Algorithms and Computation in Mathematics · Volume 17

Editors

Arjeh M. Cohen   Henri Cohen
David Eisenbud   Bernd Sturmfels

Gabriele Nebe
Eric M. Rains
Neil J.A. Sloane

Self-Dual Codes
and Invariant Theory

With 10 Figures and 34 Tables

 Springer

Authors

Gabriele Nebe

Lehrstuhl D für Mathematik
Rheinisch-Westfälische
Technische Hochschule Aachen
Templergraben 64
52062 Aachen
Germany
e-mail: nebe@math.rwth-aachen.de

Neil J.A. Sloane

Internet and Network Systems Research
AT&T Shannon Labs
180 Park Avenue
Florham Park, NJ 07932-0971
USA
e-mail: njas@research.att.com

Eric M. Rains

Department of Mathematics
University of California at Davis
1 Shields Ave
Davis, CA 95616
USA
e-mail: rains@math.ucdavis.edu

Mathematics Subject Classification (2000): 94B05, 94B60, 13A50, 16P10, 20G05; 15A66, 16D90, 68Q99, 81R99

ISSN 1431-1550

ISBN 978-3-642-06801-0   e-ISBN 978-3-540-30731-0

Springer is a part of Springer Science+Business Media
springer.com
© Springer-Verlag Berlin Heidelberg 2010
Printed in The Netherlands

Cover design: *design & production* GmbH, Heidelberg

Preface

This book has two goals. On the one hand it develops a completely new unifying theory of self-dual codes that enables us to prove a far-reaching generalization of Gleason's theorem on weight enumerators of self-dual codes. On the other hand it is an encyclopedia that gives a very extensive list of "Types" of self-dual codes and their properties—the associated Clifford-Weil groups and their invariants, in particular. For the most important Types we give bounds on their minimal distance and updated tables of the best codes.

One of the most remarkable theorems in coding theory is Gleason's 1970 theorem [191] that the weight enumerator of a binary doubly-even self-dual code is an element of the polynomial ring generated by the weight enumerators of the Hamming code of length 8 and the Golay code of length 24. In the past thirty-five years a number of different proofs of this theorem have been given, as well as many generalizations that apply to other families of self-dual codes (see for example [34], [359], [361], [383], [454], [500]). One reason for the interest in self-dual codes is that they include some of the nicest and best-known error-correcting codes, and there are strong connections with other areas of combinatorics, group theory and (as we will mention in a moment) lattices. Self-dual codes are also of considerable practical importance, although that is outside the scope of this book.

In the past, analogues of Gleason's theorem have been derived for each new family of codes on a case-by-case basis. One of the main goals of this book is to present a generalization of Gleason's theorem that applies simultaneously to weight enumerators of self-dual codes over many different alphabets. The codes we consider are linear, which for us means that the alphabet is a module V over a ring R, and a code of length N is an R-submodule of V^N. Our theorem applies to any alphabet that is a finite module over a *quasi-chain ring*—a quasi-chain ring is a product of matrix rings over chain rings, and a *chain ring* is a ring in which the left ideals are linearly ordered by inclusion. Quasi-chain rings include finite fields, the integers mod m (e.g. $\mathbb{Z}/4\mathbb{Z}$), and more generally any finite Galois ring, as well as finite quotient rings of maximal orders in quaternion algebras. It would be incorrect to say that our theory applies to self-dual codes over *any* finite ring or module, but it certainly applies to any in which the reader is likely to be interested for the foreseeable future.[1]

The weight enumerator of a classical binary code C is a homogeneous polynomial that gives the number of codewords in C of each Hamming weight. For binary doubly-even self-dual codes this polynomial belongs to the invariant ring of a certain complex matrix group of order 192, and the fact that this ring has a very simple structure leads to Gleason's theorem: the ring is a polynomial ring with two generators, and as generators one can take the weight enumerators of the Hamming and Golay codes.

Our approach provides a general setting for this connection between self-dual codes and invariant theory. To a self-dual code C over an alphabet of size v we associate a polynomial $p_C \in \mathbb{C}[x_1, \ldots, x_v]$, the "complete weight enumerator" of C. Properties of C translate into invariance properties of p_C. For example, if the length of C is even, p_C must be invariant under the transformation $x_i \mapsto -x_i$ ($i = 1, \ldots, v$). The polynomials with the given invariance properties then belong to a finitely generated graded ring. This makes it much easier to determine the possible p_C, and may allow one to deduce new properties of the codes, for instance to give bounds on the minimal distance.

We will define a general notion of Type of a self-dual code. Attached to each Type ρ is a finite complex matrix group $\mathcal{C}(\rho)$, the associated "Clifford-Weil group", and our main theorem (Theorem 5.5.7 and Corollary 5.7.5) shows that the invariant ring of $\mathcal{C}(\rho)$ is generated by the weight enumerators p_C of codes C of Type ρ. On the one hand this provides information about the possible codes of this Type (divisibility criteria for the length, bounds on the minimal distance, etc.), and on the other hand it makes it easier to compute the invariant ring of $\mathcal{C}(\rho)$. In fact our original investigations in [383] began as an attempt to generalize Sidelnikov's theorem [490], [491], [492], [493] that, for $m \geq 3$, the lowest degree harmonic invariant of the group $\mathcal{C}_m$ has degree 8. Since the invariant ring of $\mathcal{C}_m$ is spanned by the genus-m weight-enumerators of self-dual binary codes, this observation is reflected in the fact that 8 is the first length where there are two inequivalent Type I codes, i_2^4 and the Hamming code e_8 (see Chapter 6).

Our theory also applies to higher-order weight enumerators (sometimes called multiple or higher-genus weight enumerators), which consider m-tuples of codewords rather than single codewords. This leads to the higher-genus Clifford-Weil groups $\mathcal{C}_m(\rho) \leq GL_{v^m}(\mathbb{C})$. For $m = 1, 2, \ldots$ these groups form an infinite series for which the sequence of Molien series converges monotonically to the generating function $\sum_{N \geq 0} a_N t^N$ for the numbers a_N of equivalence classes of codes of Type ρ and length N (Cor. 5.7.7, Cor. 6.2.4). Note that this

---

[1] As an example, self-dual codes over the ring $\begin{pmatrix} \mathbb{Z}/4\mathbb{Z} & \mathbb{Z}/4\mathbb{Z} \\ 2\mathbb{Z}/4\mathbb{Z} & \mathbb{Z}/4\mathbb{Z} \end{pmatrix}$ are not covered by Theorem 5.5.7. Nor are codes over the group ring $\mathbb{F}_3 S_3$, where S_3 is the symmetric group of order 6.

leads to a surjection $\mathrm{Inv}(\mathcal{C}_m(\rho)) \longrightarrow \mathrm{Inv}(\mathcal{C}_{m-1}(\rho))$, analogous to the famous Siegel Φ-operator in the theory of Siegel modular forms (cf. Freitag [176]), which is presumably worth investigating further (see [381] for some initial investigations along these lines).

The Clifford-Weil groups are often very nice groups. In the case of genus-m weight enumerators (for $m \geq 1$) of self-dual binary codes, $\mathcal{C}_m(\rho)$ is the real Clifford group $\mathcal{C}_m$ of our earlier paper [383]. For the Type ρ of doubly-even self-dual binary codes, $\mathcal{C}_m(\rho)$ is the complex Clifford group $\mathcal{X}_m$ of [383]. The case $m = 1$ gives the original Gleason theorem (except for the specific identification of codes that generate the ring). In [383] we followed Bolt, Room and Wall [57], [58], [59], [536] in calling these "Clifford" groups. For self-dual codes over $\mathbb{F}_p$ containing the all-ones vector (where p is an odd prime), $\mathcal{C}_m(\rho)$ is the group $\mathcal{C}_m^{(p)}$ of [383, §7]. This is a metaplectic group, as in Weil [546], and explains why we call these "Clifford-Weil" groups in general.

These Clifford-Weil groups are also Jordan subgroups of classical Lie groups (as discussed in Alekseevskii [3], Gross and Nebe [206], Kostrikin and Tiep [334]), and provide an infinite family of examples of maximal finite matrix groups that are closely related to generalized Barnes-Wall lattices.

Besides Gleason's theorem, another remarkable fact in the background to this book is the close relationship between codes and lattices. There are some astonishing parallels between the two theories, as shown in the following list. To each of the following concepts from coding theory there is an analogue from lattice theory:

code	lattice
self-dual code	unimodular lattice
doubly-even self-dual code	even unimodular lattice
weight enumerator	theta series
invariant polynomial	modular form
MacWilliams identity	Jacobi identity
Gleason's theorem	Hecke's theorem
Molien's theorem	Selberg trace formula
Hamming code e_8	root lattice E_8
Golay code g_{24}	Leech lattice Λ_{24}

Items in the left column can be related to those in the right column by "Construction A", or one of its variants [133]. These parallels have been discussed in various articles ([500], [501], [503], Broué and Enguehard [82], [83], and most recently by Elkies [168]). One of the goals in this book, not fully realized, was to extend our main theorem to include lattices, and so to throw some additional light on the connections between codes and lattices. We were only partially successful, but the theory, presented in Chapter 9, has nevertheless led to a number of new results.

As well as lattices, another topic that has a lot in common with self-dual codes is that of quantum error-correcting codes. In fact, the construction of quantum codes was one of the initial reasons for our interest in the Clifford-Weil group. Although our main theorem does not directly apply to these codes, there are many connections to the rest of the book, and they are discussed in the final chapter.

In order to define the Type of a code in sufficient generality, we found it necessary to extend the notion of "form ring" from unitary K-theory (cf. Hahn and O'Meara [226]). In that theory, form rings are not closed under taking quotients, but with our definition, given in Chapter 1, they are. It may be worth investigating this extended notion from a K-theoretical perspective.

A note about finiteness. Although coding theory usually deals with finite alphabets (which in this book mean finite modules over finite rings), a large part of our theory is valid for arbitrary rings. In particular, the theory of form rings applies also to infinite rings. Our particular construction of the Clifford-Weil groups in Chapter 5, however, relies heavily on the finiteness of the R-module V. Consequently the proofs of the main theorems are valid only for finite form rings. On the other hand, the construction of the hyperbolic co-unitary groups applies to arbitrary form rings. We make use of this in particular in Chapter 9, where we see that the hyperbolic co-unitary groups for matrix rings over the integers coincide with Siegel modular groups.

Although this is not a textbook, our treatment is self-contained, and we have defined most of the concepts that we use, both from coding theory and invariant theory. These definitions have been kept short and expressed in our new language of form rings. As a result the book should be accessible to mathematicians, engineers and computer scientists.

The following is a brief description of the individual chapters, with emphasis on what is new. The reader is referred to the introductions to the chapters and to the table of contents for a more detailed list of what is in each chapter.

The introduction to Chapter 1 discusses how the notion of a self-dual code has been enlarged over the years. A major stimulus was the discovery in the early 1990's by Hammons, Kumar, Calderbank, Sloane and Solé [175], [91], [227] that certain notorious nonlinear binary codes could best be understood as arising from linear codes over the Galois ring $\mathbb{Z}/4\mathbb{Z}$. Our new notion of Type is defined in §1.8, after the necessary algebraic machinery has been developed in the earlier sections. In brief, a Type is a representation ρ of a form ring.

Chapter 2 begins by defining various weight enumerators associated with a code, and then follows a long section (§2.3) in which we describe all the families of self-dual codes that have been studied up to the present time as Types, using our new language of form rings. We also introduce (in §2.3.6) many new Types that treat self-dual codes over general Galois rings. Although the latter codes have so far received little attention, this may change, and in any case this section illustrates how our methods could be applied in the future if further classes of self-dual codes arise. A second long section (§2.4) then gives examples of codes and their weight enumerators for the major Types.

Chapter 2 contains two tables, Tables 2.1 (p. 78) and 2.2 (p. 79), which provide a useful list of the principal Types and the sections where they appear in the book. Another useful table appears in Chapter 11: Table 11.1 (p. 325) gives bounds on the minimal distance (used to define "extremal" codes) for the principal Types, as well as numbers ν and c such that the length must be a multiple of ν and the weights must be divisible by c. The latter property is related to the Gleason-Pierce theorem, discussed in the final section (§2.5) of Chapter 2.

Our primary interest in the book is in self-dual codes, satisfying $C^\perp = C$. Of course this implies that $C^{\perp\perp} = C$. Codes with this latter property are called *closed*. In Chapter 3 we attempt to identify just which families of codes are closed. Our main conclusion, which may be new, is that codes in certain finite representations of twisted rings are closed (see §3.3). In particular, the definition of Type given in Chapter 1 is strong enough to guarantee that all codes in a representation of a form ring are closed. Conversely, Theorem 3.2.8 shows that, while the notion of twisted rings may not be the only way to force codes to be closed, it is the only *natural* way. Our analysis in this chapter may be regarded as a continuation of the work of Wood [552], [553], [554], who concluded that quasi-Frobenius rings are the most general setting in which it makes sense to study codes over rings. Our analysis shows that one can work with the larger family of codes over twisted rings. The extra generality comes about because we consider bilinear forms taking values in a module rather than in a ring.

Chapter 4 examines the objects introduced in Chapter 1 from the point of view of category theory, and develops machinery that will be needed to prove the main theorems in the following chapter. The mathematical techniques used in this chapter are probably the most abstract in the book, and will be the least familiar to coding theorists. The Witt group of representations of a form ring, introduced in §4.6, will play an important role in several later chapters. A more detailed summary can be found in the introduction to this chapter. These results may also be of independent interest to people working in unitary K-theory.

Chapter 5 introduces the Clifford-Weil groups and their invariants. Table 5.1 on page 142 summarizes the principal Clifford-Weil groups and their structure. The main results of this book, Theorems 5.5.5 and 5.5.7, will be found in §5.5. They show that, under quite general conditions, the invariant ring of the Clifford-Weil group associated with a finite representation ρ of a form ring is spanned by the complete weight enumerators of self-dual isotropic codes of Type ρ (and arbitrary length). Although a simplified version was given in our announcement in [385], this is the first time that the complete statement of our main theorems have appeared in print. One of our two main theorems, Theorem 5.5.7 (p. 152), establishes this for self-dual codes defined over quasi-chain rings. The other main theorem, Theorem 5.5.5 (p. 150), establishes a similar result when the Type is a representation of a finite triangular form ring (defined in §1.9).

In fact we conjecture that a still more powerful theorem should hold, which would include both of the two main theorems as special cases. We state this "Weight Enumerator Conjecture" in two forms, Conjectures 5.5.2 and 5.7.2. An additional piece of evidence for this conjecture is provided by Theorem 5.5.3: an isotropic self-dual code of Type ρ and length N exists if and only if $\mathcal{C}(\rho)$ has an invariant of degree N.

Chapter 6 summarizes some of the results of our earlier paper [383] and relates them to the new situation. We can now give simpler proofs for some of the theorems in [383], including of course the main theorems, which are now special cases of the theorems in Chapter 5. The chief subjects of [383] were the real Clifford group $\mathcal{C}_m$ arising from genus-m weight enumerators of binary self-dual codes, and the complex Clifford group $\mathcal{X}_m$ arising in a similar way from doubly-even binary self-dual codes. The opening section of Chapter 6 gives some background information about the history of these groups, and the earlier work of several authors including—in roughly historical order—Barnes, Bolt, Room, Wall, Duke, Runge, Oura, Sidelnikov, Calderbank, Kantor, and Shor. This historical section concludes with the story of the amazing coincidence which led to the writing of the papers [92], [95], [96], and eventually to the present book.

In Chapter 7 we continue with the Types of codes defined in Chapter 2, and construct the associated form rings, representations, Clifford-Weil groups, and their invariants and Molien series. Chapters 6 and 7 include all the classical Types of codes.

Chapter 8 treats some further Types that were not covered in the previous two chapters, including codes over Galois rings, such as $\mathbb{Z}/4\mathbb{Z}$, and codes over $\mathbb{F}_{q^2} + \mathbb{F}_{q^2}u$ where $u^2 = 0$. The most important case of the latter family is when $q = 2$—such codes were studied by Bachoc [19] and Gaborit [178] in connection with the construction of quaternionic lattices.

Self-dual codes of many of the Types we discuss have been investigated, and their invariant rings determined, by a number of authors, including Bachoc, Bannai, Betsumiya, Bonnecaze, Choie, Conway, Dougherty, Gaborit, Gulliver, Harada, Huffman, Kim, Mallows, Munemasa, Otmani, Ozeki, Pless, Solé, and many others (as well as the present authors). However, this is the first time that these codes and their invariant rings have all been derived in a uniform way. Many of the results in Chapters 6–8 are new.

Chapter 9 presents our attempt to fit self-dual lattices into our framework of Types. The reader is referred to the long introductory section of that chapter for more information about its contents.

In Chapter 10 we apply our theory to study weight enumerators of maximally isotropic codes—that is, codes which, while not self-dual, are maximal subject to being isotropic. Note that, by definition, isotropic codes are also self-orthogonal. The weight enumerators of maximally self-orthogonal codes were first studied from this point of view by Mallows, Pless and Sloane [364], [366]. Our systematic approach enables us to correct some errors and omissions in the earlier work and to extend it to other families of codes. In particular, we describe the space of weight enumerators of maximal isotropic codes from the following families:

- doubly-even binary codes (Theorem 10.2.1)
- singly-even binary codes (Theorem 10.3.1)
- ternary codes (Theorem 10.4.1)
- ternary codes with 1 in the dual (Theorems 10.4.2 and 10.4.2)
- even additive trace-Hermitian self-orthogonal codes over $\mathbb{F}_4$ (Theorem 10.5.1)
- doubly-even codes over $\mathbb{Z}/4\mathbb{Z}$ (Theorem 10.6.1)

Almost all these results are new. In the second half of the chapter we use the results in Chapter 9 and the first half of the chapter to describe the space of modular forms spanned by the theta series of

- maximal even lattices of determinant 3^k (Corollary 10.7.7)
- maximal even lattices of determinant 2^k (Theorem 10.7.14)

Again we believe that these results are new.

One of the motivations for calculating these invariant rings is that it may then be possible to apply the linear programming method to obtain bounds on the minimal distance. The general "linear programming bound" for isotropic codes is the subject of §11.1.1 of Chapter 11. Section 11.1 summarizes the best upper bounds on codes of the principal Types that have been obtained by the linear programming and other methods; §11.2 then gives lower bounds.

We follow [454] in using the term *extremal* to indicate a code which has the highest minimal distance permitted by the appropriate linear programming bound, and *optimal* to indicate a code which has the actual highest minimal distance of any code of the given Type and length (an extremal code is automatically optimal, but in general no extremal code may exist). Table 11.1 (p. 325) summarizes what extremal means for the principal Types. The final section, §11.3, gives a summary of what is presently known about the existence of extremal and optimal codes of modest lengths. These are based on earlier tables in [454] and other sources. Although most of the material in this chapter is not new, it has not been collected in one place before. (See also the survey article of Huffman [282].)

In Chapter 12 we discuss what is presently known about the enumeration of self-dual codes of the main Types. Again this is an update of earlier tables. The main tool for these enumerations are the mass formulae given in §12.1.

The final chapter, Chapter 13, gives a brief discussion of quantum codes and their constructions and bounds. The last section, §13.6 gives a table of the best additive $[[N, k, d]]$ binary codes presently known. This is an updated version of the table in Calderbank, Rains, Shor and Sloane [96]. Again we refer the reader to the introduction to this chapter for a more detailed description of its contents and its relationship to the rest of the book.

The book concludes with an extensive bibliography. This seemed desirable, since few readers will be familiar with all the topics we mention. Furthermore, there are a large number of papers on self-dual codes, which have been scattered throughout the literature on engineering, mathematics and computer science. Besides these conventional references, we have also given cross-references to the *On-Line Encyclopedia of Integer Sequences* [504] for various number sequences that occur (coefficients of Molien series, minimal distances of optimal codes of various Types, etc.). For an example, see the reference to sequence A001399 in Eq. (5.8.1) on page 169.

A summary of some of the new results appeared in [385].

In this book we will mostly only discuss *self-dual* codes. Two topics that we will not treat are *isodual* codes, that is, codes which are equivalent to their duals under an appropriate notion of equivalence (cf. Conway and Sloane [132]), and *formally self-dual* codes, that is, possibly nonlinear codes which the property that their weight enumerator coincides with its MacWilliams transform (cf. Betsumiya, Gulliver and Harada [40], Betsumiya and Harada [44], [43], and Gulliver and Harada [210]). An isodual code is automatically formally self-dual. However, we do give a definition of formally self-dual in the language of Types at the end of §5.7.

We will also not say anything about *decoding* self-dual codes. Most of the existing work on this subject is concerned with classical codes such as the Golay and extended quadratic residue codes; little has been done on decoding self-dual codes over rings, except for the octacode of §2.4.9 (or its *alter ego* the Nordstrom-Robinson code). Readers interested in decoding are referred to the following papers: Amrani and Beéry [4], Amrani, Beéry and Vardy [5], Amrani, Beéry, Vardy, Sun and van Tilborg [6], Anderson [8], Blaum and Bruck [53], Bossert [69], Conway and Sloane [126], Dodunekov, Zinoviev and Nilsson [145], Esmaeili, Gulliver and Khandani [169], Fekri, McLaughlin, Mersereau and Schafer [171], Gaborit, Kim and Pless [184], Gordon [196], Greferath and Vellbinger [202] Greferath and Viterbo [203] Hammons, Kumar, Calderbank, Sloane and Solé [227], Higgs and Humphreys [264], Kim, Mellinger and Pless [305], Kim and Pless [306], Ping and Yeung [407], Pless [418], [421], Reed, Yin and Truong [456], Rifà [459], Solomon [509], Vardy [532], Wolfmann [549], [550], Yuan and Leung [564].

Acknowledgements

Much of this work was carried out while E.M.R. and N.J.A.S. were together at AT&T Shannon Labs. We thank AT&T Shannon Labs for making it possible for G.N. to visit the Labs in 1999 and 2002. Part of the work was done during G.N.'s visit to Harvard University with a Radcliffe Fellowship. G.N. thanks the Radcliffe Institute for its kind invitation. Since 2004, E.M.R. has been supported in part by NSF Grant DMS-0401387. Many of the calculations involving matrix groups and their Molien series and invariants were carried out with the computer algebra system MAGMA [68], [100].

We would like to thank Matthias Künzer, who made valuable comments during the course G.N. gave on the topics of this book in the summer of 2005 at RWTH Aachen. We also thank Koichi Betsumiya, Young-Ju Choie, Philippe Gaborit, Masaaki Harada, Akihiro Munemasa, Patric Östergård, Vera Pless, Heinz-Georg Quebbemann, Patrick Solé and John van Rees for providing helpful comments on the manuscript.

Although we have made every effort to be careful, it is inevitable that there will be errors in a book of this size, for which we apologize in advance. We would appreciate hearing of any corrections, as well as updates to the tables. Such items will be added to the web site for the book, which is www.research.att.com/~njas/doc/cliff2.html. They may be sent to any of the authors. Our email addresses are nebe@math.rwth-aachen.de, rains@math.ucdavis.edu and njas@research.att.com.

Aachen, Davis and Florham Park   *Gabriele Nebe ·*
October, 2005   *Eric M. Rains · N. J. A. Sloane*

General notation

Unless specified otherwise a ring (usually denoted by R) has an identity element $1 \neq 0$ and may be finite or infinite, commutative or noncommutative. Rings are always associative. Codewords are generally viewed as row vectors and the alphabet is a left R-module. The following table lists symbols that are used throughout the book.

List of Symbols.

Symbol	Meaning	See
$A.B$	group with normal subgroup isomorphic to A and quotient isomorphic to B	
$A \rtimes B$	split extension or semidirect product	
$A \wr B$	wreath product	
$A \mathsf{Y} B$	central product	
$\mathrm{Aut}(\rho)$	automorphism group	Defn 1.11.1
$C \leq V$	the code C is a submodule of V	Defn 1.2.1
$C^\perp$	dual code	Defn 1.2.1
$C \otimes R$	code C promoted to a larger ring	Rem. 2.1.10
$\mathbb{C}$	complex numbers	
$\mathcal{C}(\rho)$	Clifford-Weil group	Defn 5.3.1
$\mathcal{C}_m(\rho)$	Clifford-Weil group of genus m	Defn 5.3.4
$\mathcal{C}_m$	real Clifford group of genus m	§6.2
$\mathcal{C}_m^{(p)}$	p-Clifford group of genus m	§6.2
cwe	complete weight enumerator	Defn 2.1.2
cwe_m	genus-m complete weight enumerator	Defn 2.1.7
$e(\tau)$	$\exp(2\pi i \tau)$	Eq. (9.1.1)
$\mathrm{Ev}_n(S)$	even matrices	Defn 1.10.4
$\mathbb{F}_q$	field of order q	
fwe	full weight enumerator	Defn 2.1.3
$\widehat{G}$	character group	Defn 2.2.1
$\mathrm{GL}_n(\mathbb{F}_q)$	general linear group	
$H^\#$	dual subgroup	Defn 2.2.1
$\mathbb{H}$	real quaternions	
$H_{\iota,u_\iota,v_\iota}$	MacWilliams transform in $\mathfrak{U}(R,\Phi)$	Eq. (5.2.23)
$h_{\iota,u_\iota,v_\iota}$	MacWilliams transform in $\mathcal{C}(\rho)$	Eq. (5.3.1)
hwe	Hamming weight enumerator	Defn 2.1.2

Symbol	Meaning	See
I or I_n	$n \times n$ unit matrix	
$I \trianglelefteq R$	I is an ideal in R	
Inv	invariant ring	Eq. (5.6.3)
$\mathrm{Inv}(G, S)$	relative invariants	Defn 5.6.5
$\mathrm{Mat}_m(R)$	$m \times m$ matrices over R	
$\mathrm{Mat}_{m \times n}(R)$	$m \times n$ matrices over R	
MS	Molien series	Eq. (5.6.1)
$O_n(\mathbb{F}_q)$	orthogonal group	
$P(R, \Phi)$	parabolic subgroup of $\mathfrak{U}(R, \Phi)$	Defn 5.1.1
$P(\rho)$	parabolic subgroup of $\mathcal{C}(\rho)$	Defn 5.1.2
$\mathbb{Q}$	rational numbers	
$\mathbb{R}$	real numbers	
(R, M, ψ, Φ)	form ring	Defn 1.7.1
$\rho, (V, \rho_M, \rho_\Phi, \beta)$	representation of form ring	Defn 1.7.2
$\mathrm{Sp}_{2n}(\mathbb{F}_q)$	symplectic group	
S_N	symmetric group of order $N!$	
swe^ρ	symmetrized weight enumerator	Defn 2.1.5
swe^ρ_m	genus-m symmetrized weight enumerator	Defn 2.1.8
tr	transposed matrix	
tr, Tr	trace operators	
$T(M)$	triangular twisted ring	Defn 1.5.1
$T(M, \Phi)$	triangular form ring	§1.9
$U_n(\mathbb{F}_{q^2})$	unitary group	
$\mathfrak{U}(R, \Phi)$	hyperbolic co-unitary group	Defn 5.2.4
$\mathfrak{U}_m(R, \Phi)$	hyperbolic co-unitary group of genus m	Defn 5.2.8
V^*	dual in sense of linear algebra, space of linear functionals	
$\mathrm{WAut}(\rho)$	weak automorphism group	Defn 1.11.2
$\mathcal{X}_m$	complex Clifford group of genus m	§6.2
Z_n	cyclic group of order n	
$\mathbb{Z}_n$	n-adic integers	
$\mathbb{Z}/n\mathbb{Z}$	ring of integers mod n	
$\mathbf{1}$	all-ones vector	Ex. 1.8.4
$\{\!\!\{\ \}\!\!\}, \lambda$	structure maps	Defn 1.6.1, 4.1.1

Contents

List of Tables

List of Figures

1

The Type of a Self-Dual Code

To motivate these initial definitions, we begin by remarking that in the classical theory (cf. van Lint [350], MacWilliams and Sloane [361], Pless, Huffman and Brualdi [427], Rains and Sloane [454]) a linear error-correcting code C is a subspace of a vector space V over a finite field $\mathbb{F}$, with inner products of codewords taking values in $\mathbb{F}$ itself. The classical theory was enlarged in the early 1990's by the discovery by Hammons, Kumar, Calderbank, Sloane and Solé [175], [91], [227] that certain notorious *nonlinear* binary codes (the Nordstrom-Robinson, Kerdock and Preparata codes) could best be understood as arising from *linear* codes over the ring $\mathbb{Z}/4\mathbb{Z}$, and, in the case of the Kerdock code, from a self-dual linear code over $\mathbb{Z}/4\mathbb{Z}$.

A few years later, an important application of coding theory to quantum computers required the use of additive (but nonlinear) codes over $\mathbb{F}_4$ (Calderbank, Rains, Shor and Sloane [95], [96] and Chapter 13 below).

Furthermore, codes over rings such as $\mathbb{Z}/8\mathbb{Z}$ arise naturally in studying "Phase Shift Keying" or PSK modulation schemes—see for example Anderson [7, §3.4], Piret [408].

Thus it became clear that the theory should consider codes over rings as well as over fields, and that weaker notions of linearity should be permitted.

Concerning the weights of codewords in a self-dual code, it is easy to show that in a self-dual code over $\mathbb{F}_2$ the weight of every codeword must be even, in a self-dual code over $\mathbb{F}_3$ the weight of every codeword is a multiple of 3, and in a Hermitian self-dual code over $\mathbb{F}_4$ the weight of every codeword is even. Furthermore, there are many well-known self-dual codes over $\mathbb{F}_2$ whose weights are divisible by 4. Since these four families were the self-dual codes of main interest in the classical theory, they were called codes of *Types* I, III, IV and II respectively. In fact, as we will discuss in §2.5, a theorem of Gleason and Pierce shows that these are essentially the only possible divisibility restrictions that can be placed on the weights of self-dual codes over finite fields.

But once one allows self-dual codes to be defined over rings, there are other possible constraints that can be placed on the weights, and so in [454] we defined nine different Types of self-dual codes, each with its own separate definition.

One of the goals of this book is to introduce a more formal notion of the Type of a self-dual code, which will allow us to give a unified treatment of all the earlier definitions as well as a number of new ones. The new framework is also broad enough to include both unimodular and even–unimodular lattices, as we shall see in Chapter 9.[1]

In this framework, the *symbols* in the codewords belong to a left R-module V (the *alphabet*) where R is a ring, assumed to contain a unit 1, but which may be commutative or noncommutative, finite or infinite. A *code* C of *length* N will be an R-submodule of V^N for some positive integer N. A *codeword* $c \in C$ is an element of V^N and R is the *ground ring* underlying the code, in the sense that if $c \in C$ and $r \in R$ then $rc \in C$.

In the classical theory, inner products of codewords take values in the ground ring (which is usually the field of symbols, or a subfield if a trace is used to define the inner product). Now we allow the additional freedom that inner products of codewords will be defined by *bilinear forms* taking values in some abelian group A. For finite rings R, this abelian group A is usually a subgroup of $\mathbb{Q}/\mathbb{Z}$. This makes it possible to describe the MacWilliams transformation with respect to the $\mathbb{Q}/\mathbb{Z}$-valued bilinear forms as a complex linear transformation.

To specify additional properties of these codes, such as restrictions on the weights of codewords, or that the code contains the all-ones vector, we will use *quadratic maps* taking values in A; these are sums of quadratic forms and linear maps. We will therefore begin our discussion by defining quadratic maps in §1.1. In §1.2 we give the definition of a code and of the notions of *dual, self-orthogonal, self-dual* and *isotropic* code. To define a Type we will need the important concept of a *form ring*: this is defined in §1.7; §§1.3-1.6 contain technical material needed for this definition. Finally, the Type of a self-dual code is defined in §1.8. In brief, a Type is a *representation* ρ of a form ring (R, M, ψ, Φ). *Equivalences* and *automorphism groups* are defined using the language of Types in §1.11, and §1.12 defines the *shadow* of a code in this language.

1.1 Quadratic maps

Definition 1.1.1. Let V and A be abelian groups (see the preceding paragraphs for motivation). An *A-valued bilinear form on V* is a $\mathbb{Z}$-module homomorphism

$$\beta : V \otimes_{\mathbb{Z}} V \to A.$$

---

[1] Although so far "modular" lattices (Quebbemann [439]) do not fit into this framework.

If V is a left R-module for some ring R, then the set of all A-valued bilinear forms on V is a right $(R \otimes R)$-module, where the action is defined by

$$\beta(r \otimes s)(x, y) := \beta(rx, sy) \text{ for all } x, y \in V \text{ and all } r, s \in R.$$

This $(R \otimes R)$-module is denoted by $\mathrm{Bil}(V, A) = \mathrm{Bil}_{\mathbb{Z}}(V, A)$. An A-valued quadratic map on V is a map $\phi : V \to A$ such that

$$\phi(x+y+z)+\phi(x)+\phi(y)+\phi(z) = \phi(x+y)+\phi(x+z)+\phi(y+z)+\phi(0); \quad (1.1.1)$$

or, equivalently, such that the map $\phi : V \times V \to A$ given by

$$\phi(x, y) := \phi(x + y) - \phi(x) - \phi(y) + \phi(0) \quad (1.1.2)$$

is $\mathbb{Z}$-bilinear. A quadratic map ϕ on V is said to be *pointed* if $\phi(0) = 0$, *even* if $\phi(-x) = \phi(x)$, and *homogeneous* if it is both pointed and even. We denote the abelian group of quadratic maps from V to A by $\mathrm{Quad}(V, A)$ and the subgroup of pointed maps by $\mathrm{Quad}_0(V, A)$.

If 2 acts invertibly on A, for example, then a quadratic map ϕ is the sum of a homogeneous quadratic map (given by $x \mapsto \frac{1}{2}(\phi(x) + \phi(-x)) - \phi(0)$), a linear map (given by $x \mapsto \frac{1}{2}(\phi(x) - \phi(-x))$) and the constant $\phi(0)$.

Lemma 1.1.2. *Let* $\phi : V \to A$ *be a quadratic map. For all* $n \in \mathbb{Z}$ *and all* $x \in V$,

$$\phi(nx) = \frac{n(n + 1)}{2}\phi(x) + \frac{n(n - 1)}{2}\phi(-x) + (1 - n^2)\phi(0). \quad (1.1.3)$$

Proof. Applying (1.1.1) with $y = -x$, we find that

$$\phi(z + x) - 2\phi(z) + \phi(z - x) \quad (1.1.4)$$

is independent of z. By evaluating (1.1.4) at $z = 0, x, 2x, \ldots$ we obtain (1.1.3) for $n \geq 0$; evaluating (1.1.4) at $z = -x, -2x, \ldots$ we obtain (1.1.3) for $n < 0$. □

Corollary 1.1.3. *If the quadratic map* $\phi : V \to A$ *is homogeneous, then*

$$\phi(nx) = n^2\phi(x) \quad (1.1.5)$$

for all integers n *and all* $x \in V$.

In our applications, bilinear forms will arise from the requirement that two vectors in a self-dual code should have inner product zero. Some of the quadratic maps arise from specializations of bilinear forms, others when we impose constraints on the weights of codewords (cf. Example 1.2.2).

The reason we do not use condition (1.1.5) as well as (1.1.1) when defining a quadratic map is that (1.1.5) only applies to homogeneous quadratic functions, whereas our quadratic maps may also have a linear or constant part, for example when we study codes that must contain the all-ones vector (cf. Example 1.8.4). Furthermore, if the characteristic is 2, (1.1.5) is always satisfied.

Since the obvious action of the underlying ring R on quadratic maps is not linear, we introduce the notion of a "qmodule", generalizing the notion of a linear R-module.

Definition 1.1.4. Let R be a ring. A (right) R-*qmodule* is an abelian group Φ equipped with a pointed quadratic map $r \mapsto [r]$ from R to $\mathrm{End}(\Phi)$ (with $[r]$ acting on Φ on the right) such that $[1] = 1$, $[r][s] = [rs]$. A homomorphism between qmodules Φ_1 and Φ_2 is a map f such that $f(\phi_1 + \phi_2) = f(\phi_1) + f(\phi_2)$ and $f(\phi_1[r]_1) = f(\phi_1)[r]_2$ for all $\phi_1, \phi_2 \in \Phi, r \in R$.

Example 1.1.5. The group $\Phi = \mathrm{Quad}_0(V, A)$ of all pointed quadratic maps on a left R-module V is a right R-qmodule, with

$$(\phi[r])(v) := \phi(rv), \quad \text{for } r \in R, \phi \in \Phi, v \in V.$$

Example 1.1.6. If M is a right R-module, then $x[r] = xr$ gives M an R-qmodule structure. A qmodule obtained this way is called *linear*

Example 1.1.7. The abelian group $\mathbb{Z}/4\mathbb{Z}$ admits a natural $\mathbb{Z}/2\mathbb{Z}$-qmodule structure, given by

$$x[0] = 0, \quad x[1] = x, \quad \text{for } x \in \mathbb{Z}/4\mathbb{Z}. \tag{1.1.6}$$

1.2 Self-dual and isotropic codes

We can now define the basic coding-theoretic concepts that will be used throughout the book.

Definition 1.2.1. Let V be a left R-module, A an abelian group, $M \subset \mathrm{Bil}(V, A)$ a set of A-valued $\mathbb{Z}$-bilinear forms on V, and $\Phi \subset \mathrm{Quad}_0(V, A)$ a set of A-valued pointed quadratic maps on V. An R-submodule $C \leq V$ is called a *code*. Let $C \leq V$ be a code. The *dual* of C (with respect to M) is

$$C^\perp := \{v \in V \mid m(c, v) = 0, \text{ for all } m \in M, c \in C\}. \tag{1.2.1}$$

Generalizing the standard terminology (cf. [361], [454]), we call C *self-orthogonal* (with respect to M) if $C \subset C^\perp$ and *self-dual* if $C = C^\perp$. Furthermore, C is *isotropic* (with respect to (M, Φ)) if C is self-orthogonal with respect to M, and also $\phi(c) = 0$ for all $\phi \in \Phi, c \in C$. Hence:

$$\{ \text{ self-orthogonal codes with respect to } M \ \}$$
$$\bigcup$$
$$\{ \text{ (self-orthogonal) isotropic codes with respect to } (M, \Phi) \ \}$$
$$\bigcup$$
$$\{ \text{ self-dual isotropic codes with respect to } (M, \Phi) \ \} \ .$$

Note that according to this definition, our codes are always "linear": for us this means "an R-submodule of an R-module".

Remark. If $C \leq V$ is an R-submodule and $\beta \in \mathrm{Bil}(V, A)$ is such that $\beta(c, c') = 0$ for all $c, c' \in C$ (i.e. C is self-orthogonal with respect to β), then clearly $\beta(r \otimes s)(c, c') = \beta(rc, sc') = 0$ for all $r, s \in R$ and $c, c' \in C$, since C is an R-module. So when defining self-orthogonal codes we may as well assume that M is an $(R \otimes R)$-submodule of $\mathrm{Bil}(V, A)$.

Example 1.2.2. Classical doubly-even self-dual (or Type II) binary codes (self-dual codes in which the weight of every codeword is a multiple of 4) arise in this framework as follows. As usual, x_i denotes the i-th component of the vector $x = (x_1, \dots, x_N) \in \mathbb{F}_2^N$. We take $R := \mathbb{F}_2$, $V := \mathbb{F}_2^N$, $A := \frac{1}{4}\mathbb{Z}/\mathbb{Z}$,

$$M := \{0, m_0\} \subset \mathrm{Bil}(V, \tfrac{1}{4}\mathbb{Z}/\mathbb{Z}), \text{ where } m_0(x, y) := \sum_{i=1}^{N} \frac{1}{2} x_i y_i \, ,$$

and

$$\Phi := \{0, \phi_0, 2\phi_0, 3\phi_0\} \subset \mathrm{Quad}_0(V, \tfrac{1}{4}\mathbb{Z}/\mathbb{Z}) \text{ where } \phi_0(x) := \sum_{i=1}^{N} \frac{1}{4} x_i^2 \, .$$

Then self-dual isotropic codes with respect to (M, Φ) are precisely the doubly-even self-dual binary codes of length N (for $m_0(u, v) = 0$ ensures that the mod-2 inner product $u \cdot v$ is zero, and $\phi_0(u) = 0$ guarantees that the weight of u is a multiple of 4).

As already mentioned, our goal is to give a general definition of the "Type" of a self-dual code. Definition 1.2.1 does not quite do this, since the triple (V, M, Φ) depends on the length of the code, whereas the notion of "Type" should not. To avoid this difficulty we introduce the notion of a representation of a form ring (§1.7). Changing the length of the code will then involve changing only the representation of the form ring by adding orthogonal summands. The appropriate setting for defining isotropic codes is the notion of a "quadratic pair" (M, Φ) over R, which will be introduced in §1.6.

1.3 Twisted modules and their representations

The $(R \otimes R)$-submodules M of $\mathrm{Bil}(V, A)$ used in the previous section have a naturally defined "twist" map τ which interchanges the arguments. More generally, we have:

Definition 1.3.1. A *twisted R-module* M is a right $(R \otimes R)$-module together with an automorphism $\tau : M \to M$ such that $\tau(m(r \otimes s)) = \tau(m)(s \otimes r)$, for all $m \in M, r \in R, s \in R$, satisfying $\tau^2 = 1$.

Example 1.3.2. If V is an R-module and A an abelian group, then $M :=$ Bil(V, A) is a twisted R-module, where $\tau : M \to M$ is given by $\tau(m)(x, y) := m(y, x)$.

Definition 1.3.3. A *representation* $\rho := (V, \rho_M)$ of a twisted R-module M consists of an R-module V and a twisted R-module homomorphism $\rho_M : M \to$ Bil(V, A) (for some abelian group A) that is compatible with the twist τ in Example 1.3.2, i.e. which satisfies

$$\rho_M(\tau(m))(x, y) = \rho_M(m)(y, x), \quad \text{for } x, y \in V, m \in M. \tag{1.3.1}$$

The representation ρ is said to be *finite* if R and V are finite sets and $A = \mathbb{Q}/\mathbb{Z}$.

We generalize the notion of dual code with respect to a set of bilinear forms given in the previous section to the dual code in a representation.

Definition 1.3.4. Let $\rho = (V, \rho_M)$ be a representation of a twisted R-module M. Let $C \leq V$ be a code. The *dual of C* with respect to ρ is defined to be

$$C^\perp := \{v \in V \mid \rho_M(m)(c, v) = 0, \text{ for } m \in M, c \in C\}. \tag{1.3.2}$$

We will sometimes write $C^{\perp, \rho}$ when it is necessary to specify ρ. If $C \subset C^{\perp, \rho}$ we say that C is a *self-orthogonal code in (the representation)* ρ; if $C = C^{\perp, \rho}$ we say C is a *self-dual code in (the representation)* ρ.

1.4 Twisted rings and their representations

The case when M is *isomorphic* to R as a right R-module is especially important. One can think of this as specializing only one nonsingular bilinear form β on V and taking M to be the $1 \otimes R$-submodule of Bil(V, A) spanned by β. (Here we use "nonsingular" in its classical sense. For the formal definition see Definition 3.2.1 in Chapter 3.) If the code C is an R-submodule of V, we have

$$C^{\perp, \beta} = \{v \in V \mid \beta(v, c) = 0 \text{ for all } c \in C\}, \tag{1.4.1}$$

and for any $v \in C^{\perp, \beta}$ we have $m(v, c) = 0$ for all $m \in M$ and $c \in C$.

Definition 1.4.1. A *twisted ring* (R, M, ψ) consists of a ring R, a twisted R-module M and a right R-module isomorphism $\psi : R_R \to M_{1 \otimes R}$, such that $\epsilon := \psi^{-1}(\tau(\psi(1)))$ is a unit in R. Then ϵ is called the *associated unit* defined by the involution τ.

Definition 1.4.2. A *representation* $\rho := (V, \rho_M, \beta)$ of a twisted ring (R, M, ψ) consists of a left R-module V, an abelian group A and a twisted R-module homomorphism $\rho_M : M \to \mathrm{Bil}(V, A)$ that is compatible with the twist and such that

$$\beta := \rho_M(\psi(1)) \tag{1.4.2}$$

is nonsingular.

So a twisted ring (R, M, ψ) is just a ring with an involution (induced by τ and the isomorphism ψ). However, we prefer the clumsier notation (R, M, ψ), rather than (R, τ), since in the applications M will be identified with an R-submodule of $\mathrm{Bil}(V, A)$ and τ will be the restriction of the natural involution on $\mathrm{Bil}(V, A)$ given by interchanging the arguments in the bilinear forms. The involution on R may then be changed by varying the isomorphism ψ. We use this for instance when we rescale the twisted ring (see Remark 1.4.8). However, in most of the concrete examples we will have $R = M$ and ψ will be the identity map.

Example 1.4.3. Let V be an abelian group and $\beta \in \mathrm{Bil}_{\mathbb{Z}}(V, \mathbb{Q}/\mathbb{Z})$ a nonsingular bilinear form. Then

$$\mathrm{End}(V, \beta) := (\mathrm{End}_{\mathbb{Z}}(V), \mathrm{Bil}_{\mathbb{Z}}(V, \mathbb{Q}/\mathbb{Z}), \psi_\beta) \tag{1.4.3}$$

is a twisted ring, where $\psi_\beta(1) := \beta$ (cf. Lang [343, Ch. XIII, §5]). A representation of a twisted ring (R, M, ψ) is then a morphism[2] of twisted rings from (R, M, ψ) into $\mathrm{End}(V, \beta)$.

Definition 1.4.4. Let (R, M, ψ) be a twisted ring. By definition of ψ, $\psi(1)(1 \otimes r) = \psi(r)$ for $r \in R$. We define a map $J : R \to R$ by

$$r \mapsto r^J = \psi^{-1}(\psi(1)(r \otimes 1)). \tag{1.4.4}$$

Note in particular that $1^J = 1$. In fact J is an anti-automorphism of the ring R, as shown in the following lemma.

Lemma 1.4.5. *We have the identities*

$$(rs)^J = s^J r^J, \text{ for } r, s \in R,$$
$$\epsilon^J r^{J^2} \epsilon = r, \text{ for } r \in R, \tag{1.4.5}$$

and in particular, taking $r = 1$,

$$\epsilon^J \epsilon = 1.$$

The twisted module structure on R can be expressed in terms of J and ϵ by

---

[2] We use "morphism" in its standard categorical sense of a structure-preserving map.

$$\psi(r)(s \otimes t) = \psi(s^J rt), \text{ for } r, s, t \in R,$$
$$\tau(\psi(r)) = \psi(r^J \epsilon), \text{ for } r \in R. \tag{1.4.6}$$

The map J is bijective.

Proof. We first observe that

$$\psi(r)(s \otimes t) = \psi(1)(1 \otimes r)(s \otimes t) = \psi(1)(s \otimes 1)(1 \otimes rt) = \psi(s^J rt)$$

and

$$\psi((rs)^J) = \psi(1)(r \otimes 1)(s \otimes 1) = \psi(r^J)(s \otimes 1) = \psi(s^J r^J),$$

for all $r, s, t \in R$. Hence $(rs)^J = s^J r^J$, since ψ is an isomorphism. Furthermore,

$$\tau(\psi(r)) = \tau(\psi(1)(1 \otimes r)) = \tau(\psi(1))(r \otimes 1) = \psi(\epsilon)(r \otimes 1) = \psi(r^J \epsilon).$$

Next, since τ is an involution, we have

$$\psi(r) = \tau(\tau(\psi(r))) = \tau(\psi(r^J \epsilon)) = \psi((r^J \epsilon)^J \epsilon) = \psi(\epsilon^J r^{J^2} \epsilon).$$

In particular, taking $r = 1$, we find $\epsilon^J \epsilon = 1$, and thus $\epsilon^J = \epsilon^{-1}$ since ϵ is a unit. Moreover, taking $r = \epsilon$, we find $\epsilon^{J^2} = \epsilon$ and hence $r = (\epsilon^J r^J \epsilon)^J$. Therefore $r \mapsto \epsilon^J r^J \epsilon$ is a two-sided inverse of J and so J is bijective. $\square$

Remark. The identity $\epsilon^J \epsilon = 1$ shows that for finite rings R it is not necessary to assume that ϵ is a unit. However an example given by Loos [353, §1.3] shows that this hypothesis is needed in the case of infinite rings. (To be a unit ϵ must have both left and right inverses. The existence of a left inverse corresponds to the surjectivity of the map from R to R induced by right multiplication by ϵ, the existence of the right inverse to the injectivity of this map. If R is finite, the map is injective if and only if it is surjective.)

Remark 1.4.6. The identities in Lemma 1.4.5 may be interpreted in the context of representations of twisted rings. Let $\rho := (V, \rho_M, \beta)$ be a representation of a twisted ring (R, M, ψ). Then for all $r \in R, x, y \in V$,

$$\beta(y, rx) = \beta(x, r^J \epsilon y) \text{ and } \beta(rx, y) = \beta(x, r^J y).$$

In particular, $\beta(y, x) = \beta(x, \epsilon y)$ for all $x, y \in V$.

Proof. We have

$$\beta(y, rx) = \rho_M(\psi(r))(y, x) = \rho_M(\tau(\psi(r))(x, y)$$
$$= \rho_M(\psi(r^J \epsilon))(x, y) = \beta(x, r^J \epsilon y),$$

since $\tau(\psi(r)) = \psi(r^J \epsilon)$ by Lemma 1.4.5. Similarly,

$$\beta(rx, y) = \rho_M(\psi(1)(r \otimes 1))(x, y) = \rho_M(\psi(r^J))(x, y) = \beta(x, r^J y).   \square$$

Remark 1.4.7. A (two-sided) *ideal* in a twisted ring (R, M, ψ) is defined to be a twisted submodule M' of M, and corresponds to a ring ideal $I := \psi^{-1}(M')$ in R such that $I^J = I$. It is easily seen that the kernel of a morphism of twisted rings is an ideal, and that quotients are always defined. A twisted ring is *simple* if M and $\{0\}$ are the only twisted ideals in M.

Remark 1.4.8. Let (R, M, ψ) be a twisted ring. For any unit $u \in R$, the map

$$r \mapsto \psi(ur) \tag{1.4.7}$$

is also an isomorphism between R_R and $M_{1 \otimes R}$. We thus obtain a *rescaled* twisted ring structure on R, which we denote by R^u. The involution J' and the associated unit ϵ' of R^u are given by

$$r^{J'} = u^{-1} r^J u, \text{ for } r \in R,$$
$$\epsilon' = u^{-1} u^J \epsilon. \tag{1.4.8}$$

Note that rescaling does not change the restriction of the involution J to the center of R.

Remark. Rescaling the twisted ring $\mathrm{End}(V, \beta)$ associated with a nonsingular bilinear form $\beta = \psi(1)$ simply corresponds to choosing a different nonsingular bilinear form $\beta' = \psi(u)$ for some unit $u \in \mathrm{End}(V)$. In particular, there is a canonical bijection between the representations of the rings R and R^u. If C is a code in the representation ρ of R then C is also a code in the corresponding representation ρ^u of R^u. Rescaling does not change the notion of orthogonality for codes, because

$$C^{\perp, \rho^u} = u^{-1} C^{\perp, \rho} = C^{\perp, \rho}.$$

1.5 Triangular twisted rings

One motivation for introducing triangular rings is that representations of triangular form rings (see §1.9) are one of the two Types for which we can prove Theorem 5.5.5. Also, the notion of a triangular twisted ring is a technical construction that will help us prove the main result in Theorem 5.5.7 for direct products of matrix rings over quasi-chain rings (these include most of the rings considered in coding theory).

Definition 1.5.1. Let M be a twisted R-module. Then we define the *triangular twisted ring* $T(M) := \begin{pmatrix} R & M \\ 0 & R \end{pmatrix}$ to be the set of matrices

$$\left\{ \begin{pmatrix} a & b \\ 0 & c \end{pmatrix} \mid a, c \in R, b \in M \right\}, \tag{1.5.1}$$

with (associative) multiplication

$$\begin{pmatrix} a & b \\ 0 & c \end{pmatrix} \begin{pmatrix} a' & b' \\ 0 & c' \end{pmatrix} := \begin{pmatrix} a'a & b'(a \otimes 1) + b(1 \otimes c') \\ 0 & cc' \end{pmatrix}. \qquad (1.5.2)$$

The expression $b'(a \otimes 1) + b(1 \otimes c')$ is motivated by the usual rule for matrix multiplication, but with products involving the upper left entries being formed in the opposite ring R^{op}. Then $T(M)$ has a natural structure as a twisted ring $(T(M), T(M), \mathrm{id})$, where the twist $\tau_{T(M)} = J_{T(M)} = J$ is given by

$$\begin{pmatrix} a & b \\ 0 & c \end{pmatrix}^J := \begin{pmatrix} c & \tau(b) \\ 0 & a \end{pmatrix}, \text{ and } \epsilon := \begin{pmatrix} 1 & 0 \\ 0 & 1 \end{pmatrix}. \qquad (1.5.3)$$

Example 1.5.2. Consider the twisted ring $(R^{op} \oplus R, R^{op} \oplus R, \mathrm{id})$, where the involution $\tau = (\ )^J$ interchanges the two copies of R. This is the triangular twisted ring $T(M)$ corresponding to the trivial twisted R-module $M = \{0\}$.

Definition 1.5.3. If $\rho := (V, \rho_M)$ is a finite representation of the twisted R-module M, we may construct a representation $T(\rho) := (T(V), T(\rho_M), \beta)$ of the twisted ring $T(M)$ as follows. Define $\tilde{V} := \mathrm{Hom}(V, \mathbb{Q}/\mathbb{Z})$; thus $\tilde{V}$ is a right R-module. There is a natural action of $T(M)$ on $T(V) := \tilde{V} \oplus V$, given by

$$\begin{pmatrix} a & b \\ 0 & c \end{pmatrix} (f, v) = (x \mapsto f(ax) + \rho_M(b)(x, v), \ cv). \qquad (1.5.4)$$

Then $T(\rho_M)$ is defined by

$$T(\rho_M) \begin{pmatrix} a & b \\ 0 & c \end{pmatrix} ((f, v), (f', v')) = f(cv') + f'(av) + \rho_M(b)(v, v'), \qquad (1.5.5)$$

and

$$\beta := T(\rho_M) \begin{pmatrix} 1 & 0 \\ 0 & 1 \end{pmatrix}. \qquad (1.5.6)$$

Note that β is nonsingular.

The following results are straightforward.

Lemma 1.5.4. *Any finite representation of $T(M)$ is equivalent to one of the form $T(\rho)$, for some finite representation ρ of the twisted R-module M.*

Proof. Let $(W, \rho_{T(M)}, \beta)$ be a finite representation of the twisted ring $T(M) = (T(M), T(M), \mathrm{id})$. Let $\iota := \begin{pmatrix} 0 & 0 \\ 0 & 1 \end{pmatrix} \in T(M)$. Then ι is an idempotent in $T(M)$, $\iota \iota^J = \iota^J \iota = 0$ and $\iota + \iota^J = 1$. Moreover, $\iota T(M) \iota \cong R$ and $\iota^J T(M) \iota^J \cong R^{op}$. Define $V := \iota W$ and $V' := \iota^J W$. Then V is an R-module, V' is an R^{op}-module and $W = V \oplus V'$. For the nonsingular form β we get

$$\beta(\iota x, \iota y) = \beta(x, \iota^J \iota y) = 0 \text{ and } \beta(\iota^J x, \iota^J y) = \beta(x, \iota \iota^J y) = 0,$$

for all $x, y \in W$. Since β is nonsingular, it defines an isomorphism

$$V' \to \mathrm{Hom}(V, \mathbb{Q}/\mathbb{Z}), \ \iota^J y \mapsto (\iota x \mapsto \beta(x, \iota^J y)),$$

which is in fact an isomorphism of R^{op}-modules. $\qquad \square$

The next lemma follows immediately from Eq. (1.5.5).

Lemma 1.5.5. *The self-dual codes in the representation $T(\rho)$ take the form $C \oplus C^*$, where $C \leq V$ is a self-orthogonal code in the representation ρ and $C^* \subset \check{V}$ is the set of functionals that vanish on C.*

The following is a useful characterization of a triangular twisted ring.

Lemma 1.5.6. *A twisted ring (R, M, ψ) is a triangular twisted ring if and only if R contains an idempotent ι such that $\iota + \iota^J = 1$ and $\iota R \iota^J = 0$. Then, up to rescaling, $R \cong T(M_0)$, where $M_0 := \iota^J R \iota$ is viewed as a twisted $R_0 := \iota R \iota$-module.*

Proof. Clearly, $M_0 := \iota^J R \iota$ is a right $R_0 \otimes R_0$-module, from

$$\iota^J m \iota (\iota r \iota \otimes \iota s \iota) = \iota^J r^J \iota^J m \iota s \iota, \text{ for } r, s, m \in R.$$

The twist map $\tau_0 : M_0 \to M_0$ is given by the involution J:

$$\tau_0(\iota^J m \iota) := (\iota^J m \iota)^J = \iota^J m^J \iota, \text{ for } m \in R.$$

Every element in $r \in R$ can be written as

$$r = a_r + b_r + c_r, \text{ where } a_r = \iota^J r \iota^J \in R_0^J, \ b_r = \iota^J r \iota \in M_0, \ c_r = \iota r \iota \in R.$$

Since $\iota^J \iota = \iota \iota^J = 0$ and $\iota R \iota^J = 0$, we have

$$s := rr' = (a_r + b_r + c_r)(a_{r'} + b_{r'} + c_{r'}) = a_s + b_s + c_s,$$

with $a_s = a_r a_{r'} + b_r b_{r'} = \iota^J rr' \iota^J$, $c_s = c_r c_{r'} = \iota r \iota r' \iota = \iota rr' \iota$ (since $\iota R(1 - \iota) = 0$) and $b_s = a_r b_{r'} + b_r c_{r'}$. Hence the map

$$R \to T(M_0), \ r \mapsto \begin{pmatrix} a_r^J & b_r \\ 0 & c_r \end{pmatrix}$$

gives the desired isomorphism.   □

1.6 Quadratic pairs and their representations

When defining a "Type" of self-dual code we will need both bilinear forms and quadratic maps. The bilinear forms form an $(R \otimes R)$-module M and the quadratic maps form an R-qmodule Φ. We will also have certain mappings from the set of bilinear forms into the set of quadratic maps and vice versa. The two modules together with these "structure maps" form a "quadratic pair".

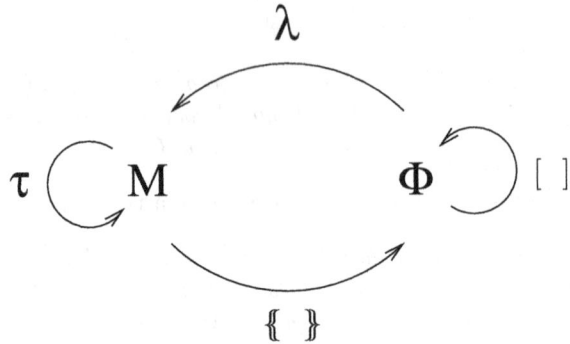

Fig. 1.1. Structure maps $\{\!\{\,\}\!\}$ and λ for a quadratic pair.

Definition 1.6.1. A *quadratic pair*[3] *over* R is a pair (M, Φ), where M is a twisted R-module and Φ is an R-qmodule, together with a pair of qmodule homomorphisms (or "structure maps") $\{\!\{\,\}\!\} : M \to \Phi$ and $\lambda : \Phi \to M$ (as in Fig. 1.1), satisfying the following conditions: for all $r, s \in R$, $m \in M$ and $\phi \in \Phi$,

$$\{\!\{\tau(m)\}\!\} = \{\!\{m\}\!\}\,,$$
$$\tau(\lambda(\phi)) = \lambda(\phi)\,,$$
$$\lambda(\{\!\{m\}\!\}) = m + \tau(m)\,,$$
$$\phi[r, s] := \phi[r + s] - \phi[r] - \phi[s] = \{\!\{\lambda(\phi)(r \otimes s)\}\!\}\,. \tag{1.6.1}$$

Note that we regard M as an R-qmodule with respect to the diagonal action $m[r] := m(r \otimes r)$. The quadratic pair (M, Φ) always has associated with it the maps $\lambda, \{\!\{\,\}\!\}, \tau$ as well as the $R \otimes R$-module structure of M and the qmodule structure of Φ.

Remark 1.6.2. The equality $\phi[r, s] = \{\!\{\lambda(\phi)(r \otimes s)\}\!\}$ implies that $\ker(\lambda)$ is a linear R-sub-qmodule of Φ.

Example 1.6.3. Let V be a left R-module and A an abelian group. Then

$$(M, \Phi) := (\mathrm{Bil}_{\mathbb{Z}}(V, A), \mathrm{Quad}_0(V, A)) \tag{1.6.2}$$

is a quadratic pair over R if we define

$$\tau(m)(v, w) = m(w, v)\,, \text{ for } m \in M, v, w \in V\,,$$
$$\{\!\{m\}\!\}(v) = m(v, v)\,, \text{ for } m \in M, v \in V\,,$$
$$\lambda(\phi)(v, w) = \phi(v, w) = \phi(v + w) - \phi(v) - \phi(w)\,, \text{ for } \phi \in \Phi, v, w \in V\,.$$

---

[3] "Quadratic pair" is used with a different meaning by Knus [324].

Remark. Let (M, Φ) be a quadratic pair. It follows from (1.6.1) that

$$\lambda(\{\!\{\lambda(\phi)\}\!\}) = 2\lambda(\phi), \text{ for } \phi \in \Phi,$$

and

$$\{\!\{\lambda(\{\!\{m\}\!\})\}\!\} = 2\{\!\{m\}\!\}, \text{ for } m \in M.$$

Definition 1.6.4. A *representation* $\rho := (V, \rho_M, \rho_\Phi)$ of a quadratic pair (M, Φ) over R consists of a representation (V, ρ_M) of the twisted module M (see Definition 1.3.3) and an additional R-qmodule homomorphism $\rho_\Phi : \Phi \to \mathrm{Quad}_0(V, A)$ that are compatible with the mappings λ and $\{\!\{\,\}\!\}$, i.e. are such that

$$\rho_M(\lambda(\phi))(x, y) = \rho_\Phi(\phi)(x + y) - \rho_\Phi(\phi)(x) - \rho_\Phi(\phi)(y), \quad \text{for } x, y \in V, \phi \in \Phi,$$
$$(1.6.3)$$

and

$$\rho_\Phi(\{\!\{m\}\!\})(x) = \rho_M(m)(x, x), \quad \text{for } x \in V, m \in M. \qquad (1.6.4)$$

The homomorphisms ρ_M and ρ_Φ map the quadratic pair (M, Φ) to the quadratic pair $(\mathrm{Bil}_\mathbb{Z}(V, A), \mathrm{Quad}_0(V, A))$ of (1.6.2). Again the representation ρ is said to be *finite* if R and V are finite sets and $A = \mathbb{Q}/\mathbb{Z}$.

Note that the function $\rho_\Phi(\phi)$ is linear if and only if the induced bilinear form is 0, i.e. if and only if $\rho_M(\lambda(\phi)) = 0$.

1.7 Form rings and their representations

The notions of "form parameter" and "form ring" were introduced by Bak [27] in order to define quadratic forms over a general class of rings (see also Hahn and O'Meara [226, §5.1], Loos [353] and W. Scharlau [475, §7.3]). Our definition of form ring, which combines the notions of twisted ring and quadratic pair, is somewhat more general than Bak's.

Definition 1.7.1. A *form ring* is a quadruple (R, M, ψ, Φ) with (R, M, ψ) a twisted ring such that (M, Φ) is a quadratic pair over R. We will often refer to this as "the form ring (R, Φ)". (Note that M is isomorphic to R via ψ.) We call (R, Φ) *finite* if $|R| < \infty$.

Definition 1.7.2. A *representation* $\rho := (V, \rho_M, \rho_\Phi, \beta)$ of a form ring (R, M, ψ, Φ) consists of a representation (V, ρ_M, β) of the twisted ring (R, M, ψ) together with an R-qmodule homomorphism

$$\rho_\Phi : \Phi \to \mathrm{Quad}_0(V, A)$$

which is compatible with ρ_M, i.e. such that (V, ρ_M, ρ_Φ) is a representation of the quadratic pair (M, Φ) over R. As usual ρ is called a *finite* representation if R and V are finite sets and $A = \mathbb{Q}/\mathbb{Z}$.

Example 1.7.3. Let V, A be abelian groups and $\beta \in \mathrm{Bil}_{\mathbb{Z}}(V, A)$ a nonsingular bilinear form. Then

$$\mathrm{End}(V; \beta, A) := (\mathrm{End}_{\mathbb{Z}}(V), \mathrm{Bil}_{\mathbb{Z}}(V, A), \psi, \mathrm{Quad}_0(V, A)) \qquad (1.7.1)$$

is a form ring, where ψ is defined by $\psi(1) := \beta$ (cf. (1.4.3), (1.6.2)). The involutions on $\mathrm{End}(V; \beta, A)$ are the obvious ones: τ is given by $\tau(m)(x, y) = m(y, x)$ for $m \in \mathrm{Bil}_{\mathbb{Z}}(V, A)$, and $q[-1](x) = q(-x)$ for $q \in \mathrm{Quad}_0(V, A)$. A representation ρ of a form ring (R, M, ψ, Φ) is then a form ring morphism from (R, M, ψ, Φ) into $\mathrm{End}(V; \beta, A)$. The representation ρ is *faithful* if the morphism is injective. For a finite representation we will omit A and simply write $\mathrm{End}(V; \beta)$ instead of $\mathrm{End}(V; \beta, \mathbb{Q}/\mathbb{Z})$.

Example 1.7.4. (Type II binary codes, cont.) As in Example 1.2.2, let $R :=$ $\mathbb{F}_2$, $V := \mathbb{F}_2$, $A := \frac{1}{4}\mathbb{Z}/\mathbb{Z}$, $M := \mathrm{Bil}(V, \frac{1}{2}\mathbb{Z}/\mathbb{Z}) = \{0, \beta\}$, where

$$\beta(x, y) := \frac{1}{2}xy, \qquad (1.7.2)$$

and

$$\Phi := \{0, \phi_0, 2\phi_0, 3\phi_0\} \subset \mathrm{Quad}_0(V, \frac{1}{4}\mathbb{Z}/\mathbb{Z}),$$

with

$$\phi_0(x) := \frac{1}{4}x^2. \qquad (1.7.3)$$

Then we have $\tau = \mathrm{id}$, $\{\!\!\{\beta\}\!\!\} = 2\phi_0$ and $\lambda(\phi_0) = \beta$ (these mappings are just the restrictions of those in the quadratic pair $(\mathrm{Bil}(V, A), \mathrm{Quad}_0(V, A))$). The equations (1.6.1) are satisfied, and so (M, Φ) is a quadratic pair over R. Moreover, the right R-module isomorphism $\psi : R \to M$ given by $\psi(1) := \beta$ makes (R, M, ψ, Φ) into a form ring.

Example 1.7.5. Bak [27] and others define a form ring to be a quadruple (R, M, ψ, Λ) such that (R, M, ψ) is a twisted ring and Λ is an R-sub-qmodule of M with $(1 + \tau)(M) \subset \Lambda \subset \ker(1 - \tau)$. Then (R, M, ψ, Λ) becomes a form ring according to our definition, with structure maps $\{\!\!\{\}\!\!\} = 1 + \tau$ and $\lambda = 1$. Similarly, $(R, M, \psi, M/\Lambda)$ becomes a form ring with the involution $\tau' = -\tau$ and with structure maps $\{\!\!\{\}\!\!\} = 1$ and $\lambda : M/\Lambda \to M$ defined by $\lambda(m + \Lambda) := m - \tau(m)$ (λ is well-defined since Λ is in the kernel of $1 - \tau$). Thus our Φ can be interpreted both as an analogue of Λ and of M/Λ. In the first case λ is injective while in the second case $\{\!\!\{\}\!\!\}$ is surjective.

Definition 1.7.6. An *ideal* in a form ring (R, Φ) is a pair (I, Γ), where I is an ideal in R and Γ is an R-sub-qmodule of Φ containing $\{\!\!\{\psi(I)\}\!\!\} + \Phi[I]$. This agrees with the usual definition of ideal when λ is injective. A form ring is *simple* if it has no nontrivial ideals. A *semisimple* form ring is a direct sum of simple form rings.

Remark 1.7.7. *Quotients of form rings.* Let $\varphi : (R_1, \Phi_1) \to (R_2, \Phi_2)$ be a morphism of form rings, and define the kernel of φ to be the pair (I, Γ) where I is the kernel of the homomorphism $R_1 \to R_2$ and Γ is the kernel of the homomorphism $\Phi_1 \to \Phi_2$. It is straightforward to verify that the kernel of a morphism is indeed an ideal. Conversely, given an ideal (I, Γ) in a form ring (R, Φ), we obtain a natural form ring structure on $(R/I, \Phi/\Gamma)$; the only nontrivial step in proving this involves the fact that for $\phi \in \Phi$, $r \in R$, $i \in I$, $\phi[r, i] = \{\!\{\lambda(\phi)(r \otimes i)\}\!\} \in \{\!\{\psi(I)\}\!\}$. Note that Bak's definition of form ring does not admit arbitrary quotients.

Remark 1.7.8. For a form ring (R, Φ), the pair $(\operatorname{rad} R, \quad \lambda^{-1}(\psi(\operatorname{rad} R)))$ (where $\operatorname{rad}(R)$ denotes the Jacobson radical of the ring R) forms an ideal which we call the *radical* of (R, Φ).

In the applications we will need the obvious notions of orthogonal sums of representations and conjugate representations, defined as follows.

Definition 1.7.9. Let ρ and ρ' be representations of a common form ring (R, M, ψ, Φ). The *orthogonal sum* $\rho + \rho'$ is the representation

$$\rho + \rho' := (V_\rho \oplus V_{\rho'}, \rho_M + \rho'_M, \rho_\Phi + \rho'_\Phi, \beta + \beta'),$$

where

$$((\rho_M + \rho'_M)(m))(v + v', w + w') = \rho_M(m)(v, w) + \rho'_M(m)(v', w')$$

and

$$((\rho_\Phi + \rho'_\Phi)(\phi))(v + v') = \rho_\Phi(\phi)(v) + \rho'_\Phi(\phi)(v'),$$

for all $v, w \in V_\rho$, $v', w' \in V_{\rho'}$, $m \in M$ and $\phi \in \Phi$. If N is a positive integer, we write

$$N\rho := \underbrace{\rho + \ldots + \rho}_{N} = (V_\rho^N, \rho_M^N, \rho_\Phi^N, \beta^N). \tag{1.7.4}$$

The *conjugate representation* of ρ is the representation

$$\bar{\rho} := (V_\rho, -\rho_M, -\rho_\Phi, -\beta).$$

1.8 The Type of a code

Definition 1.8.1. A *Type* $\mathcal{T}_\rho$ consists of a form ring (R, M, ψ, Φ) together with a representation $\rho := (V, \rho_M, \rho_\Phi, \beta)$. A *Type* $\mathcal{T}_\rho$ *code* of length N (or simply a *code of Type* ρ) is then defined to be an isotropic self-dual code in V^N (with respect to the quadratic pair $(\rho_M^N(M), \rho_\Phi^N(\Phi))$), that is, a submodule C of V^N satisfying

$$C = C^\perp := \{v \in V^N \mid \sum_{i=1}^{N} \beta(v_i, c_i) = 0, \text{ for all } c \in C\}, \tag{1.8.1}$$

and such that $\sum_{i=1}^{N} \phi(c_i) = 0$ for all $c \in C$ and $\phi \in \rho_\Phi(\Phi)$.

Since C is a submodule and the $1 \otimes R$-module $\rho_M(M)$ is spanned by β, the vector $v \in V^N$ is in the dual $C^\perp$ if and only if $\sum_{i=1}^{N} \rho_M(m)(v_i, c_i) = 0$ for all $m \in M$ and all $c \in C$.

Example 1.8.2. *Doubly-even self-dual codes over* $\mathbb{F}_2$. The appropriate form ring is

$$R(2_{\mathrm{II}}) := (\mathbb{F}_2, \mathbb{F}_2, \mathrm{id}, \mathbb{Z}/4\mathbb{Z} = \langle \phi_0 \rangle),$$

where ϕ_0 is given in (1.7.3), with structure maps defined by

$$\{\!\{1\}\!\} = 2\phi_0, \quad \lambda(\phi_0) = 1.$$

Then Example 1.7.4 gives a finite representation

$$\rho(2_{\mathrm{II}}) := (\mathbb{F}_2, \rho_M, \rho_\Phi, \beta)$$

of $R(2_{\mathrm{II}})$, defined by

$$\rho_M(1)(x,y) = \beta(x,y) := \frac{1}{2}xy \ \text{ and } \ \rho_\Phi(\phi_0)(x) := \frac{1}{4}x^2.$$

The representation $\rho(2_{\mathrm{II}})$ is the Type that defines doubly-even (or Type II) self-dual codes over $\mathbb{F}_2$.

Example 1.8.3. *Singly-even self-dual codes over* $\mathbb{F}_2$. The appropriate form ring is

$$R(2_{\mathrm{I}}) := (\mathbb{F}_2, \mathbb{F}_2, \mathrm{id}, \mathbb{F}_2),$$

with structure maps defined by

$$\{\!\{1\}\!\} = 1, \quad \lambda(1) = 0.$$

This form ring has a natural finite representation

$$\rho(2_{\mathrm{I}}) := (\mathbb{F}_2, \rho_M, \rho_\Phi, \beta)$$

defined by

$$\rho_M(1)(x,y) = \beta(x,y) := \frac{1}{2}xy \ \text{ and } \ \rho_\Phi(1)(x) := \frac{1}{2}x^2 = \frac{1}{2}x.$$

The representation $\rho(2_{\mathrm{I}})$ is the Type that defines singly-even (or Type I) self-dual codes over $\mathbb{F}_2$.

Example 1.8.4. *Self-dual codes over* $\mathbb{F}_p$ *containing the all-ones vector.* Let p be an odd prime. We use the fact that the all-ones vector $\mathbf{1}$ is in $C^\perp$ if and only if $\sum_{i=1}^{N} x_i \equiv 0 \bmod p$ for all $x \in C$. In order to apply this, we must enlarge the qmodule Φ: we take it to be $\mathbb{F}_p \oplus \mathbb{F}_p$, where the first component will correspond to the specializations of the bilinear forms in M, and the second component will correspond to the linear maps needed for our new constraint. Thus we define the form ring

$$(p)_1 := (\mathbb{F}_p, \mathbb{F}_p, \mathrm{id}, \mathbb{F}_p \oplus \mathbb{F}_p), \tag{1.8.2}$$

with structure maps given by

$$\{\!\{1\}\!\} = (1, 0), \quad \lambda((a, b)) = 2a.$$

We set $V := \mathbb{F}_p$, $A := \frac{1}{p}\mathbb{Z}/\mathbb{Z}$, and define

$$\beta : V \times V \to \frac{1}{p}\mathbb{Z}/\mathbb{Z} \quad \text{by} \quad \beta(x, y) := \frac{1}{p}xy.$$

We also define $\varphi : V \to \frac{1}{p}\mathbb{Z}/\mathbb{Z}$ by $\varphi(x) := \frac{1}{p}x = \beta(1, x)$, and set $\rho_M(a) := a\beta$ and

$$\rho_\Phi((a, b)) := \{\!\{\rho_M(a)\}\!\} + b\varphi. \tag{1.8.3}$$

The latter is the quadratic map from V to A given by

$$x \mapsto \frac{a}{p}x^2 + \frac{b}{p}x. \tag{1.8.4}$$

This defines a finite representation $\rho := (V, \rho_M, \rho_\Phi, \beta)$ of the form ring $(p)_1$. The codes of Type $\mathcal{T}_\rho$ are precisely the self-dual codes over $\mathbb{F}_p$ that contain the all-ones vector $\mathbf{1}$.

Remark 1.8.5. As the above examples illustrate, there is a technical distinction between the inner product on V^N used to define the dual code in the classical theory (which usually takes values in R) and the closely related bilinear form $\beta = \rho_M(\psi(1))$ used to define the Type (which always takes values in $\mathbb{Q}/\mathbb{Z}$).

Remark 1.8.6. A *sub-form ring* of the form ring (R, M, ψ, Φ) is a quadruple of subsets which is again a form ring with respect to the restrictions of the structure maps of (R, Φ). More precisely, a sub-form ring of (R, M, ψ, Φ) has the form $(R', \psi(R'), \psi_{|R'}, \Phi')$ where R' is a subring of R, $\psi(R')$ is closed under the involution τ, Φ' is an R' q-submodule of Φ and the restrictions of $\{\!\{\,\}\!\}$: $M' := \psi(R') \to \Phi'$ and $\lambda : \Phi' \to M'$ are well-defined.

If $\rho := (V, \rho_M, \rho_\Phi, \beta)$ is a representation of the form ring (R, Φ) and (R', Φ') is a sub-form ring of (R, Φ), then the restriction $\rho' := \rho_{|(R', \Phi')}$ is a representation of the form ring (R', Φ'), which we call a *sub-Type* of the Type ρ.

This has the somewhat unfortunate consequence that a code of Type ρ is also a code of Type ρ', whereas intuitively one expects the opposite to hold. Thus, formally, Type $\rho(2_\mathrm{I})$ is a sub-Type of Type $\rho(2_\mathrm{II})$, and additive codes over $\mathbb{F}_4$ (of Type $4^{\mathrm{H}+}$) are a sub-Type of linear codes (Type 4^{H}), whereas of course as sets the opposite is true: a Type 2_II code is also a 2_I code, and a linear code is also an additive code. The explanation for this apparent paradox is that a Type specifies constraints on a code, so a sub-Type has fewer constraints and includes more codes. We shall therefore avoid the use of "sub-Type" whenever possible.

1.9 Triangular form rings

The triangular ring construction (see Definition 1.5.1) also extends to form rings. To be precise, let (M, Φ) be a quadratic pair over R. Then we can construct a triangular form ring

$$T(M, \Phi) := (T(M), T(M), \mathrm{id}, \Phi')$$

from the triangular twisted ring $T(M)$ as follows: we set $\Phi' = R \times \Phi$, where the square bracket $[\,]$ qmodule action is given by

$$(r, \phi) \left[\begin{pmatrix} a & b \\ 0 & c \end{pmatrix} \right] = (arc, \phi[c] + \{ b(rc \otimes 1) \}). \tag{1.9.1}$$

The form maps are

$$\{ \begin{pmatrix} a & b \\ 0 & c \end{pmatrix} \} = (a + c, \{ b \}), \tag{1.9.2}$$

$$\lambda((r, \phi)) = \begin{pmatrix} r & \lambda(\phi) \\ 0 & r \end{pmatrix}. \tag{1.9.3}$$

It is easy to check that this construction defines a form ring.

Definition 1.9.1. Let $\rho := (V, \rho_M, \rho_\Phi)$ be a finite representation of a quadratic pair (M, Φ) over R. From the representation of the twisted R-module M we obtain a representation $(\tilde{V} \oplus V, T(\rho_M), \beta)$ of the twisted ring $T(M)$ (see Definition 1.5.3), where $\tilde{V} = \mathrm{Hom}(V, \mathbb{Q}/\mathbb{Z})$, which we extend to a representation

$$T(\rho) := (\tilde{V} \oplus V, T(\rho_M), T(\rho_\Phi), \beta)$$

of the triangular form ring $T(M, \Phi)$ by defining

$$T(\rho_\Phi)(r, \phi)(f, v) := f(rv) + \rho_\Phi(\phi)(v), \quad \text{for } r \in R, \phi \in \Phi, f \in \tilde{V}, v \in V.$$

We can check that $T(\rho)$ is a finite representation of the triangular form ring $T(M, \Phi)$. Just as in the case of triangular twisted rings, we have the following three easy lemmas, the proofs of which are analogous to those of Lemmas 1.5.4, 1.5.5 and 1.5.6.

Lemma 1.9.2. The finite representations of $T(M, \Phi)$ are of the form $T(\rho)$ for finite representations of ρ of (M, Φ).

Lemma 1.9.3. Let $\rho := (V, \rho_M, \rho_\Phi)$ be a finite representation of the quadratic pair (M, Φ). Then the isotropic self-dual codes in $T(\rho)$ take the form $C \oplus C^*$, where C is an isotropic self-orthogonal code in V and $C^* \subset \tilde{V}$ is the set of functionals that vanish on C.

Lemma 1.9.4. A form ring (R, M, ψ, Φ) is a triangular form ring if and only if there is an idempotent $\iota \in R$ such that $\iota + \iota^J = 1$, $\iota R \iota^J = 0$ and $\Phi[\iota] = 0$.

1.10 Matrix rings of form rings and their representations

As we will see in Chapter 2, in order to handle genus-m weight enumerators even for classical codes, we will need to allow the ground ring to be a matrix ring. For this purpose we introduce the concept of a matrix form ring. If (R, M, ψ, Φ) is a form ring, we define the *matrix form ring*

$$\mathrm{Mat}_n(R, M, \psi, \Phi) := (\mathrm{Mat}_n(R), \mathrm{Mat}_n(M), \psi_n = \mathrm{Mat}_n(\psi), \Phi_n), \qquad (1.10.1)$$

where Φ_n is the following set of upper triangular matrices:

$$\Phi_n = \left\{ \begin{pmatrix} \phi_1 & m_{12} & \cdots & m_{1n} \\ & \ddots & \ddots & \vdots \\ & & \ddots & m_{n-1,n} \\ & & & \phi_n \end{pmatrix} \; \middle| \; \phi_1, \ldots, \phi_n \in \Phi, m_{ij} \in M \right\},$$

the mapping λ_n is given by

$$\lambda_n \left(\begin{pmatrix} \phi_1 & m_{12} & \cdots & m_{1n} \\ & \ddots & \ddots & \vdots \\ & & \ddots & m_{n-1,n} \\ & & & \phi_n \end{pmatrix} \right) = \begin{pmatrix} \lambda(\phi_1) & m_{12} & \cdots & m_{1n} \\ \tau(m_{12}) & \ddots & \ddots & \vdots \\ \vdots & \ddots & \ddots & m_{n-1,n} \\ \tau(m_{1n}) & \cdots & \tau(m_{n-1,n}) & \lambda(\phi_n) \end{pmatrix},$$

$\{\!\{ \; \}\!\}_n$ is defined by

$$\left\{\!\!\left\{ \begin{pmatrix} m_{11} & \cdots & m_{1n} \\ \vdots & \cdots & \vdots \\ m_{n1} & \cdots & m_{nn} \end{pmatrix} \right\}\!\!\right\}_n = \begin{pmatrix} \{\!\{m_{11}\}\!\} & m_{12} + \tau(m_{21}) & \cdots & m_{1n} + \tau(m_{n1}) \\ & \ddots & \ddots & \vdots \\ & & \ddots & m_{n-1,n} + \tau(m_{n,n-1}) \\ & & & \{\!\{m_{nn}\}\!\} \end{pmatrix}$$

and $\psi_n = \mathrm{Mat}_n(\psi)$ is defined componentwise. Note in particular that J is given by the conjugate transpose: that is,

$$(r^J)_{ij} = (r_{ji})^J \text{ for } r \in \mathrm{Mat}_n(R).$$

The associated unit ϵ_n is the scalar matrix ϵI_n. The action of $\mathrm{Mat}_n(R)$ on $\mathrm{Mat}_n(M)$ and Φ_n is obtained by imitating matrix multiplication. (The case $n = 2$ can be seen in Eqs. (5.2.5), (5.2.6), etc., of Chapter 5.)

An equivalent definition is

$$\mathrm{Mat}_n(R, M, \psi, \Phi) := \mathrm{End}_R(R^n; \beta) \qquad (1.10.2)$$

(cf. (1.7.1)), where $\beta(v, w) = \sum_{1 \le i \le n} \psi(1)(v_i \otimes w_i)$.

Definition 1.10.1. Given a representation $\rho = (V, \rho_M, \rho_\Phi, \beta)$ of a form ring (R, M, ψ, Φ), the corresponding representation

$$\mathrm{Mat}_n(\rho) := (V^n, \rho_{\mathrm{Mat}_n(M)}, \rho_{\Phi_n}, \beta^n) \tag{1.10.3}$$

of $\mathrm{Mat}_n(R, M, \psi, \Phi)$ is defined in the obvious way. That is, V^n is a $\mathrm{Mat}_n(R)$-module via the usual matrix multiplication. The representation $\rho_{\mathrm{Mat}_n(M)}$ is defined by

$$\rho_{\mathrm{Mat}_n(M)}\left(\begin{pmatrix} m_{11} & \cdots & m_{1n} \\ \vdots & \cdots & \vdots \\ m_{n1} & \cdots & m_{nn} \end{pmatrix} \right)(x, y) := \sum_{i,j=1}^n m_{ij}(x_i, y_j),$$

for all $x = (x_1, \ldots, x_n)$ and $y = (y_1, \ldots, y_n) \in V^n$, $m_{ij} \in M$ $(1 \le i, j \le n)$. The representation ρ_{Φ_n} is defined by

$$\rho_{\Phi_n}\left(\begin{pmatrix} \phi_1 & m_{12} & \cdots & m_{1n} \\ & \ddots & \ddots & \vdots \\ & & \ddots & m_{n-1,n} \\ & & & \phi_n \end{pmatrix} \right)(x) = \sum_{i=1}^n \phi_i(x_i) + \sum_{i<j} m_{ij}(x_i, x_j),$$

for all $x = (x_1, \ldots, x_n) \in V^n$, $\phi_i \in \Phi$ and $m_{ij} \in M$.

We leave it as an exercise for the reader to show that this really does define a representation of a form ring.

In fact it follows from Morita theory (cf. Chapter 4) that all representations of $\mathrm{Mat}_n(R, M, \psi, \Phi)$ are of this form.

We illustrate the matrix ring construction by giving the matrix rings corresponding to the form rings $R(2_\mathrm{I})$ and $R(2_\mathrm{II})$ defined in Examples 1.8.3 and 1.8.2.

Example 1.10.2. We have

$$\mathrm{Mat}_n(R(2_\mathrm{II})) = (\mathrm{Mat}_n(\mathbb{F}_2), \mathrm{Mat}_n(\mathbb{F}_2), \mathrm{id}, \Phi_n),$$

where

$$\Phi_n = \left\{ \begin{pmatrix} \phi_1 & m_{12} & \cdots & m_{1n} \\ & \ddots & \ddots & \vdots \\ & & \ddots & m_{n-1,n} \\ & & & \phi_n \end{pmatrix} \mid \phi_1, \ldots, \phi_n \in \mathbb{Z}/4\mathbb{Z}, m_{ij} \in \mathbb{F}_2 \right\}.$$

The map $\{\!\!\}: \mathrm{Mat}_n(\mathbb{F}_2) \to \Phi_n$ sends a matrix (m_{ij}) to

$$\begin{pmatrix} 2m_{11} & m_{12}+m_{21} & \cdots & & m_{1n}+m_{n1} \\ & \ddots & \ddots & & \vdots \\ & & \ddots & m_{n-1,n}+m_{n,n-1} \\ & & & & 2m_{nn} \end{pmatrix}.$$

The map $\lambda : \Phi_n \to \mathrm{Mat}_n(\mathbb{F}_2)$ is defined by

$$\lambda\left(\begin{pmatrix} \phi_1 & m_{12} & \cdots & m_{1n} \\ & \ddots & \ddots & \vdots \\ & & \ddots & m_{n-1,n} \\ & & & \phi_n \end{pmatrix} \right) = \begin{pmatrix} \phi_1 \bmod 2 & m_{12} & \cdots & m_{1n} \\ m_{12} & \ddots & \ddots & \vdots \\ \vdots & \ddots & \ddots & m_{n-1,n} \\ m_{1n} & \cdots & m_{n-1,n} & \phi_n \bmod 2 \end{pmatrix}$$

and its image is the set of all symmetric matrices in $\mathrm{Mat}_n(\mathbb{F}_2)$. The involution τ is given by transposition.

Example 1.10.3. Similarly,

$$\mathrm{Mat}_n(R(2_{\mathrm{I}})) = (\mathrm{Mat}_n(\mathbb{F}_2), \mathrm{Mat}_n(\mathbb{F}_2), \mathrm{id}, \Phi_n),$$

where

$$\Phi_n = \left\{ \begin{pmatrix} \phi_1 & m_{12} & \cdots & m_{1n} \\ & \ddots & \ddots & \vdots \\ & & \ddots & m_{n-1,n} \\ & & & \phi_n \end{pmatrix} \mid \phi_1, \ldots, \phi_n \in \mathbb{F}_2, m_{ij} \in \mathbb{F}_2 \right\}.$$

The map $\{\!\!\{\ \}\!\!\} : \mathrm{Mat}_n(\mathbb{F}_2) \to \Phi_n$ sends a matrix (m_{ij}) to

$$\begin{pmatrix} m_{11} & m_{12}+m_{21} & \cdots & m_{1n}+m_{n1} \\ & \ddots & \ddots & \vdots \\ & & \ddots & m_{n-1,n}+m_{n,n-1} \\ & & & m_{nn} \end{pmatrix}.$$

The map $\lambda : \Phi_n \to \mathrm{Mat}_n(\mathbb{F}_2)$ is defined by

$$\lambda\left(\begin{pmatrix} \phi_1 & m_{12} & \cdots & m_{1n} \\ & \ddots & \ddots & \vdots \\ & & \ddots & m_{n-1,n} \\ & & & \phi_n \end{pmatrix} \right) = \begin{pmatrix} 0 & m_{12} & \cdots & m_{1n} \\ m_{12} & \ddots & \ddots & \vdots \\ \vdots & \ddots & \ddots & m_{n-1,n} \\ m_{1n} & \cdots & m_{n-1,n} & 0 \end{pmatrix},$$

and its image is the set of all symmetric matrices in $\mathrm{Mat}_n(\mathbb{F}_2)$ whose diagonal entries are 0. The latter is sometimes also denoted by $\mathrm{Alt}_n(\mathbb{F}_2)$, the set of all alternating matrices. The involution τ is given by transposition.

We will encounter symmetric matrices with even diagonal entries in several places (for instance in the definition of the theta-group in §9.1.5 and also in §8.1). They arise naturally as the image of λ_n in the matrix ring of a form ring, if the image of λ is $2M$.

Definition 1.10.4. Let S be a ring. A matrix $M \in \mathrm{Mat}_n(S)$ is called *even* if M is symmetric ($M_{ij} = M_{ji}$ for all $1 \leq i, j \leq n$) and $M_{ii} \in 2S$ for all $1 \leq i \leq n$. Let

$$\mathrm{Ev}_n(S) := \{M \in \mathrm{Mat}_n(S) \mid M \text{ is even }\}$$

denote the set of even $n \times n$ matrices over S. Then $\mathrm{Alt}_n(\mathbb{F}_2) = \mathrm{Ev}_n(\mathbb{F}_2)$.

1.11 Automorphism groups of codes

Codes that differ in only minor ways, such as the order in which the coordinates are arranged, are said to be equivalent. The precise meaning of this term depends on the particular Type of code being considered. We first define the automorphism group of a representation ρ.

Definition 1.11.1. Suppose the Type $\mathcal{T}_\rho$ is defined by the finite representation $\rho := (V, \rho_M, \rho_\Phi, \beta)$ of the form ring (R, M, ψ, Φ). An *automorphism of ρ* is an element $g \in \mathrm{End}(V)$ such that $g\rho(r) = \rho(r)g$, $\rho_M(m)(gv, gw) = \rho_M(m)(v, w)$ and $\rho_\Phi(\phi)(gv) = \rho_\Phi(\phi)(v)$, for all $r \in R, v, w \in V, m \in M$ and $\phi \in \Phi$. The set of all such automorphisms forms a subgroup of the unit group of $\mathrm{End}(V)$, which we call the *automorphism group* $\mathrm{Aut}(\rho)$.

Note that the nonsingularity of β implies that $\mathrm{Aut}(\rho)$ consists of bijective endomorphisms, and hence is indeed a subset of the unit group of $\mathrm{End}(V)$.

Automorphisms of ρ clearly preserve orthogonality and isotropy of codes of Type ρ. However, such automorphisms, together with permutations of the coordinates, do not in general give all the possible notions of code equivalence that are compatible with orthogonality and isotropy. For instance, for codes over finite fields that are not prime fields, two codes are usually considered to be equivalent if they differ by a Galois automorphism. This is only a weak automorphism, in the sense defined below, since it does not commute with the field action. (For codes over prime fields there are no such weak automorphisms.)

Definition 1.11.2. Let $\rho := (V, \rho_M, \rho_\Phi, \beta)$ be a finite representation of the form ring (R, M, ψ, Φ). A *weak automorphism of ρ* is an element $g \in \mathrm{End}(V)$ such that there exists a form-ring automorphism (defined below) $\alpha^g = (\alpha_R^g, \alpha_M^g, \alpha_\Phi^g)$ of (R, M, ψ, Φ) such that the following identities hold:

$$\begin{aligned} rgv &= g\alpha_R^g(r)v, & \text{for } r \in R, v \in V, \\ \rho_M(m)(gv, gw) &= \rho_M(\alpha_M^g(m))(v, w), & \text{for } m \in M, v, w \in V, \\ \rho_\Phi(\phi)(gv) &= \rho_\Phi(\alpha_\Phi^g(\phi))(v), & \text{for } \phi \in \Phi, v \in V. \end{aligned}$$

The set of all weak automorphisms forms the *weak automorphism group* $\mathrm{WAut}(\rho)$. A triple $\alpha := (\alpha_R, \alpha_M, \alpha_\Phi)$ of automorphisms of abelian groups $\alpha_R : R \to R$, $\alpha_M : M \to M$, $\alpha_\Phi : \Phi \to \Phi$ is called a *form ring automorphism* if α_R is a ring automorphism and the identities

$$
\begin{aligned}
\alpha_M(m(r \otimes s)) &= \alpha_M(m)(\alpha_R(r) \otimes \alpha_R(s)) \,, \\
\alpha_\Phi(\phi[r]) &= \alpha_\Phi(\phi)[\alpha_R(r)] \,, \\
\alpha_M(\lambda(\phi)) &= \lambda(\alpha_\Phi(\phi)) \,, \\
\alpha_\Phi(\{m\}) &= \{\alpha_M(m)\} \,,
\end{aligned}
$$

hold for all $r, s \in R$, $m \in M$, and $\phi \in \Phi$.

Note that we do not impose the condition that a form ring automorphism respects the chosen isomorphism ψ, and so β is not necessarily preserved under weak automorphisms. The automorphism group $\mathrm{Aut}(\rho)$ is a normal subgroup of $\mathrm{WAut}(\rho)$. One can think of weak automorphisms as being elements of the normalizer of R, whereas automorphisms centralize R. Examples of automorphism groups and weak automorphism groups of representations of form rings will be given in §2.3.

Next we define the automorphism group and weak automorphism group of a code C of Type ρ and length $N \geq 1$. For codes of length $N > 1$ we usually do not allow all automorphisms of $N\rho$, but only those that respect the chosen decomposition V^N, i.e. those that are in the wreath product

$$\mathrm{Aut}(\rho) \wr S_N := \{(a_1, \ldots, a_N)\pi \mid a_1, \ldots, a_N \in \mathrm{Aut}(\rho), \pi \in S_N\}, \quad (1.11.1)$$

which we call the *group of equivalences* of Type ρ and length N. For weak automorphisms, we consider the subgroup of the wreath product $\mathrm{WAut}(\rho) \wr S_N$ that normalizes R, that is, for which the same form-ring automorphism is applied to each coordinate. This is necessary to preserve orthogonality and isotropy of codes. So the weak automorphism group of a code is a subgroup of the *group of weak equivalences*

$$
\begin{aligned}
&(\mathrm{WAut}(\rho) \wr S_N) \cap \mathrm{WAut}(N\rho) \\
&= \{g \in \mathrm{WAut}(\rho) \wr S_N \mid \text{there is } \alpha^g \in \mathrm{Aut}((R, M, \psi, \Phi)) \\
&\qquad \text{such that } g((N\rho_X)(x)) = (N\rho_X)(\alpha_X^g(x)) \\
&\qquad\qquad \text{for } X \in \{R, M, \Phi\} \text{ and all } x \in X\}.
\end{aligned}
\quad (1.11.2)
$$

Definition 1.11.3. Let C, C' be two codes in $N\rho$.

a) C and C' are said to be *permutation-equivalent* if there is an element $g \in S_N$ with $gC = C'$.

b) C and C' are *equivalent* (resp. *weakly equivalent*) if there is an element g in the group of equivalences (resp. group of weak equivalences) with $gC = C'$.

c) The *permutation group* Perm(C) of C is the group of all $g \in S_N$ with $gC = C$. The *automorphism group* Aut(C) (resp. *weak automorphism group* WAut(C)) of C is the stabilizer of C in the group of equivalences (resp. weak equivalences).

Remark 1.11.4. 1. Usually when one speaks of the "automorphism group" of a code one is referring to the weak automorphism group WAut(C), since the larger group is to be preferred. When it is necessary to specify the smaller group Aut(C) we will call it the *strict* automorphism group. Knowledge of the smaller group is required for example when computing the group of a direct sum of codes.

2. The properties of being self-orthogonal, self-dual and isotropic are preserved under permutation equivalence, equivalence and weak equivalence.

1.12 Shadows

The notion of the shadow of a code or lattice, due to Conway and Sloane [129], [130], has been used quite fruitfully in proving nonexistence results—see also Theorem 11.1.12 of Chapter 11. This section translates this notion into our new language of form rings and so opens up the possibility of generalizing the shadow methods to new Types of codes.

Let $\rho := (V, \rho_M, \rho_\Phi, \beta)$ be a finite representation of the form ring (R, M, ψ, Φ). Let C be a self-orthogonal code in $N\rho$. Since $\rho_M(M)$ is zero on C, the mapping

$$\rho_\Phi^N(\phi) : C \to \mathbb{Q}/\mathbb{Z}$$

is linear for any $\phi \in \Phi$ (see the remark at the end of §1.6).

Definition 1.12.1. The *maximal isotropic subcode* C_0 of C is

$$C_0 := C \cap \bigcap_{\phi \in \Phi} \ker(\rho_\Phi^N(\phi)). \tag{1.12.1}$$

For $\phi \in \Phi$, we define the *ϕ-shadow* of C to be

$$S_\phi(C) := \{v \in V^N \mid \beta^N(v, c) = (\rho_\Phi^N(\phi))(c) \text{ for } c \in C\}. \tag{1.12.2}$$

Clearly, if C is isotropic, then $C_0 = C$ and $S_\phi(C) = C^\perp$ for all $\phi \in \Phi$. More generally, we have the following:

Theorem 1.12.2. The *ϕ-shadow* $S_\phi(C)$ is a coset of $C^\perp$, for any $\phi \in \Phi$.

Proof. We first show that $S_\phi(C)$ is nonempty. Since $\mathbb{Q}/\mathbb{Z}$ is injective, every linear functional on C extends to a linear functional on V^N. Since β is nonsingular, every linear functional on V^N is of the form

$$w = (w_1, \ldots, w_N) \mapsto \beta^N(v, w) = \sum_{i=1}^{N} \beta(v_i, w_i),$$

for some $v \in V^N$. Hence $S_\phi(C) \neq \emptyset$. If $v, v' \in S_\phi(C)$ then $v - v' \in C^\perp$, and conversely for all $d \in C^\perp$ and $v \in S_\phi(C)$ the vector $v + d \in S_\phi(C)$, so the ϕ-shadow is a coset of $C^\perp$.   □

Remark. The ϕ-shadow S_ϕ only depends on the coset $\phi + \{\!\{M\}\!\}$, i.e. if $\phi' = \phi + \{\!\{m\}\!\}$ then $S_{\phi'}(C) = S_\phi(C)$ for all self-orthogonal codes C in the representation $N\rho$. The number of different notions of shadow for a given Type ρ is the index $[\rho_\Phi(\Phi) : \rho_\Phi(\{\!\{M\}\!\})]$ of the subgroup $\rho_\Phi(\{\!\{M\}\!\})$ in the image of ρ_Φ.

Remark 1.12.3. The ϕ-shadow is particularly easy to describe if the quadratic map ϕ is linear and $\phi \in \ker(\lambda)$. In this case the nonsingularity of β implies that there is an element $v_0 \in V$ such that

$$\rho_\Phi(\phi)(x) = \beta(v_0, x) \text{ for } x \in V.$$

If $C \leq V^N$ is a code in $N\rho$, then

$$S_\phi(C) = \{v \in V^N \mid \beta^N(v, c) = \beta^N(v_0, c) \text{ for } c \in C\} = v_0 + C^\perp.$$

Example 1.12.4. Consider the representation $\rho(2_{\mathrm{II}})$ of the form ring $R(2_{\mathrm{II}}) = (\mathbb{F}_2, \mathbb{F}_2, \mathrm{id}, \mathbb{Z}/4\mathbb{Z} = \langle \phi_0 \rangle)$ defined in Example 1.8.2. The isotropic codes for this representation are the doubly-even binary codes. Singly-even self-dual binary codes are still self-dual for this representation (but in general are not isotropic).

Let C be a self-orthogonal code in $N\rho(2_{\mathrm{II}})$. For $\phi = 0$ or $\phi = 2\phi_0$, the ϕ-shadow $S_\phi(C) = C^\perp$. For $\phi = \pm\phi_0$, $S_\phi(C)$ is the set $S(C)$ of "parity vectors" for C:

$$S(C) = \{u \in \mathbb{F}_2^N \mid (u, c) \equiv (\frac{1}{2} \mathrm{wt}(c)) \bmod 2 \text{ for all } c \in C\}$$

$$= \{u \in V^N \mid \beta^N(u, c) \equiv (\frac{1}{4} \mathrm{wt}(c)) \bmod \mathbb{Z} \text{ for all } c \in C\}. \quad (1.12.3)$$

$S(C)$ is the usual shadow of a self-orthogonal code C, introduced in Conway and Sloane [130].[4]

In particular, if C is doubly-even, then $S(C) = C^\perp$, and the shadow of a self-dual doubly-even code is the code itself. If C is a singly-even self-dual code then C_0 is the doubly-even index 2 subcode of C, and the shadow $S(C) = C_0^\perp \setminus C$.

---

[4] A different generalization of shadow was proposed by Brualdi and Pless [86], but since it fails to possess the crucial properties given in (1.12.3) and Theorem 2.2.8 of Chapter 2 we shall not discuss it here.

Example 1.12.5. There is a second representation of the form ring $R(2_{II})$ which is of interest. This is the representation $\rho(2_{II})' := (\mathbb{F}_4, \rho'_M, \rho'_\Phi, \beta')$, where

$$\beta'(x, y) = \rho'_M(1)(x, y) = \frac{1}{2}x\bar{y}, \quad \rho'_\Phi(\phi_0)(x) = \frac{1}{2}x\bar{x},$$

and $x \mapsto \bar{x}$ denotes the nontrivial Galois automorphism of $\mathbb{F}_4$. The self-orthogonal codes in $N\rho(2_{II})'$ are the additive trace-Hermitian self-orthogonal codes over the alphabet $\mathbb{F}_4$. Codes of Type $\rho(2_{II})'$ are precisely the even additive Hermitian self-dual codes over $\mathbb{F}_4$ studied in Calderbank, Rains, Shor and Sloane [96], and which will appear as Type 4_{II}^{H+} in §2.3.4. Note however that because this representation is not faithful, the form ring $R(4_{II}^{H+})$ is a proper quotient of the form ring $R(2_{II})$. So it is a slight abuse of notation to say that $\rho(2_{II})' = \rho(4_{II}^{H+})$. These codes are important for quantum error-correction (see Chapter 13). The ϕ_0-shadow S_{ϕ_0} is exactly the shadow defined in [96]. Note that, even though these codes have $V = \mathbb{F}_4$ as their alphabet, they are not linear, only additive, and so $R = \mathbb{F}_2$ is the appropriate ground ring.

Example 1.12.6. Let m be an even integer and consider the representation $\rho(m_{II}^{\mathbb{Z}}) := (V, \rho_M, \rho_\Phi, \beta)$ of the form ring

$$R(m_{II}^{\mathbb{Z}}) := (R, M, \psi, \Phi = \langle \phi_0 \rangle) = (\mathbb{Z}/m\mathbb{Z}, \mathbb{Z}/m\mathbb{Z}, \text{id}, \mathbb{Z}/2m\mathbb{Z}),$$

where $V = \mathbb{Z}/m\mathbb{Z}$, $\rho_M(a)(x, y) = \frac{1}{m}axy$, and

$$\rho_\Phi(\phi_0)(x) := \frac{1}{2m}x^2.$$

If C is a self-dual code of Type $\rho(m_{II}^{\mathbb{Z}})$ then $S_{\phi_0}(C)$ is the usual shadow $S(C)$ (Conway and Sloane [130], [454]).

Remark. The above (by now classical) example illustrates the situation where shadows appear. Usually we begin with the smaller form ring. Given a sub-form ring $R := (R, M, \psi, \Phi)$ of a form ring $\tilde{R} := (R, M, \psi, \tilde{\Phi})$ (where the only difference between R and $\tilde{R}$ is that the form structure $\tilde{\Phi}$ of $\tilde{R}$ strictly contains that of R) and a finite representation ρ of R that extends to a representation $\tilde{\rho}$ of $\tilde{R}$ (such that ρ is the restriction $\tilde{\rho}_{|R}$), then for any $\phi_0 \in \tilde{\Phi} \setminus \Phi$ there is a ϕ_0-shadow of any code C of Type ρ, namely

$$S_{\phi_0}(C) = \{v \in V_\rho^N \mid \beta^N(v, c) = (\tilde{\rho}_\Phi^N(\phi_0))(c) \text{ for } c \in C\}.$$

We call such pairs of Types $(\tilde{\rho}, \rho)$, or the pair of form rings $(\tilde{R}, R)$, *shadow pairs*. This construction will be extensively used in Chapter 2, and Table 1.1 lists the shadow pairs that will be mentioned there.

Example 1.12.7. The condition that the usual shadow of a binary self-dual code contains the all-ones vector $\mathbf{1}$ can also be expressed in the language of

Table 1.1. The principal pairs $(\tilde{\rho}, \rho)$ of form rings used to define shadows.

ρ	$\tilde{\rho}$	Conditions	Remarks
$\rho(2_{\mathrm{I}})$	$\rho(2_{\mathrm{II}})$		
$\rho(q^{\mathrm{E}})$	$\rho(q_{\mathrm{II}}^{\mathrm{E}})$	q even prime power	
$\rho(q^{\mathrm{E}})$	$\rho(q_1^{\mathrm{E}})$	q odd prime power	ϕ linear
$\rho(q^{\mathrm{H}})$	$\rho(q_1^{\mathrm{H}})$	q arbitrary prime power	ϕ linear
$\rho(4^{\mathrm{H}+})$	$\rho(4_{\mathrm{II}}^{\mathrm{H}+})$		
$\rho(q^{\mathrm{H}+})$	$\rho(q_{\mathrm{II}}^{\mathrm{H}+})$	q even prime power	
$\rho(q^{\mathrm{H}+})$	$\rho(q_1^{\mathrm{H}+})$	q arbitrary prime power	ϕ linear
$\rho(m^{\mathbb{Z}})$	$\rho(m_{\mathrm{II}}^{\mathbb{Z}})$	m even integer	
$\rho(m^{\mathbb{Z}})$	$\rho(m_1^{\mathbb{Z}})$	m arbitrary integer	ϕ linear
$\rho(\mathrm{GR}(p^e,f)^{\mathrm{E}})$	$\rho(\mathrm{GR}(p^e,f)_{p^s}^{\mathrm{E}})$	p arbitrary prime	ϕ linear
$\rho(\mathrm{GR}(2^e,f)^{\mathrm{E}})$	$\rho(\mathrm{GR}(2^e,f)_{\mathrm{II}}^{\mathrm{E}})$		
$\rho(\mathrm{GR}(p^e,f)^{\mathrm{H}})$	$\rho(\mathrm{GR}(p^e,f)_{p^s}^{\mathrm{H}})$	p arbitrary prime	ϕ linear
$\rho(\mathrm{GR}(p^e,f)^{\mathrm{H}+})$	$\rho(\mathrm{GR}(p^e,f)_{p^s}^{\mathrm{H}+})$	p arbitrary prime	ϕ linear

form rings. This does not produce any new codes here, but the technique will be used again in the next chapter. Consider the representation

$$\rho(2_{\mathrm{S}}) := (V, \rho_M, \rho_\Phi^*, \beta)$$

of the form ring $R(2_{\mathrm{II}}) = (\mathbb{F}_2, \mathbb{F}_2, \mathrm{id}, \mathbb{Z}/4\mathbb{Z} = \langle \phi_0 \rangle)$, where $V = R = \mathbb{F}_2$, $\beta(x, y) = \rho_M(1)(x, y) = \frac{1}{2}xy$, and

$$\rho_\Phi^*(\phi_0)(x) := \frac{1}{4}x^2 + \frac{1}{2}x. \tag{1.12.4}$$

From (1.12.3), the shadow of C is the set of all $v \in \mathbb{F}_2^N$ such that, for all $c \in C$, the sum $s(c, v) := \frac{1}{2}\sum_{i=1}^N c_i^2 \equiv \sum_{i=1}^N c_i v_i \pmod{2}$. In particular, $\rho_\Phi^*(\phi_0)(c) = s(c, \mathbf{1})$, and hence an isotropic code (with respect to $\rho_\Phi^*(\phi_0)$) has the property that $\mathbf{1}$ is in the shadow. So isotropic self-dual codes of Type $\rho(2_{\mathrm{S}})$ are the self-dual binary codes C such that the shadow $S_{\phi_0}(C)$ contains $\mathbf{1}$.

Since $x = x^2 \pmod{2}$, we have

$$\rho_\Phi^*(\phi_0)(x) = -\frac{1}{4}x^2,$$

hence $\rho(2_S)$ is simply the conjugate of the representation $\rho(2_{\mathrm{II}})$ (cf. 1.8.2). Indeed, the representation $\rho(2_{\mathrm{II}}) + \rho(2_S)$ admits the isotropic code $\{00, 11\}$ (since the usual shadow of this code contains the vector 01).

Since the dual of a self-orthogonal binary code necessarily contains $\mathbf{1}$, the only way its shadow can contain $\mathbf{1}$ is if the code is doubly-even. We conclude that the codes of Type $\rho(2_S)$ are exactly the same as codes of Type $\rho(2_{\mathrm{II}})$, i.e. doubly-even self-dual binary codes.

We end this chapter with an application of the methods of Section 4.6. This is out of logical order, but it seems appropriate to give the result here.

Theorem 1.12.8. *If C is a self-dual binary code of length N, then every vector of the shadow $S(C)$ has weight congruent to $N/2$ (mod 4).*

Proof. Let v be a vector of weight l in $S(C)$; without loss of generality we may take $v = 0^k 1^l$. Then C is an isotropic code in the representation $k\rho(2_{II}) + l\rho(2_S)$, which must therefore be Witt-null (cf. Definition 4.6.6). Since $\rho(2_{II})$ has order 8 in the Witt group, this representation is Witt-null if and only if $k - l \equiv 0$ (mod 8), or in other words if $2l \equiv N$ (mod 8). $\square$

2

Weight Enumerators and Important Types

We begin this chapter by defining various weight enumerators that can be associated with a code, and then discuss the MacWilliams identity, which relates the weight enumerators of a code and its dual. In §2.3 we show how to describe the most important families of self-dual codes that have been studied to date as Types, using our new language of representations of form rings. Section §2.4 gives a number of important examples of self-dual codes of the different Types together with their weight enumerators.

In §2.3.6 we introduce many new Types that describe self-dual codes over general Galois rings. Although such codes have received little attention so far, this section is significant because it illustrates how our methods could be applied in the future if further classes of self-dual codes should arise.

2.1 Weight enumerators of codes

One important measure of the error-correcting ability of a code is its *minimal Hamming distance*, which is the minimal number of distinct components between any pair of distinct codewords. For a linear code (cf. §1.2), this is easily seen to be the minimal Hamming weight of the code, defined as follows.

Let R be a ring, V an R-module and let $C \leq V^N$ be a code over the alphabet V.

Definition 2.1.1. Let $c := (c_1, \ldots, c_N) \in V^N$. The *Hamming weight* of c is

$$\mathrm{wt}(c) := |\{i \mid c_i \neq 0\}|.$$

The *minimal Hamming weight* of C is

$$\mathrm{wt}(C) := \min\{\mathrm{wt}(c) \mid 0 \neq c \in C\}.$$

More generally, the *composition* of $c \in C$ is defined as follows: for $v \in V$, $a_v(c) := |\{i \mid c_i = v\}|$ is the number of components of c that are equal to v. Then $\mathrm{wt}(c) = N - a_0(c)$.

If $R = V = \mathbb{F}_q$ is a finite field, then a code $C \leq V^N$ of length N, dimension k and minimal Hamming weight d is sometimes referred to as an $[N, k, d]_q$ code. Here k is the dimension of C considered as a vector space over R. If $V \neq R$ or if R is not a field, then we usually define the "dimension" k to be

$$k := \frac{\log(|C|)}{\log(|V|)}.$$

We also say that C has *rate* k/N.

Two other sorts of "weight" are useful for nonbinary codes. For codes over $V = \mathbb{Z}/m\mathbb{Z} = \{0, 1, \ldots, m-1\}$ (which include codes over all finite fields of prime order) we define the *Lee weight* and *Euclidean norm* of $u \in V$ by

$$\mathrm{Lee}(u) = \min\{u, m - u\},$$
$$\mathrm{Norm}(u) = (\mathrm{Lee}(u))^2.$$

For a vector $u = (u_1, \ldots, u_N) \in V^N$, we set

$$\mathrm{Lee}(u) = \sum_{i=1}^{N} \mathrm{Lee}(u_i),$$

$$\mathrm{Norm}(u) = \sum_{i=1}^{N} \mathrm{Norm}(u_i).$$

Of course, if u is a binary vector, $\mathrm{wt}(u) = \mathrm{Lee}(u) = \mathrm{Norm}(u)$.

For example, for the ring $\mathbb{Z}/4\mathbb{Z} = \{0, 1, 2, 3\}$, the Lee weights are respectively $\{0, 1, 2, 1\}$. By analogy, Gaborit, Pless, Solé and Atkin [186] define "Lee" weights on the field $\mathbb{F}_4 = \{0, 1, \omega, \omega^2\}$ to be $\{0, 2, 1, 1\}$, respectively.

Usually the *weight enumerator* of a code is a polynomial attached to the code that gives, for example, the number of codewords of a given weight or with a given composition. We will meet several different kinds of weight enumerator.

Definition 2.1.2. Let $C \leq V^N$ be a code of length N.

- The *Hamming weight enumerator* of C is

$$\mathrm{hwe}(C)(x, y) := \sum_{c \in C} x^{N - \mathrm{wt}(c)} y^{\mathrm{wt}(c)} \in \mathbb{C}[x, y].$$

- The *complete weight enumerator* of C is

$$\mathrm{cwe}(C) := \sum_{c \in C} \prod_{i=1}^{N} x_{c_i} = \sum_{c \in C} \prod_{v \in V} x_v^{a_v(c)} \in \mathbb{C}[x_v \mid v \in V], \qquad (2.1.1)$$

again a homogeneous polynomial of degree N, which classifies the codewords by composition.

Remark. The Hamming weight enumerator $\mathrm{hwe}(C)(x, y)$ is equal to $\mathrm{cwe}(C)(x, y, \ldots, y)$, where the first argument of $\mathrm{cwe}(C)$ corresponds to $0 \in V$.

One reason for investigating weight enumerators is that, for self-dual codes, these polynomials have certain invariance properties that only depend on the Type of the code. This motivates us to study the ring of all polynomials with these same invariance properties. In many important cases it is a remarkable fact, first observed by Gleason [191], that this ring turns out to be spanned by the (complete or Hamming) weight enumerators of isotropic self-dual codes of the given Type—our main theorems, Theorems 5.5.5 and 5.5.7, provide a very general setting for this result. From this we can often obtain *a priori* information on the weight distribution of a code of given Type, for example upper bounds on the minimal distance of the code. This leads to the notion of extremal codes—see Chapter 11.

All these weight enumerators can be obtained as symmetrizations of the full weight enumerator of the code, which we define next. This is not a polynomial at all, but rather a formal sum over all codewords. While the other weight enumerators are only invariants of the (equivalence class of the) code, the full weight enumerator determines the code uniquely and in fact can be regarded as a copy of the code itself. Though full weight enumerators are not very useful from a practical point of view, we will in fact establish our main theorems (Theorems 5.5.5 and 5.5.7) for full weight enumerators. The analogous theorems for other weight enumerators can then be obtained by applying a symmetrization process (see §5.7). In particular, our main theorems also hold for complete weight enumerators.

Let R be a ring, V a finite R-module and let $C \leq V^N$ be a code.

Definition 2.1.3. Let $\mathbb{C}[V^N]$ denote the group ring of V^N, that is, the vector space generated over $\mathbb{C}$ by the symbols e_v with $v \in V^N$. The *full weight enumerator* of C is

$$\mathrm{fwe}(C) := \sum_{c \in C} e_c \in \mathbb{C}[V^N]. \qquad (2.1.2)$$

Remark 2.1.4. The complete weight enumerator of C is the image of the full weight enumerator under the projection $\mathbb{C}[V^N] \to \mathbb{C}[x_v \mid v \in V]$ defined by $e_{(v_1, \ldots, v_N)} \mapsto \prod_{i=1}^{N} x_{v_i}$.

It is sometimes useful to symmetrize $\mathrm{cwe}(C)$ in other ways, by identifying the variables x_a under the action of various permutation groups. If $R = V = \mathbb{F}_q$ is a finite field, then symmetrizing $\mathrm{cwe}(C)$ by the Galois group of $\mathbb{F}_q$ over its prime field yields the *symmetrized weight enumerator* $\mathrm{swe}(C)$. In the case $R = V = \mathbb{Z}/m\mathbb{Z}$, we may also identify x_a and x_b if $a \in V$ and $b \in V$ have the same Lee weight: this leads to the *Lee-symmetrized weight enumerator*, which we also denote by $\mathrm{swe}(C)$.

In the general situation when C is a code of Type ρ, the appropriate notion of symmetrized weight enumerator uses the automorphism group of ρ (see Definition 1.11.1). This leads to the following definition:

Definition 2.1.5. Let $\rho := (V, \rho_M, \rho_\Phi, \beta)$ be a finite representation of a form ring (R, M, ψ, Φ) and let C be a code in V^N. Let $o_1, \ldots, o_n$ denote the orbits of $\mathrm{Aut}(\rho)$ on V. For $c \in V^N$ and $i \in \{1, \ldots, n\}$, let

$$a_i(c) := |\{j \in \{1, \ldots, N\} \mid c_j \in o_i\}|$$

denote the number of components of c that belong to the orbit o_i. Then the ρ-symmetrized weight enumerator of C is

$$\mathrm{swe}^\rho(C) := \sum_{c \in C} \prod_{i=1}^n x_i^{a_i(c)} \in \mathbb{C}[x_1, \ldots, x_n].$$

Note that $\mathrm{Aut}(\rho)$ and hence also $\mathrm{swe}^\rho(C)$ depends on the representation ρ of the form ring as well as on the alphabet V. So the same code C may have different symmetrized weight enumerators when considered as a code in different representations, as the following example illustrates:

Example 2.1.6. Let $R = V := \mathbb{F}_q$, where q is an odd prime power, and consider $\mathbb{F}_q$-linear codes in V^N. There are two important Types in this family, Euclidean self-dual codes and Euclidean self-dual codes that contain the all-ones vector $\mathbf{1}$ (we have already seen a special case of the latter Type in Example 1.8.4). The corresponding form rings and their representations are discussed further in §2.3.2 below (see page 46).

For the Euclidean self-dual Type, the form ring is just

$$R(q^{\mathrm{E}}) = (\mathbb{F}_q, M = \mathbb{F}_q, \mathrm{id}, \mathbb{F}_q = \{\!\{M\}\!\}), \tag{2.1.3}$$

and the corresponding representation $\rho(q^{\mathrm{E}}) = (\mathbb{F}_q, \rho_M, \rho_\Phi, \beta)$ is defined by

$$\rho_M(a)(x, y) = \frac{1}{p} \mathrm{Tr}(axy), \quad \beta = \rho_M(1) \text{ and } \rho_\Phi(a)(x) = \frac{1}{p} \mathrm{Tr}(ax^2),$$

for $a \in M = \mathbb{F}_q$, $x, y \in V = \mathbb{F}_q$, where Tr denotes the trace from $\mathbb{F}_q$ to its prime field $\mathbb{F}_p$. We identify $\mathbb{F}_p$ with $\mathbb{Z}/p\mathbb{Z}$. The automorphism group of $\rho(q^{\mathrm{E}})$ (see Definition 1.11.1) consists of multiplication by the elements $g \in \mathbb{F}_q^*$ that satisfy $\mathrm{Tr}(gxgy) = \mathrm{Tr}(xy)$ and $\mathrm{Tr}(a(gx)^2) = \mathrm{Tr}(ax^2)$ for all $x, y \in \mathbb{F}_q$ and all $a \in \mathbb{F}_q$. Since the trace is nondegenerate, these conditions are equivalent to $g^2 = 1$, and hence $\mathrm{Aut}(\rho(q^{\mathrm{E}})) = \{1, -1\} \leq \mathbb{F}_q^*$. This group has $1 + \frac{q-1}{2}$ orbits on $V = \mathbb{F}_q$, so the $\rho(q^{\mathrm{E}})$-symmetrized weight enumerator is a polynomial in $\frac{q+1}{2}$ variables.

On the other hand, for Euclidean self-dual codes that contain $\mathbf{1}$, we may take the same R-module $M = \mathbb{F}_q$, but now the R-qmodule Φ (and hence the set of quadratic maps) must strictly contain $\{\!\{M\}\!\}$, as in Example 1.8.4. The form ring is

$$R(q_1^{\mathrm{E}}) := (\mathbb{F}_q, \mathbb{F}_q, \mathrm{id}, \mathbb{F}_q \oplus \mathbb{F}_q = \{\!\{M\}\!\} \oplus \{\phi_b \mid b \in \mathbb{F}_q\}), \tag{2.1.4}$$

with the representation $\rho(q_1^E) := (\mathbb{F}_q, \rho_M, \rho_\Phi, \beta)$ defined by $\rho_M(a)(x, y) := \frac{1}{p} \operatorname{Tr}(axy)$ and

$$\rho_\Phi((a, \phi_b))(x) := \frac{1}{p} \operatorname{Tr}(ax^2) + \frac{1}{p} \operatorname{Tr}(bx) \qquad (2.1.5)$$

(compare (1.8.3), (1.8.4)). Therefore the automorphism group of $\rho(q_1^E)$ consists of the elements $g \in \operatorname{Aut}(q^E)$ that additionally satisfy $\operatorname{Tr}(gx) = \operatorname{Tr}(x)$ for all $x \in \mathbb{F}_q$. This implies that $g = 1$ and hence $\operatorname{Aut}(\rho(q_1^E)) = \{1\}$. Therefore the $\rho(q_1^E)$-symmetrized weight enumerator is the same as the complete weight enumerator, a polynomial in q variables.

Additional information about a code can be obtained by considering more than one codeword at a time, leading to what are called *higher-genus* or *multiple weight enumerators*.

Definition 2.1.7. Let $c^{(i)} := (c_1^{(i)}, \ldots, c_N^{(i)}) \in V^N$, $i = 1, \ldots, m$, be m not necessarily distinct codewords. For $v := (v_1, \ldots, v_m) \in V^m$, let

$$a_v(c^{(1)}, \ldots, c^{(m)}) := |\{j \in \{1, \ldots, N\} \mid c_j^{(i)} = v_i \text{ for } i \in \{1, \ldots, m\}\}|.$$

The *genus-m complete weight enumerator* of C is

$$\operatorname{cwe}_m(C) := \sum_{(c^{(1)}, \ldots, c^{(m)}) \in C^m} \prod_{v \in V^m} x_v^{a_v(c^{(1)}, \ldots, c^{(m)})} \in \mathbb{C}[x_v \mid v \in V^m].$$

The genus-2 (complete) weight enumerator of a binary code is also called the *biweight enumerator*.

The analogous definition of ρ-symmetrized weight enumerator reads as follows:

Definition 2.1.8. Let ρ be a finite representation of a form ring with underlying module V and let $C \leq V^N$ be a code. For $m \in \mathbb{N}$, $\operatorname{Aut}(\rho)$ acts diagonally on V^m; let $o_1, \ldots, o_n$ denote its orbits under this action. For $c^{(i)} := (c_1^{(i)}, \ldots, c_N^{(i)}) \in V^N$, $i = 1, \ldots, m$, and $k \in \{1, \ldots, n\}$, define

$$a_k(c^{(1)}, \ldots, c^{(m)}) := |\{j \in \{1, \ldots, N\} : (c_j^{(1)}, \ldots, c_j^{(m)}) \in o_k\}|.$$

The *genus-m ρ-symmetrized weight enumerator* of C is

$$\operatorname{swe}_m^\rho(C) := \sum_{(c^{(1)}, \ldots, c^{(m)}) \in C^m} \prod_{k=1}^n x_k^{a_k(c^{(1)}, \ldots, c^{(m)})} \in \mathbb{C}[x_1, \ldots, x_n].$$

We may also define higher genus full weight enumerators, although they cannot reveal more information about the code than the full weight enumerator itself:

Definition 2.1.9. Let $\mathbb{C}[(V^m)^N] \cong \otimes^m \mathbb{C}[V^N]$ denote the group ring of V^{mN}, that is, the vector space generated over $\mathbb{C}$ by the symbols e_v with $v \in V^{mN}$. Then the *genus-m full weight enumerator* of C is

$$\text{fwe}_m(C) := \sum_{(c_1,\dots,c_m)\in C^m} e_{(c^{(1)},\dots,c^{(m)})} \in \mathbb{C}[V^{mN}].$$

As in Remark 2.1.4, the genus-m complete weight enumerator and genus-m ρ-symmetrized weight enumerator are just appropriate symmetrizations of the genus-m full weight enumerator.

Remark 2.1.10. Let $C(m) := C \otimes R^m \le V^N \otimes R^m$. We can identify $V^N \otimes R^m$ with V^{mN} via a suitable isomorphism, and regard $C(m)$ as a code in $(V^m)^N$ with ground ring $\text{Mat}_m(R)$, the ring of $m \times m$-matrices with entries in R. Thus $C(m)$ is C "promoted" to a code over V^m (column vectors of length m over V with componentwise addition and scalar multiplication). Then

$$\text{fwe}_m(C) = \text{fwe}(C(m)), \text{cwe}_m(C) = \text{cwe}(C(m)), \text{ etc.}$$

An analogous remark also holds for ρ-symmetrized weight enumerators, since $\text{Aut}(\text{Mat}_m(\rho)) = \text{Aut}(\rho)$, where the latter acts diagonally on V^m. Let ρ be a finite representation of a form ring with underlying module V, and let $C \le V^N$ be a code. Then

$$\text{swe}_m^\rho(C) = \text{swe}^{\text{Mat}_m(\rho)}(C(m)).$$

This is one of the main reasons why we allow noncommutative ground rings: even when we consider genus-m weight enumerators for classical binary codes, $\text{Mat}_m(\mathbb{F}_2)$ arises naturally as the ground ring.

We close this section with an example. Further examples of codes and their weight enumerators are given in §2.4.

Example 2.1.11. Let $R = \mathbb{F}_4 = \{0, 1, \omega, \omega^2\}$ be the field with four elements, where $\omega \in \mathbb{F}_4 \backslash \mathbb{F}_2$ satisfies $\omega^2 + \omega + 1 = 0$. This will be our standard notation for this field. For codes over $\mathbb{F}_4$ we will normally use the indeterminates x, y for the Hamming weight enumerator, x, y, z for the symmetrized weight enumerator and x, y, z, t (instead of $x_0, x_1, x_\omega, x_{\omega^2}$) for the complete weight enumerator, with $\text{swe}(x, y, z) = \text{cwe}(x, y, z, z)$.

We take the alphabet V to be $\mathbb{F}_4$, and let $C \le \mathbb{F}_4^2$ be the code with generator matrix $[1, \omega]$. This is an $\mathbb{F}_4$-linear code of length 2 that is self-dual with respect to the Hermitian inner product, that is, a code of Type 4^{H} (see (2.3.24) below). The code C has four codewords:

$$C = \{(0,0), (1,\omega), (\omega,\omega^2), (\omega^2,1)\}.$$

The Hamming weight enumerator is

$$\text{hwe}(C)(x, y) = x^2 + 3y^2 \,, \tag{2.1.6}$$

the complete weight enumerator is

$$\text{cwe}(C)(x, y, z, t) = x^2 + yz + zt + yt \,, \tag{2.1.7}$$

and the symmetrized weight enumerator is

$$\text{swe}(C)(x, y, z) = \text{cwe}(C)(x, y, z, z) = x^2 + 2xy + z^2 \,. \tag{2.1.8}$$

These weight enumerators can all be obtained as symmetrizations of the full weight enumerator of C, which is

$$\text{fwe}(C) = e_{(0,0)} + e_{(1,\omega)} + e_{(\omega,\omega^2)} + e_{(\omega^2,1)} \,.$$

Remark 2.1.12. The work of Jaffe [292], [293] on proving nonexistence of linear codes uses certain refined weight enumerators. For example, he might use linear programming to prove that a codeword of weight w exists, and then without loss of generality assume a particular vector of that weight. This corresponds to looking at isotropic self-dual codes in a representation of the form $w\rho_1 + (N - w)\rho_2$; and the weight enumerators he considers are the maximal symmetrizations of full weight enumerators in that representation. Although we will not pursue this point of view further, we do note that our main theorems are sufficiently general to cover this approach.

Remark 2.1.13. As a practical matter, it is worth remarking that one can compute the genus-m complete weight enumerator for a code over a field $\mathbb{F}_q$ with symbolic algebra programs such as MAGMA [100] by changing the field to $\mathbb{F}_{q^m}$ and asking for the ordinary complete weight enumerator.

2.2 MacWilliams identity and generalizations

The fundamental property of weight enumerators of linear codes, which underlies all the invariance properties, is the MacWilliams identity (MacWilliams [356]; [361, Chap. 5]). This identity expresses the weight enumerator of the dual code as a linear transformation of the weight enumerator of the original code. In Theorem 2.2.4 we give a general setting for this identity. As usual the proof depends on the Poisson summation formula for finite abelian groups (which will be V^N in our case).

Definition 2.2.1. Let G be a finite abelian group. Then the *character group*

$$\widehat{G} := \{\chi : G \to \mathbb{C}^* \mid \chi(a + b) = \chi(a)\chi(b) \text{ for } a, b \in G\} = \text{Hom}(G, \mathbb{C}^*)$$

is the group of group homomorphism of G into the multiplicative group $\mathbb{C}^*$ of nonzero complex numbers. The *dual* of a subgroup $H \leq G$ is

$$H^\# := \{\chi \in \widehat{G} \mid \chi(h) = 1 \text{ for all } h \in H\} \,.$$

Remark. $H^\#$ is a subgroup of $\widehat{G}$ which is canonically isomorphic to the character group of G/H.

Theorem 2.2.2. *Let G be a finite abelian group. The order of $\widehat{G}$ is*

$$|\widehat{G}| = |G|.$$

Moreover, the elements of $\widehat{G}$ form an orthonormal basis for the space of all functions $f : G \to \mathbb{C}$ with respect to the Hermitian inner product

$$\langle f_1, f_2 \rangle_G := \frac{1}{|G|} \sum_{g \in G} f_1(g)\overline{f_2(g)}.$$

Remark. Theorem 2.2.2 is a standard result in representation theory (see for instance Serre [479]). It also holds for finite nonabelian groups if $\widehat{G}$ is replaced by the set of characters of irreducible representations of G, and the space of functions from G to $\mathbb{C}$ by the space of "class-functions" on G, i.e. those functions that are constant on the conjugacy classes of G. One may also replace "finite" by "compact" (or even "locally compact"), provided the sum is replaced by the appropriate integral. Only the finite abelian case will be proved here.

Proof. Let $G = \langle g_1 \rangle \times \cdots \times \langle g_m \rangle$ be a decomposition of G into a direct product of cyclic groups, and let $d_j := |\langle g_j \rangle|$ be the order of g_j. Any element $f \in \widehat{G}$ is uniquely determined by its values on the g_j. The only restriction is that $f(g_j)$ must be a d_j-th root of 1. Therefore

$$|\widehat{G}| = d_1 \cdots d_m = |G|.$$

Let $f_1, f_2 \in \widehat{G}$. The product $f_1\overline{f_2} \in \widehat{G}$, and $\langle f_1, f_2 \rangle_G = \langle f_1\overline{f_2}, 1 \rangle_G$ where 1 is the *trivial character* of G, i.e. the constant function 1. It only remains to show that, for $f \in \widehat{G}$,

$$\langle f, 1 \rangle_G = \frac{1}{|G|} \sum_{g \in G} f(g) = \begin{cases} 0 & \text{if } f \neq 1, \\ 1 & \text{if } f = 1. \end{cases}$$

This is clear if $f = 1$. If $f \neq 1$ there is some j with $f(g_j) \neq 1$. Then

$$\sum_{g \in G} f(g) = \left(\sum_{i=0}^{d_j - 1} f(g_j)^i \right) \sum_{g \in G_j} f(g),$$

where G_j is the subgroup of G generated by all the g_i except g_j. Since $\sum_{i=0}^{d_j - 1} f(g_j)^i$ is a multiple of a sum over all powers of a nontrivial root of unity, it is 0, and therefore $\langle f, 1 \rangle_G = 0$. $\square$

Remark. Let $C \leq V$ be a code and let $\beta \in \mathrm{Bil}(V, \mathbb{Q}/\mathbb{Z})$ be a nonsingular bilinear form on V. Then β induces an isomorphism between V and $\widehat{V} = \mathrm{Hom}(V, \mathbb{C}^*) \cong \mathrm{Hom}(V, \mathbb{Q}/Z) = V^*$ by

$$v \mapsto (w \mapsto \exp(2\pi i \beta(v, w))).$$

Under this isomorphism the dual code $C^{\perp}$ is mapped onto $C^{\#}$.

Theorem 2.2.3. (Cf. Dym and McKean [163, §4.5], Terras [520, Chapter 12].) *The Poisson summation formula for finite abelian groups. Let $H \leq G$ be a subgroup of a finite abelian group G and let $f : G \to \mathbb{C}$. Then for any $g \in G$,*

$$\frac{1}{|H|} \sum_{h \in H} f(g + h) = \frac{1}{|G|} \sum_{\chi \in H^{\#}} \widehat{f}(\chi)\chi(g), \qquad (2.2.1)$$

where

$$\widehat{f}(\chi) := \sum_{a \in G} f(a)\overline{\chi(a)} = |G|\langle f, \chi \rangle_G. \qquad (2.2.2)$$

Proof. Let $f' : G/H \to \mathbb{C}$ be defined by

$$f'(g + H) = \sum_{h \in H} f(g + h).$$

Calculating the coefficients of f' with respect to the orthonormal basis $\widehat{G/H}$ we find

$$f' = \sum_{\chi \in \widehat{G/H}} \langle f', \chi \rangle \chi.$$

Then identifying $\widehat{G/H}$ with $H^{\#}$ we have

$$\langle f', \chi \rangle = \frac{|H|}{|G|} \sum_{\gamma \in G/H} f'(\gamma)\overline{\chi(\gamma)} = \frac{|H|}{|G|} \sum_{g \in G} f(g)\overline{\chi(g)} = |H|\langle f, \chi \rangle,$$

which completes the proof. $\qquad\qquad\qquad\square$

Applying the Poisson summation formula to a code equipped with a bilinear form we obtain the following result, which may be regarded as a very general version of the MacWilliams identity.

Theorem 2.2.4. *Let $C \leq V$ be a code, $\beta \in \mathrm{Bil}(V, \mathbb{Q}/Z)$ a nonsingular bilinear form, and $C^{\perp} = C^{\perp,\beta}$ (cf. (1.4.1)). Let f be any function $f : V \to \mathbb{C}$. Then*

$$\sum_{c \in C} f(c) = \frac{1}{|C^{\perp}|} \sum_{c' \in C^{\perp}} \widehat{f}(c'), \qquad (2.2.3)$$

where

$$\widehat{f}(c') = \sum_{v \in V} f(v) \exp(-2\pi i \beta(c', v)). \qquad (2.2.4)$$

Proof. Apply Eq. (2.2.1) with $G = V, H = C, g = 0$. Then

$$\sum_{c \in C} f(c) = \frac{|C|}{|V|} \sum_{\chi \in C^{\perp}} \widehat{f}(\chi)\chi(0) = \frac{1}{|C^{\perp}|} \sum_{c' \in C^{\perp}} \widehat{f}(c') . \qquad \square$$

Example 2.2.5. *MacWilliams identity for Hamming weight enumerators.* We apply Theorem 2.2.4 to the case when $C \leq V^N$ is a code of length N and

$$f(v) := x^{N-\mathrm{wt}(v)}y^{\mathrm{wt}(v)} ,$$

where we think of x, y as arbitrary complex numbers. Then

$$\widehat{f}(v) = \sum_{w \in V^N} (\prod_{j=1}^{N} \exp(-2\pi i\beta(v_j, w_j)))x^{N-\mathrm{wt}(w)}y^{\mathrm{wt}(w)} = \prod_{j=1}^{N} (q(v_j)y + x) ,$$

where

$$q(b) := \sum_{0 \neq a \in V} \exp(-2\pi i\beta(b, a)) \text{ for } b \in V .$$

If $v_j = 0$, $q(v_j) = |V| - 1$. We claim that $q(v_j) = -1$ if $v_j \neq 0$. The function $\chi_b : a \mapsto \exp(-2\pi i\beta(b, a))$ is a character on V. If $b \neq 0$ then by the non-degeneracy of β this character is not the trivial character, hence

$$\langle \chi_b, 1 \rangle_V = \frac{1}{|V|} \sum_{a \in V} \exp(-2\pi i\beta(b, a)) = 0 ,$$

and so $q(b) = -1$. Therefore $\widehat{f}(v) = (x - y)^{\mathrm{wt}(v)}(x + (|V| - 1)y)^{N-\mathrm{wt}(v)}$, and the formula above becomes the familiar MacWilliams identity for Hamming weight enumerators [361, Chap. 5, Theorem 13]:

$$\mathrm{hwe}(C)(x, y) = \frac{1}{|C^{\perp}|} \mathrm{hwe}(C^{\perp})(x + (|V| - 1)y, x - y) . \qquad (2.2.5)$$

Example 2.2.6. *MacWilliams identity for full weight enumerators.* This is even easier to deal with. Let $\beta \in \mathrm{Bil}(V, \mathbb{Q}/\mathbb{Z})$ be nonsingular. Then for any code $C \leq V$,

$$\sum_{w \in V} \sum_{v \in C} \exp(2\pi i\beta(w, v)) e_w = |C| \sum_{w \in C^{\perp}} e_w = |C| \mathrm{fwe}(C^{\perp}) , \qquad (2.2.6)$$

since for $w \in V$ the sum

$$\sum_{v \in C} \exp(2\pi i\beta(w, v)) = |C|\langle \chi_w, 1 \rangle = \begin{cases} |C| & \text{if } w \in C^{\perp} , \\ 0 & \text{otherwise} , \end{cases}$$

where χ_w is the character $\chi_w : C \to \mathbb{C}^*, v \mapsto \exp(2\pi i\beta(w, v))$. This is the trivial character if and only if $w \in C^{\perp}$. In particular, if $C = C^{\perp}$ is a self-dual code in V, then the full weight enumerator of C is invariant under the "change of variables"

$$e_v \mapsto |V|^{-1/2} \sum_{w \in V} \exp(2\pi i\beta(w, v)) e_w . \qquad (2.2.7)$$

Example 2.2.7. *MacWilliams identity for complete weight enumerators.* Similarly (cf. Remark 2.1.4), the complete weight enumerator of $C^\perp$ is the image of the complete weight enumerator of C under the change of variables $x_v \mapsto \sum_{w \in V} \exp(2\pi i \beta(w,v))\, x_w$, divided by $|C|$.

2.2.1 The weight enumerator of the shadow

The full (and also the complete) weight enumerator of the ϕ-shadow of a self-orthogonal code C can be obtained in a similar way. To shorten notation, we write $\phi^N := \perp^N \rho_\Phi(\phi) = \rho_\Phi^N(\phi)$. The ϕ-shadow (see (1.12.2)) is

$$S_\phi(C) := \{v \in V \mid \beta^N(v,c) = \phi^N(c) \text{ for } c \in C\}. \qquad (2.2.8)$$

The restriction of ϕ^N to C is linear on any self-orthogonal code C. We can apply the above theory, replacing β by $\beta^N - \phi^N$.

Theorem 2.2.8. *Let $C \le V^N$ be a self-orthogonal code. Then the full weight enumerator of the shadow is given by*

$$\mathrm{fwe}(S_\phi) = \frac{1}{|C|} \sum_{c \in C} \sum_{w \in V^N} \exp(2\pi i(\beta^N(w,c) - \phi^N(c)))\, e_w\,. \qquad (2.2.9)$$

This is the image of the full weight enumerator of C under the change of variables $e_v \mapsto \sum_{w \in V^N} \exp(2\pi i(\beta^N(w,v) - \phi^N(v)))\, e_w$, divided by $|C|$.

Proof. The proof is similar to that in Example 2.2.6. For $w \in V$, let

$$\chi_w : C \to \mathbb{C}^*, c \mapsto \exp(2\pi i(\beta^N(w,c) - \phi^N(c)))\,.$$

Since ϕ^N is linear on C, the mapping χ_w belongs to $\widehat{C}$. Clearly $\chi_w = 1$ is the trivial character if and only if $w \in S_\phi(C)$. Therefore the coefficient of e_w in the transformed full weight enumerator is

$$\sum_{c \in C} \exp(2\pi i(\beta^N(w,c) - \phi^N(c))) = |C|\langle \chi_w, 1 \rangle = \begin{cases} |C| & \text{if } w \in S_\phi(C)\,, \\ 0 & \text{otherwise}\,. \end{cases} \qquad \square$$

Corollary 2.2.9. *Let $C \le V^N$ be a self-orthogonal code. The complete weight enumerator of the ϕ-shadow of C is the image of the complete weight enumerator of C under the change of variables $x_v \mapsto \sum_{w \in V} \exp(2\pi i(\beta(w,v) - \rho_\Phi(\phi)(v)))\, x_w$, divided by $|C|$.*

2.3 Catalogue of important types

This long section describes the Types of the most important families of self-dual codes in our new language as representations of form rings. We will use the same notation for these families that was introduced in [454], and give analogous names to the corresponding form rings. Tables 2.1, 2.2 at the end of the chapter list the Types and give the sections where the definitions and examples can be found. The representations of the form rings here will always be faithful (cf. Example 1.7.3), and therefore define the involutions τ and $[-1]$ as well as the structure maps $\{\!\{\ \}\!\}$ and λ. If $\rho(\mathcal{T})$ is a a representation of a form ring $R(\mathcal{T})$, we will sometimes say "codes of Type $\mathcal{T}$" rather than the clumsier "codes of Type $\rho(\mathcal{T})$".

Remark 2.3.1. We begin with some general remarks on the meaning of Type I and Type II codes.   A binary self-dual code C with all weights divisible by 4 has classically been called *doubly-even* or of *Type* II; if we do not impose this restriction then C is *singly-even* or of *Type* I. A Type I code may or may not also be of Type II: the classes are not mutually exclusive. We say a code is *strictly* Type I if it is not of Type II.

We use a similar terminology when $R = \mathbb{F}_2$ and V is an arbitrary finite R-module; that is, we say that a code $C \leq V^N$ is *doubly-even* or of Type II if all its Hamming weights are divisible by 4, or *singly-even* or of Type I if its Hamming weights are even.

Similarly, we say that a self-dual code over $\mathbb{Z}/m\mathbb{Z}$, m even, is of *Type* II if the Euclidean norms are divisible by $2m$, or of *Type* I if they are divisible by m. (This terminology was used by Bannai et al. [28], Bonnecaze et al. [66] and Dougherty et al. [154].)

There is one other situation where a similar distinction can be made. An additive trace-Hermitian self-dual code over $\mathbb{F}_4$ from the family $4^{\mathrm{H}+}$ is of *Type* II if the Hamming weights are even, or of *Type* I if odd weights may occur (if odd weights do occur then the code cannot be linear).

Informally, one may think of Type II as generally meaning a code which is "more even than it needs to be". Note that the cases where there is a notion of Type II code are also cases where there is a nontrivial shadow (compare §1.12).

In the following sections each paragraph is labeled with the symbol for the family (following [454]).

Note that $\Phi = \{\!\{M\}\!\}$ indicates that no additional quadratic maps are needed beyond the specializations of bilinear forms in M.

We begin with binary codes.

2.3.1 Binary codes

2

Binary linear codes: $R = \mathbb{F}_2 = \{0, 1\}$, $V = \mathbb{F}_2$ with the standard inner product

$$(x, y) := xy, \tag{2.3.1}$$

C = subspace of V^N. Here we distinguish two important Types of self-dual codes, doubly-even and singly-even codes, as follows.

2$_{\mathrm{I}}$

Singly-even self-dual binary codes (the classical Type I codes).   These are self-dual codes in $\mathbb{F}_2^N$ with the standard inner product (2.3.1). Such codes exist if and only if the length N is even. For the formal definition of the form ring (R, M, ψ, Φ) we take $\mathbb{F}_2 = \{0, 1\}$, $R = M = \mathbb{F}_2$, $\psi : R \xrightarrow{\cong} M$ to be the identity map, and $\Phi = \{\!\{M\}\!\}$; that is, we use the form ring

$$R(2_{\mathrm{I}}) := (R, M, \psi, \Phi) = (\mathbb{F}_2, \mathbb{F}_2, \mathrm{id}, \{\!\{M\}\!\})\,. \tag{2.3.2}$$

The Type for singly-even self-dual binary codes is then the representation

$$\rho(2_{\mathrm{I}}) := (V, \rho_M, \rho_\Phi, \beta)$$

of this form ring, where $V = \mathbb{F}_2$, $\beta := \rho_M(1)$ is the bilinear form

$$\beta : (x, y) \mapsto \frac{1}{2}xy \;\in\; \mathrm{Bil}(V, \tfrac{1}{2}\mathbb{Z}/\mathbb{Z})\,, \tag{2.3.3}$$

and $\rho_\Phi(\{\!\{\beta\}\!\})(x) = \frac{1}{2}x^2$ for all $x \in V$.

2$_{\mathrm{II}}$

Doubly-even self-dual binary codes (the classical Type II codes).   These are singly-even self-dual codes in $\mathbb{F}_2^N$ satisfying the extra condition that the weight of every codeword is divisible by 4,

$$\mathrm{wt}(c) \in 4\mathbb{Z} \text{ for all } c \in C\,.$$

Such codes exist if and only if the length N is a multiple of 8. As already discussed in Example 1.8.2, the Type for doubly-even self-dual binary codes is the representation $\rho(2_{\mathrm{II}}) := (V, \rho_M, \rho_\Phi, \beta)$ of the form ring

$$R(2_{\mathrm{II}}) := (R, M, \psi, \Phi = \langle \phi_0 \rangle) = (\mathbb{F}_2, \mathbb{F}_2, \mathrm{id}, \mathbb{Z}/4\mathbb{Z})\,, \tag{2.3.4}$$

where $V = \mathbb{F}_2$, again $\rho_M(1)(x, y) = \frac{1}{2}xy$ and $\rho_\Phi(\phi_0)(x) = \frac{1}{4}x^2$ for all $x, y \in V$.

2$_{\mathrm{S}}$

These are singly-even self-dual codes C in $\mathbb{F}_2^N$ such that the shadow contains the all-ones vector, $\mathbf{1} \in S_{\phi_0}(C)$. As already discussed in Example 1.12.7, the Type of these codes is the representation $\rho(2_{\mathrm{S}}) := (V, \rho_M, \rho_\Phi, \beta)$ of the form ring $R(2_{\mathrm{II}})$, where $V = \mathbb{F}_2$, $\rho_M(1)(x, y) = \frac{1}{2}xy$, and $\rho_\Phi(\phi_0)(x) = \frac{1}{4}x^2 + \frac{1}{2}x$ for all $x, y \in V$.

For all three Types, the automorphism groups and weak automorphism groups of the representations are trivial:

$$\mathrm{Aut}(\rho(2_{\mathrm{I}})) = \mathrm{WAut}(\rho(2_{\mathrm{I}})) = \mathrm{Aut}(\rho(2_{\mathrm{II}})) = \mathrm{WAut}(\rho(2_{\mathrm{II}}))$$
$$= \mathrm{Aut}(\rho(2_{\mathrm{S}})) = \mathrm{WAut}(\rho(2_{\mathrm{S}})) = \mathbb{F}_2^* = \{1\}.$$

So in particular, the groups of equivalences and weak equivalences in all three cases are just the symmetric group S_N. Thus two codes are equivalent if they differ only by a permutation of the coordinates, and the automorphisms of a code are those permutations that preserve the code. This is the usual notion of the automorphism group of a binary code.

The following are three equivalent formulations of the MacWilliams identity for binary codes (see (2.2.5)):

$$\mathrm{hwe}(C^\perp)(x,y) = \frac{1}{|C|}\,\mathrm{hwe}(C)(x+y, x-y)\,, \tag{2.3.5}$$

which can also be written as

$$\sum_{u \in C^\perp} x^{N-\mathrm{wt}(u)}y^{\mathrm{wt}(u)} = \frac{1}{|C|}\sum_{u \in C}(x+y)^{N-\mathrm{wt}(u)}(x-y)^{\mathrm{wt}(u)}\,. \tag{2.3.6}$$

If A_i (resp. $A_i^\perp$) is the number of codewords of weight i in C (resp. $C^\perp$), so that

$$\mathrm{hwe}(C)(x,y) = \sum_{i=0}^{N} A_i x^{N-i}y^i\,, \tag{2.3.7}$$

then (2.3.5) is equivalent to

$$A_k^\perp = \frac{1}{|C|}\sum_{i=0}^{N} A_i P_k(i)\,, \tag{2.3.8}$$

where

$$P_k(x) = \sum_{j=0}^{k}(-1)^j \binom{x}{j}\binom{N-x}{k-j}, \quad k = 0, \ldots, N\,,$$

is a Krawtchouk polynomial ([361, Chap. 5]). There are analogous Krawtchouk polynomials for any alphabet, see [361, p. 151]. For the other families we will give just the formulation in terms of weight enumerators. From Corollary 2.2.9, the weight enumerator of the shadow is (Conway and Sloane [130]):

$$\mathrm{hwe}(S_{\phi_0}(C))(x,y) = \frac{1}{|C|}\,\mathrm{hwe}(C)(x+y, i(x-y))\,. \tag{2.3.9}$$

2.3.2 Euclidean codes

4^{E}

Euclidean linear[1] codes over $\mathbb{F}_4$ (or Euclidean quaternary codes): $R = \mathbb{F}_4 = \{0, 1, \omega, \omega^2\}$, where $\omega^2 + \omega + 1 = 0$, $\omega^3 = 1$, with the Euclidean inner product

---

[1] At first glance the adjective "linear" may appear superfluous, but we use it to signify a code which is a vector space over the ground field, as opposed to one which is merely additive (cf. Type $4^{\mathrm{H}+}$ below).

$$(x, y) = xy \, ; \tag{2.3.10}$$

$C =$ subspace of $\mathbb{F}_4^N$. This is a special case of the next family.

The MacWilliams identities for codes from the family 4^{E} are:

$$\mathrm{hwe}(C^\perp)(x, y) = \frac{1}{|C|} \, \mathrm{hwe}(C)(x + 3y, x - y) \, ,$$

$$\mathrm{swe}(C^\perp)(x, y, z) = \frac{1}{|C|} \, \mathrm{swe}(C)(x + y + 2z, x + y - 2z, x - y) \, ,$$

$$\mathrm{cwe}(C^\perp)(x, y, z, t) = \frac{1}{|C|} \, \mathrm{cwe}(C^\perp)(x + y + z + t, x + y - z - t,$$

$$x - y - z + t, x - y + z - t) \, . \tag{2.3.11}$$

q^{E} (even)

Euclidean $\mathbb{F}_q$-linear codes, where $q = 2^f$ is a power of 2: $R = \mathbb{F}_q$, with the Euclidean inner product $(x, y) = xy$; $C =$ subspace of $\mathbb{F}_q^N$. In our notation the Type of Euclidean self-dual linear codes over $\mathbb{F}_q$ is the representation $\rho(q^{\mathrm{E}}) := (V, \rho_M, \rho_\Phi, \beta)$ of the form ring

$$R(q^{\mathrm{E}}) := (R, M, \psi, \Phi = \{\!\{M\}\!\}) = (\mathbb{F}_q, \mathbb{F}_q, \mathrm{id}, \mathbb{F}_q) \, , \tag{2.3.12}$$

where $V = \mathbb{F}_q$,

$$\rho_M(a)(x, y) = \frac{1}{2} \, \mathrm{Tr}(axy) \text{ for } a \in M = \mathbb{F}_q, x, y \in V = \mathbb{F}_q \, ,$$

and Tr denotes the trace of $\mathbb{F}_q$ over the prime field $\mathbb{F}_2$. This defines the representation ρ_Φ, since

$$\rho_\Phi(\{\!\{a\}\!\})(x) = \{\!\{\rho_\Phi(a)\}\!\}(x) = \frac{1}{2} \, \mathrm{Tr}(ax^2) \, .$$

Since squaring is a Galois automorphism, $\mathrm{Tr}(x^2) = \mathrm{Tr}(x)$. Hence for any self-orthogonal code C the all-ones vector $\mathbf{1}$ lies in $C^\perp$. In particular, self-dual codes contain $\mathbf{1}$. We have $\mathrm{Aut}(\rho(q^{\mathrm{E}})) = \{g \in \mathbb{F}_q \mid g^2 = 1\} = \{1\}$ and $\mathrm{WAut}(\rho(q^{\mathrm{E}})) = \mathbb{F}_q^* \rtimes \mathrm{Gal}(\mathbb{F}_q/\mathbb{F}_2) \cong Z_{q-1} \rtimes Z_f$. The group of equivalences is S_N. The group of weak equivalences is $(Z_{q-1} \rtimes Z_f) \times S_N$, since we also allow *global scalar multiplication*, that is, multiplication of every coordinate by the same scalar, as well as *global conjugation*, simultaneous application of any element of $\mathrm{Gal}(\mathbb{F}_q/\mathbb{F}_2) \cong Z_f$ to every coordinate. Of course, global scalar multiplication is an automorphism of every code, so has no effect on whether codes are weakly equivalent.

The MacWilliams identities for codes from the family q^{E} are as follows:

$$\mathrm{hwe}(C^\perp)(x, y) = \frac{1}{|C|} \, \mathrm{hwe}(C)(x + (q - 1)y, x - y) \, . \tag{2.3.13}$$

The cwe for $C^\perp$ is obtained from the cwe for C by replacing each x_ξ, $\xi \in \mathbb{F}_q$, by

$$\sum_{\mu \in \mathbb{F}_q} \exp(\pi i \operatorname{Tr}(\mu\xi)) \, x_\mu \qquad (2.3.14)$$

and dividing by $|C|$. We omit discussion of the swe, since there are several possible symmetrizations.

$q_{\mathrm{II}}^{\mathrm{E}}$

Generalized doubly-even self-dual codes. Quebbemann [438] extended the notion of doubly-even code to codes over $\mathbb{F}_q$, where $q = 2^f$, $f \geq 1$. A linear code $C \leq \mathbb{F}_q^N$ is called *doubly-even* if

$$\sum_{i=1}^{N} c_i = 0 \text{ and } \sum_{1 \leq i < j \leq N} c_i c_j = 0, \text{ for all } c \in C. \qquad (2.3.15)$$

Doubly-even codes are self-orthogonal with respect to the usual bilinear form $\sum_{i=1}^{N} c_i c_i'$. This follows from the identity

$$\sum_{i<j}(c_i + c_i')(c_j + c_j') = \sum_{i<j} c_i c_j + \sum_{i<j} c_i' c_j' + \sum_{i=1}^{N} c_i \sum_{i=1}^{N} c_i' - \sum_{i=1}^{N} c_i c_i'.$$

If $f = 1$ these are precisely the usual doubly-even binary codes (if c has even weight w, then $\sum_{i<j} c_i c_j \equiv w(w-1)/2 \bmod 2$, which is $0 \bmod 2$ if and only if $w \equiv 0 \pmod 4$ by Kummer's theorem.)

Identifying $\mathbb{F}_q$ with $\mathbb{F}_2^f$ using a self-complementary basis (that is, an $\mathbb{F}_2$-basis for $\mathbb{F}_{2^f}$ which is orthonormal with respect to the trace bilinear form), a doubly-even code of length N in the new sense becomes a doubly-even code of length fN over $\mathbb{F}_2$, cf. Corollary 2.4.5 below.

A self-dual doubly-even code in the new sense is called a *generalized doubly-even self-dual code*. For $q = 4$ we obtain exactly the Type II codes over $\mathbb{F}_4$ considered by Gaborit et al. [186]; for us this is the Type $4_{\mathrm{II}}^{\mathrm{E}}$ of Euclidean self-dual codes over $\mathbb{F}_4$ in which the Lee weight of every codeword is a multiple of 4.

The Type of these codes can be specified as follows. Let $R = \mathbb{F}_{2^f}$ and let $O := \mathbb{Z}_2[\zeta_{2^f - 1}]$ be the ring of integers in the unramified extension of degree f of the 2-adic numbers, so that $R \cong O/2O$. If $x \in R$, x^2 is uniquely determined modulo 4, so squares of elements of R can be considered as elements of $O/4O$. The usual trace $\operatorname{Tr} : O \to \mathbb{Z}$ maps $4O$ into $4\mathbb{Z}$. The Type we want is the representation $\rho(q_{\mathrm{II}}^{\mathrm{E}}) := (V, \rho_M, \rho_\Phi, \beta)$ of the form ring

$$R(q_{\mathrm{II}}^{\mathrm{E}}) := (R, M, \psi, \Phi) = (\mathbb{F}_q, \mathbb{F}_q, \mathrm{id}, O/4O), \qquad (2.3.16)$$

given by $V = \mathbb{F}_q$ and

$$\rho_M(a)(x, y) = \frac{1}{2} \operatorname{Tr}(axy),$$

for all $a \in M = \mathbb{F}_q$, $x, y \in V = \mathbb{F}_q$ (as in the family q^{E}). Then $\beta = \rho_M(1)$ is defined by $\beta(x, y) := \frac{1}{2} \operatorname{Tr}(xy)$ for all $x, y \in V$. The representation ρ_Φ is given by[2]

$$\rho_\Phi(a)(x) := \frac{1}{4} \operatorname{Tr}(ax^2), \tag{2.3.17}$$

for $a \in \Phi = O/4O$ and $x \in V$. Since the Galois automorphisms of $\mathbb{F}_q$ lift to Galois automorphisms of O, and similarly for scalar multiplication, we again have $\operatorname{Aut}(\rho(q_{\mathrm{II}}^{\mathrm{E}})) = \{1\}$ and $\operatorname{WAut}(\rho(q_{\mathrm{II}}^{\mathrm{E}})) = \mathbb{F}_q^* \rtimes \operatorname{Gal}(\mathbb{F}_q/\mathbb{F}_2) \cong Z_{q-1} \rtimes Z_f$. The groups of equivalences and weak equivalences are as for Type q^{E}.

Theorem 2.3.2. *Codes of Type $q_{\mathrm{II}}^{\mathrm{E}}$ are exactly the generalized doubly-even self-dual codes.*

Proof. (i) Let $C \leq \mathbb{F}_{2^f}^N$ be a generalized doubly-even self-dual code. Since λ is surjective, it is enough to show that $\sum_{i=1}^N \rho_\Phi(a)(c_i) = 0$ for all $c \in C$ and $a \in \Phi = O/4O$. Now $\sum_{i=1}^N c_i = 0$ in $O/2O$, so as an element of $O/4O$ the square

$$\left(\sum_{i=1}^N c_i\right)^2 = \sum_{i=1}^N c_i^2 + 2\sum_{i<j} c_i c_j = 0 \in O/4O. \tag{2.3.18}$$

Since $\sum_{i<j} c_i c_j = 0$ it follows that $\sum_{i=1}^N c_i^2 = 0 \in O/4O$. (ii) To obtain the other inclusion let C be a code of Type $\rho(q_{\mathrm{II}}^{\mathrm{E}})$. By the non-degeneracy of the trace form, $\sum_{i=1}^N c_i^2 = 0$ in $O/4O$. Therefore by (2.3.18), $(\sum_{i=1}^N c_i)^2 \equiv 0$ (mod $2O$) and hence also $\sum_{i=1}^N c_i = 0 \in \mathbb{F}_{2^f}$. Then $(\sum_{i=1}^N c_i)^2 \equiv 0$ (mod $4O$) and (2.3.18) implies that $\sum_{i<j} c_i c_j = 0$. $\qquad\square$

The above Type will be described in a more sophisticated way (using Witt vectors) in §7.6.4 (see also Nebe, Quebbemann, Rains and Sloane [382]).

3

Ternary codes: $R = \mathbb{F}_3 = \{0, 1, 2\}$, $(x, y) = xy$; C = subspace of $\mathbb{F}_3^N$. This is a special case of the next family. The MacWilliams identities for ternary codes are:

$$\operatorname{hwe}(C^\perp)(x, y) = \frac{1}{|C|} \operatorname{hwe}(C)(x + 2y, x - y), \tag{2.3.19}$$

$$\operatorname{cwe}(C^\perp)(x, y, z) = \frac{1}{|C|} \operatorname{cwe}(C)(x + y + z, x + \omega y + \overline{\omega} z,$$

$$x + \overline{\omega} y + \omega z), \tag{2.3.20}$$

where $\omega = \exp(2\pi i/3)$. The automorphism groups $\operatorname{Aut}(\rho(3))$ and $\operatorname{WAut}(\rho(3))$ are both $\{\pm 1\}$.

---

[2] This was incorrectly written as $\frac{1}{4} \operatorname{Tr}(a^2 x^2)$ in [385].

q^{E} (odd)

Linear codes over $\mathbb{F}_q$, where $q = p^f$ is a power of an odd prime p, with the Euclidean inner product $(x, y) = xy$; $C =$ subspace of $\mathbb{F}_q^N$. If q is a square, the Hermitian family q^{H} below is generally preferred to q^{E}, since the codes have more interesting properties.

The Type of Euclidean self-dual linear codes over $\mathbb{F}_q$ is the representation $\rho(q^{\mathrm{E}}) := (V, \rho_M, \rho_\Phi, \beta)$ of the form ring

$$R(q^{\mathrm{E}}) := (R, M, \psi, \Phi = \{\!\{M\}\!\}) = (\mathbb{F}_q, \mathbb{F}_q, \mathrm{id}, \mathbb{F}_q), \qquad (2.3.21)$$

where $V = \mathbb{F}_q$, $\rho_M(a)(x, y) = \frac{1}{p}\mathrm{Tr}(axy)$ for all $a \in M = \mathbb{F}_q$, $x, y \in V = \mathbb{F}_q$, and Tr denotes the trace of $\mathbb{F}_q$ over the prime field $\mathbb{F}_p$. For odd q we find

$$\mathrm{Aut}(\rho(q^{\mathrm{E}})) = \{g \in \mathbb{F}_q^* \mid g^2 = 1\} = \{1, -1\} \cong Z_2,$$

$$\mathrm{WAut}(\rho(q^{\mathrm{E}})) = \mathbb{F}_q^* \rtimes \mathrm{Gal}(\mathbb{F}_q/\mathbb{F}_p) \cong Z_{q-1} \rtimes Z_f.$$

The group of equivalences consists of all permutations and sign changes of the coordinates; for the weak equivalences we also allow global conjugation and global scalar multiplication. Of course, when q is a prime, two codes are equivalent if and only if they are weakly equivalent, since global scalar multiplication preserves every code.

The MacWilliams identity for Hamming weight enumerators is (2.3.13) again. The cwe for $C^\perp$ is obtained from the cwe for C by replacing each x_ξ, $\xi \in \mathbb{F}_q$, by

$$\sum_{\mu \in \mathbb{F}_q} \exp(2\pi i \, \mathrm{Tr}(\mu\xi)/p) \, x_\mu \qquad (2.3.22)$$

and dividing by $|C|$.

q_1^{E} (odd)

Codes belonging to the family q^{E} (odd) that contain the all-ones vector $\mathbf{1}$. For our purposes Euclidean self-dual codes that contain $\mathbf{1}$ play an important role, since their weight enumerators are invariant under a much larger group (and so invariant theory leads to stronger restrictions on these codes). Although codes from the family q^{E} automatically contain $\mathbf{1}$ if q is even, this is an additional restriction when q is odd. To obtain the Type of these codes we must enlarge the set of quadratic maps in $R(q^{\mathrm{E}})$, as we have already seen in §2.1.6. The form ring is now

$$R(q_1^{\mathrm{E}}) := (R, M, \psi, \Phi = \{\!\{M\}\!\} \oplus \langle\varphi\rangle) = (\mathbb{F}_q, \mathbb{F}_q, \mathrm{id}, \mathbb{F}_q \oplus \mathbb{F}_q), \qquad (2.3.23)$$

where $\ker(\lambda) = \langle\varphi\rangle$. The representation $\rho(q_1^{\mathrm{E}})$ is defined as in the previous family q^{E}, together with

$$\rho_\Phi(\varphi)(x) := \beta(1, x) = \frac{1}{p} \operatorname{Tr}(x)$$

for all $x \in V$. Now -1 is not an automorphism of $\rho(q_1^{\mathrm{E}})$, so $\operatorname{Aut}(\rho(q_1^{\mathrm{E}})) = \{1\}$ and $\operatorname{WAut}(\rho(q_1^{\mathrm{E}})) = \mathbb{F}_q^* \rtimes \operatorname{Gal}(\mathbb{F}_q/\mathbb{F}_p) \cong Z_{q-1} \rtimes Z_f$. (Note that $\operatorname{Aut}(\rho(q^{\mathrm{E}}))$ is the stabilizer of $\mathbf{1}$ in the group $\operatorname{Aut}(\rho(q^{\mathrm{E}}))$.) The group of equivalences is thus just S_N, while the group of weak equivalences is still $(Z_{q-1} \rtimes Z_f) \times S_N$.

2.3.3 Hermitian codes

4^H

Linear codes over $\mathbb{F}_4$ (or Hermitian quaternary codes): $R = \mathbb{F}_4$, $\overline{x} = x^2$ for $x \in \mathbb{F}_4$, with the Hermitian inner product

$$(x, y) = x\overline{y}; \tag{2.3.24}$$

$C =$ subspace of $\mathbb{F}_4^N$. Note that for $x, y \in \mathbb{F}_4$, $(x + y)^2 = x^2 + y^2$, $x^4 = x$. This is a special case of the next family. The MacWilliams identities for codes from the family(4^H) are as follows.

$$\operatorname{cwe}(C^\perp)(x, y, z, t) = \frac{1}{|C|} \operatorname{cwe}(C)(x + y + z + t, x + y - z - t,$$
$$x - y + z - t, x - y - z + t). \tag{2.3.25}$$

The hwe and swe are given by (2.3.11).

q^H

Linear codes over $\mathbb{F}_q$ (or q-ary linear codes), where $q = r^2 = p^f$ is an even power of an arbitrary prime p, with $\overline{x} = x^r$ for $x \in \mathbb{F}_q$, $(x, y) = x\overline{y}$; $C =$ subspace of $\mathbb{F}_q^N$. Note that for $x, y \in \mathbb{F}_q$, $(x + y)^r = x^r + y^r$, $x^q = x$ and $^-$ generates the Galois group of $\mathbb{F}_q$ over its subfield $\mathbb{F}_r$.

The Type of these codes is the representation $\rho(q^{\mathrm{H}}) := (V, \rho_M, \rho_\Phi, \beta)$ of the form ring

$$R(q^{\mathrm{H}}) := (R, M, \psi, \Phi = \{\!\{M\}\!\}) = (\mathbb{F}_q, \mathbb{F}_q, \operatorname{id}, \mathbb{F}_q), \tag{2.3.26}$$

where $V = \mathbb{F}_q$, $\rho_M(a)(x, y) = \frac{1}{p} \operatorname{Tr}(ax\overline{y})$ for all $a \in M = \mathbb{F}_q$, $x, y \in V = \mathbb{F}_q$, and Tr denotes the trace of $\mathbb{F}_q$ over the prime field $\mathbb{F}_p$. This specifies ρ_Φ since $\{\!\{\}\!\}$ is surjective. The automorphism group of $\rho(q^{\mathrm{H}})$ is the full unitary group of $\mathbb{F}_q$:

$$\operatorname{Aut}(\rho(q^{\mathrm{H}})) = \{g \in \mathbb{F}_q^* \mid g^{r+1} = g\overline{g} = 1\} \cong Z_{r+1}. \tag{2.3.27}$$

Since the Galois group of $\mathbb{F}_q$ is abelian,

$$\operatorname{WAut}(\rho(q^{\mathrm{H}})) = \mathbb{F}_q^* \rtimes \operatorname{Gal}(\mathbb{F}_q/\mathbb{F}_p) \cong Z_{q-1} \rtimes Z_f. \tag{2.3.28}$$

The group of equivalences consists of monomial matrices with nonzero entries $g \in \mathbb{F}_q^*$ with $g\bar{g} = 1$; for the weak equivalences we also allow global multiplication by elements of $\mathbb{F}_q^*$ and global conjugation by elements of $\mathrm{Gal}(\mathbb{F}_q/\mathbb{F}_p)$.

The MacWilliams identities for codes from the family q^{H} are as follows. For the hwe see (2.3.13) again. The cwe for $C^{\perp}$ is obtained from the cwe for C by replacing each x_ξ, $\xi \in \mathbb{F}_q$, by

$$\sum_{\mu \in \mathbb{F}_q} \exp(2\pi i\, \mathrm{Tr}(\mu\bar{\xi})/p)\, x_\mu \qquad (2.3.29)$$

and dividing by $|C|$.

q_1^{H}

Codes in q^{H} that contain $\mathbf{1}$. As in the Euclidean case, it is often helpful to consider only those codes in q^{H} that contain the all-ones vector. This extra condition can be expressed by adding certain linear quadratic maps to the R-qmodule Φ.

The Type of Hermitian self-dual $\mathbb{F}_q$-codes that contain $\mathbf{1}$ is the representation $\rho(q_1^{\mathrm{H}}) := (V, \rho_M, \rho_\Phi, \beta)$ of the form ring

$$R(q_1^{\mathrm{H}}) := (R, M, \psi, \Phi = \{\!\{M\}\!\} \oplus \{\varphi_a \mid a \in \mathbb{F}_q\}) = (\mathbb{F}_q, \mathbb{F}_q, \mathrm{id}, \mathbb{F}_q \oplus \mathbb{F}_q)\,, \qquad (2.3.30)$$

where $V = \mathbb{F}_q$ and ρ_M are defined as in q^{H} and

$$\rho_\Phi(\varphi_a)(x) := \frac{1}{p}\,\mathrm{Tr}(ax)\,.$$

If C is an isotropic code in $N\rho(q_1^{\mathrm{H}})$, then $\mathrm{Tr}(a\sum_{i=1}^N c_i)$ is divisible by p for all $c \in C$ and $a \in \mathbb{F}_q$. Therefore $\mathbf{1} \in C^{\perp}$ and hence isotropic self-dual codes contain the all-ones vector. The automorphism group of $\rho(q_1^{\mathrm{H}})$ is trivial and the weak automorphism group coincides with that of $\rho(q^{\mathrm{H}})$.

2.3.4 Additive codes

$4^{\mathrm{H}+}$

Trace-Hermitian additive codes over $\mathbb{F}_4$ (or additive quaternary codes): $R = \mathbb{F}_2$, $V = \mathbb{F}_4$, with

$$(x, y) = xy^2 + x^2y = \mathrm{Tr}(x\bar{y})\,; \qquad (2.3.31)$$

$C = $ additive subgroup of $\mathbb{F}_4^N$. These codes will play an important role in Chapter 13. This is a special case of the next family. The MacWilliams identities are the same as for the family 4^{H}. For the ϕ_0-shadow $S := S_{\phi_0}(C)$, where ϕ_0 is the quadratic map from the form ring $R(4_{\mathrm{II}}^{\mathrm{H}+})$ defined below, given by

$$\rho_\Phi(\phi_0)(x) := \frac{1}{2}x^3 = \begin{cases} 0 & \text{if } x = 0 \\ \frac{1}{2} & \text{else} \end{cases}$$

we have:

$$\text{hwe}(S)(x,y) = \frac{1}{|C|}\,\text{hwe}(C)(x+3y, y-x)\,,$$

$$\text{swe}(S)(x,y,z) = \frac{1}{|C|}\,\text{swe}(C)(x+y+2z, -x-y+2z, y-x)\,,$$

$$\text{cwe}(S)(x,y,z,t) = \frac{1}{|C|}\,\text{cwe}(C)(x+y+z+t, -x-y+z+t,$$

$$-x+y-z+t, -x+y+z-t)\,. \tag{2.3.32}$$

For this Type, $\text{Aut}(\rho(4^{\text{H}+})) = \text{WAut}(\rho(4^{\text{H}+})) = \text{SL}_2(\mathbb{F}_2) \cong S_3$. Note in particular therefore that the true symmetrized weight enumerator is the same as the Hamming weight enumerator, and what we call the symmetrized weight enumerator above is only partially symmetrized.

$q^{\text{H}+}$ (even)

Trace-Hermitian codes over $\mathbb{F}_q$ which are linear over $\mathbb{F}_r$, where $r^2 = q$ is even: $R = \mathbb{F}_r$, $V = \mathbb{F}_q$, with $(x,y) = xy^r + x^r y = \text{Tr}(x\bar{y})$ where $\bar{x} = x^r$ for $x, y \in V$; $C = \mathbb{F}_r$-linear subspace of $\mathbb{F}_q^N$. The Type of these codes is the representation $\rho(q^{\text{H}+}) := (V, \rho_M, \rho_\Phi, \beta)$ of the form ring

$$R(q^{\text{H}+}) := (R, M, \psi, \Phi = \{\!\{M\}\!\} = \{0\}) = (\mathbb{F}_r, \mathbb{F}_r, \text{id}, \{0\})\,, \tag{2.3.33}$$

where $V = \mathbb{F}_q$ and $\rho_M(a)(x,y) = \frac{1}{2}\text{Tr}(ax\bar{y})$ for all $a \in M = \mathbb{F}_r$ and $x, y \in V = \mathbb{F}_q$. Since $ax\bar{x} \in \mathbb{F}_r$, $\text{Tr}(ax\bar{x}) = 2\,\text{Tr}_{\mathbb{F}_r/\mathbb{F}_2}(ax\bar{x}) = 0$ and so $\{\!\{M\}\!\} = \{0\}$. The automorphism group $\text{Aut}(\rho(q^{\text{H}+}))$ is the stabilizer $\text{SL}_2(\mathbb{F}_r)$ of the symplectic inner product (x,y); the weak automorphism group is the group $\text{GL}_2(\mathbb{F}_r) \rtimes \text{Gal}(\mathbb{F}_r/\mathbb{F}_p)$. The MacWilliams identities are the same as for q^{H}.

$q_1^{\text{H}+}$ (even)

Codes in family $q^{\text{H}+}$ (even) that contain the all-ones vector. The Type of these codes is the representation $\rho(q_1^{\text{H}+}) := (V, \rho_M, \rho_\Phi, \beta)$ of the form ring

$$R(q_1^{\text{H}+}) := (R, M, \psi, \Phi = \langle \varphi \rangle) = (\mathbb{F}_r, \mathbb{F}_r, \text{id}, \mathbb{F}_r)\,, \tag{2.3.34}$$

where $V = \mathbb{F}_q$ and ρ_M are as in $\rho(q^{\text{H}+})$ and ρ_Φ is defined by

$$\rho_\Phi(\varphi)(x) = \beta(x,1) = \frac{1}{2}\text{Tr}_{\mathbb{F}_r/\mathbb{F}_2}(x+\bar{x}) = \frac{1}{2}\text{Tr}(x)\,.$$

The automorphism group $\text{Aut}(\rho(q_1^{\text{H}+}))$ is the stabilizer in $\text{Aut}(\rho(q^{\text{H}+}))$ of $\mathbf{1}$, isomorphic to the *additive* group $\mathbb{F}_r$. The weak automorphism group is the stabilizer in $\text{WAut}(\rho(q^{\text{H}+}))$ of the $\mathbb{F}_r$-span of $\mathbf{1}$, so is isomorphic to the semidirect product of the group of upper-triangular matrices in $\text{GL}_2(\mathbb{F}_r)$ by the Galois group $\text{Gal}(\mathbb{F}_r/\mathbb{F}_p)$.

$q_{\mathrm{II}}^{\mathrm{H+}}$ (even)

Type II or even trace-Hermitian codes over $\mathbb{F}_q$ which are linear over $\mathbb{F}_r$, where $r^2 = q$ is even: $R = \mathbb{F}_r$, $V = \mathbb{F}_q$, with $(x, y) = xy^r + x^r y = \mathrm{Tr}(x\overline{y})$, where $\overline{x} = x^r$, for $x, y \in V$; $C = \mathbb{F}_r$-linear subspace of $\mathbb{F}_q^N$. The Type of these codes is the representation $\rho(q_{\mathrm{II}}^{\mathrm{H+}}) := (V, \rho_M, \rho_\Phi, \beta)$ of the form ring

$$R(q_{\mathrm{II}}^{\mathrm{H+}}) := (R, M, \psi, \Phi = \langle \phi_0 \rangle) = (\mathbb{F}_r, \mathbb{F}_r, \mathrm{id}, \mathbb{F}_r), \qquad (2.3.35)$$

where $V = \mathbb{F}_q$, ρ_M is defined as in $\rho(q^{\mathrm{H+}})$ and ρ_Φ is defined by

$$\rho_\Phi(\phi_0)(x) = \frac{1}{2} \mathrm{Tr}_{\mathbb{F}_r/\mathbb{F}_2}(x\overline{x}).$$

The automorphism group $\mathrm{Aut}(\rho(q_{\mathrm{II}}^{\mathrm{H+}}))$ is the stabilizer of the $\mathbb{F}_r$-valued quadratic form $x \mapsto x\overline{x} = x^{r+1}$, so is given by the dihedral group $Z_{r+1} \rtimes Z_2$, where $Z_{r+1} \subset \mathbb{F}_q^*$ is the group of elements of norm 1, and Z_2 is $\mathrm{Gal}(\mathbb{F}_q/\mathbb{F}_r)$. Similarly, the weak automorphism group $\mathrm{WAut}(\rho(q_{\mathrm{II}}^{\mathrm{H+}})) = \mathbb{F}_q^* \rtimes \mathrm{Gal}(\mathbb{F}_q/\mathbb{F}_p)$.

Remark. If $q = 4$, then $x\overline{x} = 1$ for all $x \in \mathbb{F}_4 \setminus \{0\}$. Hence in an isotropic code from the family $4^{\mathrm{H+}}$, even if it is not self-dual, the Hamming weight of every codeword is even. (This is not true in general. Let $a, b, c \in \mathbb{F}_{16}$ have traces $1, \omega, \omega^2 \in \mathbb{F}_4$, respectively. Then the word (a, b, c) is isotropic but has odd Hamming weight.)

$q_{\mathrm{II},1}^{\mathrm{H+}}$ (even)

Codes in family $q_{\mathrm{II}}^{\mathrm{H+}}$ (even) that contain the all-ones vector. The Type of these codes is the representation $\rho(q_{\mathrm{II},1}^{\mathrm{H+}}) := (V, \rho_M, \rho_\Phi, \beta)$ of the form ring

$$R(q_{\mathrm{II},1}^{\mathrm{H+}}) := (R, M, \psi, \Phi = \langle \phi_0 \rangle \oplus \langle \varphi \rangle) = (\mathbb{F}_r, \mathbb{F}_r, \mathrm{id}, \mathbb{F}_r \oplus \mathbb{F}_r), \qquad (2.3.36)$$

where $V = \mathbb{F}_q$, ρ_M and $\rho_\Phi(\phi_0)$ are as in $\rho(q_{\mathrm{II}}^{\mathrm{H+}})$, and $\rho_\Phi(\varphi) = \beta(1, \ )$ is as in $\rho(q_1^{\mathrm{H+}})$. The automorphism group $\mathrm{Aut}(\rho(q_{\mathrm{II},1}^{\mathrm{H+}})) = \mathrm{Gal}(\mathbb{F}_q/\mathbb{F}_r)$, and the weak automorphism group $\mathrm{WAut}(\rho(q_{\mathrm{II},1}^{\mathrm{H+}})) = \mathbb{F}_r^* \rtimes \mathrm{Gal}(\mathbb{F}_q/\mathbb{F}_p)$.

$q^{\mathrm{H+}}$ (odd)

Trace-Hermitian codes over $\mathbb{F}_q$ which are linear over $\mathbb{F}_r$, where $r^2 = q = p^f$ is odd: $R = \mathbb{F}_r$, $V = \mathbb{F}_q$, $\overline{x} = x^r$, for all $x \in V$. The codes C of length N are $\mathbb{F}_r$-linear subspaces of $\mathbb{F}_q^N$. The obvious inner product to use would appear to be the skew-symmetric $\mathbb{F}_r$-bilinear form $(x, y) = x\overline{y} - \overline{x}y = xy^r - x^r y$. However, this takes values in $\mathbb{F}_q$, which is undesirable for an $\mathbb{F}_r$-linear code. We will change it by multiplying by some nonzero element $\alpha \in \mathbb{F}_q$ with $\alpha^r = -\alpha$. Since $(x, y)^r = -(x, y)$ for all $x, y \in \mathbb{F}_q$, the values $\alpha(x, y)$ lie in $\mathbb{F}_r$. We will therefore require that the codes be self-dual with respect to the new inner product $(x, y) = x\alpha y^r - x^r \alpha y = \text{Tr}(\alpha x \bar{y})$.

The Type of these codes is defined by the representation $\rho(q^{\text{H}+}) := (V, \rho_M, \rho_\Phi, \beta)$ of the form ring

$$R(q^{\text{H}+}) := (R, M, \psi, \Phi = \{\!\{M\}\!\} = \{0\}) = (\mathbb{F}_r, \mathbb{F}_r, \text{id}, \{0\}), \qquad (2.3.37)$$

where $V = \mathbb{F}_q$ and $\rho_M(a)(x, y) = \frac{1}{p} \text{Tr}(\alpha a x \bar{y})$, for all $a \in M = \mathbb{F}_r$ and $x, y \in \mathbb{F}_q$. Since $\bar{\alpha} = -\alpha$, $\text{Tr}(\alpha a x \bar{x}) = 0$ for all $a \in \mathbb{F}_r$, and so $\{\!\{M\}\!\} = \{0\}$.

The image of $\rho(q^{\text{H}+})$ does not depend on the choice of α, since any other α' with $\bar{\alpha}' = -\alpha'$ is of the form αa for some $a \in \mathbb{F}_r$. The representation depends on α, but we get the same notions of duality and isotropy in all cases. Note that, for any choice of α, codes of Type q^{H} are also codes of Type $q^{\text{H}+}$.

The automorphism and weak automorphism groups are analogous to those in the even characteristic case.

For the hwe see (2.3.13) again. The cwe for $C^\perp$ is obtained from the cwe for C by replacing each x_ξ, $\xi \in \mathbb{F}_q$, by

$$\sum_{\mu \in \mathbb{F}_q} \exp(2\pi i \, \text{Tr}(\alpha \mu \bar{\xi})/p) \, x_\mu \qquad (2.3.38)$$

and dividing by $|C|$.

$q_1^{\text{H}+}$ (odd)

Codes in family $q^{\text{H}+}$ that contain the all-ones vector. The Type of these codes is the representation $\rho(q_1^{\text{H}+}) := (V, \rho_M, \rho_\Phi, \beta)$ of the form ring

$$R(q_1^{\text{H}+}) := (R, M, \psi, \Phi = \langle \varphi \rangle) = (\mathbb{F}_r, \mathbb{F}_r, \text{id}, \mathbb{F}_r), \qquad (2.3.39)$$

where $V = \mathbb{F}_q$ and ρ_M are as in $\rho(q^{\text{H}+})$ and ρ_Φ is defined by

$$\rho_\Phi(\varphi)(x) = \beta(1, x) = \frac{1}{p} \text{Tr}_{\mathbb{F}_r/\mathbb{F}_p}(\alpha x - \alpha \bar{x}) = \frac{1}{p} \text{Tr}(\alpha x).$$

Again, the automorphism and weak automorphism groups agree with the even characteristic case.

2.3.5 Codes over Galois rings $\mathbb{Z}/m\mathbb{Z}$

Let $q = p^e$ be a prime power. A *Galois ring* $\text{GR}(q, f)$ is an unramified Galois extension of degree f of $\mathbb{Z}/q\mathbb{Z}$ (see Helleseth and Kumar [262, §81], McDonald [369], for further information, references, etc.). Let K denote the unique unramified extension of degree f of the p-adic numbers, and let $O := \mathbb{Z}_p[\zeta_{p^f-1}]$ be the ring of integers in K. Then pO is the unique maximal ideal of O and the Galois ring is defined by

$$\mathrm{GR}(p^e, f) := O/p^e O = (\mathbb{Z}/p^e\mathbb{Z})[\zeta_{p^f-1}]. \tag{2.3.40}$$

Since Galois automorphisms of K preserve the ideal $p^e O$, they can be viewed as Galois automorphisms of $\mathrm{GR}(p^e, f)$. Let $\zeta := \zeta_{p^f-1} + p^e O \in \mathrm{GR}(p^e, f)$. Then $\mathrm{GR}(p^e, f) = (\mathbb{Z}/p^e\mathbb{Z})[\zeta]$ and any element $x \in \mathrm{GR}(p^e, f)$ has a unique representation as

$$x = x_0 + px_1 + \ldots + p^{e-1}x_{e-1}, \tag{2.3.41}$$

with $x_i \in \{0, \zeta^0 = 1, \zeta, \ldots, \zeta^{p^f-2}\}$. The Galois group of $\mathrm{GR}(p^e, f)$ over $\mathbb{Z}/p^e\mathbb{Z}$ is generated by the Frobenius automorphism Frob that maps ζ to ζ^p, and hence is a cyclic group of order f. In the notation of (2.3.41), we have $\mathrm{Frob}(x) = x_0^p + px_1^p + \ldots + p^{e-1}x_{e-1}^p$. This defines a trace function

$$\mathrm{Tr} : \mathrm{GR}(p^e, f) \to \mathbb{Z}/p^e\mathbb{Z}, \quad \mathrm{Tr}(x) := \sum_{j=0}^{f-1} \mathrm{Frob}^j(x). \tag{2.3.42}$$

The simplest examples of Galois rings are the rings $\mathbb{Z}/q\mathbb{Z}$ themselves, where q is a prime power. All finite quotients of $\mathbb{Z}$ are of the form $\mathbb{Z}/m\mathbb{Z}$ for some $m \in \mathbb{N}$. Suppose $m = \prod_{i=1}^{s} p_i^{n_i}$, where the primes p_i are distinct. Then $\mathbb{Z}/m\mathbb{Z} = \oplus_{i=1}^{s}\mathbb{Z}/(p_i^{n_i}\mathbb{Z})$ is a direct product of Galois rings of prime power characteristic. So in principle it would be enough to treat only the case of prime powers $m = p^n$, but since it introduces no additional complication, we will define the Types for arbitrary finite rings $R = \mathbb{Z}/m\mathbb{Z}$.

$4^{\mathbb{Z}}$

$\mathbb{Z}/4\mathbb{Z}$-linear codes: $R = \mathbb{Z}/4\mathbb{Z} = \{0, 1, 2, 3\}$, with $(x, y) = xy \pmod 4$; $C = \mathbb{Z}/4\mathbb{Z}$-submodule of $(\mathbb{Z}/4\mathbb{Z})^N$. This is a special case of the next family.

The MacWilliams identities for codes from the family $4^{\mathbb{Z}}$ are:

$$\mathrm{hwe}(C^\perp)(x, y) = \frac{1}{|C|}\,\mathrm{hwe}(C)(x + 3y, x - y),$$

$$\mathrm{swe}(C^\perp)(x, y, z) = \frac{1}{|C|}\,\mathrm{swe}(C)(x + 2y + z, x - y, x - 2y + z),$$

$$\mathrm{cwe}(C^\perp)(x, y, z, t) = \frac{1}{|C|}\,\mathrm{cwe}(C)(x + y + z + t, x + iy - z - it,$$

$$x - y + z - t, x - iy - z + it). \tag{2.3.43}$$

The variables (x, y, z, t) in the complete weight enumerator correspond to (x_0, x_1, x_2, x_3); we obtain the symmetrized weight enumerator by setting $t = y$

For the non-trivial shadow $S := S_{\phi_0}(C)$ where ϕ_0 is the quadratic map in $4_{\mathrm{II}}^{\mathbb{Z}}$ defined in (2.3.52) below, we have:

2.3 Catalogue of important types   53

$$\text{swe}(S)(x,y,z) = \frac{1}{|C|}\,\text{swe}(C)(x+2y+z, \eta(x-y), -x+2y-z)\,,$$

$$\text{cwe}(S)(x,y,z,t) = \frac{1}{|C|}\,\text{swe}(C)(x+y+z+t, \eta(x+iy-z-it),$$
$$-(x-y+z-t), \eta(x-iy-z+it))\,, \qquad (2.3.44)$$

where $\eta = \exp(\pi i/4)$.

$m^{\mathbb{Z}}$

$R = \mathbb{Z}/m\mathbb{Z}$, where m is an integer ≥ 2, with $(x,y) = xy \pmod m$; $C = \mathbb{Z}/m\mathbb{Z}$-submodule of $(\mathbb{Z}/m\mathbb{Z})^N$. The Type of these codes is given by the representation $\rho(m^{\mathbb{Z}}) := (V, \rho_M, \rho_\Phi, \beta)$ of the form ring

$$R(m^{\mathbb{Z}}) := (R, M, \psi, \Phi = \{\!\{M\}\!\}) = (\mathbb{Z}/m\mathbb{Z}, \mathbb{Z}/m\mathbb{Z}, \text{id}, \mathbb{Z}/m\mathbb{Z})\,, \qquad (2.3.45)$$

where $V = \mathbb{Z}/m\mathbb{Z}$, $\rho_M(a)(x,y) = \frac{1}{m}axy$ for all $a \in M = \mathbb{Z}/m\mathbb{Z}$ and $x,y \in V = \mathbb{Z}/m\mathbb{Z}$, and

$$\rho_\Phi(\{\!\{a\}\!\})(x) = \{\!\{\rho_\Phi(a)\}\!\}(x) = \frac{1}{m}ax^2\,.$$

In particular, isotropic codes in the family $m^{\mathbb{Z}}$ have the property that the Euclidean norms of codewords are divisible by m. Then

$$\text{Aut}(\rho(m^{\mathbb{Z}})) = \{g \in (\mathbb{Z}/m\mathbb{Z})^* \mid g^2 = 1\} \qquad (2.3.46)$$

is the 2-torsion subgroup of the unit group $(\mathbb{Z}/m\mathbb{Z})^* = \text{WAut}(\rho(m^{\mathbb{Z}}))$. This subgroup has order 2^{o+e}, where o denotes the number of distinct odd prime divisors of m, and $e = 2$ if m is divisible by 8, $e = 1$ if 4 but not 8 divides m, and $e = 0$ if m is not divisible by 4. The group of equivalences consists of monomial matrices over (2.3.46); for the weak equivalences we also allow global multiplication by units of $\mathbb{Z}/m\mathbb{Z}$.

The MacWilliams identities for codes from the family $m^{\mathbb{Z}}$ are as follows:

$$\text{hwe}(C^\perp)(x,y) = \frac{1}{|C|}\,\text{hwe}(C)(x+(m-1)y, x-y)\,. \qquad (2.3.47)$$

The cwe for $C^\perp$ is obtained from the cwe for C by replacing each x_j, $0 \leq j \leq m-1$, by

$$\sum_{k=0}^{m-1} \exp(2\pi ijk/m)\,x_k \qquad (2.3.48)$$

and dividing by $|C|$. If m is even, then $(m^{\mathbb{Z}}, m_{\text{II}}^{\mathbb{Z}})$ is a shadow pair and the non-trivial shadow of a code of Type $\rho(m^{\mathbb{Z}})$ is defined by the quadratic map $\rho_\Phi(m_{\text{II}}^{\mathbb{Z}})(\phi_0)$ given in (2.3.52). The cwe of the ϕ_0-shadow is obtained from the cwe for C by replacing each x_j, $0 \leq j \leq m-1$, by

$$\sum_{k=0}^{m-1} \exp(\pi i(j^2 + 2jk)/m)\, x_k \qquad (2.3.49)$$

and dividing by $|C|$.

$m_1^{\mathbb{Z}}$

Codes in family $m^{\mathbb{Z}}$ that contain $\mathbf{1}$. The Type of these codes is given by the representation $\rho(m_1^{\mathbb{Z}}) := (V, \rho_M, \rho_\Phi, \beta)$ of the form ring

$$R(m_1^{\mathbb{Z}}) := (R, M, \psi, \Phi = \langle\{\!\{M\}\!\}, \varphi\rangle)$$
$$= (\mathbb{Z}/m\mathbb{Z}, \mathbb{Z}/m\mathbb{Z}, \mathrm{id}, \mathbb{Z}/m\mathbb{Z} \oplus \mathbb{Z}/m'\mathbb{Z}), \qquad (2.3.50)$$

where $V = \mathbb{Z}/m\mathbb{Z}$ and ρ_M are as in $\rho(m^{\mathbb{Z}})$, and ρ_Φ is defined by

$$\rho_\Phi(\varphi)(x) = \beta(1, x) = \frac{1}{m}x,$$

where $m' = m$ if m is odd, $m' = m/2$ if m is even. Note that $m'\varphi \in \{\!\{M\}\!\}$, and hence $\Phi \cong \mathbb{Z}/m\mathbb{Z} \oplus \mathbb{Z}/m'\mathbb{Z}$. Also $\mathrm{Aut}(\rho(m_1^{\mathbb{Z}})) = \{1\}$, $\mathrm{WAut}(\rho(m_1^{\mathbb{Z}})) = \mathrm{WAut}(\rho(m^{\mathbb{Z}}))$.

$m_{\mathrm{II}}^{\mathbb{Z}}$

These are the Type II codes in the family $m^{\mathbb{Z}}$, where m is even, that is, they have the additional property that the Euclidean norm of every codeword is divisible by $2m$. The Type of these codes is given by the representation $\rho(m_{\mathrm{II}}^{\mathbb{Z}}) := (V, \rho_M, \rho_\Phi, \beta)$ of the form ring

$$R(m_{\mathrm{II}}^{\mathbb{Z}}) := (R, M, \psi, \Phi = \langle \phi_0 \rangle) = (\mathbb{Z}/m\mathbb{Z}, \mathbb{Z}/m\mathbb{Z}, \mathrm{id}, \mathbb{Z}/2m\mathbb{Z}), \qquad (2.3.51)$$

where $V = \mathbb{Z}/m\mathbb{Z}$ and $\rho_M(a)(x, y) = \frac{1}{m}axy$ as in the family $m^{\mathbb{Z}}$, and

$$\rho_\Phi(\phi_0)(x) := \frac{1}{2m}x^2. \qquad (2.3.52)$$

Also

$$\mathrm{Aut}(\rho(m_{\mathrm{II}}^{\mathbb{Z}})) = \{g \in \mathbb{Z}/m\mathbb{Z}^* \mid g^2 \equiv 1 \pmod{2m}\} \qquad (2.3.53)$$

is of order 2^{o+e}, where o denotes the number of distinct odd prime divisors of m and $e = 1$ if m is divisible by 4, $e = 0$ otherwise. The weak automorphism group again is the entire unit group $\mathbb{Z}/m\mathbb{Z}^*$. The group of equivalences consists of monomial matrices over (2.3.53); for the weak equivalences we also allow global multiplication by units of $\mathbb{Z}/m\mathbb{Z}$.

$m_{\mathrm{II},1}^{\mathbb{Z}}$

These are codes in the family $m_{\mathrm{II}}^{\mathbb{Z}}$ that contain $\mathbf{1}$. The Type of these codes is given by the representation $\rho(m_{\mathrm{II},1}^{\mathbb{Z}}) := (V, \rho_M, \rho_\Phi, \beta)$ of the form ring

$$R(m_{\mathrm{II},1}^{\mathbb{Z}}) := (R, M, \psi, \Phi = \langle \phi_0, \varphi \rangle)$$
$$= (\mathbb{Z}/m\mathbb{Z}, \mathbb{Z}/m\mathbb{Z}, \mathrm{id}, \mathbb{Z}/2m\mathbb{Z} \times \mathbb{Z}/(m/2)\mathbb{Z}), \qquad (2.3.54)$$

where $V = \mathbb{Z}/m\mathbb{Z}$, ρ_M and $\rho_\Phi(\phi_0)$ are as in the family $m_{\mathrm{II}}^{\mathbb{Z}}$, and $\rho_\Phi(\varphi)(x) = \beta(1, x) = \frac{1}{m}x$. Note that $\frac{m}{2}\varphi = m\phi_0$. As in the case $m_1^{\mathbb{Z}}$ we have $\mathrm{Aut}(\rho(m_{\mathrm{II},1}^{\mathbb{Z}})) = \{1\}$, $\mathrm{WAut}(\rho(m_{\mathrm{II},1}^{\mathbb{Z}})) = (\mathbb{Z}/m\mathbb{Z})^*$.

$m_{\mathrm{S}}^{\mathbb{Z}}$

These are codes in the family $m^{\mathbb{Z}}$, where m is even, whose shadow contains $\mathbf{1}$. The Type of these codes is given by the representation $\rho(m_{\mathrm{S}}^{\mathbb{Z}}) := (V, \rho_M, \rho_\Phi, \beta)$ of the form ring $R(m_{\mathrm{II}}^{\mathbb{Z}})$, where $V = \mathbb{Z}/m\mathbb{Z}$, $\rho_M(a)(x, y) = \frac{1}{m}axy$ are as in the family $m^{\mathbb{Z}}$, and

$$\rho_\Phi(\phi_0)(x) := \frac{1}{2m}x^2 + \frac{1}{m}x\,.$$

2.3.6 Codes over more general Galois rings

We now consider more general Galois rings. Of course the Types $m^{\mathbb{Z}}$, etc. defined above as well as the Types over finite fields are all special cases of Types over general Galois rings. However, following the historical development—and also because codes over general Galois rings are less well known—we will retain the special names used above for the Types of codes over fields and the rings $\mathbb{Z}/m\mathbb{Z}$.

Arbitrary Galois rings have the same relationship to the rings $\mathbb{Z}/q\mathbb{Z}$ as arbitrary finite fields have to prime fields. Thus there are Types of codes over $\mathrm{GR}(p^e, f)$ that are parallel to all the Types over $\mathbb{F}_{p^f}$.

$\mathrm{GR}(p^e, f)^{\mathbf{E}}$

Euclidean self-dual $\mathrm{GR}(p^e, f)$-linear codes. The Type of these codes is the representation $\rho(\mathrm{GR}(p^e, f)^{\mathrm{E}}) := (V, \rho_M, \rho_\Phi, \beta)$ of the form ring

$$R(\mathrm{GR}(p^e, f)^{\mathrm{E}}) := (R, M, \psi, \Phi = \{\!\{M\}\!\})$$
$$= (\mathrm{GR}(p^e, f), \mathrm{GR}(p^e, f), \mathrm{id}, \mathrm{GR}(p^e, f)), \qquad (2.3.55)$$

where $V = \mathrm{GR}(p^e, f)$, $\rho_M(a)(x, y) = \frac{1}{p^e}\mathrm{Tr}(axy)$ for $a \in M = \mathrm{GR}(p^e, f)$, $x, y \in V = \mathrm{GR}(p^e, f)$, and Tr denotes the trace of $\mathrm{GR}(p^e, f)$ to $\mathbb{Z}/p^e\mathbb{Z}$ defined in (2.3.42). This defines the representation ρ_Φ, since

$$\rho_\Phi(\{\!\{a\}\!\})(x) = \{\!\{\rho_\Phi(a)\}\!\}(x) = \frac{1}{p^e}\,\mathrm{Tr}(ax^2)\,.$$

Then $\mathrm{Aut}(\rho(\mathrm{GR}(p^e,f)^{\mathrm{E}})) = \{g \in \mathrm{GR}(p^e,f) \mid g^2 = 1\} = \{1,-1\}$ if $p \neq 2$; and $\mathrm{Aut}(\rho(\mathrm{GR}(2^e,f)^{\mathrm{E}})) = 1 + 2^{e-1}\,\mathrm{GR}(2^e,f)$ is isomorphic to Z_2^f if $p = 2$ and $e > 1$ and is trivial if $p = 2$ and $e = 1$. The weak automorphism group is $\mathrm{WAut}(\rho(\mathrm{GR}(p^e,f)^{\mathrm{E}})) = \mathrm{GR}(p^e,f)^* \rtimes \mathrm{Gal}(\mathrm{GR}(p^e,f),\mathbb{Z}/p^e\mathbb{Z})$.

$\mathbf{GR}(p^e,f)_1^{\mathbf{E}}$

Assume that p is an odd prime. The Type of Euclidean self-dual linear codes over $\mathrm{GR}(p^e,f)$ containing the all-ones vector $\mathbf{1}$ is the representation $\rho(\mathrm{GR}(p^e,f)_1^{\mathrm{E}}) := (V,\rho_M,\rho_\Phi,\beta)$ of the form ring

$$\begin{aligned}
R(\mathrm{GR}(p^e,f)_1^{\mathrm{E}}) &:= (R,M,\psi,\Phi = \{\!\{M\}\!\} \oplus \ker(\lambda))\\
&= (\mathrm{GR}(p^e,f),\mathrm{GR}(p^e,f),\mathrm{id},\mathrm{GR}(p^e,f) \oplus \mathrm{GR}(p^e,f))\,,
\end{aligned}$$
$$(2.3.56)$$

where $V = \mathrm{GR}(p^e,f)$, $\rho_M(a)(x,y) = \frac{1}{p^e}\,\mathrm{Tr}(axy)$ for $a \in M = \mathrm{GR}(p^e,f)$ and $x,y \in V = \mathrm{GR}(p^e,f)$ are as above, and

$$\rho_\Phi(a,b)(x) = \frac{1}{p^e}\,\mathrm{Tr}(ax^2 + bx)\,.$$

We have $\mathrm{Aut}(\rho(\mathrm{GR}(p^e,f)_1^{\mathrm{E}})) = 1$ and

$$\mathrm{WAut}(\rho(\mathrm{GR}(p^e,f)_1^{\mathrm{E}})) = \mathrm{WAut}(\rho(\mathrm{GR}(p^e,f)^{\mathrm{E}}))\,.$$

$\mathbf{GR}(p^e,f)_{p^s}^{\mathbf{E}}$

In the previous Type we considered codes containing the all-ones vector $\mathbf{1}$. Such codes also contain $a\mathbf{1}$ for all $a \in \mathrm{GR}(p^e,f)$. As a weaker requirement we may specify that C should contain $p^s\mathbf{1}$ for some $0 \leq s < e$ (for $s = 0$ this coincides with the previous Type). For odd primes p, the Type of these codes is the representation $\rho(\mathrm{GR}(p^e,f)_{p^s}^{\mathrm{E}}) := (V,\rho_M,\rho_\Phi,\beta)$ of the form ring

$$\begin{aligned}
R(\mathrm{GR}(p^e,f)_{p^s}^{\mathrm{E}}) &:= (R,M,\psi,\Phi = \{\!\{M\}\!\} \oplus \ker(\lambda))\\
&= (\mathrm{GR}(p^e,f),\mathrm{GR}(p^e,f),\mathrm{id},\mathrm{GR}(p^e,f) \oplus \mathrm{GR}(p^{e-s},f))\,,
\end{aligned}$$
$$(2.3.57)$$

where $V = \mathrm{GR}(p^e,f)$, $\rho_M(a)(x,y) = \frac{1}{p^e}\,\mathrm{Tr}(axy)$ for $a \in M = \mathrm{GR}(p^e,f)$ and $x,y \in V = \mathrm{GR}(p^e,f)$ are as above, and

$$\rho_\Phi(a,b)(x) = \frac{1}{p^e}\,\mathrm{Tr}(ax^2) + \frac{1}{p^{e-s}}\,\mathrm{Tr}(bx)\,.$$

$\mathbf{GR(2^e, f)^E_{2^s}}$

For $p = 2$, squaring is a Galois automorphism modulo 2, hence $\mathrm{Tr}(x) \equiv \mathrm{Tr}(x^2)$ (mod 2). Therefore the Type of Euclidean self-dual $GF(2^e, f)$-linear codes that contain $2^s \mathbf{1}$ for some $0 \le s < e$ is the representation $\rho(\mathrm{GR}(2^e, f)^E_{2^s}) := (V, \rho_M, \rho_\Phi, \beta)$ of the form ring

$$
\begin{aligned}
R(\mathrm{GR}(2^e, f)^E_{2^s}) &:= (R, M, \psi, \Phi = \langle \phi = \{\mathbf{1}\}, \varphi_s \rangle) \\
&= (\mathrm{GR}(2^e, f), \mathrm{GR}(2^e, f), \mathrm{id}, \mathrm{GR}(2^e, f) \oplus \mathrm{GR}(2^{e-s-1}, f)),
\end{aligned}
\tag{2.3.58}
$$

where $V = \mathrm{GR}(2^e, f)$, $\rho_M(a)(x, y) = \frac{1}{2^e} \mathrm{Tr}(axy)$ for $a \in M = \mathrm{GR}(2^e, f)$ and $x, y \in V = \mathrm{GR}(2^e, f)$, $\rho_\Phi(\phi)(x) = \frac{1}{2^e} \mathrm{Tr}(x^2)$ for all $x \in V$ are as above, and

$$
\rho_\Phi(\varphi_s)(x) = 2^s \beta(1, x) = \frac{1}{2^{e-s}} \mathrm{Tr}(x) \text{ for all } x \in V.
$$

Note that $\rho_\Phi(2^{e-s-1}\varphi_s) = \rho_\Phi(2^{e-1}\phi)$.

$\mathbf{GR(2^e, f)^E_{II}}$

The notion of generalized doubly-even self-dual codes over fields of characteristic 2 given above (see family q^E_{II}) can also be extended to Galois rings of characteristic a power of 2. Equation (2.3.15) is replaced by:

$$
\sum_{i=1}^N c_i^2 = 0 \text{ for all } c \in C
\tag{2.3.59}
$$

(viewing squaring as a map from $\mathrm{GR}(2^e, f)$ to $\mathrm{GR}(2^{e+1}, f)$). The Type of these codes is the representation $\rho(\mathrm{GR}(2^e, f)^E_{II}) := (V, \rho_M, \rho_\Phi, \beta)$ of the form ring

$$
\begin{aligned}
R(\mathrm{GR}(2^e, f)^E_{II}) &:= (R, M, \psi, \Phi = \langle \phi_0 \rangle) \\
&= (\mathrm{GR}(2^e, f), \mathrm{GR}(2^e, f), \mathrm{id}, \mathrm{GR}(2^{e+1}, f)),
\end{aligned}
\tag{2.3.60}
$$

where $V = \mathrm{GR}(2^e, f)$,

$$
\rho_M(a)(x, y) = \frac{1}{2^e} \mathrm{Tr}(axy),
$$

for all $a \in M = \mathrm{GR}(2^e, f)$ and $x, y \in V = \mathrm{GR}(2^e, f)$ (as in the family $\mathrm{GR}(2^e, f)^E$), and

$$
\rho_\Phi(\phi_0)(x) := \frac{1}{2^{e+1}} \mathrm{Tr}(x^2),
$$

for all $x \in V$. Since the Galois automorphisms of $\mathrm{GR}(2^e, f)$ lift to Galois automorphisms of $\mathrm{GR}(2^{e+1}, f)$, we have

$$
\mathrm{Aut}(\rho(\mathrm{GR}(2^e, f)^E_{II})) = \{g \in \mathrm{GR}(2^e, f) \mid g^2 \equiv 1 \pmod{2^{e+1}}\} = \{\pm 1\},
$$

$$
\mathrm{WAut}(\rho(\mathrm{GR}(2^e, f)^E_{II})) = \mathrm{GR}(2^e, f)^* \rtimes \mathrm{Gal}(\mathrm{GR}(2^e, f), \mathbb{Z}/2^e\mathbb{Z}).
$$

$\mathbf{GR}(2^e, f)_{\mathrm{II}, 2^s}^{\mathrm{E}}$

Codes of Type $\mathrm{GR}(2^e, f)_{\mathrm{II}}^{\mathrm{E}}$ that contain $2^s \mathbf{1}$ (and hence the all-ones vector if $s = 0$). If $s = e - 1$, this Type coincides with $\rho(\mathrm{GR}(2^e, f)_{\mathrm{II}}^{\mathrm{E}})$ above, so we assume that $s \leq e - 2$. The Type of these codes is given by the representation $\rho(\mathrm{GR}(2^e, f)_{\mathrm{II}, 2^s}^{\mathrm{E}}) := (V, \rho_M, \rho_\Phi, \beta)$ of the form ring

$$R(\mathrm{GR}(2^e, f)_{\mathrm{II}, 2^s}^{\mathrm{E}}) := (R, M, \psi, \Phi = \langle \phi_0, \varphi_s \rangle)$$
$$= (\mathrm{GR}(2^e, f), \mathrm{GR}(2^e, f), \mathrm{id}, \mathrm{GR}(2^{e+1}, f) \times \mathrm{GR}(2^{e-s-1}, f)),$$
$$(2.3.61)$$

where $V = \mathrm{GR}(2^e, f)$,

$$\rho_M(a)(x, y) = \frac{1}{2^e} \operatorname{Tr}(axy),$$

for all $a \in M = \mathrm{GR}(2^e, f)$ and $x, y \in V = \mathrm{GR}(2^e, f)$ (as in the family $\mathrm{GR}(2^e, f)^{\mathrm{E}}$), and

$$\rho_\Phi(\phi_0)(x) := \frac{1}{2^{e+1}} \operatorname{Tr}(x^2), \text{ and } \rho_\Phi(\varphi_s)(x) := \frac{1}{2^{e-s}} \operatorname{Tr}(x).$$

Since $x \mapsto x^2$ is a Galois automorphism of $\mathbb{F}_{2^f}$, we have $\rho_\Phi(2^{e-s-1} \varphi_s) = \rho_\Phi(2^e \phi_0)$ and the complement of $\langle \phi_0 \rangle$ in Φ is generated by $2^{s+1} \phi_0 + \varphi_s$. As in the case $\mathrm{GR}(2^e, f)_{\mathrm{II}}^{\mathrm{E}}$, the mapping λ is surjective. Now the kernel of λ is generated by φ_s (since $2^e \phi_0 = 2^{e-s-1} \varphi_s$) and is isomorphic to $\mathrm{GR}(2^{e-s}, f)$.

$\mathbf{GR}(p^e, f)^{\mathrm{H}}$

Linear codes over $\mathrm{GR}(p^e, f)$, for p odd and $f = 2\ell$ even, which are Hermitian self-dual with respect to the inner product $(x, y) = x\bar{y}$, where $^- = \mathrm{Frob}^\ell$ generates the Galois group $\mathrm{Gal}(\mathrm{GR}(p^e, f)/\mathrm{GR}(p^e, \ell))$. Note that if $x = \sum_{i=0}^{e-1} x_i p^i$ as in (2.3.41), $\bar{x} = \sum_{i=0}^{e-1} \bar{x_i} p^i = \sum_{i=0}^{e-1} x_i^{p^\ell} p^i$. The Type of these codes is the representation $\rho(\mathrm{GR}(p^e, f)^{\mathrm{H}}) := (V, \rho_M, \rho_\Phi, \beta)$ of the form ring

$$R(\mathrm{GR}(p^e, f)^{\mathrm{H}}) := (R, M, \psi, \Phi = \{\!\{ M \}\!\})$$
$$= (\mathrm{GR}(p^e, f), \mathrm{GR}(p^e, f), \mathrm{id}, \mathrm{GR}(p^e, f)),$$
$$(2.3.62)$$

where $V = \mathrm{GR}(p^e, f)$, $\rho_M(a)(x, y) = \frac{1}{p^e} \operatorname{Tr}(ax\bar{y})$ for all $a \in M = \mathrm{GR}(p^e, f)$ and $x, y \in V = \mathrm{GR}(p^e, f)$, and Tr denotes the trace of $\mathrm{GR}(p^e, f)$ over $\mathbb{Z}/p^e\mathbb{Z}$ defined in (2.3.42). This specifies ρ_Φ since $\{\!\{ \}\!\}$ is surjective.

$\mathbf{GR}(p^e, f)_{p^s}^{\mathrm{H}}$

Codes in $\mathrm{GR}(p^e, f)^{\mathrm{H}}$ that contain $p^s \mathbf{1}$ for some $0 \leq s < e$. The Type of these codes is the representation $\rho(\mathrm{GR}(p^e, f)_{p^s}^{\mathrm{H}}) := (V, \rho_M, \rho_\Phi, \beta)$ of the form ring

$$R(\mathrm{GR}(p^e, f)^{\mathrm{H}}_{p^s}) := (R, M, \psi, \Phi = \{\!\{M\}\!\} \oplus \langle \varphi_s \rangle)$$
$$= (\mathrm{GR}(p^e, f), \mathrm{GR}(p^e, f), \mathrm{id}, \mathrm{GR}(p^e, f) \oplus \mathrm{GR}(p^{e-s}, f)), \quad (2.3.63)$$

where $V = \mathrm{GR}(p^e, f)$ and ρ_M are defined as in $\mathrm{GR}(p^e, f)^{\mathrm{H}}$, and

$$\rho_\Phi(\varphi)(x) = p^s \beta(1, x) = \frac{1}{p^{e-s}} \mathrm{Tr}(x).$$

The automorphism group is

$$\mathrm{Aut}(\rho(\mathrm{GR}(p^e, f)^{\mathrm{H}}_{p^s})) = \{ g \in 1 + p^{e-s} \, \mathrm{GR}(p^e, f) \mid g\bar{g} = 1 \},$$

of order $p^{\ell s}$, and the weak automorphism group is the split extension of $\mathrm{Aut}(\rho(\mathrm{GR}(p^e, f)^{\mathrm{H}}_{p^s}))$ by $Z_f \cong \mathrm{Gal}(\mathrm{GR}(p^e, f), \mathbb{Z}/p^e\mathbb{Z})$.

$\mathrm{GR}(p^e, f)^{\mathrm{H}+}$

Trace-Hermitian codes over $\mathrm{GR}(p^e, f)$, p odd, which are linear over $\mathrm{GR}(p^e, \ell)$, where $f = 2\ell$ and $^-$ denotes the Galois automorphism as in the case $\mathrm{GR}(p^e, f)^{\mathrm{H}}$. These are codes $C \leq \mathrm{GR}(p^e, f)^N$ which are $\mathrm{GR}(p^e, \ell)$-submodules and are self-dual with respect to the non-standard Hermitian inner product $(x, y) = x\alpha\bar{y} - \bar{x}\alpha y$, where α is a nonzero element in $\mathrm{GR}(p^e, f)$ with $\bar{\alpha} = -\alpha$ (compare family $q^{\mathrm{H}+}$ on page 50). Then the Type of these codes is the representation $\rho(\mathrm{GR}(p^e, f)^{\mathrm{H}+}) := (V, \rho_M, \rho_\Phi, \beta)$ of the form ring

$$R(\mathrm{GR}(p^e, f)^{\mathrm{H}+}) := (R, M, \psi, \Phi = \{\!\{M\}\!\} = \{0\})$$
$$= (\mathrm{GR}(p^e, \ell), \mathrm{GR}(p^e, \ell), \mathrm{id}, \{0\}), \quad (2.3.64)$$

where $V = \mathrm{GR}(p^e, f)$ and $\rho_M(a)(x, y) = \frac{1}{p^e} \mathrm{Tr}(a\alpha x\bar{y})$ for all $a \in M = \mathrm{GR}(p^e, \ell)$ and $x, y \in V = \mathrm{GR}(p^e, f)$. Since $ax\bar{x} \in \mathrm{GR}(p^e, \ell)$ for all $x \in V$ and $a \in \mathrm{GR}(p^e, \ell)$, the trace

$$\mathrm{Tr}(\alpha ax\bar{x}) = \mathrm{Tr}_{\mathrm{GR}(p^e, \ell)/(\mathbb{Z}/p^e\mathbb{Z})}(ax\bar{x}(\alpha + \bar{\alpha})) = 0,$$

and hence $\{\!\{M\}\!\} = \{0\}$. Again the automorphism group is the full unitary group of $\mathrm{GR}(p^e, f)$, of order $p^{\ell e}(p^\ell + 1)$.

$\mathrm{GR}(p^e, f)^{\mathrm{H}+}_{p^s}$

Codes in family $\mathrm{GR}(p^e, f)^{\mathrm{H}+}$ that contain $p^s 1$ (for some $0 \leq s < e$). The Type of these codes is the representation $\rho(\mathrm{GR}(p^e, f)^{\mathrm{H}+}_{p^s}) := (V, \rho_M, \rho_\Phi, \beta)$ of the form ring

$$R(\mathrm{GR}(p^e, f)^{\mathrm{H}+}_{p^s}) := (R, M, \psi, \Phi = \langle \varphi_s \rangle)$$
$$= (\mathrm{GR}(p^e, \ell), \mathrm{GR}(p^e, \ell), \mathrm{id}, \mathrm{GR}(p^{e-s}, \ell)), \quad (2.3.65)$$

where $V = \mathrm{GR}(p^e, f)$ and ρ_M are as in $\rho(\mathrm{GR}(p^e, f)^{\mathrm{H}+})$, and ρ_Φ is defined by

$$\rho_\Phi(\varphi_s)(x) = p^s \beta(1, x) = \frac{1}{p^{e-s}} \mathrm{Tr}(\alpha x).$$

Here the automorphism group consists of those unitary elements in $\mathrm{GR}(p^e, 2\ell)$ that are 1 modulo $p^s \, \mathrm{GR}(p^e, 2\ell)$.

2.3.7 Linear codes over p-adic integers

$\mathbb{Z}_p$

Linear codes over the p-adic integers: $R = \mathbb{Z}_p$ is the ring of p-adic integers and $(x, y) := xy$; $C = \mathbb{Z}_p$-submodule of $\mathbb{Z}_p^N$. The Type of these codes is given by the representation $\rho(\mathbb{Z}_p) := (V, \rho_M, \rho_\Phi, \beta)$ of the form ring

$$R(\mathbb{Z}_p) := (R, M, \psi, \Phi = \{\!\{M\}\!\}) = (\mathbb{Z}_p, \mathbb{Z}_p, \text{id}, \mathbb{Z}_p) , \qquad (2.3.66)$$

where $V = \mathbb{Z}_p$ and $\rho_M(a)(x, y) = axy$. Since R is infinite, this is not a finite representation.

More general p-adic integers

More generally, let R be the ring of integers in a finite extension of the rational p-adic field $\mathbb{Q}_p$. Let $\mathbb{F} = R/\pi R$ be the residue class field of R, where $\pi \in R$ is any nonzero prime. There are Types over R analogous to the Types over Galois rings, namely Euclidean self-dual codes (possibly containing $\mathbf{1}$), Type II codes if $p = 2$, Hermitian self-dual codes (possibly containing 1) and additive Hermitian self-dual codes if $|\mathbb{F}|$ is a square. One may view these p-adic codes as limits of codes over Galois rings. The form rings are completely analogous to those for Galois rings. Since they will not be mentioned elsewhere in this book, we omit the details.

2.4 Examples of self-dual codes

This section will describe some important examples of self-dual codes. The following notation will be used when specifying generators for codes. If $a_1 \ldots a_N$ is a vector in V^N then

$$a_1 \ldots a_m (a_{m+1} \ldots a_N)$$

will abbreviate the set of $N - m$ vectors in V^N whose first m coordinates are $a_1 \ldots a_m$ and whose last $N - m$ coordinates are obtained by cyclic permutations of $a_{m+1} \ldots a_N$. Similarly,

$$(a_1 \ldots a_m) a_{m+1} \ldots a_N$$

stands for the set of m vectors obtained by cyclically permuting the first m coordinates.

We begin with codes over finite fields.

2.4.1 2: Binary codes

We start with the families 2_{I} and 2_{II}. A third family will be described in §7.2.1. Of course for binary codes the automorphism group and weak automorphism group coincide.

2_I: Singly-even binary self-dual codes

The codes of Type 2_I are subspaces $C \leq \mathbb{F}_2^N$ such that

$$\sum_{i=1}^{N} c_i c_i' = 0 \text{ for } c, c' \in C. \tag{2.4.1}$$

The first example of such a code is the $[2,1,2]$ repetition code $i_2 := \{00, 11\}$, with Hamming weight enumerator

$$\mathrm{hwe}(i_2)(x,y) = x^2 + y^2, \tag{2.4.2}$$

This is also the complete weight enumerator, since the alphabet has only two elements. The full weight enumerator of i_2 is $e_{00} + e_{11} \in \mathbb{C}[e_{00}, e_{01}, e_{10}, e_{11}]$. To obtain the genus-2 weight enumerator of i_2, we consider pairs of codewords. Thus

$$\mathrm{cwe}_2(i_2)(x_{00}, x_{01}, x_{10}, x_{11}) = x_{00}^2 + x_{01}^2 + x_{10}^2 + x_{11}^2. \tag{2.4.3}$$

Similarly the genus-m weight enumerator is

$$\mathrm{cwe}_m(i_2) = \sum_{w \in \mathbb{F}_2^m} x_w^2. \tag{2.4.4}$$

The automorphism group $\mathrm{Aut}(i_2)$ of i_2 is the cyclic group $Z_2 \cong S_2$ (see Definition 1.11.3).

2_{II}: Doubly-even binary self-dual codes

These are perhaps the most famous self-dual codes.

The $[8,4,4]$ *Hamming* code e_8 (see [133, p. 80], [361]), generated by $1(1101000)$, or with generator matrix

$$\begin{bmatrix} 0 & 0 & 0 & 0 & 1 & 1 & 1 & 1 \\ 0 & 0 & 1 & 1 & 0 & 0 & 1 & 1 \\ 0 & 1 & 0 & 1 & 0 & 1 & 0 & 1 \\ 1 & 1 & 1 & 1 & 1 & 1 & 1 & 1 \end{bmatrix}, \tag{2.4.5}$$

or alternatively

$$\begin{bmatrix} 1 & 0 & 0 & 0 & 0 & 1 & 1 & 1 \\ 0 & 1 & 0 & 0 & 1 & 0 & 1 & 1 \\ 0 & 0 & 1 & 0 & 1 & 1 & 0 & 1 \\ 0 & 0 & 0 & 1 & 1 & 1 & 1 & 0 \end{bmatrix}, \tag{2.4.6}$$

is self-dual with Hamming weight enumerator

$$\mathrm{hwe}(e_8)(x,y) = x^8 + 14x^4y^4 + y^8. \tag{2.4.7}$$

The genus-m complete weight enumerator of e_8 was calculated in [383, Theorem 4.14]. Let $G(m, k)$ denote the set of k-dimensional subspaces of $\mathbb{F}_2^m$. Then

$$\mathrm{cwe}_m(e_8) = \sum_{v \in \mathbb{F}_2^m} x_v^8 + 14 \sum_{U \in G(m,1)} \sum_{d \in \mathbb{F}_2^m/U} \prod_{v \in d+U} x_v^4$$
$$+ 168 \sum_{U \in G(m,2)} \sum_{d \in \mathbb{F}_2^m/U} \prod_{v \in d+U} x_v^2$$
$$+ 1344 \sum_{U \in G(m,3)} \sum_{d \in \mathbb{F}_2^m/U} \prod_{v \in d+U} x_v . \quad (2.4.8)$$

The second term on the right-hand side is equal to $14 \sum_{\{u,v\}} x_u^4 x_v^4$, where $\{u, v\}$ runs through unordered pairs of distinct elements of $\mathbb{F}_2^m$. To check the value at $x_v \equiv 1$, we observe that

$$2^m + 14 \begin{bmatrix} m \\ 1 \end{bmatrix} 2^{m-1} + 168 \begin{bmatrix} m \\ 2 \end{bmatrix} 2^{m-2} + 1344 \begin{bmatrix} m \\ 3 \end{bmatrix} 2^{m-3} = 2^{4m} ,$$

where $\begin{bmatrix} m \\ k \end{bmatrix} = |G(m, k)|$ is a Gaussian binomial coefficient.

For genus-2 complete weight enumerators of binary codes (the word 'complete' could of course be omitted), we will use the variables w, x, y, z instead of $x_{00}, x_{01}, x_{10}, x_{11}$. It will also be convenient to use standard symmetric function notation (as in David, Kendall and Barton [140]), so that

$$\sigma_i = w^i + x^i + y^i + z^i ,$$
$$\sigma_{i,i} = w^i x^i + w^i y^i + w^i z^i + x^i y^i + x^i z^i + y^i z^i ,$$
$$\sigma_{i,j} = w^i(x^j + y^j + z^j) + x^i(w^j + y^j + z^j)$$
$$+ y^i(w^j + x^j + z^j) + z^i(w^j + x^j + y^j) \ (\text{for } i \neq j), \quad (2.4.9)$$

and so on. Note that the genus-2 complete weight enumerator of any binary code containing $\mathbf{1}$ is a symmetric function of its four variables, since $\mathrm{AGL}_2(2) \cong S_4$.

Setting $m = 2$ in (2.4.8) we obtain

$$\mathrm{cwe}_2(e_8) = \sigma_8 + 14\,\sigma_{4,4} + 168\,\sigma_{2,2,2} . \quad (2.4.10)$$

This is also the complete weight enumerator of the code $e_8 \otimes \mathbb{F}_4$ (cf. Remark 2.1.10). Equation (2.4.10) will be mentioned again in Chapters 6 and 7.

The automorphism group of e_8 is

$$\mathrm{Aut}(e_8) \cong \mathrm{AGL}_3(2) ,$$

the general affine linear group of degree 3 over $\mathbb{F}_2$ of order $1344 = 8.7.6.4$.

The $[24, 12, 8]$ *binary Golay* code g_{24} ([133, Chaps. 3,11], [361]) is generated by

$$1(101011100011000000000000),\tag{2.4.11}$$

or equivalently by the "idempotent" generator

$$1(000001010011001101010111).\tag{2.4.12}$$

The Hamming weight enumerator is

$$\mathrm{hwe}(g_{24})(x,y) = x^{24} + 759x^{16}y^8 + 2576x^{12}y^{12} + 759x^8y^{16} + y^{24}.\tag{2.4.13}$$

The genus-2 Hamming weight enumerator (see [359]) is

$$\begin{aligned}
\sigma_{24} &+ 759\,\sigma_{16,8} + 2576\,\sigma_{12,12} + 212520\,\sigma_{12,4,4,4} + 340032\,\sigma_{10,6,6,2} \\
&+ 22770\,\sigma_{8,8,8} + 1275120\,\sigma_{8,8,4,4} + 4080384\,\sigma_{6,6,6,6}.
\end{aligned}\tag{2.4.14}$$

The automorphism group $\mathrm{Aut}(g_{24})$ is the Mathieu group M_{24}, of order

$$24.23.22.21.20.48 = 244823040.$$

There are two closely related codes: the $[7,3,4]$ Hamming code e_7 and the $[23,11,8]$ Golay code g_{23}, obtained by deleting (say) the first coordinate from (2.4.5) or (2.4.6), (2.4.11) or (2.4.12), These are maximally self-orthogonal, but not self-dual; they will appear again in Chapter 10. They are also duals of quadratic residue codes, which over a general ground field $\mathbb{F}_q$ are defined as follows. Let p be a prime not dividing q, and let ζ be a primitive pth root of unity in the algebraic closure $\overline{\mathbb{F}_q}$. Let

$$g(x) := \prod_{a\in(\mathbb{F}_p^*)^2} (x - \zeta^a) \in \overline{\mathbb{F}_q}[x],\tag{2.4.15}$$

where a runs through the nonzero squares in $\mathbb{F}_p$. If q is a square modulo p, then the Frobenius automorphism $\zeta \mapsto \zeta^q$ permutes the factors of $g(x)$ and therefore preserves $g(x)$. Hence $g(x) \in \mathbb{F}_q[x]$. We define the *quadratic residue code* $QR(\mathbb{F}_q, p)$ of length p to be the cyclic code over $\mathbb{F}_q$ with generator polynomial $g(x)$. Then $QR(\mathbb{F}_q, p)$ has dimension $p - \deg g(x) = (p+1)/2$. In the binary ($q = 2$) case, we have:

Theorem 2.4.1. ([361, p. 490]) *If p is a prime $\equiv 7$ (mod 8) then $QR(\mathbb{F}_2, p) \supset QR(\mathbb{F}_2, p)^\perp$. Moreover, the* extended *quadratic residue code $XQ(\mathbb{F}_2, p)$ of length $p + 1$, obtained by appending a zero-sum check symbol, is of Type 2_{II}.*

In particular, $QR(\mathbb{F}_2, 7) = e_7^\perp$, $XQ(\mathbb{F}_2, 7) = e_8$, $QR(\mathbb{F}_2, 23) = g_{23}^\perp$ and $XQ(\mathbb{F}_2, 23) = g_{24}$.

For $N \geq 4$, let d_N denote the $[N, (N-1)/2, 4]$ code generated by "tetrads":

$$\begin{bmatrix} 1\,1\,1\,1\,0\,0\,0\,0\,0\,0\cdots \\ 0\,0\,1\,1\,1\,1\,0\,0\,0\,0\cdots \\ 0\,0\,0\,0\,1\,1\,1\,1\,0\,0\cdots \\ 0\,0\,0\,0\,0\,0\,1\,1\,1\,1\cdots \\ \cdot\,\cdot\,\cdot\,\cdot\,\cdot\,\cdot\,\cdot\,\cdot\,\cdot\,\cdot\,\cdot\,\cdot \end{bmatrix}. \qquad (2.4.16)$$

By convention, d_N is the zero code of length N for $N \leq 3$. The weight enumerator of d_N is

$$x^\varepsilon ((x^2 + y^2)^{\lfloor N/2 \rfloor} + (x^2 - y^2)^{\lfloor N/2 \rfloor})/2, \qquad (2.4.17)$$

where $\varepsilon = 0$ if N is even, 1 if N is odd. For $N \equiv 0 \pmod 4$, with $N \geq 4$, adjoining the "glue word" $0101\ldots01$ (cf. §9.6) produces the self-dual code d_N^+. For $N \geq 8$ the code d_N^+ has minimal weight 4, and $d_8^+ = e_8$.

All the codes $i_2, e_7, e_8, g_{23}, g_{24}$ are unique in the sense that any linear or nonlinear code with the same length, size, minimal distance and containing the zero vector is linear and equivalent to the code given above (cf. Pless [412], [423], Snover [507]).

2.4.2 4^{E}: Euclidean self-dual codes over $\mathbb{F}_4$

The smallest example of a code of Type 4^{E} (self-dual with respect to the Euclidean inner product) is the $[2, 1, 2]_4$ repetition code $\mathbb{F}_4 \otimes i_2$ with generator matrix $[1, 1]$ (cf. (2.4.2)). The symbol $[2, 1, 2]_4$ implies that this is a linear code of length 2, dimension 1 and minimal Hamming distance 2 over $\mathbb{F}_4$.

The smallest Type II example is the $[4, 2, 3]_4$ Reed-Solomon code RS_4 with generator matrix

$$\begin{bmatrix} 1\,1\,1\,\,1 \\ 0\,1\,\omega\,\omega^2 \end{bmatrix}, \qquad (2.4.18)$$

and

$$\begin{aligned} \mathrm{hwe}(RS_4) &= x^4 + 12xy^3 + 3y^4, \\ \mathrm{swe}(RS_4) &= x^4 + y^4 + 2z^4 + 12xyz^2, \\ \mathrm{cwe}(RS_4) &= x^4 + y^4 + z^4 + t^4 + 12xyzt. \qquad (2.4.19) \end{aligned}$$

The weak automorphism group is $3.S_4$, of order 72. As can be seen from the swe, the Lee weights are indeed divisible by 4, so this is of Type $4_{\mathrm{II}}^{\mathrm{E}}$. This example shows that the Hamming weights in a Type $4_{\mathrm{II}}^{\mathrm{E}}$ code need not be even. RS_4 is also the smallest example of an extended quadratic residue code over $\mathbb{F}_4$: $RS_4 \cong XQ(\mathbb{F}_4, 3)$.

Theorem 2.4.2. ([361, p. 508]) *If p is a prime $\equiv 3 \pmod 4$ then $QR(\mathbb{F}_4, p) \supset QR(\mathbb{F}_4, p)^\perp$, and appending a zero-sum check symbol gives a Type $4_{\mathrm{II}}^{\mathrm{E}}$ code $XQ(\mathbb{F}_4, p)$ of length $p + 1$.*

Note that if $p \equiv 7 \pmod 8$ then $XQ(\mathbb{F}_4, p) = XQ(\mathbb{F}_2, p) \otimes \mathbb{F}_4$, so only the $p \equiv 3 \pmod 8$ case is truly interesting.

If C is a binary self-dual code, then $C \otimes \mathbb{F}_4$ is a self-dual code belonging to both families 4^E and 4^H. Conversely, it is not difficult to show that if C is self-dual over $\mathbb{F}_4$ with respect to both the Euclidean and Hermitian inner products, then $C = B \otimes \mathbb{F}_4$ for some self-dual binary code B.

2.4.3 q^E (even or odd): Euclidean self-dual codes over $\mathbb{F}_q$

There is a restriction on the length N: if $q \equiv 3 \pmod 4$ then self-dual codes exist if and only if N is a multiple of 4; for other values of q, N need only be even (cf. Pless [411] and Theorem 7.4.1).

Provided $q \not\equiv 3 \pmod 4$, $\mathbb{F}_q$ contains an element α such that $\alpha^2 = -1$ and then $[1, \alpha]$ generates a self-dual code of Type q^E for q even or odd.

The general condition for an extended quadratic residue code to be self-dual is as follows:

Theorem 2.4.3. *If p is a prime $\equiv 3 \pmod 4$ and q is a nonzero square mod p then $QR(\mathbb{F}_q, p) \supset QR(\mathbb{F}_q, p)^\perp$, and the code $XQ^E(\mathbb{F}_q, p)$ obtained by appending a zero-sum check symbol and then dividing the new symbol by $\sqrt{-p}$ is of Type q^E. The code $XQ^E(\mathbb{F}_q, p)$ is called a Euclidean extended quadratic residue code of length $p + 1$.*

Note that the existence of a square root of $-p$ is guaranteed by quadratic reciprocity, because q is a square mod p. Note also that if $q = 4$ then $-p = \sqrt{-p} = 1 \in \mathbb{F}_4$ and $XQ^E(\mathbb{F}_4, p) = XQ(\mathbb{F}_4, p)$ as defined in Theorem 2.4.2. We omit the proof of Theorem 2.4.3, since it is similar to that of Theorem 2.4.11 given below.

2.4.4 q_{II}^E: Generalized doubly-even self-dual codes

The code RS_4 of §2.4.2 is of Type 4_{II}^E. Before giving further examples of codes of this Type, we establish some general properties of these codes.

One interesting property of generalized doubly-even self-dual codes is that this family is closed both under field extensions (this is immediate from the definition) and also under restriction of scalars, as we now show.

Let $q = 2^f$ and $\mathbb{F} := \mathbb{F}_q$. Let $B = (b_1, \ldots, b_f)$ be an $\mathbb{F}_2$-basis of $\mathbb{F}$ such that $\mathrm{Tr}(b_i b_j) = \delta_{ij}$ for $i, j = 1, \ldots, f$, where Tr denotes the trace of $\mathbb{F}$ over $\mathbb{F}_2$. Then B is called a *self-complementary* basis for $\mathbb{F}$. Using such a basis we identify $\mathbb{F}$ with $\mathbb{F}_2^f$ and define

$$\varphi : \mathbb{F} \to \mathbb{Z}/4\mathbb{Z}, \quad \varphi\left(\sum_{i=1}^{f} a_i b_i\right) := \mathrm{wt}(a_1, \ldots, a_f) + 4\mathbb{Z}$$

to be the weight modulo 4. Since $\mathrm{Tr}(b_i) = \mathrm{Tr}(b_i^2) = 1$, we have

$$\varphi(a) + 2\mathbb{Z} = \text{Tr}(a),$$

and (considering $2\,\text{Tr}$ as a map onto $2\mathbb{Z}/4\mathbb{Z}$),

$$\varphi(a+b) = \varphi(a) + \varphi(b) + 2\,\text{Tr}(ab),$$

for all $a, b \in \mathbb{F}$. More generally,

$$\varphi(\sum_{i=1}^{N} c_i) = \sum_{i=1}^{N} \varphi(c_i) + 2\,\text{Tr}(\sum_{i<j} c_i c_j).$$

We extend φ to a quadratic function on $\mathbb{F}^N$:

$$\phi : \mathbb{F}^N \to \mathbb{Z}/4\mathbb{Z}, \quad \phi(c) := \sum_{i=1}^{N} \varphi(c_i).$$

Proposition 2.4.4. *A code $C \leq \mathbb{F}^N$ is doubly-even if and only if $\phi(C) = \{0\}$.*

Proof. For $r \in \mathbb{F}, c \in \mathbb{F}^N$,

$$\phi(rc) = \varphi(\sum_{i=1}^{N} rc_i) - 2\,\text{Tr}(\sum_{i<j} r^2 c_i c_j).$$

This equation shows in particular that $\phi(C) = \{0\}$ if C is doubly-even. Conversely, if $\phi(C) = \{0\}$ then the same equation shows that $\text{Tr}(r \sum_{i=1}^{N} c_i) = \varphi(\sum_{i=1}^{N} rc_i) + 2\mathbb{Z} = 0$ for all $r \in \mathbb{F}$, $c \in C$. Since the trace bilinear form is non-degenerate, this implies that $\sum_{i=1}^{N} c_i = 0$ for all $c \in C$. The same equality then implies that $\text{Tr}(r^2 \sum_{i<j} c_i c_j) = 0$ for all $r \in \mathbb{F}$ and $c \in C$. The mapping $r \mapsto r^2$ is an automorphism of $\mathbb{F}$, so again the non-degeneracy of the trace bilinear form yields $\sum_{i<j} c_i c_j = 0$ for all $c \in C$. $\square$

Corollary 2.4.5. *Let $\mathbb{F}^N$ be identified with $\mathbb{F}_2^{Nf}$ via a self-complementary basis. Then a doubly-even code $C \leq \mathbb{F}^N$ becomes a doubly-even binary code $C_{\mathbb{F}_2} \leq \mathbb{F}_2^{Nf}$.*

For example, we may regard RS_4 as a linear code of dimension 4 and length 8 over $\mathbb{F}_2$ via the self-complementary basis (ω, ω^2). Thus we apply the mapping (used in [186])

$$0 \mapsto 0, 0, \quad 1 \mapsto 1, 1, \quad \omega \mapsto 1, 0, \quad \omega^2 \mapsto 0, 1 \tag{2.4.20}$$

(compare the Gray map of (2.4.35)), and find that $(RS_4)_{\mathbb{F}_2}$ is the Hamming code e_8.

Remark 2.4.6. Let $C \leq \mathbb{F}^N$ be a doubly-even code. Then $1 \in C^\perp$. Hence if C is self-dual then 4 divides N.

In the following remark we use the fact that the length of a doubly-even self-dual binary code is divisible by 8.

Remark 2.4.7. If $f \equiv 1 \pmod 2$ then the length of a Type $q_{\mathrm{II}}^{\mathrm{E}}$ code over $\mathbb{F}$ is divisible by 8. If $f \equiv 0 \pmod 2$ then $\mathbb{F} \otimes_{\mathbb{F}_4} RS_4$ is a Type $q_{\mathrm{II}}^{\mathrm{E}}$ code of length 4 over $\mathbb{F}$.

Further examples of doubly-even self-dual codes are provided by extended quadratic-residue codes (cf. [361]). By combining Theorems 2.4.1 and 2.4.2 with Proposition 2.4.4, we obtain the following:

Theorem 2.4.8. Let p be a prime $\equiv 3 \pmod 4$. Then if either $p \equiv 7 \pmod 8$ or if f is even, the Euclidean extended quadratic residue code $XQ^E(\mathbb{F}_q, p)$ is a Type $q_{\mathrm{II}}^{\mathrm{E}}$ code of length $p+1$ over $\mathbb{F}_q$, where $q = 2^f$.

2.4.5 3: Euclidean self-dual codes over $\mathbb{F}_3$

Self-dual ternary codes are codes of Type 3 (sometimes called Type III codes, although we will not use that notation here). They exist if and only if the length N is a multiple of 4. Here again the automorphism group and weak automorphism group coincide.

The $[4, 2, 3]_3$ *tetracode* t_4, generated by $\{1110, 0121\}$ ([133, p. 81]) has Hamming weight enumerator

$$\mathrm{hwe}(t_4)(x, y) = x^4 + 8xy^3, \tag{2.4.21}$$

and complete weight enumerator $\mathrm{cwe}(t_4) = x_0\{x_0^3 + (x_1 + x_2)^3\}$. $\mathrm{Aut}(t_4) = 2.S_4$.

The $[12, 6, 6]_3$ *ternary Golay* code g_{12} ([133], p. 85), generated by $1(11210200000)$, has

$$\mathrm{hwe}(g_{12})(x, y) = x^{12} + 264x^6 y^6 + 440 x^3 y^9 + 24 y^{12}, \tag{2.4.22}$$

and (assuming the all-ones codeword is present)

$$\begin{aligned}
\mathrm{cwe}(g_{12})(x_0, x_1, x_2) &= x_0^{12} + x_1^{12} + x_2^{12} + 22(x_0^6 x_1^6 + x_1^6 x_2^6 + x_2^6 x_0^6) \\
&\quad + 220(x_0^6 x_1^3 x_2^3 + x_0^3 x_1^6 x_2^3 + x_0^3 x_1^3 x_2^6). \tag{2.4.23}
\end{aligned}$$

$\mathrm{Aut}(g_{12}) = 2.M_{12}$ (where M_{12} is a Mathieu group), of order 190080.

These two codes are unique in the same sense as our binary examples (Pless [412], [423]).

By Theorem 2.4.3, the quadratic residue construction also gives self-dual codes in this case.

Theorem 2.4.9. If p is a prime $\equiv 11 \pmod{12}$ then $QR(\mathbb{F}_3, p) \supset QR(\mathbb{F}_3, p)^\perp$, and appending a zero-sum check symbol gives a Type 3 code $XQ(\mathbb{F}_3, p) = XQ^E(\mathbb{F}_3, p)$ of length $p + 1$.

In particular, $XQ(\mathbb{F}_3, 11) = g_{12}$.

2.4.6 4^{H}: Hermitian self-dual codes over $\mathbb{F}_4$

The $[2,1,2]_4$ repetition code $i_2 = \{00, 11, \omega\omega, \overline{\omega}\,\overline{\omega}\}$ has

$$
\begin{aligned}
\mathrm{hwe}(i_2) &= x^2 + 3y^2\,, \\
\mathrm{swe}(i_2) &= x^2 + y^2 + 2z^2\,, \\
\mathrm{cwe}(i_2) &= x^2 + y^2 + z^2 + t^2\,.
\end{aligned}
\tag{2.4.24}
$$

The automorphism group of i_2 is

$$
\mathrm{Aut}(i_2) \cong \mathbb{F}_4^* \times Z_2 \cong Z_6,\ \text{of order }6\,,
$$

and the weak automorphism group is

$$
\mathrm{WAut}(i_2) \cong \mathbb{F}_4^* \rtimes \mathrm{Gal}(\mathbb{F}_4/\mathbb{F}_2) \times Z_2 \cong S_3 \times Z_2,\ \text{of order }12\,.
$$

Of course i_2 is simply the $\mathbb{F}_4$-span of the binary code i_2 defined above. The code C in Example 2.1.11 is equivalent to i_2.

The $[6,3,4]_4$ *hexacode* h_6 ([133, p. 82], [427, §12]) in the form with generator matrix

$$
\begin{bmatrix}
1 & 0 & 0 & 1 & \omega & \omega \\
0 & 1 & 0 & \omega & 1 & \omega \\
0 & 0 & 1 & \omega & \omega & 1
\end{bmatrix}
\tag{2.4.25}
$$

has

$$
\begin{aligned}
\mathrm{hwe}(h_6) &= x^6 + 45x^2y^4 + 18y^6\,, \\
\mathrm{swe}(h_6) &= x^6 + y^6 + 2z^6 + 15(2x^2y^2z^2 + x^2z^4 + y^2z^4)\,, \\
\mathrm{cwe}(h_6) &= x^6 + y^6 + z^6 + t^6 \\
&\quad + 15(x^2y^2z^2 + x^2y^2t^2 + x^2z^2t^2 + y^2z^2t^2)\,,
\end{aligned}
\tag{2.4.26}
$$

and $\mathrm{WAut}(h_6) = 3.S_6$, of order 2160. Again this code is unique (cf. §12.4).

The hexacode h_6 is also the smallest example of an extended quadratic residue code of Type 4^{H}.

Theorem 2.4.10. *If p is a prime $\equiv 5$ or $7 \pmod 8$ then $QR(\mathbb{F}_4, p)$ contains its Hermitian dual, and appending a zero-sum check symbol gives a Type 4^{H} code $XQ(\mathbb{F}_4, p)$ of length $p + 1$.*

Again, if $p \equiv 7 \pmod 8$ then $XQ(\mathbb{F}_4, p) = XQ(\mathbb{F}_2, p) \otimes \mathbb{F}_4$.

2.4.7 q^{H}: Hermitian self-dual linear codes over $\mathbb{F}_q$

Since the norm map from $\mathbb{F}_q$ to $\mathbb{F}_{\sqrt{q}}$ is surjective, there is an element $a \in \mathbb{F}_q$ with $a\bar{a} = -1$. Then $[1\,a]$ is self-dual.

In some cases it is possible to construct an extended quadratic residue of Type q^{H}.

Theorem 2.4.11. *Suppose $-\sqrt{q}$ is a non-square modulo the prime p, and let $\alpha \in \mathbb{F}_q$ be chosen so that $\alpha\bar{\alpha} = -p$. Then $QR(\mathbb{F}_q, p)$ contains its Hermitian dual, and the code $XQ^H(\mathbb{F}_q, p)$ obtained by appending a zero-sum check symbol to $QR(\mathbb{F}_q, p)$ and then dividing the new symbol by α is of Type q^H and length $p+1$. The code $XQ^H(\mathbb{F}_q, p)$ is called a Hermitian extended quadratic residue code of length $p+1$.*

Note that again the code $XQ(\mathbb{F}_4, p)$ from Theorem 2.4.10 is equivalent to $XQ^H(\mathbb{F}_4, p)$, since α may be chosen to be 1 in these cases.

Proof. Let $g(x)$ be the generator polynomial for the cyclic code $QR(\mathbb{F}_q, p)$, as in (2.4.15). Then the generator polynomial for the Euclidean dual is $\prod_{a \in \mathbb{F}_p \setminus (\mathbb{F}_p^*)^2}(x - \zeta^{-a})$. Applying the Galois automorphism $v \mapsto v^{\sqrt{q}}$, which extends to a ring automorphism of $\mathbb{F}_q[x]$, we see that the generator polynomial for the Hermitian dual is $g_H(x) := \prod_{a \in \mathbb{F}_p \setminus (\mathbb{F}_p^*)^2}(x - \zeta^{-\sqrt{q}a})$. Since $-\sqrt{q} \notin (\mathbb{F}_p^*)^2$, we obtain $g_H(x) = (x - 1)g(x)$, and hence $QR(\mathbb{F}_q, p)$ contains its Hermitian dual $QR(\mathbb{F}_q, p)^{\perp, \beta_H}$ as a subcode of codimension 1. Since the all-ones vector $\mathbf{1}$ (corresponding to the polynomial $(x^p - 1)/(x - 1) \in (g(x)) \setminus (g_H(x))$) lies in $QR(\mathbb{F}_q, p)$ but not in $QR(\mathbb{F}_q, p)^{\perp, \beta_H}$, we have

$$QR(\mathbb{F}_q, p) = \langle \mathbf{1}, QR(\mathbb{F}_q, p)^{\perp, \beta_H} \rangle.$$

In particular, $QR(\mathbb{F}_q, p)^{\perp, \beta_H}$ is a Hermitian self-orthogonal code whose codewords have zero-sum check symbol 0 (since they are orthogonal to the all-ones vector). To see that the extended code is Hermitian self-dual, it therefore suffices to calculate the norm of $c := (\mathbf{1}, -p\alpha^{-1})$, which is $\beta_H(c, c) = p - p^2 \frac{1}{p} = 0$. $\qquad\square$

2.4.8 4^{H+}: Trace-Hermitian additive codes over $\mathbb{F}_4$

The smallest example of a code of Type 4^{H+} is the $[1, \frac{1}{2}, 1]_{4+}$ code $i_1 = \{0, 1\}$, with trivial automorphism group. The symbol $[1, \frac{1}{2}, 1]_{4+}$ implies that this is an additive code over $\mathbb{F}_4$ of length 1, with $4^{1/2} = 2$ codewords and minimal Hamming distance 1.

Another nice example with odd length is the $[5, 2.5, 3]_{4+}$ shortened hexacode h_5, generated as an additive code by $(01\omega\omega 1)$, with weight enumerator $x^5 + 10x^2y^3 + 15xy^4 + 6y^5$ and $|\text{WAut}(h_5)| = 120$.

The $[12, 6, 6]_{4+}$ *dodecacode* z_{12} can be defined as the cyclic code with generator $(\omega 10100100101)$ ([96], see also Höhn [266]) and

$$\text{hwe}(z_{12}) = x^{12} + 396x^6y^6 + 1485x^4y^8 + 1980x^2y^{10} + 234y^{12}.$$

$\text{Aut}(z_{12}) = \text{WAut}(z_{12})$ is a semi-direct product of Z_3^3 with S_4, and has order 648. It is noteworthy that there is no linear $[12, 6, 6]_4$ code (Conway, Pless and Sloane [123]). Since z_{12} has only even weights, it is of Type 4_{II}^{H+}.

2.4.9 $4^{\mathbb{Z}}$: Self-dual codes over $\mathbb{Z}/4\mathbb{Z}$

Codes over rings are probably less familiar to the reader than codes over fields, and so we will add some remarks here about the first such case, codes over $\mathbb{Z}/4\mathbb{Z}$. General references for these codes are: [454, §§7.10, 7.11], [446], Conway and Sloane [131], [133], Dougherty, Harada and Oura [160], Fields, Gaborit, Leon and Pless [174]; Gaborit [178], Hammons, Kumar, Calderbank, Sloane and Solé [227], Huffman [281], Klemm, [315], [316], Pless, Leon and Fields [428] and Z.-X. Wan [539].

Any code (self-dual or not) over $\mathbb{Z}/4\mathbb{Z}$ is equivalent to one with generator matrix of the form

$$\begin{bmatrix} I_{k_1} & X & Y_1 + 2Y_2 \\ 0 & 2I_{k_2} & 2Z \end{bmatrix}, \qquad (2.4.27)$$

where X, Y_1, Y_2, Z are $\{0,1\}$-matrices. Then C is an elementary abelian group of type $4^{k_1}2^{k_2}$, containing $2^{2k_1+k_2}$ words. We indicate this by writing $|C| = 4^{k_1}2^{k_2}$. The dual code $C^{\perp}$ has generator matrix

$$\begin{bmatrix} (-Y_1 + 2Y_2)^{tr} - Z^{tr}X^{tr} & Z^{tr} & I_{n-k_1-k_2} \\ 2X^{tr} & 2I_{k_2} & 0 \end{bmatrix},$$

and $|C^{\perp}| = 4^{N-k_1-k_2}2^{k_2}$.

There are two binary codes $C^{(1)} = C/(2\mathbb{Z})^N \subset (\mathbb{Z}/2\mathbb{Z})^N$ and $C^{(2)} = C^{(1)} + \frac{1}{2}(C \cap 2\mathbb{Z}^N)$ associated with C, having generator matrices

$$[I_{k_1}\ X\ Y_1] \quad \text{and} \quad \begin{bmatrix} I_{k_1} & X & Y_1 \\ 0 & I_{k_2} & Z \end{bmatrix}. \qquad (2.4.28)$$

If C is self-orthogonal (with respect to the mod 4 inner product) then $C^{(1)}$ is doubly-even and $C^{(1)} \subseteq C^{(2)} \subseteq C^{(1)\perp}$. If C is self-dual then $C^{(2)} = C^{(1)\perp}$. The next two theorems give the converse assertions.

Theorem 2.4.12. *If A, B are binary codes with $A \subseteq B$ then there is a code C over $\mathbb{Z}/4\mathbb{Z}$ with $C^{(1)} = A$, $C^{(2)} = B$. If in addition A is doubly-even and $B \subseteq A^{\perp}$ then C can be made self-orthogonal. If $B = A^{\perp}$ then C is self-dual.*

Proof. Suppose A, B have generator matrices as shown in (2.4.28). Then

$$\begin{bmatrix} I_{k_1} & X & Y_1 \\ 0 & 2I_{k_2} & 2Z \end{bmatrix} \qquad (2.4.29)$$

is a generator matrix for a code C with $C^{(1)} = A$, $C^{(2)} = B$. To establish the second assertion we must modify (2.4.29) to make C self-orthogonal. We do this by adding a lower triangular $\{0,2\}$-matrix to I_{k_1}: we replace the (j,i)th entry of (2.4.29) by the inner product modulo 4 of rows i and j, for $1 \le i < j \le k_1$. □

Remark 2.4.13. (i) The examples below show that this construction is typically not unique.

(ii) Question: What are necessary and sufficient conditions under which C can be made isotropic? See Theorem 2.4.15 for the case $B = A = A^{\perp}$.

(iii) Via Theorem 2.4.12, every self-*orthogonal* doubly-even binary code corresponds to one or more self-*dual* codes over $\mathbb{Z}/4\mathbb{Z}$.

Theorem 2.4.14. (Gaborit [178]) *A code C over $\mathbb{Z}/4\mathbb{Z}$ with generator matrix (2.4.27) is self-dual if and only if $C^{(1)}$ is doubly-even, $C^{(2)} = C^{(1)\perp}$, and Y_2 is chosen so that if $M = Y_1 Y_2^{tr}$, then $M_{ij} + M_{ji} \equiv \frac{1}{2} \operatorname{wt}(v_i \cap v_j)$, where $v_1, \ldots, v_{k_1}$ are the generators of $C^{(1)}$.*

In contrast to self-dual codes over fields, self-dual codes over $\mathbb{Z}/4\mathbb{Z}$ exist for all lengths, even or odd. Furthermore, a self-dual code C over $\mathbb{Z}/4\mathbb{Z}$ of length N can be shortened to a self-dual code of length $N - 1$ by deleting any one of its coordinates. This is accomplished as follows. If the projection of C onto the ith coordinate contains all of $\mathbb{Z}/4\mathbb{Z}$, the shortened code is obtained by taking those words of C that are 0 or 2 in the ith coordinate and omitting that coordinate. If the projection of C onto the ith coordinate contains only 0 and 2, we take the words of C that are 0 on the ith coordinate and omit that coordinate.

In this way all self-dual codes over $\mathbb{Z}/4\mathbb{Z}$ belong to a common "family tree". The beginning of this tree, showing all self-dual codes of lengths $N \leq 8$ up to weak equivalence, is shown in Fig. 2.1.

For codes over $\mathbb{Z}/4\mathbb{Z}$ we will use x_0, x_1, x_2, x_3 as the variables in complete weight enumerators, and we set $x_0 = x$, $x_1 = x_3 = y$, $x_2 = z$ to get the Lee-symmetrized weight enumerators.

The root of the tree, the simplest self-dual code over $\mathbb{Z}/4\mathbb{Z}$, is the code

$$\mathcal{A}_1 := \{0, 2\}$$

of length 1. There are two weakly equivalent, but (permutation) inequivalent self-dual codes $\mathcal{D}_4^{\oplus}$ and $\mathcal{D}_4^{\oplus'}$, with generator matrices

$$\begin{bmatrix} 1 & 1 & 1 & 1 \\ 0 & 2 & 0 & 2 \\ 0 & 0 & 2 & 2 \end{bmatrix}, \quad \begin{bmatrix} 1 & 3 & 3 & 3 \\ 0 & 2 & 0 & 2 \\ 0 & 0 & 2 & 2 \end{bmatrix}. \tag{2.4.30}$$

Both $\mathcal{D}_4^{\oplus}$ and $\mathcal{D}_4^{\oplus'}$ have Lee-symmetrized weight enumerator

$$\operatorname{swe}(\mathcal{D}_4^{\oplus}) = \operatorname{swe}(\mathcal{D}_4^{\oplus'}) = x^4 + 6x^2 z^2 + z^4 + 8y^4. \tag{2.4.31}$$

Their weak automorphism groups are

$$\operatorname{WAut}(\mathcal{D}_4^{\oplus}) = \operatorname{WAut}(\mathcal{D}_4^{\oplus'}) \cong Z_2^3 \rtimes S_4,$$

of order $2^3 4!$.

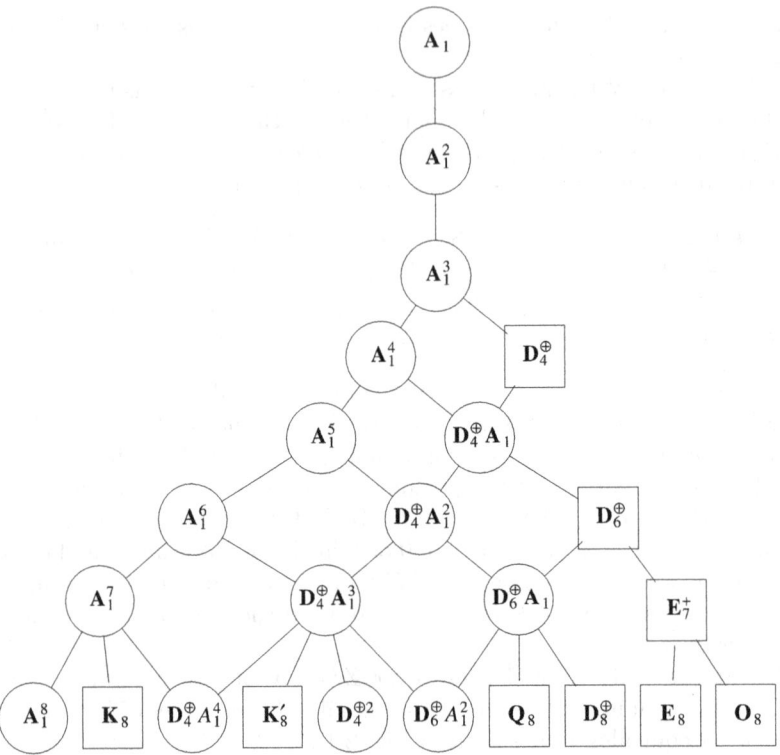

Fig. 2.1. All self-dual codes over $\mathbb{Z}/4\mathbb{Z}$ of length $N \leq 8$. In this figure, taken from [131], codes that are denoted by calligraphic letters in the text appear as bold letters: thus for example $\mathcal{O}_8$ appears as $\mathbf{O}_8$. A vertical or sloping line indicates that the upper code can be obtained by shortening the lower code. The indecomposable codes are indicated by squares, the others by circles. There are no indecomposable codes of length 9, so that the codes of length 9 are obtainable by adjoining the code $\mathcal{A}_1 \ (= \mathbf{A}_1)$ to the codes of length 8.

The *octacode* $\mathcal{O}_8$ (Conway and Sloane [131], [133], [454])[3] is the length 8 self-dual code over $\mathbb{Z}/4\mathbb{Z}$ generated by the vectors $3(2001011)$, or equivalently with generator matrix

$$\begin{bmatrix} 1\,0\,0\,0\,2\,3\,3\,3 \\ 0\,1\,0\,0\,1\,2\,3\,1 \\ 0\,0\,1\,0\,1\,1\,2\,3 \\ 0\,0\,0\,1\,1\,3\,1\,2 \end{bmatrix}, \qquad (2.4.32)$$

having minimal Lee weight 6 and minimal norm 8. This is the smallest example of a self-dual code of Type $4_{\mathrm{II}}^{\mathbb{Z}}$: that is, all norms are divisible by 8. The complete weight enumerator of $\mathcal{O}_8$ is

---

[3] Eq. (2.4.32) differs from the generator matrix given in [454], which did not contain the all-ones vector.

$$\mathrm{cwe}(\mathcal{O}_8) = x_0^8 + x_1^8 + x_2^8 + x_3^8 + 14x_1^4x_3^4 + 14x_0^4x_2^4 + 56x_0^3x_1^3x_2x_3$$
$$+ 56x_0^3x_1x_2x_3^3 + 56x_0x_1^3x_2^3x_3 + 56x_0x_1x_2^3x_3^3 \, . \qquad (2.4.33)$$

The Lee-symmetrized weight enumerator of $\mathcal{O}_8$ is

$$\mathrm{swe}(\mathcal{O}_8) = \mathrm{cwe}(\mathcal{O}_8)(x, z, y, z) = x^8 + 16y^8 + z^8 + 14x^4z^4 + 112xy^4z(x^2 + z^2) \, ,$$
$$(2.4.34)$$

and $|\,\mathrm{WAut}(\mathcal{O}_8)| = 2 \cdot 1344$.

The most interesting property of the octacode is that, when mapped to a binary code under the *Gray map*

$$0 \mapsto 0, 0, \quad 1 \mapsto 0, 1, \quad 2 \mapsto 1, 1, \quad 3 \mapsto 1, 0 \, , \qquad (2.4.35)$$

$\mathcal{O}_8$ becomes the notorious Nordstrom-Robinson code, a nonlinear binary code of length 16, minimal distance 6, containing 256 words (Forney, Sloane and Trott [175], Hammons, Kumar, Calderbank, Sloane and Solé [227]; see also Carlet [102], Greferath and Schmidt [200], Tapia-Recillas and G. Vega [519]).

The octacode reduces mod 2 to the Hamming code e_8. There is another lift of e_8 to $\mathbb{Z}/4\mathbb{Z}$, namely the code $\mathcal{E}_8$, with generator matrix

$$\begin{bmatrix} 1 & 0 & 0 & 0 & 0 & 1 & 1 & 1 \\ 0 & 1 & 0 & 0 & 3 & 0 & 1 & 3 \\ 0 & 0 & 1 & 0 & 3 & 3 & 0 & 1 \\ 0 & 0 & 0 & 1 & 3 & 1 & 3 & 0 \end{bmatrix} \, , \qquad (2.4.36)$$

but the minimal Lee weight and norm are now both only 4. The Lee-symmetrized weight enumerator of $\mathcal{E}_8$ is

$$\mathrm{swe}(\mathcal{E}_8) = x^8 + 16x^4y^4 + 14x^4z^4 + 48x^3y^4z$$
$$+ 96x^2y^4z^2 + 48xy^4z^3 + 16y^8 + 16y^4z^4 + z^8 \, . \qquad (2.4.37)$$

However, not all binary self-dual codes lift to self-dual codes over $\mathbb{Z}/4\mathbb{Z}$, e.g. $\{00, 11\}$ does not.

Theorem 2.4.15. (*a*) *Let C be a binary self-dual code of length N. A necessary and sufficient condition for C to be liftable to a self-dual code $\widehat{C}$ over $\mathbb{Z}/4\mathbb{Z}$ is that C be doubly-even.*

(*b*) *If this condition is satisfied, $\widehat{C}$ can be chosen so that all norms are divisible by 8.*

(*c*) *More generally, a self-dual code over $\mathbb{Z}/m\mathbb{Z}$, m even, that reduces to a self-dual code mod 2 lifts to $\mathbb{Z}/2m\mathbb{Z}$ precisely when all norms are divisible by $2m$, and in that case all norms in the lifted code can be arranged to be divisible by $4m$. Thus if a code lifts from $\mathbb{Z}/m\mathbb{Z}$ to $\mathbb{Z}/2m\mathbb{Z}$ then it lifts to $\mathbb{Z}/2^k m\mathbb{Z}$ for all k. In particular, if a binary code lifts to $\mathbb{Z}/4\mathbb{Z}$ then it lifts to a self-dual code over the 2-adic integers.*

Proof. (a) (Necessity) Suppose $v \in C$ has weight $\text{wt}(v) \not\equiv 0 \pmod 4$ and let $\widehat{v} \in \widehat{C}$ be any lift of v. Then $\text{Norm}(\widehat{v}) \equiv \text{Norm}(v) \pmod 4$ because for integers x, y if $x \equiv y \pmod 2$ then $x^2 \equiv y^2 \pmod 4$.

(Sufficiency) Without loss of generality C has a generator matrix of the form $[IA]$ where $AA^{tr} \equiv -I \pmod 2$. Let B be any lift of A to $\mathbb{Z}/4\mathbb{Z}$. We wish to find $\widehat{A} = B + 2M$ such that $\widehat{A}\widehat{A}^{tr} \equiv -I \pmod 4$, since then we can take $\widehat{C} = [I\widehat{A}]$. We have

$$\widehat{A}\widehat{A}^{tr} \equiv BB^{tr} + 2(MB^{tr} + BM^{tr}) \pmod 4.$$

The condition on C implies that $BB^{tr} + I$ has even coefficients and is zero on the diagonal. But then there exists a binary matrix M' such that $2(M' + (M')^{tr}) = BB^{tr} + I$, and we take $M = M'(B^{-1})^{tr}$. This completes the proof of (a).

(b) We must show that we can choose $\widehat{A}$ so the diagonal entries of $\widehat{A}\widehat{A}^{tr} + I$ are zero mod 8. Set $\widehat{A}' = \widehat{A} - 2L\widehat{A}$, where L is symmetric, so that

$$\widehat{A}'(\widehat{A}')^{tr} = \hat{A}\hat{A}^{tr} + 4L + 4L^2 \pmod 8.$$

Let $\Delta = \frac{1}{4}(\hat{A}\hat{A}^{tr} + I)$. Then we need $L^2 + L + \Delta \pmod 2$ to be symmetric with zero diagonal. It is easy to see that we can accomplish this provided trace $\Delta \equiv 0 \pmod 2$ (consider, for instance, $L = \begin{pmatrix} 1 & 1 \\ 1 & 0 \end{pmatrix}$.) In fact, we have

$$1 \equiv \det(\hat{A}\hat{A}^{tr}) \equiv 1 + 4 \text{ trace } \Delta \pmod 8$$

so trace Δ is even.

The proof of (c) is analogous. $\qquad \square$

It follows from Theorem 2.4.15 that the Golay code g_{24} can be lifted to $\mathbb{Z}/4\mathbb{Z}$. Since g_{24} is an extended cyclic code, the lift can be easily performed by Graeffe's method (Conway and Sloane [131], Uspensky [531]). Suppose $g_2(x)$ divides $x^n - 1 \pmod 2$ and we wish to find a monic polynomial $g(x)$ over $\mathbb{Z}/4\mathbb{Z}$ such that $g(x) \equiv g_2(x) \pmod 2$ and $g(x)$ divides $x^n - 1 \pmod 4$. Let $g_2(x) = e(x) - d(x)$, where $e(x)$ contains only even powers and $d(x)$ only odd powers. Then $g(x)$ is given by $g(x^2) = \pm(e^2(x) - d^2(x))$. Applying this technique to the generator polynomial for g_{24}, that is, to $g_2(x) = 1 + x + x^5 + x^6 + x^7 + x^9 + x^{11}$, we obtain $g(x) = -1 + x + 2x^4 - x^5 - x^6 - x^7 - x^9 + 2x^{10} + x^{11}$ and so

$$3(310023330321000000000000) \tag{2.4.38}$$

generates a self-dual code G_{24} of length 24 which is the Golay code lifted to $\mathbb{Z}/4\mathbb{Z}$. Iterating this process enables us to lift cyclic or extended cyclic codes to $\mathbb{Z}_{2^m}$ for arbitrarily large m.

For use in Chapter 8 we list some further examples.

$\mathcal{J}_{10}$ is the self-dual code with generator matrix

$$\begin{bmatrix} 1 & 0 & 1 & 1 & 1 & 1 & 1 & 0 & 1 & 1 \\ 0 & 1 & 0 & 0 & 0 & 0 & 3 & 3 & 1 & 0 \\ 0 & 0 & 2 & 0 & 0 & 0 & 0 & 0 & 0 & 2 \\ 0 & 0 & 0 & 2 & 0 & 0 & 0 & 0 & 0 & 2 \\ 0 & 0 & 0 & 0 & 2 & 0 & 0 & 0 & 0 & 2 \\ 0 & 0 & 0 & 0 & 0 & 2 & 0 & 0 & 0 & 2 \\ 0 & 0 & 0 & 0 & 0 & 0 & 2 & 0 & 2 & 0 \\ 0 & 0 & 0 & 0 & 0 & 0 & 0 & 2 & 2 & 2 \end{bmatrix}$$

and $|\mathcal{J}_{10}| = 4^2 2^6$ ([131]). $\mathcal{J}_{16}$ has generator matrix

$$\begin{bmatrix} 1 & 0 & 0 & 0 & 0 & 0 & 1 & 1 & 1 & 0 & 3 & 3 & 1 & 0 & 3 & 2 \\ 0 & 1 & 0 & 0 & 0 & 0 & 1 & 0 & 0 & 1 & 3 & 3 & 1 & 1 & 2 & 3 \\ 0 & 0 & 1 & 0 & 0 & 0 & 1 & 0 & 0 & 0 & 0 & 2 & 2 & 0 & 3 & 3 \\ 0 & 0 & 0 & 1 & 0 & 0 & 0 & 1 & 1 & 1 & 3 & 0 & 2 & 3 & 3 & 1 \\ 0 & 0 & 0 & 0 & 1 & 0 & 0 & 1 & 1 & 1 & 0 & 0 & 1 & 1 & 1 & 1 \\ 0 & 0 & 0 & 0 & 0 & 1 & 0 & 0 & 0 & 0 & 3 & 2 & 0 & 1 & 1 \\ 0 & 0 & 0 & 0 & 0 & 0 & 2 & 0 & 0 & 0 & 2 & 0 & 2 & 0 & 2 \\ 0 & 0 & 0 & 0 & 0 & 0 & 0 & 2 & 0 & 0 & 0 & 0 & 2 & 2 & 2 \\ 0 & 0 & 0 & 0 & 0 & 0 & 0 & 0 & 2 & 0 & 0 & 0 & 2 & 2 & 2 \\ 0 & 0 & 0 & 0 & 0 & 0 & 0 & 0 & 0 & 2 & 0 & 0 & 0 & 2 & 0 & 0 \end{bmatrix}$$

and $|\mathcal{J}_{16}| = 4^6 2^4$. Note that the codes $\mathcal{J}_{10}$ and $\mathcal{J}_{16}$ are not Type II codes. We will also need the Type II code C_{16} with generator matrix

$$\begin{bmatrix} 1 & 1 & 1 & 1 & 1 & 1 & 1 & 1 & 1 & 1 & 1 & 1 & 1 & 1 & 1 & 1 \\ 1 & 0 & 1 & 1 & 1 & 1 & 1 & 1 & 0 & 0 & 0 & 0 & 1 & 0 & 0 & 0 \\ 1 & 1 & 0 & 1 & 0 & 0 & 1 & 1 & 1 & 1 & 0 & 0 & 0 & 1 & 0 & 0 \\ 1 & 1 & 1 & 0 & 1 & 0 & 1 & 0 & 1 & 0 & 1 & 0 & 0 & 0 & 1 & 0 \\ 0 & 0 & 0 & 0 & 1 & 1 & 1 & 1 & 1 & 1 & 1 & 0 & 0 & 0 & 0 & 1 \\ 0 & 0 & 0 & 0 & 0 & 2 & 0 & 0 & 0 & 0 & 2 & 2 & 0 & 0 & 2 \\ 0 & 0 & 0 & 0 & 0 & 0 & 2 & 0 & 0 & 0 & 2 & 2 & 2 & 2 \\ 0 & 0 & 0 & 0 & 0 & 0 & 0 & 2 & 0 & 0 & 0 & 2 & 2 & 0 & 2 \\ 0 & 0 & 0 & 0 & 0 & 0 & 0 & 0 & 2 & 0 & 0 & 0 & 2 & 2 & 2 \\ 0 & 0 & 0 & 0 & 0 & 0 & 0 & 0 & 0 & 2 & 0 & 2 & 0 & 2 & 0 & 2 \\ 0 & 0 & 0 & 0 & 0 & 0 & 0 & 0 & 0 & 0 & 2 & 2 & 0 & 0 & 2 & 2 \end{bmatrix}$$

and $|C_{16}| = 4^5 2^6$.

The code $\mathcal{K}_{4m}$ ($m \geq 1$, but note that $\mathcal{K}_4 \cong \mathcal{D}_4^\oplus$) is a self-dual code introduced by Klemm [315], having generator matrix

$$\begin{bmatrix} 1 & 1 & 1 & \ldots & 1 & 1 \\ 0 & 2 & 0 & \ldots & 0 & 2 \\ 0 & 0 & 2 & \ldots & 0 & 2 \\ \cdot & \cdot & \cdot & \ldots & \cdot & \cdot \\ 0 & 0 & 0 & \ldots & 2 & 2 \end{bmatrix} \; ;$$

$$|\mathcal{K}_{4m}| = 4^1 2^{4m-2}, \ \text{cwe}(\mathcal{K}_{4m}) = (A^{4m} + B^{4m} + C^{4m} + D^{4m})/2, \text{ where}$$

$$A = x_0 + x_2, \ B = x_1 + x_3, \ C = x_0 - x_2, \ D = x_1 - x_3, \qquad (2.4.39)$$

and $\mathrm{WAut}(\mathcal{K}_{4m}) = Z_2^{4m-1} : S_{4m}$.

2.4.10 Codes over other Galois rings

Research Problem 2.4.16. *Find interesting (nontrivial) examples of self-dual codes over more general Galois rings of the Types described in §2.3.6.*

Of course one may lift self-dual codes over fields $\mathbb{F}_{p^f}$ to self-dual codes over $\mathrm{GR}(p^e, f)$, as shown in Theorems 2.4.15 and 2.4.17 for the case $f = 1$. As an example we take the hexacode h_6 defined in §2.4.6. If ω is a primitive cube root of unity in $\mathrm{GR}(2^e, 2)$ and $\alpha = a + b\omega \in \mathrm{GR}(2^e, 2)$, $a, b \in \mathbb{Z}/2^e\mathbb{Z}$, is an element such that $\alpha\bar{\alpha} = a^2 - ab + b^2 = -3$ (the existence of such an α is guaranteed by the surjectivity of the norm form), then

$$\begin{bmatrix} \alpha & 0 & 0 & 1 & \omega & \omega \\ 0 & \alpha & 0 & \omega & 1 & \omega \\ 0 & 0 & \alpha & \omega & \omega & 1 \end{bmatrix} \qquad (2.4.40)$$

generates a self-dual isotropic code $h_6(\alpha, e)$ of Type $\rho(\mathrm{GR}(2^e, 2)^\mathrm{H})$. Note that, for $e > 1$, this is not a code of Type $\mathrm{GR}(2^e, 2)_1^\mathrm{H}$, since the length of such a code is necessarily divisible by 2^e.

For a concrete example we may take $e = 2$ and $\alpha = 3 - \omega$. Then the symmetrized, Lee and Hamming weight enumerators of $h_6((3 - \omega), 2)$ are respectively

$$\begin{aligned} \mathrm{swe}(h_6(3 - \omega, 2)) = \ &x^6 + 45x^2y^4 + 18y^6 + 216z^5t + 720z^3t^3 + 216zt^5 \\ &+ (36x^2 + 108xy + 216y^2)(z^4 + t^4) \\ &+ (108x^2 + 864xy + 1188y^2)z^2t^2, \\ \mathrm{lwe}(h_6(3 - \omega, 2)) = \ &x^6 + 45x^2y^4 + 180x^2z^4 + 1080xyz^4 + 18y^6 \\ &+ 1620y^2z^4 + 1152z^6, \\ \mathrm{hwe}(h_6(3 - \omega, 2)) = \ &x^6 + 225x^2y^4 + 1080xy^5 + 2790y^6. \end{aligned}$$

Note that in the symmetrized weight enumerator, the variable x corresponds to 0's, y to elements in $2\,\mathrm{GR}(4, 2) \setminus \{0\}$, z to the $\langle-\omega\rangle$ orbit of 1, and t to the orbit of $1 + 2\omega$. Also $\mathrm{lwe}(x, y, z) = \mathrm{swe}(x, y, z, z)$ and $\mathrm{hwe}(x, y) = \mathrm{lwe}(x, y, y)$. Since $(3-\omega)(3-\bar{\omega}) \equiv -3 \pmod 8$, we also obtain a code $h_6(3-\omega, 3)$ of Type $\mathrm{GR}(2^3, 2)^\mathrm{H}$ and length 6.

2.4.11 $\mathbb{Z}_p$: Codes over the p-adic numbers

References: Calderbank and Sloane [98], Calderbank, Li and Poonen [94]. The 2-adic Hamming code [98] is the self-dual code of length 8 with generator matrix

$$\begin{bmatrix} 1 & \lambda & \lambda-1 & -1 & 0 & 0 & 0 & 1 \\ 0 & 1 & \lambda & \lambda-1 & -1 & 0 & 0 & 1 \\ 0 & 0 & 1 & \lambda & \lambda-1 & -1 & 0 & 1 \\ 0 & 0 & 0 & 1 & \lambda & \lambda-1 & -1 & 1 \end{bmatrix}, \qquad (2.4.41)$$

where λ is the 2-adic integer $(1+\sqrt{-7})/2$. The 2-adic expansion of λ is

$$\lambda = 2 + 4 + 32 + 128 + 256 + 512 + 1024 + 2048 + 4096 + 32768 + \cdots.$$

This is the cyclic code with generator

$$(1, \lambda, \lambda - 1, -1, 0, 0, 0) \qquad (2.4.42)$$

with a 1 appended to each of the generators.

Similarly, the 2-adic self-dual Golay code [98] of length 24 is the cyclic code with generator

$$(1, 1 - \lambda, -2 - \lambda, -4, \lambda - 4, 2\lambda - 3, 2\lambda + 1, \lambda + 3, 4, 3 - \lambda,$$
$$-\lambda, -1, 0, 0, 0, 0, 0, 0, 0, 0, 0, 0, 0, 0), \qquad (2.4.43)$$

where now $\lambda = (1+\sqrt{-23})/2$, with a 1 appended to each of the 12 generators.

The 3-adic self-dual Golay code [98] of length 12 is the cyclic code with generator

$$(1, \lambda, -1, 1, \lambda - 1, -1, 0, 0, 0, 0, 0, 0), \qquad (2.4.44)$$

where $\lambda = (1+\sqrt{-11})/2$, again with a 1 appended to each generator.

For $p = 2$, Theorem 2.4.15 gives a necessary and sufficient condition for when a binary self-dual code can be lifted to a 2-adic self-dual code. For odd primes p, self-dual codes over $\mathbb{F}_p$ always lift to self-dual p-adic codes:

Theorem 2.4.17. *Let p be an odd prime and C a self-dual code over $\mathbb{F}_p$. Then there is a p-adic self-dual code $\tilde{C}$ with $C = \tilde{C}$ (mod p).*

Proof. The proof is similar to that of Theorem 2.4.15. We show that a self-dual code C of Type $m^{\mathbb{Z}}$ with $m = p^k$ that reduces to a self-dual code modulo p can be lifted to a self-dual code $\hat{C}$ of Type $n^{\mathbb{Z}}$ with $n = p^{k+1}$. Then the theorem follows by induction.

Without loss of generality, C has a generator matrix of the form $[I A]$ where $AA^{tr} \equiv -I$ (mod p^k). Let B be any lift of A to $\mathbb{Z}/p^{k+1}\mathbb{Z}$ and put $\hat{A} = B + p^k M$. We must find an integral matrix M (in fact it is enough to find a matrix M mod p) such that

$$\hat{A}\hat{A}^{tr} \equiv BB^{tr} + p^k(MB^{tr} + BM^{tr}) \,(\text{mod } p^{k+1}) \equiv -I \,(\text{mod } p^{k+1}).$$

Table 2.1. The principal Types and the sections where their definitions and further information such as examples, Clifford-Weil groups, Molien series, etc. can be found. The words "even" and "odd" in parentheses refer to the parity of q, not of the weights: see the Notes at the end of the table.

Type	Definition	See also Sections:
2	2.3.1	2.4.1, Tables 11.2, 12.2–12.8
2_I	2.3.1	2.4.1, 6.3.1, Tables 11.2, 12.2–12.8
2_II	1.8.2, 2.3.1	1.12.4, 2.4.1, 6.4.1, Tables 11.2, 12.2, 12.3–12.8
2_S	1.12.7, 2.3.1	
$2^\mathrm{lin}, 2^\mathrm{lin}_1$	7.2.1	7.2.1
$2^\mathrm{lin}_{1'}, 2^\mathrm{lin}_{1,1'}$	7.2.1	7.2.1
4^E	2.3.2	2.4.2, 7.6.2
4^E_II	2.3.2	2.4.2, 7.6.3
q^E (even)	2.3.2	2.4.3, 7.6.2
q^E_II	2.3.2	2.4.4, 7.6.4
3	2.3.2	2.4.5, 7.4.1, Tables 11.4, 12.9
q^E (odd)	2.3.2	2.4.3, 7.4.1
q^E_1 (odd)	1.8.4, 2.3.2	7.4.1
4^H	2.3.3	2.4.6, 7.3.1, Tables 11.7, 12.10
q^H	2.3.3	2.4.7, 7.3.1
q^H_1	2.3.3	7.3.1
$4^{\mathrm{H}+}$	2.3.4	2.4.8, 7.6.1, Table 11.8
$4^{\mathrm{H}+}_\mathrm{II}$	2.3.4	7.6.3
$q^{\mathrm{H}+}$ (even)	2.3.4	7.6.1
$q^{\mathrm{H}+}_1$ (even)	2.3.4	7.6.1
$q^{\mathrm{H}+}_\mathrm{II}$ (even)	2.3.4	7.6.3
$q^{\mathrm{H}+}_{\mathrm{II},1}$ (even)	2.3.4	7.6.4
$q^{\mathrm{H}+}$ (odd)	2.3.4	7.5.1
$q^{\mathrm{H}+}_1$ (odd)	2.3.4	7.5.1
$q^\mathrm{lin}, q^\mathrm{lin}_1$	7.2.1	7.2.1
$q^\mathrm{lin}_{1'}, q^\mathrm{lin}_{1,1'}$	7.2.1	7.2.1

The condition on C implies that the entries of $BB^{tr} \equiv -I \pmod{p^k}$. In particular B is invertible modulo p^{k+1}. Let $B' = B^{-1} \pmod{p^{k+1}}$. Then $M := -\frac{1}{2p^k}(BB^{tr} + I)(B')^{tr} \pmod p$ is the desired matrix.   □

Table 2.2. The principal Types (cont.)

Type	Definition	Further information
$4^{\mathbb{Z}}$	2.3.5	$2.4.9, 8.1, 8.2.1,$ Table 11.11
$4_{\mathrm{II}}^{\mathbb{Z}}$	2.3.5	$2.4.9, 8.1, 8.2.4$
$m^{\mathbb{Z}}$	2.3.5	$8.1, 8.2.1, 8.3$
$m_1^{\mathbb{Z}}$	2.3.5	$8.1, 8.2.2$
$m_{\mathrm{II}}^{\mathbb{Z}}$	$1.12.6, 2.3.5$	$8.1, 8.2.4, 8.3$
$m_{\mathrm{II},1}^{\mathbb{Z}}$	2.3.5	$8.1, 8.2.5, 8.3$
$m_{\mathrm{S}}^{\mathbb{Z}}$	2.3.5	$8.2.3$
$\mathrm{GR}(p^e, f)^{\mathrm{E}}$	2.3.5	$8.4.1, 8.4.4$
$\mathrm{GR}(p^e, f)_1^{\mathrm{E}}$	2.3.5	$8.4.1, 8.4.4$
$\mathrm{GR}(p^e, f)_{p^s}^{\mathrm{E}}$	2.3.5	$8.4.1, 8.4.4$
$\mathrm{GR}(2^e, f)_{2^s}^{\mathrm{E}}$	2.3.5	$8.4.1, 8.4.4$
$\mathrm{GR}(2^e, f)_{\mathrm{II}}^{\mathrm{E}}$	2.3.5	$8.4.1, 8.4.4$
$\mathrm{GR}(2^e, f)_{\mathrm{II},2^s}^{\mathrm{E}}$	2.3.5	$8.4.1, 8.4.4$
$\mathrm{GR}(p^e, f)^{\mathrm{H}}$	2.3.5	$8.4.2, 8.4.4$
$\mathrm{GR}(p^e, f)_{p^s}^{\mathrm{H}}$	2.3.5	$8.4.2, 8.4.4$
$\mathrm{GR}(p^e, f)^{\mathrm{H}+}$	2.3.5	$8.4.3, 8.4.4$
$\mathrm{GR}(p^e, f)_{p^s}^{\mathrm{H}+}$	2.3.5	$8.4.3, 8.4.4$
$\mathbb{Z}_p$	2.3.7	$2.4.11$
$\mathbb{F}_{q^2} + \mathbb{F}_{q^2} u$	8.5	8.5

Notes: q or m can be even or odd: $q^{\mathrm{E}}, q^{\mathrm{H}}, q_1^{\mathrm{H}}, q^{\mathrm{H}+}, q_1^{\mathrm{H}+}, m^{\mathbb{Z}}, m_1^{\mathbb{Z}}$;
q or m must be even: $q_{\mathrm{II}}^{\mathrm{E}}, q_{\mathrm{II}}^{\mathrm{H}+}, q_{\mathrm{II},1}^{\mathrm{H}+}, m_{\mathrm{II}}^{\mathbb{Z}}, m_{\mathrm{II},1}^{\mathbb{Z}}, m_{\mathrm{S}}^{\mathbb{Z}}$;
for q_1^{E}, q is assumed to be odd, since otherwise $q_1^{\mathrm{E}} = q^{\mathrm{E}}$.

Remark 2.4.18. Let q be an odd prime power. There exist elements $\alpha_1, \alpha_2 \in \mathbb{F}_q$ with $\alpha_1^2 + \alpha_2^2 = -1$. Then

$$\begin{bmatrix} 1 & 0 & \alpha_1 & \alpha_2 \\ 0 & 1 & \alpha_2 & -\alpha_1 \end{bmatrix} \tag{2.4.45}$$

generates a Euclidean self-dual $\mathbb{F}_q$-code of length 4.

Corollary 2.4.19. *Let p be an odd prime. Then there is a self-dual p-adic code of length 4. Reducing modulo p^k, this gives a code of Type $(p^k)^{\mathbb{Z}}$ of length 4.*

Remark. The ring $\mathbb{Z}_p$ is just one example of a local ring. One could also consider codes over other local rings, for example the ring O of integers in a finite dimensional extension of the p-adic rationals $\mathbb{Q}_p$. Codes over such rings O are similar to codes over $\mathbb{Z}_p$, and there are analogues of the above theorems and constructions.

2.5 The Gleason-Pierce Theorem

It is elementary that in a binary self-orthogonal code the weight of every vector is even, in a ternary self-dual code the weight of every vector is a multiple of 3, and in a Hermitian self-dual code over $\mathbb{F}_4$ the weight of every vector is even. Furthermore, there are many well-known binary self-dual codes whose weights are divisible by 4—see §2.4.1. The following theorem, due to Gleason and Pierce, shows that these four are essentially the only possible nontrivial divisibility restrictions that can be imposed on the weights of self-dual codes.

Theorem 2.5.1. (Gleason and Pierce, quoted in Assmus, Mattson and Turyn [15]; see also [503]) *If C is a self-dual code belonging to any of the families 2, 3, q^{E} (even or odd), q^{H}, $q^{\mathrm{H}+}$ (even or odd) or $m^{\mathbb{Z}}$ which has all its Hamming weights divisible by an integer $c > 1$, then one of the following holds:*
(a) $|\mathbb{F}| = 2$, $\quad c = 2$ *(so Type 2_I)*
(b) $|\mathbb{F}| = 2$, $\quad c = 4$ *(so Type 2_II)*
(c) $|\mathbb{F}| = 3$, $\quad c = 3$ *(so Type 3)*
(d) $|\mathbb{F}| = 4$, $\quad c = 2$ *(so Types 4^H, 4^E, $4^{\mathrm{H}+}$, $4^{\mathbb{Z}}$ – but only Type 4^H has the property that* all *codes have all weights divisible by 2)*
(e) $|\mathbb{F}| = q$, q arbitrary, $\quad c = 2$, $\mathrm{hwe}(C) = (x^2 + (q-1)y^2)^{N/2}$.

Remark 2.5.2. 1. The theorem may be proved by considering how the Hamming weight enumerator behaves under the MacWilliams transform — see [503] for details. An alternative proof of a somewhat more general result is given by Ward [541], [544, Theorem 13.5].

2. The same conclusion holds if "C is self-dual" is replaced by "C is formally self-dual". (See also Kennedy [299].)

3. Note that there are no nontrivial examples from families q^H ($q > 4$), q^E ($q > 4$) or $m^{\mathbb{Z}}$ ($m > 4$).

4. There are several points to be mentioned concerning case (e). *Linear* self-dual codes with Hamming weight enumerator $(x^2 + (q-1)y^2)^{N/2}$ always exist in families 2, 4^H, 4^E, $4^{\mathrm{H}+}$, q^H; and exist in families q^E and $m^{\mathbb{Z}}$ precisely when there is a square root of -1 in $\mathbb{F}_q$ or $\mathbb{Z}/m\mathbb{Z}$ respectively; in particular, they never exist in families 3 or $4^{\mathbb{Z}}$.

Furthermore, it is easy to see that any linear code over $\mathbb{F}_q$ for $q > 2$ with weight enumerator $(x^2 + (q-1)y^2)^{N/2}$ is a direct sum of codes of length 2. However, in the binary case there are many examples of linear codes with weight enumerator $(x^2 + y^2)^{N/2}$ that are not self-dual: these have been classified for $N \leq 16$, see [503]. These are examples of formally self-dual codes. There are also examples from family $4^{\mathrm{H}+}$, e.g. the additive code $[1100, 0110, 0011, \omega\omega\omega\omega]$ with weight enumerator $(x^2 + 3y^2)^2$.

5. In some cases, analogous restrictions can be imposed on Euclidean norms of codewords. In particular, suppose C is a self-dual code over $\mathbb{Z}/m\mathbb{Z}$ (that is, a code from families $4^{\mathbb{Z}}$ or $m^{\mathbb{Z}}$) where m is even. Then the Euclidean norms of the codewords *must* be divisible by m, and *may* be divisible by $2m$ (Bannai et al. [28], Bonnecaze et al. [66], Dougherty et al. [154]; see also Theorem 2.4.15 above).

3

Closed Codes

Our primary interest is in self-dual codes, satisfying $C^\perp = C$. Of course this implies that $C^{\perp\perp} = C$. Codes with the latter property are called *closed*. In this chapter we attempt to answer the question: which families of codes are closed?

In Corollary 3.3.3 we show that codes in certain finite representations ρ of twisted rings are closed. In fact, the discussions in this chapter were our original motivation for the definition of twisted ring given in §1.4; see in particular Theorem 3.2.8. Example 3.1.7 shows that some restrictions are certainly needed to obtain a reasonable theory of duality for codes. Our approach can be regarded as a continuation of the work of Wood [552], [553], [554], who concluded that quasi-Frobenius rings are the natural setting for studying certain properties of codes (in particular, the MacWilliams extension theorem). This chapter shows that one can work with the larger family of codes over twisted rings. The extra generality comes about because we consider bilinear forms taking values in a module rather than a ring. Also Wood considers only finite rings.

The reader who is only interested in the main results presented in Chapter 5 may omit this chapter. However, certain results on the behavior of duality under multiplication by idempotents or under the formation of tensor products (in particular, Theorem 3.5.9, Theorem 3.4.1 and Corollary 3.4.2) will be required for the proofs of the main theorems in Chapter 5.

The main results of this chapter are Theorem 3.3.2 and Corollary 3.3.3, where the abstract algebra introduced in §§3.1–3.4 is applied.

3.1 Bilinear forms and closed codes

We begin by reconsidering the definition of duality for codes. The definition of dual code given in §1.2 involved choosing a set of bilinear forms M. For the discussion in this chapter it will suffice to replace M by a single bilinear form β.

Definition 3.1.1. Let R and S be rings, V be a left R-module, W a left S-module and A a left $R \otimes_{\mathbb{Z}} S$-module. Then an A-*valued bilinear form* on $V \otimes W$ is an $R \otimes_{\mathbb{Z}} S$-module homomorphism

$$\beta : {}_R V \otimes_{\mathbb{Z}} {}_S W \to {}_{R \otimes S} A.$$

We freely write $\beta(v, w)$ for $\beta(v \otimes w)$.

Note that, in this generality, the restriction to a single form is no restriction; associated with any submodule M of the space of A-valued bilinear forms is a natural $\mathrm{Hom}(M, A)$-valued bilinear form

$$\beta : V \times W \to \mathrm{Hom}(M, A), \ \beta(v, w)(m) := m(v, w), \ \text{for } m \in M, \ v \in V, \ w \in W.$$

The way that representations of twisted rings fit into this context will appear in §3.3.

We proceed to define three notions of equivalence for bilinear forms.

(i) Suppose β' is an A-valued bilinear form on the $R \otimes S$-module $V' \otimes W'$. Then β and β' are said to be *equivalent* if there exist isomorphisms $\varphi_V : V \to V'$, $\varphi_W : W \to W'$ such that the diagram

$$
\begin{array}{ccc}
V \otimes W & \xrightarrow{\ \varphi_V \otimes \varphi_W\ } & V' \otimes W' \\[4pt]
\beta \downarrow & & \downarrow \beta' \\[4pt]
A & =\!=\!=\!=\!= & A
\end{array}
\qquad (3.1.1)
$$

commutes.

(ii) More generally, if β' is an A'-valued bilinear form on $V' \otimes W'$, where A' is another $(R \otimes S)$-module, then β and β' are said to be *weakly equivalent* if there are:

   – automorphisms $\sigma_R : R \to R$, $\sigma_S : S \to S$,
   – a σ_R-semilinear isomorphism $\varphi_V : V \to V'$,
   – a σ_S-semilinear isomorphism $\varphi_W : W \to W'$, and
   – a $\sigma_R \otimes \sigma_S$-semilinear isomorphism $\sigma_A : A \to A'$,

such that the diagram

$$
\begin{array}{ccc}
V \otimes W & \xrightarrow{\ \varphi_V \otimes \varphi_W\ } & V' \otimes W' \\[4pt]
\beta \downarrow & & \downarrow \beta' \\[4pt]
A & \xrightarrow{\ \sigma_A\ } & A'
\end{array}
\qquad (3.1.2)
$$

commutes.

(iii) We call β and β' *similar* if they are weakly equivalent with $\sigma_S = \mathrm{id}$ and $\sigma_R = \mathrm{id}$.

We will say that the pair (φ_V, φ_W) is respectively an *equivalence*, a *weak equivalence* or a *similitude* in the three cases.

Definition 3.1.2. Let A be a left $(R \otimes S)$-module, V a left R-module, W a left S-module and β an A-valued bilinear form on V and W. If C is a subgroup of V, we define the *dual subgroup* by

$$C^{\perp} = \{x \in W \mid \beta(c, x) = 0 \text{ for all } c \in C\}. \qquad (3.1.3)$$

Similarly if D is a subgroup of W we define its dual by

$$D^{\perp} = \{x \in V \mid \beta(x, d) = 0 \text{ for all } d \in D\}.$$

If $C \subset V$ and $D \subset W$ are subgroups with $C \subset D^{\perp}$ we write $C \perp D$ (see Fig. 3.1).

$$
\begin{array}{ccc}
C & \Longrightarrow & C^{\perp} \\
\cap & & \cup \\
D^{\perp} & \Longleftarrow & D \\
\end{array}
$$

$$V \qquad W$$

Fig. 3.1. In this situation we say $C \perp D$.

Since $C \subset D^{\perp}$ if and only if $D \subset C^{\perp}$, $C \perp D$ if and only if $D \perp C$. Note that $C \perp D$ means $D \subset C^{\perp}$, not necessarily that $D = C^{\perp}$.

Note also that $C^{\perp}$ is the union of all subgroups $D \subset W$ with $D \perp C$.

Remark. Weak equivalence preserves orthogonality: if (φ_V, φ_W) is a weak equivalence from β to β', then

$$\varphi_V(C)^{\perp} = \varphi_W(C^{\perp}), \text{ for } C \subset V.$$

Definition 3.1.3. With the same hypotheses as in Definition 3.1.2, a *code* C is either an R-submodule of V or an S-submodule of W. We will sometimes say simply that C is a code *with respect to* β, the other data — R, S, A, V, W — usually being clear from the context.

If C is a subgroup of V (resp. W) we define the code *generated by* C to be RC (resp. SC), the smallest submodule containing C.

Since the dual is just the intersection of preimages of the submodule $\{0\} \le A$ under module homomorphisms, we see that $C^{\perp}$ is a code. More precisely:

Lemma 3.1.4. *For any subgroup $C \subset V$, the dual subgroup $C^{\perp} \subset W$ is a code, and furthermore $C^{\perp}$ is equal to the dual of the code generated by C. In particular, if C is a code, so is $C^{\perp}$ (and then we call $C^{\perp}$ the* dual code *to C).*

Corollary 3.1.5. *For any pair of subgroups $C \subset V$ and $D \subset W$, $C \perp D$ if and only if $RC \perp SD$.*

In view of Corollary 3.1.5 we will henceforth restrict our attention to codes, i.e. submodules rather than subgroups. Note that if $C \subset V$ and $D \subset W$ are codes with $C \perp D$ and (φ_V, φ_W) is a weak equivalence, then $\varphi_V(C) \perp \varphi_W(D)$.

Definition 3.1.6. A code $C \subset V$ is called *closed* (with respect to β) if $C^{\perp\perp} = C$.

Since this property is essential for the investigation of self-dual codes, it would be nice to have necessary and sufficient conditions (on R, S, A, V, W, β) which imply that all codes in V are closed. At present we only know certain sufficient conditions. These are, however, general enough to include a large number of examples. Indeed, in §3.3 we will see that our notion of Type introduced in §1.8 satisfies these conditions, and so, in particular, for every representation of the form rings mentioned in the previous chapter and in Chapters 7 and 8 (see Tables 2.1, 2.2), all codes are closed.

It is clear that the left radical of β, namely the set $\{v \in V \mid \beta(v, W) = \{0\}\}$ is contained in $C^{\perp\perp}$ for all $C \subset V$, and so must also be contained in any closed code C. The following example shows that still further conditions are needed.

Example 3.1.7. Let $R = S = \mathbb{F}_2$ and $V = W = \mathbb{F}_2^N$. Choose a basis $(\beta_1, \ldots, \beta_d)$ for the space of all $\mathbb{F}_2$-valued alternating bilinear forms on V (so $d = \binom{N}{2}$), and define

$$\beta := (\beta_1, \ldots, \beta_d) : V \times V \to \mathbb{F}_2^d .$$

Then $C \leq V$ is closed if and only if $\dim(C)$ equals $0, 1$ or N.

For example, take $N = 3$, and let $\beta = (\beta_1, \beta_2, \beta_3)$, where

$$\beta_1 := \begin{pmatrix} 0 & 1 & 0 \\ 1 & 0 & 0 \\ 0 & 0 & 0 \end{pmatrix}, \quad \beta_2 := \begin{pmatrix} 0 & 0 & 1 \\ 0 & 0 & 0 \\ 1 & 0 & 0 \end{pmatrix}, \quad \beta_3 := \begin{pmatrix} 0 & 0 & 0 \\ 0 & 0 & 1 \\ 0 & 1 & 0 \end{pmatrix} .$$

Then $(x, y, z) \in C^{\perp}$ if and only if $c \cdot (y, x, 0) = c \cdot (z, 0, x) = c \cdot (0, z, y) = 0$ for all $c \in C$, where the dot indicates the ordinary mod-2 scalar product. One can check that any one-dimensional code is self-dual with respect to β, but if C is the two-dimensional code generated by $(1, 1, 0), (1, 0, 1)$ then $C^{\perp} = \{0\}, C^{\perp\perp} = \mathbb{F}_2^3$, and C is not closed.

3.2 Families of closed codes

Theorem 3.2.2 below will allow us to show that many important families of codes are closed. In particular it will allow us to show in §3.3 that all codes in a representation of a twisted ring are closed.

Definition 3.2.1. An A-valued bilinear form $\beta : V \otimes W \to A$ is *left nonsingular* if the induced homomorphism from V to $\mathrm{Hom}_S(W, A)$ is an isomorphism. Similarly, β is *right nonsingular* if the induced homomorphism from W to $\mathrm{Hom}_R(V, A)$ is an isomorphism. If both conditions hold β is called *nonsingular* and V is said to be A-*reflexive* (cf. Lam [341]). (If β is nonsingular the canonical map $V \to \mathrm{Hom}_S(\mathrm{Hom}_R(V, A), A)$ is an isomorphism.)

An R-module is said to have *finite length* if it possesses a composition series; that is, if there exists a finite chain

$$0 = V_0 \subset V_1 \subset \cdots \subset V_\ell = V \tag{3.2.1}$$

of R-submodules such that each quotient V_i/V_{i-1} is a simple R-module; in this case we say that V has length ℓ.

Theorem 3.2.2. *Let R_0 and S_0 be rings and let R and S be algebras over R_0 and S_0 respectively. Suppose U is a left $(R \otimes S)$-module with the following properties:*

- *U is injective as an R-module; that is, for any injection $f : V \to V'$ of left R-modules, $f^* : \mathrm{Hom}_R(V', U) \to \mathrm{Hom}_R(V, U)$ is a surjection.*
- *Any simple R-module of finite length over R_0 is U-reflexive.*
- *U is injective as an S-module and any simple S-module of finite length over S_0 is U-reflexive.*

Then any R-module V of finite length over R_0 is U-reflexive.

Proof. Note that an R-module V has finite length over R_0 if and only if it has finite length over R and all of the simple quotients in a composition series for V have finite length over R_0.

We show that V is U-reflexive by induction on the R-composition length of V. When V has length 1 (i.e. V is a simple R-module), this is true by assumption. Otherwise, let C be a submodule of V and consider the quotient V/C. Writing $V^* = \mathrm{Hom}_R(V, U)$ for the U-dual of V, we have the following commutative diagram:

$$
\begin{array}{ccccccccc}
0 & \longrightarrow & C & \xrightarrow{\iota} & V & \xrightarrow{\pi} & V/C & \longrightarrow & 0 \\
& & \downarrow & & \downarrow & & \downarrow & & \\
0 & \longrightarrow & C^{**} & \xrightarrow{\iota^{**}} & V^{**} & \xrightarrow{\pi^{**}} & (V/C)^{**} & \longrightarrow & 0
\end{array}
\tag{3.2.2}
$$

where the vertical arrows are the natural maps to the double-dual. Now the first row is clearly exact; since U is injective, the second row is also exact. Furthermore, since C and V/C have lengths smaller than that of V, we may assume by induction that C and V/C are U-reflexive. But then two of the vertical arrows are isomorphisms; it follows that the third is also an isomorphism and thus V is U-reflexive. $\qquad\square$

We now apply this theorem to codes.

Corollary 3.2.3. *Suppose:*

- *R_0 and S_0 are rings,*
- *R and S are algebras over R_0 and S_0 respectively,*
- *U is a left $(R \otimes S)$-module satisfying the hypotheses of Theorem 3.2.2,*
- *$_RV$ is an R-module of finite length over R_0,*
- *$_SW$ is an S-module of finite length over S_0,*
- *β is a nonsingular bilinear form on $_RV \otimes _SW$ taking values in U.*

Then all codes C in V or W are closed (with respect to β).

Proof. Suppose β is a nonsingular form on V and W. Without loss of generality, we may take $W = \mathrm{Hom}_R(V, U)$ and

$$\beta(v, w) = w(v).$$

Now, essentially by definition, $C^\perp$ is the kernel of the restriction map ι^* : $V^* \to C^*$, $f \mapsto f|_C$; from the exact sequence

$$0 \longrightarrow (V/C)^* \xrightarrow{\;\pi^*\;} V^* \xrightarrow{\;\iota^*\;} C^* \longrightarrow 0,$$

we see that $C^\perp \cong (V/C)^*$, injected by π^*. Dualizing again, we see that $C^{\perp\perp}$ corresponds to the image of $\iota^{**} : C^{**} \to V^{**}$; that is, $C^{\perp\perp} = C$. □

We will give several applications.

3.2.1 Codes over commutative rings

Associated with any ring R is a certain injective left R-module called its "minimal injective cogenerator" $U^0(R)$ (Lam [341, §19, p. 510]). Let k be a commutative ring. We will apply Theorem 3.2.2 with $R_0 = S_0 = R = S = k$ and $U = U^0(k)$. Since k is commutative, the k-module $U^0(k)$ can be regarded as a $k \otimes k$-module in the obvious way, and does satisfy the hypotheses of Theorem 3.2.2 (see Exercise 19.21 in [341]). A few special cases are of particular importance:

- $k = \mathbb{Z}$, $U^0(k) \cong \mathbb{Q}/\mathbb{Z}$,
- $k = \mathbb{Z}_p$, $U^0(k) \cong \mathbb{Q}_p/\mathbb{Z}_p$,
- $k = $ a field, $U^0(k) \cong k$.

Then if β takes values in $U^0(k)$, all codes over the ring k are closed.

3.2.2 Codes over quasi-Frobenius rings

Quasi-Frobenius rings are an important generalization of semisimple rings. Their key property is that they are "self-injective". Technically, a ring is *quasi-Frobenius* if it is left and right Artinian and left and right self-injective. There are several equivalent definitions, and the standard references (Faith [170], Kasch [297], Lam [341], Nicholson and Yousif [387], Rowen [460]) give much more information.

Examples of quasi-Frobenius rings include the following:

- any field (finite or infinite),
- $\mathbb{Z}/n\mathbb{Z}$,
- Galois rings (see §2.3.5),
- semisimple rings,
- $\mathrm{Mat}_n(R)$ if R is quasi-Frobenius,
- the group ring RG if R is quasi-Frobenius and G is a finite group.

The above references give further examples of quasi-Frobenius rings.

Examples of rings that are not quasi-Frobenius include $\mathbb{Z}$ and the ring $R := \begin{pmatrix} k & k \\ 0 & k \end{pmatrix}$, where k is a division ring. (The latter is not right self-injective: consider the ideal $I := \begin{pmatrix} 0 & k \\ 0 & 0 \end{pmatrix}$. The right R-homomorphism from I to R given by $\begin{pmatrix} 0 & a \\ 0 & 0 \end{pmatrix} \mapsto \begin{pmatrix} 0 & 0 \\ 0 & a \end{pmatrix}$ cannot be extended to all of R.)

Codes that are defined specifically over quasi-Frobenius rings have been studied by Horimoto and Shiromoto [268] and Shiromoto and Storme [485].

A somewhat stronger notion is that of a Frobenius ring. A quasi-Frobenius ring is *Frobenius* if it satisfies the additional condition that

$$\mathrm{soc}(R_R) \cong \overline{R}_R,$$

where soc is the socle of the R-module R, that is, the largest semi-simple submodule of R_R, and $\overline{R} = R/J$ denotes the largest semi-simple quotient module of R_R. Here J is the Jacobson radical (Lam, [341, § 16]).

It is not so easy to find examples of rings that are quasi-Frobenius but not Frobenius. Lam [341], following Nakayama [378], gives a certain 6-dimensional matrix ring which has this property.

Codes over Frobenius rings seem to have been studied for the first time in 1995, when Wood [554] showed that finite commutative Frobenius rings are an appropriate setting for generalizing the MacWilliams extension theorem to codes over rings. (See also Wood [552], [553], Ward and Wood [545], Kheifets [300].)

Theorem 3.2.4. *Let T be a quasi-Frobenius ring. If β takes values in T then all codes over T are closed (with respect to β).*

Proof. We take $R_0 = S_0 = R = S = T$. It follows immediately from the definition that $U = {}_{T^{op}\otimes T}T$ satisfies the hypotheses of Theorem 3.2.2. $\square$

Quasi-Frobenius rings are the largest class of rings R having the property that codes over R are closed with respect to a bilinear form taking values in R.

The minimal injective cogenerator of a quasi-Frobenius ring R is R itself; thus if R is commutative, the situation considered in this subsection is a special case of that in §3.2.1.

3.2.3 Algebras over a commutative ring

The commutative case (§3.2.1) and the quasi-Frobenius case (§3.2.2) when the ring is finite or commutative have a further common generalization.

Recall that a module $F = F_R$ is called a *progenerator* if it is finitely generated, projective and such that

$$R \cong \operatorname{Hom}_R(F, R) \otimes_{\operatorname{End}_R(F)} F \text{ as an } (R, R)\text{-bimodule ;} \qquad (3.2.3)$$

or equivalently if R_R is a direct summand of F^n for some n. Note that then the $\operatorname{End}_R(F)$ right module $\operatorname{Hom}_R(F, R)_{\operatorname{End}_R(F)}$ is also a progenerator.

An $R \otimes S$ right module $F = F_{R \otimes S}$ is a *faithfully balanced* progenerator if F is a progenerator both as an R-module and as an S-module, and if in addition there exists an $R \otimes S$ left module $F^{-1} = {}_{R \otimes S}F^{-1}$ such that the following S- resp. R-bimodules are isomorphic

$$F \otimes_R F^{-1} \cong S, \quad F \otimes_S F^{-1} \cong R. \qquad (3.2.4)$$

Using such modules F, a third way to generate modules U that satisfy the hypotheses of Theorem 3.2.2 is the following:

Lemma 3.2.5. *Fix a commutative ring k and k-algebras R and S. Let $F_{R \otimes_k S}$ be a faithfully balanced progenerator. Then $U = \operatorname{Hom}_k(F_{R \otimes S}, U^0(k))$ satisfies the hypotheses of Theorem 3.2.2 for R and S with $R_0 = S_0 = k$.*

Proof. By the "Injective Producing Lemma" (Lam [341, Chap. 1]), U is injective over both R and S; it remains to show that simple R-modules of finite length over k are U-reflexive.

Indeed, for any R-module V of finite length over k, we have

$$
\begin{aligned}
V^{**} &= \operatorname{Hom}_S(\operatorname{Hom}_R(V, U), U) \\
&= \operatorname{Hom}_S(\operatorname{Hom}_R(V, U), \operatorname{Hom}_k(F, U^0(k))) \\
&\cong \operatorname{Hom}_S(F, \operatorname{Hom}_k(\operatorname{Hom}_R(V, U), U^0(k))) \\
&= \operatorname{Hom}_S(F, \operatorname{Hom}_k(\operatorname{Hom}_R(V, \operatorname{Hom}_k(F, U^0(k))), U^0(k))) \\
&\cong \operatorname{Hom}_S(F, \operatorname{Hom}_k(\operatorname{Hom}_k(F \otimes_R V, U^0(k)), U^0(k))) . \qquad (3.2.5)
\end{aligned}
$$

The first isomorphism in (3.2.5) is obtained by mapping any element

$$X \in \operatorname{Hom}_S(\operatorname{Hom}_R(V, U), \operatorname{Hom}_k(F, U^0(k)))$$

to the element in $\operatorname{Hom}_S(F, \operatorname{Hom}_k(\operatorname{Hom}_R(V, U), U^0(k)))$ that sends $a \in F$ to the k-homomorphism

$$\operatorname{Hom}_R(V, U) \to U^0(k), \ f \mapsto X(f)(a).$$

The inverse isomorphism sends any $\Psi \in \operatorname{Hom}_S(F, \operatorname{Hom}_k(\operatorname{Hom}_R(V, U), U^0(k)))$ to the S-homomorphism

$$\operatorname{Hom}_R(V, U) \to \operatorname{Hom}_k(F, U^0(k)), \ f \mapsto (a \mapsto \Psi(a)(f)).$$

The second isomorphism in (3.2.5), $\operatorname{Hom}_R(V, \operatorname{Hom}_k(F, U^0(k))) \cong \operatorname{Hom}_k(F \otimes_R V, U^0(k))$, is the usual one sending $X \in \operatorname{Hom}_R(V, \operatorname{Hom}_k(F, U^0(k)))$ to the k-homomorphism

$$F \otimes_R V \to U^0(k), \ (f \otimes v) \mapsto X(v)(f).$$

Now $F \otimes_R V$ has finite length over k (F is a direct summand of R^n, so $F \otimes_R V$ is a direct summand of $R^n \otimes_R V = V^n$), so the $U^0(k)$-bidual of $F \otimes_R V$ is

$$\operatorname{Hom}_k(\operatorname{Hom}_k(F \otimes_R V, U^0(k)), U^0(k)) \cong F \otimes_R V$$

and hence

$$V^{**} \cong \operatorname{Hom}_S(F, F \otimes_R V) \cong V. \qquad \square$$

Since for all k-algebras S, the module $S_{S^{op} \otimes_k S}$ is a faithfully balanced progenerator, Lemma 3.2.5 constructs a module "U" such that for U-valued bilinear forms all codes are closed.

Corollary 3.2.6. *Fix a commutative ring k and a k-algebra S. Then $U = \operatorname{Hom}_k(S_{S^{op} \otimes S}, U^0(k))$ satisfies the hypotheses of Theorem 3.2.2 for $R = S$, $R_0 = S_0 = k$.*

In this case we can say more:

Theorem 3.2.7. *Fix a commutative ring k. Let S be a k-algebra, let V be an S^{op}-module, let W be an S-module and let $A = \operatorname{Hom}(S_{S^{op} \otimes S}, A_0)$ for some k-module A_0. For an A-valued bilinear form β on V and W, define an A_0-valued form β_0 on V and W (viewed as k-modules) by*

$$\beta_0(v, w) = (\beta(v, w))(1_S).$$

If β is nonsingular, then so is β_0; moreover, for any code for β, we have

$$C^{\perp_\beta} = C^{\perp_{\beta_0}}.$$

Proof. For the second claim, we take $C \subset W$ and observe:

$$
\begin{aligned}
C^{\perp_{\beta_0}} &= \{v \in V \mid \beta_0(v, C) = 0\} \\
&= \{v \in V \mid \beta(v, C)(1) = 0\} \\
&= \{v \in V \mid \beta(v, SC)(1) = 0\} \\
&= \{v \in V \mid \beta(v, C)(S) = 0\} \\
&= \{v \in V \mid \beta(v, C) = 0\} \\
&= C^{\perp_\beta} .
\end{aligned}
$$

It remains to show that β_0 is nonsingular. By symmetry we need only show it is right nonsingular, which means by definition that the induced map

$$
\tilde{\beta}_0 : {}_kW \to \mathrm{Hom}({}_kV, A_0), \quad w \mapsto (v \mapsto \beta_0(v, w))
$$

is an isomorphism. This map $\tilde{\beta}_0$ is the image of the isomorphism

$$
\tilde{\beta} : {}_SW \to \mathrm{Hom}_{S^{op}}({}_{S^{op}}V, A), \quad w \mapsto (v \mapsto \beta(v, w))
$$

under the canonical isomorphism

$$
{}_kS \otimes_S \mathrm{Hom}_{S^{op}}({}_{S^{op}}V, {}_{S^{op}\otimes S}A) \to \mathrm{Hom}({}_kV, A_0), \quad (s \otimes \varphi) \mapsto (v \mapsto (\varphi(v)(s))
$$

and thus must itself be an isomorphism. $\qquad \square$

When $A_0 = U^0(k)$, we have the following converse, the gist of which is this. Consider a duality on codes given by some collection of inner products, with its induced notion of closed codes, and suppose this duality agrees on closed codes with that given by a single nonsingular form. Then the duality is of the kind considered in Theorem 3.2.7; in particular, there exists a natural module structure on V and W such that all closed codes are submodules, and all submodules are closed. Note that if we insist that the duality given by a twisted module representation have this form, we immediately find that the twisted module can be given a twisted ring structure (see §3.3).

Theorem 3.2.8. *Fix a commutative ring k, let V and W be k-modules of finite length and let β be an A-valued form on V and W for some k-module A. If there exists a nonsingular $U^0(k)$-valued form β_0 on V and W such that*

$$
C^{\perp_\beta} = C^{\perp_{\beta_0}}
$$

for all β-closed codes C, then there exists a k-algebra S, an S^{op}-module structure on V and an S-module structure on W such that a code is β-closed if and only if it is an S^{op} or S-submodule, as appropriate. Furthermore, if we set $A' = \mathrm{Hom}_k(S_{S^{op}\otimes S}, U^0(k))$, there exists a k-module map $\eta : A' \to A$ and a nonsingular $S^{op} \otimes S$-linear form $\beta' : V \otimes W \to A'$ satisfying $\beta(v, w) = \eta(\beta'(v, w))$ and $\beta_0(v, w) = \beta'(v, w)(1_S)$.

Proof. Replacing A by a submodule as necessary, we may assume that $\beta(V \otimes W) = A$. Then A is a quotient of a finite length module, so must itself have finite length. Thus A is $U^0(k)$-reflexive, and we may therefore assume that A has the form $A = \mathrm{Hom}_k(N, U^0(k))$ for some finite length k-module N. Using the canonical isomorphism

$$\mathrm{Hom}_k(V \otimes W, \mathrm{Hom}_k(N, U^0(k))) \cong \mathrm{Hom}_k(N, \mathrm{Hom}_k(V \otimes W, U^0(k))),$$

we see that we may identify N with a submodule of $\mathrm{Hom}_k(V \otimes W, U^0(k))$. Furthermore, if C and D are codes with $C \perp D$, then we have

$$\gamma(C \otimes D) = 0,$$

for all $\gamma \in N$. We may further assume that all such $U^0(k)$-valued bilinear forms are in N; otherwise, we can simply enlarge N as necessary; note that β factors through the resulting surjection and the resulting form gives rise to the same duality.

In particular, we find $\beta_0 \in N$. To any element $\gamma \in N$ and any element $w \in W$ we associate an element $\gamma \cdot w \in W$ by

$$\beta_0(v, \gamma \cdot w) = \gamma(v, w);$$

since β_0 is nonsingular, $\gamma \cdot w$ exists uniquely. For any closed code $C \subset V$, we have $C \perp_\gamma C^{\perp_\beta}$ and thus $C \perp_{\beta_0} \gamma \cdot C^{\perp_\beta}$, so

$$\gamma \cdot C^{\perp_\beta} \subset C^{\perp_{\beta_0}} = C^{\perp_\beta}.$$

Since $C^{\perp\perp\perp} = C^\perp$, we have

$$\gamma \cdot C^{\perp_\beta} \subset C^{\perp_\beta},$$

even when C is not closed.

We similarly define $v \cdot \gamma$ and observe

$$C \perp D \implies C \cdot \gamma_1 \perp \gamma_2 \cdot D,$$

for codes $C \subset V$, $D \subset W$ and elements $\gamma_1, \gamma_2 \in N$.

Now, given $\gamma_1, \gamma_2 \in N$, we define a bilinear form

$$(\gamma_1 \cdot \gamma_2)(v, w) = \beta_0(v \cdot \gamma_1, \gamma_2 \cdot w).$$

But then $(\gamma_1 \cdot \gamma_2)(C, D) = 0$ whenever $C \perp_\beta D$, so $\gamma_1 \cdot \gamma_2 \in N$. Furthermore, we readily verify $\gamma \cdot \beta_0 = \beta_0 \cdot \gamma = \gamma$. We thus obtain a ring structure on N compatible with the k-module structure and may thus identify N as a k-algebra S and have already defined the S^{op}-module structure on V and the S-module structure on W. The fact that $\beta_0(v, w) = \beta(v, w)(1_S)$ is immediate.

It remains to show that β is nonsingular. Choose

$$\phi \in \mathrm{Hom}_S({}_S W, \mathrm{Hom}_k(S_{S^{op} \otimes S}, U^0(k)));$$

we need to find $v \in V$ such that $\beta(v, w) = \phi(w)$. Since β_0 is nonsingular, we may certainly find $v \in V$ with $\beta_0(v, w) = \phi(w)(1_S)$. But then

$$\beta(v, w)(s) = \beta_0(v, ws) = \phi(ws)(1_S) = \phi(w)(s),$$

as required. □

3.2.4 Direct summands

In the next theorem, note that, if R is a field, then V and W are simply vector spaces over R, any subspace is a direct summand and the result is immediate.

Theorem 3.2.9. *Let R be a ring and let V be a finitely generated projective left R-module and W a finitely generated projective left R^{op}-module. If β is a nonsingular $_{R \otimes R^{op}}R$-valued bilinear form on V and W then any code which is a direct summand is closed.*

Proof. If $V = V_1 \oplus V_2$, β induces an isomorphism

$$W \cong V^* = \operatorname{Hom}_R(V, R) = V_1^* \oplus V_2^*.$$

Then $V_1^\perp = V_2^*$ and hence $(V_1^\perp)^\perp = V_1$ as required. □

Remark 3.2.10. Being a direct summand is a sufficient but not necessary condition for a code to be closed—see for instance Theorem 3.2.4 for the case where R is a quasi-Frobenius ring.

3.3 Representations of twisted rings and closed codes

We next show that the definition of Type given in §1.8 is strong enough to guarantee that all codes are closed.

Definition 3.3.1. Let k be a commutative ring and let (R, M, ψ) be a *twisted k-algebra* (i.e. a twisted ring such that R is a k-algebra). Then a representation $\rho := (V, \rho_M, \beta)$ is called a *k-representation* of R if V is a k-module of finite length, β is a nonsingular $U^0(k)$-valued k-bilinear form on V, and ρ is a homomorphism from R to $\operatorname{End}_k(V; \beta)$.

In particular, any twisted ring is a twisted $\mathbb{Z}$-algebra and finite representations are $\mathbb{Z}$-representations.

Note that a k-representation of R induces a natural left R-module structure on V. Note also that the $U^0(k)$-valued form β is not necessarily R-bilinear. In general it does not make sense to talk about R-bilinearity, since $U^0(k)$ is not an $R^{op} \otimes R$-module. However, our notion of twisted ring allows us to construct an $R^{op} \otimes R$-bilinear form from β, which enables us to apply the general theory developed at the beginning of this chapter to our notion of Type.

Given a left twisted R-module M and a left R-module V, an M-valued *Hermitian form* on V is an M-valued bilinear form β on $V \times V$ such that

$$\beta(w, v) = \tau(\beta(v, w)) \text{ for all } v, w \in V.$$

Theorem 3.3.2. *Fix k, R, V, and let M be the underlying twisted module of R. Furthermore, fix an R-module structure on V. Then the k-representations of R on V compatible with the given R-module structure are in natural correspondence with the nonsingular, $(R \otimes R)$-linear, $\mathrm{Hom}_k(M, U^0(k))$-valued Hermitian forms on V.*

Proof. Clearly any such k-representation is uniquely determined by the nonsingular $U^0(k)$-valued k-bilinear form β. Then $\beta_0 : V \times V \to \mathrm{Hom}_k(M, U^0(k))$ defined by

$$\beta_0(v, w)(\psi(r)) := \beta(v, rw), \text{ for } v, w \in V, r \in R$$

is a nonsingular $(R \otimes R)$-linear, $\mathrm{Hom}_k(M, U^0(k))$-valued Hermitian form on V. On the other hand, given such a form β_0, then

$$\beta : V \times V \to U^0(k), (v, w) \mapsto \beta_0(v, w)(\psi(1))$$

is a nonsingular $U^0(k)$-valued k-bilinear form on V. This gives the desired natural bijection. $\square$

Corollary 3.3.3. *Let $\rho = (V, \rho_M, \beta)$ be a k-representation of the twisted k-algebra (R, M, ψ). Then any submodule $C \subset V$ is closed.*

Proof. Apply Theorem 3.2.7. $\square$

Note in particular that the dual code (cf. Definition 1.3.4) of a code C is given by

$$C^\perp = \{v \in V \mid \beta(v, C) = \{0\}\}. \tag{3.3.1}$$

Thus the notion of twisted rings suffices to make all codes closed; conversely, by Theorem 3.2.8, we see that, while this may not be the only way to force codes to be closed, it is the only *natural* way.

The next lemma is also a corollary of Theorem 3.3.2.

Lemma 3.3.4. *Let $\rho = (V, \rho_M, \beta)$ be a k-representation of the twisted k-algebra R. Then for any code C, we have the exact sequence*

$$0 \longrightarrow C \longrightarrow V \longrightarrow \mathrm{Hom}(C^\perp, U^0(k)) \longrightarrow 0 \, ;$$

thus for any maximal ideal $\mathfrak{p}$ of k, we find that the number $\ell_\mathfrak{p}(V)$ of k-composition factors of V isomorphic to $k/\mathfrak{p}$ is given by

$$\ell_\mathfrak{p}(C) + \ell_\mathfrak{p}(C^\perp) = \ell_\mathfrak{p}(V) \, .$$

Remark. In particular, when $k = \mathbb{Z}$, the last statement becomes $|V| = |C||C^\perp|$.

3.4 Morita theory

We now consider how Morita equivalences between rings affect the duality of codes.

Let $_RV$ and $_SW$ be left modules and let $_{R'}V'_R$ and $_{S'}W'_S$ be bimodules. We observe that

$$(V' \otimes W') \otimes_{R \otimes S} (V \otimes W) \cong (V' \otimes_R V) \otimes (W' \otimes_S W).$$

Thus, if A is an $(R \otimes S)$-module and β is an A-valued form on $V \times W$, tensoring with $V' \otimes W'$ gives us a $(V' \otimes W') \otimes_{R \otimes S} A$-valued bilinear form

$$\tilde{\beta} = (\mathrm{id}_{V'} \otimes \mathrm{id}_{W'}) \otimes \beta \; : \; (V' \otimes_R V) \times (W' \otimes_S W) \to (V' \otimes W') \otimes_{R \otimes S} A.$$

Theorem 3.4.1. (i) *With notation as above, if codes $C \subset V$ and $D \subset W$ satisfy $C \perp D$ with respect to β, then the codes $C' := V' \otimes_R C \subset V' \otimes_R V$ and $D' := W' \otimes_S D \subset W' \otimes_S W$ satisfy*

$$C' \perp D' \text{ with respect to } \tilde{\beta}.$$

(ii) *If V' and W' are progenerators over R and S, then the converse is true, and moreover*

$$(C')^{\perp, \tilde{\beta}} = W' \otimes_S C^{\perp, \beta}.$$

Proof. (i) We observe that, for $v' \in V'$, $w' \in W'$, $c \in C$, $d \in D$,

$$\tilde{\beta}((v' \otimes_R c) \otimes (w' \otimes_S d)) = (v' \otimes w') \otimes \beta(c, d) = 0.$$

(ii) To prove the second assertion, we first note that enlarging R' and S' will not affect the relation $\perp_\beta$; we may therefore take $R' = \mathrm{End}_R(V')$, $S' = \mathrm{End}_S(W')$. We write $V'' = \mathrm{Hom}(V', R)$, $W'' = \mathrm{Hom}(W', S)$ and observe that the canonical isomorphism

$$(V'' \otimes W'') \otimes_{R' \otimes S'} (V' \otimes W') \cong R \otimes S,$$

so the associativity of $\otimes$ tells us that

$$(V'' \otimes W'') \otimes_{R' \otimes S'} ((V' \otimes W') \otimes_{R \otimes S} \beta) \cong \beta,$$

up to canonical similitude. But then

$$(V' \otimes C) \perp (W' \otimes D) \implies V'' \otimes (V' \otimes C) \perp W'' \otimes (W' \otimes D) \implies C \perp D.$$

The remaining assertion follows from the fact that $C^\perp$ is the union of all codes D with $C \perp D$.   □

Remark. Note that if A has the form $\mathrm{Hom}_k(A_0, U^0(k))$, then when V' and W' are projective over R and S, we obtain a canonical isomorphism

$$(V' \otimes W') \otimes_{R \otimes S} A = \mathrm{Hom}_k(\mathrm{Hom}_{R \otimes S}(V' \otimes W', A_0), U^0(k)).$$

There are two special cases of Theorem 3.4.1 that are of particular importance.

Corollary 3.4.2. *Let $\iota \in R$ and $\iota' \in S$ be idempotents. Then*

$$C \perp D \implies \iota C \perp \iota' D.$$

If ι and ι' are basic idempotents *(Lam [342, p. 372]), that is, $R\iota R = R$ and $S\iota' S = S$, then*

$$C \perp D \iff \iota C \perp \iota' D.$$

Proof. Define

$$R' = \iota R\iota, \ V' = \iota R, \ S' = \iota' S\iota', \ W' = \iota' S.$$

Then we obtain an $(R \otimes S)$-module $(\iota \otimes \iota')A$ and a form $(\iota \otimes \iota')\beta$, and the statements follow from Theorem 3.4.1. $\qquad\qquad\square$

Example 3.4.3. Let $R := \mathbb{Z}/6\mathbb{Z}$. Then R contains the two idempotents $\iota := 4 + 6\mathbb{Z}$ and $\iota' := 1 - \iota := 3 + 6\mathbb{Z}$. We have $\iota R = \iota R\iota \cong \mathbb{Z}/3\mathbb{Z}$ and $\iota' R \cong \mathbb{Z}/2\mathbb{Z}$. A code $C \le R^N$ is self-dual if and only if ιC is a self-dual ternary code and $\iota' C$ is a self-dual binary code.

The second special case is motivated by the notion of higher-genus weight enumerators. As already pointed out in Remark 2.1.10, if C is a code over R then its genus-m weight enumerator is essentially the weight enumerator of $C \otimes R^m \cong C^m$. Since this is a module over the noncommutative ring $\mathrm{Mat}_m(R)$, the study of codes over commutative rings leads naturally to the noncommutative theory.

Corollary 3.4.4. *Let m and l be positive integers. We may take*

$$V' = R^m, \ R' = \mathrm{Mat}_m(R), \ W' = S^l, \ S' = \mathrm{Mat}_l(S)$$

in Theorem 3.4.1, obtaining a new bilinear module and form denoted by $A_{m \times l}$ and $\beta_{m \times l}$. We can think of the elements of $A_{m \times l}$ as $m \times l$ matrices with entries in A.

Continuing with the notation of Theorem 3.4.1, if V' and W' are finitely generated projective modules and $R' = \mathrm{End}_R(V')$, $S' = \mathrm{End}_S(W')$, then any code with respect to the tensor product form is itself a tensor product. The two special cases mentioned above have this property.

Besides tensor products, the notion of (direct) product also extends to forms, using the map

$$\mathrm{Bil}(V, W; A) \times \mathrm{Bil}(V', W'; A') \ \to \ \mathrm{Bil}(V \times V', W \times W'; A \times A')$$

given by

$$(\beta, \beta')(v, v'; w, w') = (\beta(v, w), \beta'(v', w')).$$

Corollary 3.4.5. *Let β be an $_{R \otimes S} A$-valued form on $_R V$ and $_S W$, and let β' be an $_{R' \otimes S'} A'$-valued form on $_{R'} V'$ and $_{S'} W'$. Then any code $C^+ \subset V \times V'$ can be expressed in the form $C \times C'$ with $C \subset V$, $C' \subset V'$, and we have the relation*

$$(C \times C')^\perp = C^\perp \times C'^\perp.$$

Proof. Let

$$C = (1,0)C^+, \quad C' = (0,1)C^+.$$

By Corollary 3.4.2 we get

$$(1,0)(C \times C')^\perp \subset C^\perp, \quad (0,1)(C \times C')^\perp \subset C'^\perp,$$

and thus

$$(C \times C')^\perp \subset C^\perp \times C'^\perp.$$

Thus it remains to show

$$(C \times C') \perp (C^\perp \times C'^\perp).$$

But

$$\beta(C \times C', C^\perp \times C'^\perp) = (\beta(C, C^\perp), \beta'(C', C'^\perp)) = 0. \qquad \square$$

3.5 New representations from old

3.5.1 Subquotients and quotients

In this section, we let R and S be rings, $_R V$ and $_S W$ left R- resp. S-modules, and we fix a bilinear form $\beta : V \times W \to A$, where A is a left $(R \otimes S)$-module.

Definition 3.5.1. Let $C \subset C' \subset V$ and $D \subset D' \subset W$ be codes with $C \perp D'$ and $C' \perp D$. Then for $v \in C'$, $w \in D'$ we have

$$\beta(v + c, w + d) = \beta(v, w) \text{ for } c \in C, d \in D. \tag{3.5.1}$$

We thus obtain a new A-valued bilinear form on $(C'/C) \times (D'/D)$, which we call a *subquotient form*. If C and D are closed and $C' = D^\perp$, $D' = C^\perp$ then we denote the new form by $\beta/(C, D)$ and call it a *quotient form* (rather than subquotient).

The significance of the subquotient construction is that orthogonal codes with respect to a subquotient form lift to orthogonal codes with respect to the original form. Dual codes can likewise be lifted from quotient forms.

Theorem 3.5.2. *Suppose A satisfies the three conditions on U in the statement of Theorem 3.2.2, and assume $\beta : V \times W \to A$ is nonsingular. Then for any codes $C \leq V$ and $D \leq W$ with $C \perp D$, the quotient form $\beta/(C, D)$ is nonsingular.*

Proof. We may as well assume $W = V^* := \operatorname{Hom}_R(V, A)$, and recall that V is A-reflexive. Furthermore, since

$$\beta/(C, D) = (\beta/(C, 0))/(0, D),$$

it suffices to prove the theorem when $D = 0$.

Now $C^\perp \cong (V/C)^* \subset V^*$. It follows that $\beta/(C, 0)$ is left nonsingular. Since V/C is reflexive, we conclude that $\beta/(C, 0)$ is nonsingular, as required.   □

Remark. Similarly, if β satisfies the hypotheses of Theorem 3.2.9, then $\beta/(C, D)$ is nonsingular for any direct summands C and D.

Remark. Taking the image under the natural epimorphism $V \times W \to V/C \times W/D$ yields a one-to-one correspondence between the set of pairs of orthogonal codes containing $C \times D$:

$$\{(\tilde{C}, \tilde{D}) \leq V \times W \mid \tilde{C} \supseteq C, \tilde{D} \supseteq D \text{ and } \tilde{C} \perp_\beta \tilde{D}\},$$

and the set of pairs of orthogonal codes for $\beta/(C, D)$. The latter are called *glue codes*, and we obtain $\tilde{C}$ and $\tilde{D}$ by *gluing* codes for $\beta/(C, D)$ to C and D; this is an important construction for self-orthogonal and self-dual codes, as well as a major tool for their classification. We will return to gluing theory in §§9.6, 9.7.

For representations of twisted rings, we may take $S = R$, $W = V$ and C a self-orthogonal code in V to obtain a new representation with underlying module $C^\perp/C$, called the quotient representation.

Definition 3.5.3. Let (R, M, ψ) be a twisted k-algebra over a commutative ring k, let $\rho := (V, \rho_M, \beta)$ be a k-representation of (R, M, ψ), and let $C \subset V$ be a self-orthogonal code. Then the *quotient representation* ρ/C is given by $(C^\perp/C, \widetilde{\rho_M}, \tilde{\beta})$, where

$$\widetilde{\rho_M}(m)(v + C, w + C) = \rho_M(m)(v, w), \text{ for } v, w \in C^\perp.$$

3.5.2 Direct sums and products

This section defines the sum and product of two representations and the conjugate of a representation. In all these definitions we will leave to the reader the verification that the new "β" is always nonsingular.

Definition 3.5.4. Let (R, M, ψ) be a twisted k-algebra over a commutative ring k. Let $\rho := (V, \rho_M, \beta)$ and $\rho' := (V', \rho'_M, \beta')$ be k-representations of this twisted k-algebra. Then the *orthogonal sum* of these representations is given by

$$\rho + \rho' := (V \oplus V', \rho_M + \rho'_M, \beta \perp \beta'),$$

where

$$(\rho_M + \rho'_M)(m)(v_1 + v'_1, v_2 + v'_2) = \rho_M(m)(v_1, v_2) + \rho'_M(m)(v'_1, v'_2)$$

and

$$(\beta + \beta')(v + v', w + w') = \beta(v, w) + \beta'(v', w').$$

As above, if N is a positive integer, we write

$$N\rho := \underbrace{\rho + \ldots + \rho}_{N} = (V^N, \rho_M^N, \beta^N).$$

Definition 3.5.5. The *conjugate representation* $\bar{\rho}$ is defined by

$$\bar{\rho} := (V, -\rho_M, -\beta). \tag{3.5.2}$$

In contrast, the product of two representations involves representations of different twisted k-algebras, and is a representation of the product twisted k-algebra. (The product should not be confused with the tensor product, which will be discussed in §3.5.3.)

Definition 3.5.6. Let (R, M, ψ) and (R', M', ψ') be twisted k-algebras over the same commutative ring k. Their *product* $(R \times R', M \times M', \psi \times \psi')$ is also a twisted k-algebra, where the involution is $\tau \times \tau'$. Let $\rho := (V, \rho_M, \beta)$ and $\rho' := (V', \rho'_{M'}, \beta')$ be k-representations of (R, M, ψ) and (R', M', ψ') respectively. Then the *product* of the representations ρ and ρ' is the representation

$$\rho \times \rho' := (V \times V', \rho_M \times \rho'_{M'}, \beta \times \beta')$$

of $R \times R'$, where $V \times V'$ is the obvious $(R \times R')$-module and

$$(\rho_M \times \rho'_{M'})(m, m')((v, v'), (w, w')) = \rho_M(m)(v, w) + \rho'_{M'}(m')(v', w').$$

3.5.3 Tensor products

The construction of the tensor product of two representations is more complicated, however. The result is again a representation of a different twisted ring. Let (R, M, ψ) be a twisted k-algebra and $\rho = (V, \rho_M, \beta)$ a k-representation. The tensor product $V' \otimes_R V$ of V with a right R-module V' is a left $\mathrm{End}_R(V')$-module. To construct a twisted ring of which $V' \otimes_R V$ is a representation, assume that V' is a finitely generated projective right R-module and let $\beta' : V' \times V' \to M$ be a nonsingular M-valued R-bilinear form on V'. The desired twisted ring is then

$$\mathrm{End}_R(V', \beta') := (R', M', \psi') := (\mathrm{End}_R(V'), \mathrm{Bil}_R(V', M), \psi_{\beta'}),$$

where

$$\psi_{\beta'} : \mathrm{End}_R(V') \to \mathrm{Bil}_R(V', M), \quad f \mapsto (\beta')^f,$$

with $(\beta')^f(v', w') = \beta'(v', f(w'))$ for all $v', w' \in V'$ and $f \in \mathrm{End}_R(V')$. This twisted ring has the representation

$$(V', \beta') \otimes_R \rho := (V' \otimes_R V, \rho_{M'}, \beta'),$$

where

$$\rho_{M'}(m')(v' \otimes v, w' \otimes w) := (m'(v', w'))(v, w).$$

This is called the *tensor product* of (V', β') with the representation ρ.

Note that the notion of M-valued form is equivalent to the notion of sesquilinear R-valued form $\beta' : V' \times V' \to R$:

$$\beta'(xr, ys) = r^J \beta'(x, y)s, \text{ for } x, y \in V', \ r, s \in R.$$

However, the latter notion is not invariant under rescaling.

Remark 3.5.7. Let (R, M, ψ) be a twisted ring and let $V' = \iota R$ for some idempotent ι. Then $\mathrm{End}_R(V') = \iota R \iota$, and any $\beta' \in \mathrm{Bil}_R(V', M)$ is uniquely determined by $\beta'(\iota, \iota) = m \in M(\iota \otimes \iota) = \psi(\iota^J R \iota)$ (note that $\psi(r)(\iota \otimes \iota) = \psi(\iota^J r \iota)$ for $r \in R$). Now

$$\mathrm{Hom}_R(\iota R, M_{R \otimes 1}) \cong \mathrm{Hom}_R(\iota R, \ _{R^J}R) \cong \iota^J R,$$

as right R-modules, hence the module V' admits a nonsingular M-valued form if and only if there is an isomorphism of right R-modules

$$\iota R \cong \iota^J R,$$

in which case we say that the idempotent is *symmetric*. This isomorphism then induces an isomorphism

$$\mathrm{End}_R(\iota R) = \iota R \iota \cong \iota^J R \iota \cong M(\iota \otimes \iota)$$

of right $\iota R \iota$-modules, as required. Note that any isomorphism

$$\kappa : \iota R \cong \iota^J R$$

takes the form

$$\kappa(\iota x) = v_\iota x, \ \kappa^{-1}(\iota^J x) = u_\iota x,$$

where $u_\iota \in \iota R \iota^J$ and $v_\iota \in \iota^J R \iota$ satisfy

$$u_\iota v_\iota = \iota, \ v_\iota u_\iota = \iota^J.$$

Then $(R' := \iota R \iota, M' := \psi(\iota^J R \iota), \psi')$ is a twisted ring with structure maps

$$\psi'(\iota r \iota) = \psi(v_\iota r \iota),$$
$$\tau'(\iota r \iota) = u_\iota r^J v_\iota^J \epsilon,$$
$$\psi'(\iota t \iota)(\iota r \iota \otimes \iota s \iota) = \psi(u_\iota r^J v_\iota t \iota s \iota).$$

Example 3.5.8. If $V' = R^n$, then we may take $\beta' : V' \times V' \to M$ to be defined by

$$\beta'(v, w) = \sum_{1 \leq i \leq n} \psi(1)(v_i \otimes w_i).$$

This is nonsingular, so gives a twisted ring structure on $\mathrm{Mat}_n(R)$. The resulting twist is obtained by transposing the matrix and then twisting element-wise.

In order for this construction (and in particular the two special cases above) to be useful, it needs to preserve the class of self-dual codes, which indeed it does:

Theorem 3.5.9. *Let ρ be a finite representation of the twisted ring (R, M, ψ), and let V' be a finitely generated projective right R-module admitting a nonsingular M-valued R-bilinear form. Then if a code C is self-dual, so is $V' \otimes_R C$.*

Proof. Since V' is finitely generated and projective, we may write $V' = \iota R^n$ for some $n > 0$ and some idempotent $\iota \in \mathrm{Mat}_n(R)$. The module

$$\iota \, \mathrm{Mat}_n(R) = \iota R^n \otimes_R R^n$$

admits a nonsingular $\mathrm{Mat}_n(R)$-valued sesquilinear form, so ι must be symmetric. It thus suffices to consider the cases $V' = R^n$ and $V' = \iota R$. The first case follows immediately from Theorem 3.4.1, since R^n is a progenerator.

It remains to consider the second case. Let $I \subset R$ be the ideal annihilated by ρ. Then $\iota + I$ is still symmetric in R/I. In particular, we may assume $I = 0$, and thus R is finite. In particular, R is semilocal, and thus $\iota R \cong \iota^J R$ implies $(1 - \iota)R \cong (1 - \iota^J)R$ (see Lam [342, §20]), so $1 - \iota$ is also symmetric. Thus both ιC and $(1 - \iota)C$ are self-orthogonal. We thus find that

$$|\iota C|^2 \leq |\iota V|,$$
$$|(1 - \iota)C|^2 \leq |(1 - \iota)V|.$$

Taking products, we obtain $|C|^2 \leq |V|$. Since equality holds here, we must have $|\iota C|^2 = |\iota V|$, and thus ιC is self-dual, as required. $\square$

4

The Category **Quad**

This chapter studies the objects introduced in Chapter 1 from the point of view of category theory. One of the main reasons for doing this is so that we can introduce the notion of a matrix ring over a form ring (see §1.10), which will be used to define the hyperbolic co-unitary group (see (5.2.5)) and in the proofs of the main theorems in Chapter 5. Another application will be the definition of the Witt group of representations of a form ring (§4.6). This will be used to define the universal Clifford-Weil group associated with a finite form ring (see §5.4).

There are also two other reasons why we adopt this categorical approach. The first is to justify the definition of a quadratic pair by deducing it from a simpler notion. Secondly, we want to extend Morita theory to the quadratic case.

For both of these purposes, we need a notion of tensor product of spaces of quadratic forms. Since we would like the tensor product of the spaces of quadratic forms on V and W to be the space of quadratic forms on $V \otimes W$, we need to add some additional structure, just as when dealing with Hermitian forms one needs to consider the ambient space of all bilinear forms. Since the definition of a quadratic form is that it induces a bilinear form, this suggests that the correct abstraction of "space of quadratic forms" should also include a space of bilinear forms.

Thus we need a space M (of bilinear forms) and a space Φ (of quadratic forms); we also need a map $\lambda : \Phi \to M$ that corresponds to taking the induced bilinear form. Moreover, since a bilinear form induces a quadratic form by setting both arguments equal, we need a map $\{\!\{\,\}\!\} : M \to \Phi$. This leads us to Definition 4.1.1.

One indication that Definition 4.1.1 is the correct one is that there are two functors F and Q (the "free" functor and the "squaring" functor, see Lemma 4.1.5) from the category of abelian groups to the category of quadratic groups, such that the space of morphisms from $Q(A)$ to $F(B)$ is simply the space of pointed quadratic maps from A to B. Thus, in addition to the objects being models of spaces of quadratic forms, the morphisms in the category are themselves analogues of quadratic forms. And indeed, the space of morphisms between two quadratic groups can be extended to a quadratic group structure, giving an "internal hom" functor on Quad (see Definition 4.2.1). This forces a corresponding notion of tensor product, making Quad a closed category, with associated notions of ring, module, etc., all of which behave as expected. The resulting structure is rich enough to give analogues of Morita theory; in addition, our notions of quadratic pairs, form rings, and representations thereof, all have direct categorical meanings.

4.1 The category of quadratic groups

One of the main concepts introduced in Chapter 1 was the notion of a quadratic pair over a ring R (§1.1). As usual when taking a categorical approach, it is helpful to consider the special case $R = \mathbb{Z}$. In this case the definition can be considerably simplified, and we obtain the following notion, which is equivalent to a "quadratic pair over $\mathbb{Z}$".

Definition 4.1.1. A *quadratic group* is a quadruple $(M, \Phi, \{\!\{\,\}\!\}, \lambda)$, where M and Φ are abelian groups and $\{\!\{\,\}\!\} : M \to \Phi$, $\lambda : \Phi \to M$ are homomorphisms satisfying

$$\{\!\{\,\}\!\} \circ \lambda \circ \{\!\{\,\}\!\} = 2\{\!\{\,\}\!\} \,, \tag{4.1.1}$$

$$\lambda \circ \{\!\{\,\}\!\} \circ \lambda = 2\lambda . \tag{4.1.2}$$

Remark. If no confusion will arise we will sometimes refer to a quadratic group as (M, Φ), leaving the maps implicit.

Given a quadratic group $(M, \Phi, \{\!\{\,\}\!\}, \lambda)$, we define maps $\tau : M \to M$ and $[-1] : \Phi \to \Phi$ by

$$\tau = \lambda \circ \{\!\{\,\}\!\} - \mathrm{id}_M \,, \tag{4.1.3}$$

$$[-1] = \{\!\{\,\}\!\} \circ \lambda - \mathrm{id}_\Phi \,, \tag{4.1.4}$$

(compare Figure 1.1).

The following is a straightforward calculation.

Lemma 4.1.2. *The maps* τ, $\{\!\{\,\}\!\}$, λ, $[-1]$ *satisfy*

$$\tau \circ \tau = \mathrm{id}_M \,, \tag{4.1.5}$$

$$[-1] \circ [-1] = \mathrm{id}_\Phi \,, \tag{4.1.6}$$

$$\tau \circ \lambda = \lambda \,, \tag{4.1.7}$$

$$\lambda \circ [-1] = \lambda \,, \tag{4.1.8}$$

$$\{\!\{\,\}\!\} \circ \tau = \{\!\{\,\}\!\} \,, \tag{4.1.9}$$

$$[-1] \circ \{\!\{\,\}\!\} = \{\!\{\,\}\!\} \,. \tag{4.1.10}$$

Conversely, if $\{\!\!\{\,\}\!\!\} : M \to \varPhi$ *and* $\lambda : \varPhi \to M$ *are homomorphisms such that the above identities hold, then* $(M, \varPhi, \{\!\!\{\,\}\!\!\}, \lambda)$ *is a quadratic group.* (*In other words,* (4.1.3)–(4.1.10) *imply* (4.1.1) *and* (4.1.2).)

As we mentioned above, a quadratic group is simply a quadratic pair over $\mathbb{Z}$; the $\mathbb{Z}$-actions on M and $\varPhi$ are given by

$$m(j \otimes k) = jkm\,,$$
$$\phi[j] = \frac{j(j+1)}{2}\phi + \frac{j(j-1)}{2}[-1](\phi)\,, \tag{4.1.11}$$

for all $m \in M$, $\phi \in \varPhi$, $j, k \in \mathbb{Z}$. In particular, $\phi[-1] = [-1](\phi)$, as we would expect.

Definition 4.1.3. A *morphism* of quadratic groups is a pair of group homomorphisms $\varphi : M \to M'$, $\chi : \varPhi \to \varPhi'$ such that

$$\varphi(\lambda(\phi)) = \lambda'(\chi(\phi)), \quad \text{for } \phi \in \varPhi,$$
$$\chi(\{\!\!\{ x \}\!\!\}) = \{\!\!\{ \varphi(x) \}\!\!\}', \quad \text{for } x \in M\,.$$

We denote the category of quadratic groups and morphisms by Quad.

Remark. A quadratic morphism is *monic* if and only if φ and χ are both injective, and *epic* if and only if φ and χ are both surjective; indeed, a sequence of morphisms is exact if and only if the φ and χ sequences are exact. In particular, products, direct sums, kernels, images, etc. can all be defined "element-wise", i.e. on M and $\varPhi$ independently. This follows from the proof of the next lemma.

We recall that a *bicomplete* abelian category is an abelian category in which all small limits exist (i.e. the category is complete) and all small colimits exist (i.e. the category is cocomplete)—see e.g. MacLane [355, Chapter III, V]. For example, module categories over rings are bicomplete [355, Theorem V.1.2, Exercise V.1.8]. In plain language this means that products, direct sums, kernels, images, etc. all exist and behave reasonably in a bicomplete abelian category, and in particular in Quad.

Lemma 4.1.4. *The category* Quad *is a bicomplete abelian category.*

Proof. Indeed, Quad is isomorphic to the category of (left) modules over the (noetherian) ring with generators x, y, and relations $x^2 = x$, $y^3 = 2y$, $xy + yx = y$. Here x corresponds to $\mathrm{id}_M + 0$ and y corresponds to $\{\!\!\{\,\}\!\!\} + \lambda$, both acting on $M \times \varPhi$. $\qquad\square$

Remark. As an abelian group, the above ring is free on six generators, corresponding to id_M, $\mathrm{id}_\varPhi$, λ, $\{\!\!\{\,\}\!\!\}$, τ and $[-1]$.

Lemma 4.1.5. *Given a quadratic group $(M, \Phi, \{\!\{\ \}\!\}, \lambda)$, we define its underlying abelian group to be Φ; this defines a functor*

$$U : \mathsf{Quad} \to \mathsf{Ab}, \ U((M, \Phi, \{\!\{\ \}\!\}, \lambda)) := \Phi.$$

The functor $F : \mathsf{Ab} \to \mathsf{Quad}$ given by

$$F(A) = (A, \qquad A \times A, \ x \mapsto (x, x), \ (x, y) \mapsto x + y)$$
$$= (F(A)_M, F(A)_\Phi, \{\!\{\ \}\!\}_{F(A)}, \quad \lambda_{F(A)} \ )$$

is both left adjoint and right adjoint to U. The functor F is a full, faithful, exact embedding of Ab in Quad. We call F the free functor *and U the* underlying group functor.

Proof. We first need to produce a natural isomorphism between $\mathrm{Hom}(A, \Phi)$ and $\mathrm{Hom}(F(A), (M, \Phi))$, where A is an abelian group and (M, Φ) is a quadratic group. If $\varphi \in \mathrm{Hom}(A, \Phi)$, then

$$(x \mapsto \lambda(\varphi(x)), \ (x, y) \mapsto \varphi(x) + [-1]\varphi(y)) \in \mathrm{Hom}(F(A), (M, \Phi));$$

the inverse map is given by $(\eta, \chi) \mapsto (x \mapsto \chi(x, 0))$.

Similarly, we obtain a natural isomorphism between $\mathrm{Hom}(\Phi, A)$ and $\mathrm{Hom}((M, \Phi), F(A))$ by taking $\varphi \in \mathrm{Hom}(\Phi, A)$ to

$$(x \mapsto \varphi(\{\!\{x\}\!\}), \ \phi \mapsto (\varphi(\phi), \varphi([-1]\phi))). \qquad \square$$

One easily calculates that the involutions on $F(A)$ are given by

$$\tau(x) = x \text{ for all } x \in A,$$
$$[-1](x, y) = (y, x) \text{ for all } x, y \in A.$$

Remark. U and F should not be confused with the underlying group and free functors corresponding to the structure of Quad as a module category.

Of equal importance for our purposes is the *squaring functor* Q from Ab to Quad. Given an abelian group A, we set $Q(A)_M = A \otimes_{\mathbb{Z}} A$, and define $Q(A)_\Phi$ to be the abelian group generated by elements $[x]$ for $x \in A$ modulo the relations

$$[0] = 0,$$
$$[x + y + z] + [x] + [y] + [z] = [x + y] + [x + z] + [y + z], \quad \text{for } x, y, z \in A.$$

The maps $\{\!\{\ \}\!\}$ and λ are given by

$$\{\!\{x \otimes y\}\!\} = [x, y] := [x + y] - [x] - [y], \quad \text{for } x, y \in A,$$
$$\lambda([x]) = x \otimes x, \quad \text{for } x \in A.$$

Note that $[x, y] = [y, x]$ and $[x, y + z] = [x, y] + [x, z]$, so $[x, y]$ is a symmetric biadditive form. We observe that

$$\tau(x \otimes y) = y \otimes x, \quad \text{for } x, y \in A \,,$$
$$[-1][x] = [-x], \quad \text{for } x \in A \,;$$

for the latter identity, note that $[x, x] = -[x, -x] = [x] + [-x]$. If A, B are abelian groups and $\varphi \in \text{Hom}(A, B)$, then $Q(\varphi) \in \text{Hom}(Q(A), Q(B))$ is defined by

$$Q(\varphi)(x \otimes y) = \varphi(x) \otimes \varphi(y), \quad \text{for } x, y \in A \,,$$
$$Q(\varphi)([x]) = [\varphi(x)], \quad \text{for } x \in A \,.$$

Remark. Unlike the underlying and free functors, the squaring functor does not preserve the addition of morphisms (indeed, it is quadratic, rather than linear). In particular, it does not preserve products and coproducts, so is neither left nor right exact, and does not admit an adjoint. It does, however, preserve the zero object, as well as epimorphisms, but in general not monomorphisms.

Lemma 4.1.6. *Let A and B be abelian groups. A morphism from $Q(A)$ to $F(B)$ is equivalent to a pointed quadratic map $\phi : A \to B$; that is, a function ϕ such that*

$$\phi(0) = 0 \,,$$
$$\phi(x + y + z) + \phi(x) + \phi(y) + \phi(z) = \phi(x + y) + \phi(x + z) + \phi(y + z) \,,$$

for all $x, y, z \in A$ (see Definition 1.1.1).

Proof. Given a morphism $(\varphi, \chi) : Q(A) \to F(B)$, define a quadratic map $\phi : A \to B$ by

$$\phi(a) = \pi_1(\chi([a])) \,,$$

where π_1 is the projection onto the first component of $F(B)_\Phi = B \times B$. Since

$$\chi([-a]) = \chi([-1][a]) = [-1]\chi([a]) \,,$$

we find that $\pi_2(\chi([a])) = \phi(-a)$, and thus

$$\chi([a]) = (\phi(a), \phi(-a)) \,.$$

Similarly, we have

$$\{\varphi(a \otimes b)\} = \chi([a + b] - [a] - [b]) = (\phi(a, b), \phi(a, b)) = \{\phi(a, b)\} \,,$$

where we recall from (1.1.2) that $\phi(a, b) = \phi(a + b) - \phi(a) - \phi(b)$. Since $\{\}$ is injective for $F(B)$, we conclude that

$$\varphi(a \otimes b) = \phi(a, b) \,.$$

Thus the morphism (φ, χ) is determined by the quadratic map ϕ.

Conversely, if ϕ is any quadratic map from A to B, then

$$(\varphi : a \otimes b \mapsto \phi(a,b), \ \chi : [a] \mapsto (\phi(a), \phi(-a)))$$

is a morphism from $Q(A)$ to $F(B)$. Indeed, we find that

$$\varphi(\lambda([a])) = \varphi(a \otimes a) = \phi(a,a) = \phi(a) + \phi(-a) = \lambda(\chi([a]))$$

and

$$\{\!\{\varphi(a \otimes b)\}\!\} = \chi(\{\!\{a \otimes b\}\!\})$$

as above.   □

4.2 The internal hom-functor IHom

Given a quadratic map $\phi : A \to B$, we obtain a $\mathbb{Z}$-bilinear function $\lambda(\phi)$ by taking

$$\lambda(\phi)(x,y) = \phi(x+y) - \phi(x) - \phi(y) \, ;$$

similarly, given a bilinear function β from $A \times A$ to B, we obtain a pointed quadratic map $\{\!\{\beta\}\!\}$ by taking

$$\{\!\{\beta\}\!\}(x) = \beta(x,x) \, .$$

We thus obtain a quadratic group, with $M = \mathrm{Bil}(A,B)$ given by the group of bilinear functions from $A \times A$ to B, and $\Phi = \mathrm{Quad}_0(A,B)$ given by the group of pointed quadratic maps from A to B.

This construction may be generalized.

Definition 4.2.1. Given two quadratic groups (M, Φ) and (M', Φ'), we define a new quadratic group $\mathrm{IHom}((M, \Phi), (M', \Phi'))$, the *internal hom* of (M, Φ) and (M', Φ'), as follows. We take

$$\mathrm{IHom}((M, \Phi), (M', \Phi')) = (M'', \Phi'') \, ,$$

where $M'' = \mathrm{Hom}(M, M')$ and $\Phi'' = \mathrm{Hom}((M, \Phi), (M', \Phi'))$; the maps λ'' and $\{\!\{\ \}\!\}''$ are given by

$$\lambda''(\varphi, \chi) = \varphi \, ,$$
$$\{\!\{\varphi\}\!\}'' = (x \mapsto \varphi(x) + \tau'(\varphi(\tau(x)))), \ \phi \mapsto \{\!\{\varphi(\lambda(\phi))\}\!\}') \, .$$

We observe that $\mathrm{IHom}(Q(A), F(B))$ is precisely the quadratic group constructed above. This suggests the following definition.

Definition 4.2.2. Let (M, Φ) be a quadratic group and A an abelian group. An (M, Φ)-*valued quadratic form on* A is a homomorphism from $Q(A)$ to (M, Φ).

Recall that a (symmetric monoidal) *closed* category is a category admitting a bifunctor $\square$ which up to coherent isomorphisms is associative and symmetric, i.e. satisfies

$$A \square B \cong B \square A \text{ and } (A \square B) \square C \cong A \square (B \square C),$$

and for which there is an identity E in the category (i.e. $_\_ \square E$ and $E \square _\_$ are equivalent to the identity functor) such that each functor $_\_ \square B$ has a right adjoint (MacLane [355, p. 180]).

Theorem 4.2.3. *For any object* $(M, \Phi) \in$ Quad, *the functor* IHom$((M, \Phi), _\_)$ *admits a left adjoint, denoted by* $_\_ \otimes (M, \Phi)$ *(the* tensor product*). Together these make* Quad *a symmetric monoidal closed category.*

Proof. We define the tensor product $((M, \Phi) \otimes (M', \Phi'))$ to be $(\tilde{M}, \tilde{\Phi})$, where $\tilde{M} = M \otimes M'$ and $\tilde{\Phi}$ is the quotient of $(M \otimes M') \times (\Phi \otimes_{\mathbb{Z}[Z_2]} \Phi')$ by the relations

$$(0, \{\!\{m\}\!\} \otimes \phi') = (m \otimes \lambda'(\phi'), 0), \quad \text{for } m \in M, \phi' \in \Phi',$$
$$(0, \phi \otimes \{\!\{m'\}\!\}') = (\lambda(\phi) \otimes m', 0), \quad \text{for } m' \in M', \phi \in \Phi,$$

and the group Z_2 acts as $[-1]$. The structure maps are given by

$$\{\!\{m \otimes m'\}\!\} = (m \otimes m', 0), \quad \text{for } m \in M, m' \in M',$$
$$\tilde{\lambda}((m \otimes m', 0)) = m \otimes m' + \tau(m) \otimes \tau'(m'), \quad \text{for } m \in M, m' \in M',$$
$$\tilde{\lambda}((0, \phi \otimes \phi')) = \lambda(\phi) \otimes \lambda'(\phi'), \quad \text{for } \phi \in \Phi, \phi' \in \Phi'.$$

We need to verify that this is adjoint to IHom, that the tensor product is commutative and associative (up to coherent isomorphisms) and that

$$I := F(\mathbb{Z}) = (\mathbb{Z}, \mathbb{Z} \times \mathbb{Z} \cong \mathbb{Z}[Z_2])$$

is an identity for the tensor product.

We first observe that to specify a morphism $(M, \Phi) \otimes (M', \Phi') \to (M'', \Phi'')$, where (M'', Φ'') is now an arbitrary form ring, it suffices to give the maps $\varphi : M \otimes M' \to M''$, $\chi : \Phi \otimes_{\mathbb{Z}[Z_2]} \Phi' \to \Phi''$, since then the map $\chi : M \otimes M' \to \Phi''$ is determined by

$$\chi(m \otimes m') = \chi(\{\!\{m \otimes m'\}\!\}) = \{\!\{\varphi(m \otimes m')\}\!\}, \quad \text{for } m \in M, m' \in M'.$$

(Here and in the sequel we omit the superscripts $'$ and $''$ on the structure maps.) Conversely, two such maps (φ, χ) determine a morphism from the tensor product if and only if they satisfy the relations:

$$\varphi(\lambda(\phi) \otimes \lambda(\phi')) = \lambda(\chi(\phi \otimes \phi')),$$
$$\varphi(\tau(m) \otimes \tau(m')) = \tau(\varphi(m \otimes m')),$$
$$\chi(\{\!\{m\}\!\} \otimes \phi') = \{\!\{\varphi(m \otimes \lambda'(\phi'))\}\!\},$$
$$\chi(\phi \otimes \{\!\{m'\}\!\}) = \{\!\{\varphi(\phi \otimes \lambda'(m'))\}\!\};$$

for all $\phi \in \Phi, \phi' \in \Phi', m \in M, m' \in M'$. The first two relations arise from the requirement that φ and χ respect λ, while the last two relations (and the sufficiency of the relations) arise from the fact that

$$((M, \Phi) \otimes (M', \Phi'))_\Phi$$

is defined as a quotient.

Similarly, a morphism $(\varphi, \chi) : (M, \Phi) \rightarrow \mathrm{IHom}((M', \Phi'), (M'', \Phi''))$ is determined by $\varphi(m)(m')$ and $\chi(\phi)(\phi')$, since we must have $\chi(\phi)(m') = \varphi(\lambda(\phi))(m')$. The isomorphism

$$\mathrm{Hom}((M, \Phi) \otimes (M', \Phi'), (M'', \Phi'')) \rightarrow \mathrm{Hom}((M, \Phi), \mathrm{IHom}((M', \Phi'), (M'', \Phi'')))$$

required by adjointness is then induced by the corresponding isomorphisms for M and Φ; that is, we take the pair of morphisms $(\varphi : M \otimes M' \rightarrow M'', \chi : \Phi \otimes_{\mathbb{Z}[Z_2]} \Phi' \rightarrow \Phi'')$ specifying a map on the left to the corresponding morphisms $(\varphi' : M \rightarrow \mathrm{Hom}(M', M''), \chi' : \Phi \rightarrow \mathrm{Hom}_{\mathbb{Z}[Z_2]}(\Phi', \Phi''))$ specifying a map on the right; the identities required for φ and χ to give a map from the tensor product translate directly into the identities required for φ' and χ' to be morphisms.

For commutativity, the required isomorphism

$$(M, \Phi) \otimes (M', \Phi') \cong (M', \Phi') \otimes (M, \Phi)$$

is simply specified by the isomorphisms $M \otimes M' \cong M' \otimes M$, $\Phi \otimes_{\mathbb{Z}[Z_2]} \Phi' \cong \Phi' \otimes_{\mathbb{Z}[Z_2]} \Phi$; similarly we have the isomorphisms $I \otimes (M, \Phi) \cong (M, \Phi)$ and $(M, \Phi) \otimes I \cong (M, \Phi)$. For associativity, the isomorphism

$$(M, \Phi) \otimes ((M', \Phi') \otimes (M'', \Phi'')) \rightarrow ((M, \Phi) \otimes (M', \Phi')) \otimes (M'', \Phi'')$$

follows from the corresponding isomorphisms for the M and Φ groups, once we observe that in the Φ group of $(M, \Phi) \otimes ((M', \Phi') \otimes (M'', \Phi''))$ we have the relation

$$\phi \otimes (m' \otimes m'') = \lambda(\phi) \otimes (m' \otimes m''),$$

and thus the group is naturally isomorphic to a quotient of

$$M \otimes (M' \otimes M'') \times \Phi \otimes (\Phi' \otimes \Phi'') ;$$

and similarly for the other triple tensor product.

The coherence conditions are automatic, since the canonical isomorphisms are directly induced by the corresponding isomorphisms for the usual tensor product. $\square$

Remark. The relation

$$\phi \otimes [-1](\phi') = [-1](\phi) \otimes \phi',$$

implicit in taking $\Phi \otimes_{\mathbb{Z}[Z_2]} \Phi'$ instead of $\Phi \otimes \Phi'$ above, is redundant, since in any case

$$\phi \otimes [-1](\phi') = \phi \otimes \{\lambda(\phi')\} - \phi \otimes \phi' = \lambda(\phi) \otimes \lambda(\phi') - \phi \otimes \phi' = [-1](\phi) \otimes \phi'.$$

Remark 4.2.4. Since the above canonical morphisms are all induced by the canonical morphisms for the usual tensor product, the same is true for all of the other canonical morphisms induced by these morphisms and adjointness; thus, for instance, the "evaluation" map

$$\mathrm{IHom}((M, \Phi), (M', \Phi')) \otimes (M, \Phi) \to (M', \Phi')$$

is specified by the maps

$$\varphi \otimes m \mapsto \varphi(m),$$
$$(\varphi, \chi) \otimes \phi \mapsto \chi(\phi).$$

Lemma 4.2.5. *For any pair of abelian groups* $A, B \in \mathsf{Ab}$, *we have canonical morphisms*

$$\mathrm{IHom}(F(A), F(B)) \cong F(\mathrm{Hom}(A, B)),$$
$$F(A) \otimes F(B) \cong F(A \otimes B),$$
$$\mathrm{IHom}(Q(A), Q(B)) \cong Q(\mathrm{Hom}(A, B)),$$
$$Q(A) \otimes Q(B) \cong Q(A \otimes B).$$

Furthermore $F(\mathbb{Z}) = Q(\mathbb{Z}) = I$. *Thus* F *and* Q *are functors of closed categories.*

Proof. For instance, the canonical morphism (φ, χ) from $Q(A) \otimes Q(B)$ to $Q(A \otimes B)$ is defined by

$$\varphi((x \otimes x') \otimes (y \otimes y')) = (x \otimes y) \otimes (x' \otimes y'), \quad \text{for } x, x' \in A, y, y' \in B,$$
$$\chi([x] \otimes [y]) = [x \otimes y], \quad \text{for } x \in A, y \in B.$$

We find that

$$\varphi(\tau((x \otimes x')) \otimes \tau((y \otimes y'))) = (x' \otimes y') \otimes (x \otimes y)$$
$$= \tau(\varphi((x \otimes x') \otimes (y \otimes y'))),$$
$$\varphi(\lambda([x]) \otimes \lambda([y])) = (x \otimes y) \otimes (x \otimes y)$$
$$= \lambda(\chi([x] \otimes [y])),$$
$$\chi(\{\!\{x \otimes x'\}\!\} \otimes [y]) = [x \otimes y, x' \otimes y]$$
$$= \{\!\{\varphi((x \otimes x') \otimes \lambda([y]))\}\!\},$$
$$\chi([x] \otimes \{\!\{y \otimes y'\}\!\}) = \{\!\{\varphi(\lambda([x]) \otimes (y \otimes y'))\}\!\},$$

and thus (φ, χ) determines a morphism from the tensor product. Moreover, φ and χ are both bijective and thus (φ, χ) is an isomorphism.

As above, the required coherence conditions are automatic. Clearly $F(\mathbb{Z}) = I$ (see proof of Theorem 4.2.3 above). The isomorphism (φ, χ) from $F(\mathbb{Z})$ to $Q(\mathbb{Z})$ is given by

$$\varphi(a) = 1 \otimes a = a(1 \otimes 1) \text{ for } a \in \mathbb{Z},$$
$$\chi(a, b) = a[1] + b[1].$$

We check that

$$\varphi(\lambda(a, b)) = (a + b)(1 \otimes 1),$$
$$\lambda(\chi(a, b)) = a\lambda([1]) + b\lambda([-1])$$
$$= a(1 \otimes 1) + b(-1 \otimes -1)$$
$$= (a + b)(1 \otimes 1),$$

so this is indeed a morphism. The inverse is given by

$$a \otimes b \mapsto ab,$$
$$[a] \mapsto (a(a + 1)/2, a(a - 1)/2)$$

(recall from (1.1.3) that

$$[k] = \frac{k(k + 1)}{2}[1] + \frac{k(k - 1)}{2}[-1]$$

for any integer k).   □

Since Quad is a bicomplete symmetric monoidal closed category, there is an associated notion of a Quad-based category, or Quad-category (cf. MacLane [355, p. 180]): a category in which the Homs are elements of Quad, with identity and composition given by Quad-morphisms

$$I \to \mathrm{Hom}(A, A),$$
$$\mathrm{Hom}(A, B) \otimes \mathrm{Hom}(B, C) \to \mathrm{Hom}(A, C),$$

with the obvious associativity and identity requirements. Since the functors F and Q are functors of closed categories, they induce actions on the corresponding enriched categories; more precisely, F takes a Quad-category to an Ab-category (the "underlying preadditive category") and Q takes an Ab-category to a Quad-category.

We also obtain a notion of a quadratic ring (a Quad-ring, which is a Quad-category with one object), as well as a corresponding notion of bimodule. By tracing through the definitions, we obtain the following more concrete formulation.

Definition 4.2.6. A *quadratic ring* R is a quadruple $R = (R_M, R_\Phi, \{\}, \lambda)$, where R_M and R_Φ are (associative) rings with identity, $\{\} : R_M \to R_\Phi$ and $\lambda : R_\Phi \to R_M$ are ring homomorphisms, and the following identities hold:

$$\{r\}\phi = \{r\lambda(\phi)\}, \quad \text{for } r \in R_M, \phi \in R_\Phi,$$
$$\phi\{r\} = \{\lambda(\phi)r\}, \quad \text{for } r \in R_M, \phi \in R_\Phi.$$

If $R := (R_M, R_\Phi, \{\!\{\}\!\}, \lambda)$ and $S := (S_M, S_\Phi, \{\!\{\}\!\}, \lambda)$ are quadratic rings, an (R, S)-*bimodule* is a quadratic group $(M, \Phi, \{\!\{\}\!\}, \lambda)$ such that M is a (R_M, S_M)-bimodule, Φ is a (R_Φ, S_Φ)-bimodule and the following identities hold:

$$\tau(rms) = \tau(r)\tau(m)\tau(s), \quad r \in R_M, s \in S_M, m \in M,$$
$$\lambda(\phi_R \phi \phi_S) = \lambda(\phi_R)\lambda(\phi)\lambda(\phi_S), \quad \text{for } \phi_R \in R_\Phi, \phi_S \in S_\Phi, \phi \in \Phi,$$
$$\phi_R \{\!\{m\}\!\} \phi_S = \{\!\{\lambda(\phi_R) m \lambda(\phi_S)\}\!\}, \quad \text{for } \phi_R \in R_\Phi, \phi_S \in S_\Phi, m \in M,$$
$$\{\!\{r\}\!\} \phi = \{\!\{r\lambda(\phi)\}\!\}, \quad \text{for } r \in R_M, \phi \in \Phi,$$
$$\phi \{\!\{s\}\!\} = \{\!\{\lambda(\phi)s\}\!\}, \quad \text{for } s \in S_M, \phi \in \Phi.$$

A *morphism of* (R, S)-*bimodules* is a morphism of quadratic groups that consists of a pair of (R_M, S_M)- and (R_Φ, S_Φ)-bimodule morphisms.

Remark. We have omitted the redundant identity discussed above for the tensor product; this simply has the effect of requiring that R_Φ be a $\mathbb{Z}[Z_2]$-algebra with Z_2 acting as $[-1]$ on R_Φ and Φ.

4.3 Properties of quadratic rings

For the basic properties of rings and bimodules over closed categories we refer to Pareigis [403] (where such rings are called *monoids*). We briefly mention the most important facts. The identity I for the tensor product has a canonical quadratic ring structure, and every quadratic group inherits an (I, I)-bimodule structure; there is thus a notion of left- and right-modules, namely (R, I)- and (I, S)-bimodules. For every pair (R, S) of quadratic rings, the (R, S)-bimodules and bimodule morphisms form an abelian category $R - \text{Mod} - S$. The tensor product of quadratic rings is itself a quadratic ring; there are then equivalences

$$R - \text{Mod} - S \cong \text{Mod} - (R^{op} \otimes S) \cong (R \otimes S^{op}) - \text{Mod}.$$

If R, S and T are quadratic rings and $_R(M, \Phi)_S$, $_S(M', \Phi')_T$ are (R, S)- and (S, T)-bimodules, then the (R, T)-bimodule $(M, \Phi) \otimes_S (M', \Phi')$ is defined to be the quotient of $(M, \Phi) \otimes (M', \Phi')$ by the relations

$$ms \otimes m' = m \otimes sm' \quad \text{for } m \in M, m' \in M', s \in S_M, \tag{4.3.1}$$
$$\phi \phi_S \otimes \phi' = \phi \otimes \phi_S \phi' \quad \text{for } \phi \in \Phi, \phi' \in \Phi', \phi_S \in S_\Phi. \tag{4.3.2}$$

Similarly, if (M, Φ) is instead an (S, R)-bimodule, then $\text{IHom}_S((M, \Phi), (M', \Phi'))$ is the (R, T)-bimodule which is the submodule of $\text{IHom}((M, \Phi), (M', \Phi'))$ for which the corresponding maps on M and Φ are S_M- and S_Φ-module morphisms. The various canonical morphisms and isomorphisms between Homs and tensor products of bimodules over ordinary rings then all have direct analogues for bimodules over quadratic rings. In particular, $\text{IHom}_S((M, \Phi), \_)$ is right adjoint to $\_ \otimes_S (M, \Phi)$ and is therefore left exact.

In our case, since kernels and cokernels can be computed componentwise, and the canonical maps for the unadorned IHom and $\otimes$ are induced by the canonical maps for their ordinary analogues, it follows that the various canonical maps for IHom_R and $\otimes_R$ are also directly induced by their counterparts for ordinary modules. This is also the reason why R_M and $R_\varPhi$ are ordinary rings and M and $\varPhi$ are ordinary bimodules in the first place.

Lemma 4.3.1. *Let R and S be (ordinary) rings and let M be an (R, S)-bimodule. Then $F(R)$ and $F(S)$ are quadratic rings, and $F(M)$ is an $(F(R), F(S))$-bimodule; similarly, $Q(R)$ and $Q(S)$ are quadratic rings, and $Q(M)$ is a $(Q(R), Q(S))$-bimodule. If ${}_R M_S$ and ${}_S M'_T$ are bimodules over the ordinary rings R, S, T, then there are canonical morphisms*

$$F(M) \otimes_{F(S)} F(M') \cong F(M \otimes_S M'),$$
$$Q(M) \otimes_{Q(S)} Q(M') \cong Q(M \otimes_S M').$$

Proof. This is immediate from the fact that F and Q are functors of closed categories.   □

The ring structure on $Q(R)$ is given by:

$$(r \otimes r')(s \otimes s') = (rs \otimes r's'),$$
$$[r][s] = [rs], \tag{4.3.3}$$

for all $r, r', s, s' \in R$. A $(Q(R), Q(S))$-bimodule is then a quadratic group $(M, \varPhi, \{\!\{\,\}\!\}, \lambda)$, where M is a $(Q(R)_M, Q(S)_M)$-bimodule, $\varPhi$ is a $(Q(R)_\varPhi, Q(S)_\varPhi)$-bimodule and

$$\tau((r \otimes r')x(s \otimes s')) = (r' \otimes r)\tau(x)(s' \otimes s),$$
$$\lambda([r]\phi[s]) = (r \otimes r)\lambda(\phi)(s \otimes s),$$
$$[r]\{\!\{x\}\!\}[s] = \{\!\{(r \otimes r)x(s \otimes s)\}\!\},$$
$$[r, r']\phi = \{\!\{(r \otimes r')\lambda(\phi)\}\!\},$$
$$\phi[s, s'] = \{\!\{\lambda(\phi)(s \otimes s')\}\!\},$$

for $x \in M$, $\phi \in \varPhi$, $r, r' \in R$, $s, s' \in S$. In particular, we find that a right $Q(R)$-module is simply what we called a quadratic pair over R above.

Definition 4.3.2. Let R be an ordinary ring, let $(M, \varPhi)$ be a left $Q(R)$-module, and let V be a left R-module. An $(M, \varPhi)$-*valued quadratic form* is defined to be a $Q(R)$-linear map from $Q(V)$ to $(M, \varPhi)$. A *morphism* of $(M, \varPhi)$-valued quadratic forms is an R-linear map $V \to W$ making the obvious diagram (shown in Fig. 4.1(a)) commute; a *weak morphism* of quadratic forms is a homomorphism $R \to S$ together with semilinear maps $V \to W$ and $(M, \varPhi) \to (M', \varPhi')$ making the diagram in Fig. 4.1(b) commute. *Isomorphism* and *weak isomorphism* of quadratic forms are defined analogously.

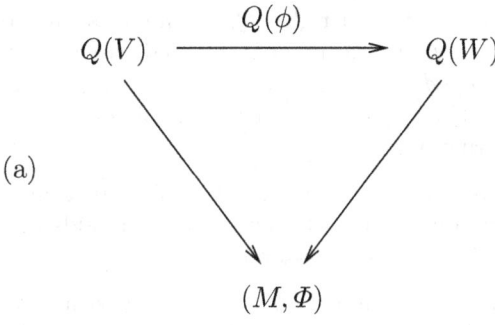

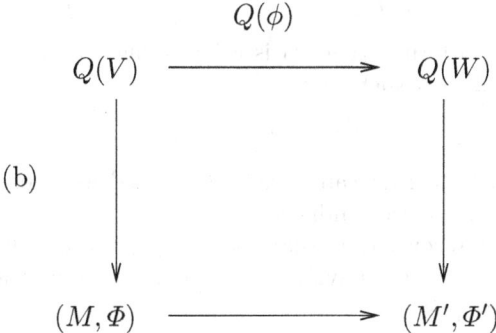

Fig. 4.1. (a) Morphism and (b) weak morphism of quadratic forms.

Remark. In this context we think of "isomorphism" as meaning "isometry" and "weak isomorphism" as meaning "semilinear similarity". The automorphism group is a normal subgroup of the weak automorphism group.

Definition 4.3.3. Let k be a commutative ring equipped with an involution ($x \mapsto \overline{x}$ such that $\overline{xy} = \overline{x}\,\overline{y}$, $\overline{\overline{x}} = x$). Then we define the quadratic ring $Q_k(k)$ to be the quotient of the quadratic ring $Q(k)$ obtained by imposing the additional relations

$$x \otimes y = 1 \otimes \overline{x}y = x\overline{y} \otimes 1,$$
$$[x, y] = [1, \overline{x}y] = [x\overline{y}, 1].$$

We then obtain a *functor* Q_k from the category of k-modules to the category of $Q_k(k)$-modules by defining

$$Q_k(V) = Q(V) \otimes_{Q(k)} Q_k(k),$$

viewing $Q_k(k)$ as a $Q(k)$-module in the obvious way. Since $Q_k(k)$ is commutative (i.e. both $Q_k(k)_M$ and $Q_k(k)_\Phi$ are commutative), there is no need to distinguish between left- and right- $Q_k(k)$-modules; moreover, the category $Q_k(k)$ − Mod is a symmetric monoidal closed category. The corresponding notion of ring, a *quadratic k-algebra*, is simply a quadratic ring R equipped with a morphism $Q_k(k) \to R$; the notion of bimodule is inherited from the quadratic ring structure.

Remark. Just as tensor products linearize bilinear forms, we can view Q as linearizing the notion of quadratic form; Q_k then adds the requirement that the associated bilinear form be k-semilinear.

Definition 4.3.4. Let k be a commutative ring with involution. A *form k-algebra* is a quadruple (R, M, ψ, Φ), where R is a k-algebra, (M, Φ) a right $Q_k(R)$-module and $\psi : R \to M$ is an abelian group isomorphism such that

$$\psi(rs) = \psi(r)(1 \otimes s)\text{for } r, s \in R.$$

Equivalently, a form k-algebra is a form ring (R, M, ψ, Φ) equipped with a morphism $\gamma : k \to R$ such that

$$\gamma(\overline{x}) = \gamma(x)^J \quad \text{for } x \in k.$$

In particular, a form ring can always be viewed as a form algebra over its center (with the involution induced by J).

Of course, we recover the usual notions of quadratic rings and form rings by taking $k = \mathbb{Z}$, with the trivial involution, so nothing is lost by considering algebras in the sequel.

4.4 Morita theory for quadratic rings

Ordinary Morita theory classifies equivalences between (ordinary) module categories. As we have seen in Chapter 3, this has important consequences for duality: Morita equivalences act naturally on codes, preserving duality. To extend this theory to include quadratic forms, we will need an analogue of Morita theory for categories of modules over quadratic rings. In fact there are two such analogues, depending on whether we view the categories as Ab-categories or Quad-categories. In both cases, equivalences are classified by progenerators (finitely generated projective generators), but the notions of "projective" and "generators" differ.

The usual notions of "projective" and "generator" can be defined in terms of the Hom-functor as follows: an R-module M is *projective* if and only if the functor $\text{Hom}(M, \_)$ is exact, and a *generator* if and only if the functor is faithful. In our case we have two Hom-functors, Hom and IHom, and thus two notions.

Definition 4.4.1. Let R be a quadratic ring. An R-module (M, Φ) is Ab-*projective* if the functor $\text{Hom}_R((M, \Phi), \_)$ is exact, and an Ab-*generator* if the functor $\text{Hom}((M, \Phi), \_)$ is faithful. Similarly, an R-module (M, Φ) is Quad-*projective* (a Quad-*generator*) if $\text{IHom}_R((M, \Phi), \_)$ is exact (resp. faithful).

The relevant notions of finite generation (of modules) are the same; see Lam [341, (18.2) and (18.3)].

Before considering these notions in detail, we first introduce a functor that will turn out to control the relation between the Ab- and Quad- notions.

Definition 4.4.2. Let $R = (R_M, R_\Phi, \{\!\{\,\}\!\}, \lambda)$ be a quadratic ring. For an R_M-module M, define an R-module $D(M)$ by $D(M) = (M \oplus M, M)$, with structure maps

$$\{\!\{(m, m')\}\!\} = m + m', \text{ for } m, m' \in M,$$
$$\lambda(m) = (m, m), \text{ for } m \in M,$$
$$r(m, m') = (rm, \tau(r)m'), \text{ for } r \in R_M, m, m' \in M,$$
$$\phi_R m = \lambda(\phi_R)m, \text{ for } \phi_R \in R_\Phi, m \in M.$$

Lemma 4.4.3. Let R be a quadratic ring, M a right R_M-module, and (M', Φ') a right R-module. Then

$$\mathrm{IHom}_R(D(M), (M', \Phi')) \cong D(\mathrm{Hom}_{R_M}(M, M')),$$
$$\mathrm{IHom}_R((M', \Phi'), D(M)) \cong D(\mathrm{Hom}_{R_M}(M', M)).$$

Similarly, if (M', Φ') is a left R-module, then

$$D(M) \otimes_R (M', \Phi') \cong D(M \otimes_{R_M} M').$$

Proof. The map $(\psi, \chi) : D(\mathrm{Hom}_{R_M}(M, M')) \to \mathrm{IHom}_R(D(M), (M', \Phi'))$ is given by

$$\psi(\psi_0, \psi_1)(m, m') = \psi_0(m) + \tau(\psi_1(m')),$$
$$\chi(\psi_0)(m, m') = \psi_0(m) + \tau(\psi_0(m')),$$
$$\chi(\psi_0)(m) = \{\!\{\psi_0(m)\}\!\}.$$

It is straightforward to verify that this is an isomorphism.

Similarly, the isomorphism

$$(\psi, \chi) : D(\mathrm{Hom}_{R_M}(M', M)) \to \mathrm{IHom}_R((M', \Phi'), D(M))$$

is given by

$$\psi(\psi_0, \psi_1)(m) = (\psi_0(m), \psi_1(m)),$$
$$\chi(\psi_0)(m) = (\psi_0(m), \psi_0(\tau(m))),$$
$$\chi(\psi_0)(\phi) = \psi_0(\lambda(\phi));$$

while the isomorphism $(\psi, \chi) : D(M) \otimes_R D(M') \to D(M \otimes_{R_M} M')$ is given by

$$\psi((m_1, m_2) \otimes (m_1', m_2')) = (m_1 \otimes m_1', m_2 \otimes m_2'),$$
$$\chi(m \otimes \phi') = m \otimes \lambda(\phi').$$

□

Theorem 4.4.4. *Let R be a quadratic ring. An R-module (M,Φ) is* Quad-*projective, respectively a* Quad-*generator, if and only if the R-module $(M,\Phi)\oplus D(M)$ is* Ab-*projective, respectively an* Ab-*generator.*

Proof. Let (ψ,χ) be an R-module morphism. The morphism $\mathrm{IHom}_R((M,\Phi), (\psi,\chi))$ is surjective, respectively nonzero, if and only if the same is true for the two abelian group morphisms

$$\mathrm{Hom}_R((M,\Phi),(\psi,\chi)) \text{ and } \mathrm{Hom}_{R_M}(M,\psi).$$

But $\mathrm{Hom}_{R_M}(M,\psi)$ is surjective or nonzero if and only if the same is true for

$$\mathrm{Hom}_R(D(M),(\psi,\chi)).$$

The theorem follows.   □

Theorem 4.4.5. *Let R be a quadratic ring. Every R-module (M,Φ) is a surjective image of a direct sum*

$$\bigoplus_I (R_M, R_\Phi) \oplus \bigoplus_{I'} D(R_M),$$

for some sets I, I', which can be taken to be finite if (M,Φ) is finitely generated. The R-module (M,Φ) is Ab-*projective if and only if it is a direct summand of such a module, and an* Ab-*generator if and only if both (R_M, R_Φ) and $D(R_M)$ are surjective images of a direct sum $\bigoplus_I (M,\Phi)$.*

Proof. Observe that (R_M, R_Φ) is the free R-module on one generator in Φ, while $D(R_M)$ is the free R-module on one generator in M. Thus in the first claim, we need simply take $I \subseteq M$, $I' \subseteq \Phi$ to be generating sets, and the other claims follow similarly.   □

Corollary 4.4.6. *An R-module (M,Φ) is* Quad-*projective if and only if it is* Ab-*projective.*

Proof. Indeed, both notions are preserved under direct sums and taking direct summands; moreover, both (R_M, R_Φ) and $D(R_M)$ are Quad-projective.   □

In particular, the module $(R_M, R_\Phi) \oplus D(R_M)$ is an Ab-progenerator, and thus induces a faithful, exact Ab-functor from $R - \mathsf{Mod}$ to $\widehat{R} - \mathsf{Mod}$, where $\widehat{R}$ is the *ordinary* ring

$$\mathrm{End}_R((R_M, R_\Phi) \oplus D(R_M)).$$

In fact, this functor is an equivalence, and thus $R - \mathsf{Mod}$ can be viewed as a module category, and ordinary Morita theory applies. The ring $\widehat{R}$ can be viewed as the ring generated by $R_M \oplus R_\Phi$, λ and $\{\!\{\ \}\!\}$, with the necessary relations. The details are omitted, as this version of Morita theory is of little interest for present purposes.

Morita theory for Quad-categories of modules is a straightforward analogue of ordinary Morita theory. An equivalence of Quad-categories preserves Quad-progenerators; since the module (R_M, R_Φ) is a Quad-progenerator, an equivalence of module Quad-categories determines Quad-progenerators in each category, and thus has the form

$$\mathrm{IHom}_R((M, \Phi), \_) : R - \mathsf{Mod} \to \mathrm{IEnd}_R((M, \Phi)) - \mathsf{Mod},$$

for some Quad-progenerator (M, Φ) in (R_M, R_Φ).

For the converse, we will need the following lemma. Given an R-module (M, Φ), define the dual R-module $(M, \Phi)^* := \mathrm{IHom}_R((M, \Phi), R)$.

Lemma 4.4.7. *Let R be a quadratic ring, and let (P_M, P_Φ) be a finitely generated projective left R-module. Then the canonical map $(P_M, P_\Phi) \to (P_M, P_\Phi)^{**}$ is an isomorphism. For any other left R-module (M, Φ), the canonical maps*

$$(M, \Phi)^* \otimes_R (P_M, P_\Phi) \to \mathrm{IHom}_R((M, \Phi), (P_M, P_\Phi)),$$
$$(P_M, P_\Phi)^* \otimes_R (M, \Phi) \to \mathrm{IHom}_R((P_M, P_\Phi), (M, \Phi)),$$

are isomorphisms.

Proof. Each statement is preserved by finite direct sums and the taking of direct summands, so it only remains to verify them when $(P_M, P_\Phi) = (R_M, R_\Phi)$ (trivially true) and when $(P_M, P_\Phi) = D(R_M)$. The latter case follows from Lemma 4.4.3, recalling that the above canonical maps are induced by the corresponding canonical maps over Ab.   □

Corollary 4.4.8. *Let k be a commutative ring with involution, and let R be a k-algebra. If V is a finitely generated projective R-module, then the canonical map*

$$Q_k(V^*) \to Q_k(V)^*$$

is an isomorphism. If (M, Φ) is an arbitrary $Q_k(R)$-module, then

$$\mathrm{IHom}_{Q_k(R)}(Q_k(V), (M, \Phi)) \cong Q_k(V^*) \otimes (M, \Phi).$$

In particular,

$$\mathrm{IEnd}_{Q_k(R)}(Q_k(V)) \cong Q_k(\mathrm{End}_R(V)).$$

Proof. As above, the first claim is true for $V = R$, and is preserved under finite direct sums and the taking of direct summands, so is true for all finitely generated projective modules. The remaining claims are then immediate.   □

In particular, if (P_M, P_Φ) is a finitely generated projective module, then

$$(P_M, P_\Phi)^* \otimes_R (P_M, P_\Phi) \cong \mathrm{IEnd}_R((P_M, P_\Phi)),$$

and so, setting $S := \mathrm{IEnd}_R((P_M, P_\Phi))$, we have natural isomorphisms of functors:

$$\mathrm{IHom}_R((P_M, P_\Phi), \mathrm{IHom}_S((P_M, P_\Phi)^*, \text{-}))$$
$$\cong \mathrm{IHom}_S((P_M, P_\Phi)^* \otimes_R (P_M, P_\Phi), \text{-}) \cong \mathrm{Id} \ . \qquad (4.4.1)$$

If (P_M, P_Φ) is a Quad-progenerator in $R - \mathrm{Mod}$, then $(P_M, P_\Phi)^*$ is a Quad-progenerator in $S - \mathrm{Mod}$ (see e.g. Lam [341, Prop. 18.22]), and thus $\mathrm{IHom}_R((P_M, P_\Phi), \text{-})$ and $\mathrm{IHom}_S((P_M, P_\Phi)^*, \text{-})$ are inverse functors.

In particular, we have the following special case.

Theorem 4.4.9. *Let k be a commutative ring with involution, and let R be a k-algebra. If $_RV$ is a progenerator in $R - \mathrm{Mod}$, then $Q_k(V)$ is a Quad-progenerator in $Q_k(R) - \mathrm{Mod}$. Thus if the k-algebras R and S are Morita equivalent, then the $Q_k(k)$-algebras $Q_k(R)$ and $Q_k(S)$ are Quad-Morita equivalent.*

4.5 Morita theory for form rings

For form rings, Morita theory has the following consequences.

Theorem 4.5.1. *Let (R, M, ψ, Φ) be a form ring. Let V be a finitely generated projective module in $\mathrm{Mod} - R$, and let $\beta : V \otimes V \to M$ be a nonsingular $R \otimes R$-linear form. Then $S := \mathrm{End}_R(V)$ inherits a form ring structure with quadratic pair $\mathrm{IHom}_{Q(R)}(Q(V), (M, \Phi))$ and a structure map ψ such that*

$$\psi(1) = \beta \ .$$

Note that the underlying twisted ring is precisely the twisted ring $\mathrm{End}_R(V; \beta)$ defined above; we therefore also denote this form algebra by $\mathrm{End}_R(V; \beta)$. When V is a progenerator, this construction is invertible, taking the inner product on V^* to be the composition

$$V^* \otimes V^* \xrightarrow{\psi(1) \otimes 1} M \otimes_{R \otimes R} (V^* \otimes V^*) \to \mathrm{Hom}_{R \otimes R}(V \otimes V, M) \ .$$

Theorem 4.5.2. *Take (R, M, ψ, Φ) as above. If ρ is a finite representation of (R, M, ψ, Φ) on the left R-module V_ρ, then the composition*

$$\mathrm{IHom}_R(Q(V), (M, \Phi)) \xrightarrow{\varphi} \mathrm{IHom}_R(Q(V), \mathrm{IHom}(Q(V_\rho), F(\mathbb{Q}/\mathbb{Z})))$$
$$\cong \mathrm{IHom}(Q(V \otimes_R V_\rho), F(\mathbb{Q}/\mathbb{Z})) \ ,$$

where $\varphi = \mathrm{IHom}_{Q(R)}(Q(V), \rho)$, defines a finite representation of $\mathrm{End}_R(V; \beta)$ on $V \otimes_R V_\rho$. If a code C is isotropic or self-dual, then the same is true of $V \otimes_R C$. If V is a progenerator, then all finite representations of $\mathrm{End}_R(V; \beta)$ are of this form, and a code C is isotropic or self-dual if and only if the same is true for $V \otimes_R C$.

Proof. That this gives a representation of the quadratic pair $\mathrm{IHom}_R(Q(V),$ $(M, \Phi))$ is straightforward; that it gives a finite representation of the form ring follows from the corresponding statement for twisted rings. The statement for codes is also straightforward. When V is a progenerator, the construction is invertible, so the remaining claims follow. $\square$

Example 4.5.3. Let ι be a symmetric idempotent in R, with corresponding elements $u_\iota \in \iota R \iota^J$, $v_\iota \in \iota^J R \iota$. Then v_ι induces a nonsingular form on $R\iota$ via

$$\beta(r \otimes s) = \psi(v_\iota)(r \otimes s),$$

and we obtain a form ring structure on $\iota R \iota$.

Of particular interest are the possible form structures on matrix rings $\mathrm{Mat}_n(R)$. If (R, M, ψ, Φ) is a form ring, then one can define a form ring

$$\mathrm{Mat}_n(R, M, \psi, \Phi) := \mathrm{End}_R(R^n; \beta) = (\mathrm{Mat}_n(R), \mathrm{Mat}_n(M), \mathrm{Mat}_n(\psi), \Phi_n)$$

as in §1.10. In certain situations, this is essentially the only way to give $\mathrm{Mat}_n(R)$ the structure of a form ring:

Theorem 4.5.4. *Let R_0 be a semiperfect ring such that 0 and 1 are the only idempotents in R_0. Then all form rings with underlying ring $R := \mathrm{Mat}_n(R_0)$ are isomorphic to a form ring $\mathrm{Mat}_n(R_0, M_0, \psi_0, \Phi_0)$ for some form structure (M_0, Φ_0) on R_0.*

Proof. Let ι be a primitive idempotent in R. Then ι reduces to a primitive idempotent in $R/\mathrm{rad}\, R$, and thus $\iota R \iota \cong R_0$. In particular, ι is a basic idempotent, and thus ιR is a progenerator for R. Now, all primitive idempotents in $R/\mathrm{rad}\, R$ are equivalent, and thus all primitive idempotents in R are equivalent (see Zassenhaus [565]); in particular, it follows that ι is symmetric. So R is Morita-equivalent to a form ring $(R_0, M_0, \psi_0, \Phi_0)$. In other words, R is isomorphic to $\mathrm{End}_{R_0}(R_0^n; \beta)$ for some nonsingular M_0-valued form β. Changing β simply rescales R; in particular, up to rescaling, we may take $\beta(x, y) = \sum_{i=1}^n \psi_0(1)(x_i, y_i)$. In other words, up to rescaling, $R \cong \mathrm{Mat}_n(R_0, \Phi_0)$. $\square$

4.6 Witt rings, groups and modules

In this section, k is a fixed commutative ring with involution denoted by $^-$.

Definition 4.6.1. A *projective representation* of the quadratic ring $Q_k(k)$ is a pair (V, β), where V is a finitely generated projective k-module and β is a morphism $\beta : Q_k(V) \to Q_k(k)$ such that the induced k-valued bilinear form on V is nonsingular. A projective representation is *metabolic* if there is a self-dual, isotropic direct summand $C \subset V$.

We define the direct sum of two projective representations just as we did for finite representations. This induces a commutative semigroup structure on the set of isomorphism classes of projective representations; the metabolic representations form a subsemigroup. The quotient is a group, with inverse given by the conjugate representation.

Definition 4.6.2. The *projective Witt ring* $W_\pi(Q_k(k))$ is the above group, with multiplication defined by $(V, \beta)(W, \beta') = (V \otimes W, \beta \otimes \beta')$, where $\beta \otimes \beta'$ is defined by:

$$(\beta \otimes \beta')((v \otimes w) \otimes (v' \otimes w')) = \beta(v \otimes v')\beta'(w \otimes w'),$$
$$(\beta \otimes \beta')([v] \otimes [w]) = \beta([v])\beta'([w]).$$

Theorem 4.6.3. *The Witt ring is well-defined.*

Proof. That the given product takes projective representations to projective representations is straightforward; it remains to show that if either representation is metabolic, then the product is as well. Indeed, if $C \subset V$ is a self-dual, isotropic, direct summand, then so is $C \otimes_k W$.   □

Example 4.6.4. Let p be an odd prime, and let $\mathbb{Z}_p$ be the ring of p-adic integers. A projective representation of $Q_{\mathbb{Z}_p}(\mathbb{Z}_p)$ corresponds to a triple $(V \cong \mathbb{Z}_p^n, \beta, v_0 \in V)$, where β is a nonsingular $\mathbb{Z}_p$-valued symmetric bilinear form; an isotropic self-dual direct summand is a direct summand $C \subset V$ containing v_0, self-dual with respect to V. In particular, a metabolic representation has $\beta(v_0) = 0$, so $\beta(v_0)$ gives a homomorphism from $W_{\mathbb{Z}_p}(\mathbb{Z}_p)$ to $\mathbb{Z}_p$. Conversely, if $\beta(v_0) = 0$, then $v_0 = p^k w$ for some $k \in \mathbb{N}$, $w \in V$ such that w generates an isotropic direct summand; thus the kernel of the homomorphism reduces to the case $v_0 = 0$, and is therefore the projective Witt group of $\mathbb{Z}_p$ (cf. W. Scharlau [475]). Thus

$$W_\pi(Q_{\mathbb{Z}_p}(\mathbb{Z}_p)) \cong \begin{cases} \mathbb{Z}/2\mathbb{Z} \times \mathbb{Z}/2\mathbb{Z} \times \mathbb{Z}_p, & p \equiv 1 \pmod 4, \\ \mathbb{Z}/4\mathbb{Z} \times \mathbb{Z}_p, & p \equiv 3 \pmod 4. \end{cases}$$

Example 4.6.5. Similarly, taking $\mathbb{Z}_2$ to be the ring of 2-adic integers, a representation of $Q_{\mathbb{Z}_2}(\mathbb{Z}_2)$ corresponds to a triple $(V \cong \mathbb{Z}_2^n, \beta, v_0 \in V)$, where β is a nonsingular $\mathbb{Z}_2$-valued symmetric bilinear form, and v_0 satisfies

$$\beta(v_0, w) \equiv \beta(w, w) \pmod 2, \text{ for } w \in V.$$

We find that

$$W_\pi(Q_{\mathbb{Z}_2}(\mathbb{Z}_2)) \cong \mathbb{Z}/2\mathbb{Z} \times \mathbb{Z}_2.$$

Indeed, $\beta(v_0, v_0) \pmod 8$ induces the standard map (W. Scharlau [475]) from $W_q(\mathbb{Z}_2)$ to $\mathbb{Z}/8\mathbb{Z}$, so the kernel of the homomorphism $\beta(v_0, v_0)$ from $W_\pi(Q_{\mathbb{Z}_2}(\mathbb{Z}_2)) \to \mathbb{Z}_2$ precisely corresponds to the $\mathbb{Z}/2\mathbb{Z}$ factor of $W_q(\mathbb{Z}_2)$.

There is a similar notion for finite representations of form rings.

Definition 4.6.6. Let (R, M, ψ, Φ) be a form k-algebra over the commutative ring k. A k-representation $\rho := (V, \rho_M, \rho_\Phi, \beta)$ of (R, M, ψ, Φ) is *Witt-null* if it contains an isotropic self-dual code.

Lemma 4.6.7. *If a representation ρ is Witt-null then so is any quotient of ρ.*

Proof. Let C_0 be an isotropic self-dual code in ρ, and let C be any isotropic code. Then the code

$$C + (C_0 \cap C^\perp)$$

is self-dual and isotropic. Indeed, it is a sum of orthogonal isotropic submodules, and its dual is

$$(C + (C_0 \cap C^\perp))^\perp = C^\perp \cap (C_0 + C) = (C^\perp \cap C_0) + C \,. \qquad \square$$

Definition 4.6.8. Let (R, M, ψ, Φ) be a form k-algebra. Two k-representations of (R, M, ψ, Φ) are *Witt-equivalent* if they admit isomorphic quotients.

Thus a representation is Witt-null if and only if it is Witt-equivalent to the trivial (zero) representation.

Lemma 4.6.9. *Two representations ρ and ρ' are Witt-equivalent if and only if the representation $\rho \oplus \bar{\rho}'$ is Witt-null.*

Proof. Suppose the representations are Witt-equivalent; let $\gamma : \rho/C \to \rho'/C'$ be an equivalence. Then the code

$$C_\gamma = \{(v, w) \in V_\rho \oplus V'_\rho \mid \rho(v + C) = w + C'\}$$

is readily verified to be isotropic and self-dual with respect to $\rho \oplus \bar{\rho}'$. Conversely, an isotropic self-dual code in $\rho \oplus \bar{\rho}'$ is necessarily of this form. $\square$

In particular, the representation $\rho \oplus \bar{\rho}$ is Witt-null.

Theorem 4.6.10. *Witt-equivalence is an equivalence relation.*

Proof. Let $\rho_1 \sim \rho_2$ and $\rho_2 \sim \rho_3$. Then the representation

$$(\rho_1 \oplus \overline{\rho_2}) \oplus (\rho_2 \oplus \overline{\rho_3}) \,,$$

as a sum of Witt-null representations, is Witt-null. On the other hand, it is equivalent to the representation

$$(\rho_1 \oplus \overline{\rho_3}) \oplus (\rho_2 \oplus \overline{\rho_2}) \,.$$

Thus $\rho_1 \oplus \overline{\rho_3}$ is a quotient of a Witt-null representation, and is Witt-null as required. $\square$

Definition 4.6.11. The *Witt group* $W(R, M, \psi, \Phi)$ of a form k-algebra (R, M, ψ, Φ) is the abelian group formed by the Witt-equivalence classes of k-representations of (R, M, ψ, Φ), with addition given by direct sums.

An element of the Witt group is canonically determined (up to ordinary equivalence) by its quotient by a maximal isotropic subcode. In other words, the Witt group is in one-to-one correspondence with the set of equivalence classes of *anisotropic* representations, that is, representations containing no nontrivial isotropic code.

Example 4.6.12. We calculate the Witt group of the form ring $R(2_{II}) = (\mathbb{F}_2, \mathbb{F}_2, \mathrm{id}, \mathbb{Z}/4\mathbb{Z} = \langle \phi_0 \rangle)$ (see Example 1.8.2, p. 16; §2.3.1, p. 41). There are two natural representations $\rho(2_{II})$ and $\rho(2_{II})'$ of $R(2_{II})$, defined by

$$\rho(2_{II}) := (\mathbb{F}_2, \rho_M, \rho_\Phi, \beta),$$

$$\rho_M(1)(x, y) = \frac{1}{2}xy,$$

$$\rho_\Phi(\phi_0)(x) = \frac{1}{4}x^2,$$

and

$$\rho(2_{II})' := (\mathbb{F}_4, \rho'_M, \rho'_\Phi, \beta'),$$

$$\rho'_M(1)(x, y) = \frac{1}{2}\mathrm{Tr}(x\bar{y}),$$

$$\rho'_\Phi(\phi_0)(x) = \frac{1}{2}x\bar{x}.$$

The codes of Type $\rho(2_{II})$ are precisely the doubly-even binary self-dual codes, while the codes of Type $\rho(2_{II})'$ are precisely the even additive trace-Hermitian self-dual codes over $\mathbb{F}_4$ (see family q_{II}^{H+}, p. 50). Note that $\rho(2_{II})'$ is Witt-equivalent to $4\rho(2_{II})$; in fact, $\rho(2_{II})' \cong 4\rho(2_{II})/C$, where $C = \langle 1111 \rangle$. (The point is that $4\rho(2_{II})$ contains the one-dimensional isotropic code $C = \{0000, 1111\}$, for which $C^\perp/C$ contains four elements, namely the cosets $s_0 = C, s_1 = \{1100, 0011\}, s_2 = \{1010, 0101\}, s_3 = \{1001, 0110\}$, so that $S := \{s_0, s_1, s_2, s_3\}$ has the same structure as $\mathbb{F}_4$. More precisely, S with $\rho_M(1)(s_i, s_j) = \frac{1}{2}(i \neq j)$, $\rho_\Phi(\phi_0)(s_i) = \frac{1}{2}(i \neq 0)$ has the same structure as $\mathbb{F}_4$ with $\rho'_M(1)(x, y) = \frac{1}{2}(x \neq y)$, $\rho'_\Phi(\phi_0)(x) = \frac{1}{2}(x \neq 0)$.)

We claim that $W(R(2_{II})) \cong \mathbb{Z}/8\mathbb{Z}$ and is generated by $\rho(2_{II})$.

Proof. Since there exists a doubly-even binary self-dual code of length 8, $\rho(2_{II})$ has order dividing 8 in the Witt group; since $\rho(2_{II})'$ is not Witt-null, $\rho(2_{II})$ has order exactly 8. Thus it remains only to show that $\rho(2_{II})$ generates the Witt group.

Any representation $\rho = (V, \rho_M, \rho_\Phi, \beta)$ of $R(2_{II})$ is given by a $\frac{1}{4}\mathbb{Z}/\mathbb{Z}$-valued quadratic form $\rho_\Phi(\phi_0)$ (since λ is surjective). If there is a $v \in V$ such that $\rho_\Phi(\phi_0)(v) \in \{\pm\frac{1}{4}\}$ then

$$V = \langle v \rangle \oplus \langle v \rangle^\perp,$$

and the representation splits accordingly. The representation on $\langle v \rangle$ is either $\rho(2_{II})$ or $\bar{\rho}(2_{II})$. Thus it remains to consider $\frac{1}{2}\mathbb{Z}/\mathbb{Z}$-valued quadratic forms.

Since β is nonsingular, V must be even-dimensional, and thus by standard results on quadratic forms, ρ is a direct sum of 2-dimensional representations. The unique anisotropic 2-dimensional representation is $\rho(2_{\mathrm{II}})'$ (cf. Kitaoka [312, Prop. 1.3.1]). □

Example 4.6.13. The following are some further examples of Witt ring computations. (In general one can find the Witt group by following the proof of Theorem 4.6.15 below.)

$R(q^{\mathrm{E}}), q$ odd: A representation of q^{E} is specified precisely by a nonsingular $\mathbb{F}_q$-valued quadratic form on a vector space $\mathbb{F}_q^k$. Thus an anisotropic representation corresponds to an anisotropic form; since there are four equivalence classes of such forms, $|W(R(q^{\mathrm{E}}))| = 4$. The group structure is $Z_2 \times Z_2$ when $q \equiv 1 \pmod 4$ and Z_4 when $q \equiv 3 \pmod 4$.

$R(q_1^{\mathrm{E}}), q$ odd: An anisotropic representation of $R(q_1^{\mathrm{E}})$ restricts to a representation of $R(q^{\mathrm{E}})$, and any representation of $R(q^{\mathrm{E}})$ lifts to a representation of $R(q_1^{\mathrm{E}})$ trivial on $\ker(\lambda)$. Thus $W(R(q_1^{\mathrm{E}}))$ is a split extension of $W(R(q^{\mathrm{E}}))$; i.e. $W(R(q^{\mathrm{E}}))$ is a direct summand of $W(R(q_1^{\mathrm{E}}))$. There is also (as in the proof of Theorem 4.6.15) a set of natural characters of $W(R(q_1^{\mathrm{E}}))$ given by $\beta(v, \alpha v)$, where v is such that $\beta(v, w) = \psi((0, 1))(w)$. A Witt-equivalence class is in the kernel of this collection of characters if and only if it is induced from $W(R(q^{\mathrm{E}}))$; we conclude that $W(R(q_1^{\mathrm{E}})) = W(R(q^{\mathrm{E}})) \oplus \mathbb{F}_q$.

$R(q^{\mathrm{H}})$: A representation of q^{H} is given by a nonsingular Hermitian form over $\mathbb{F}_q$. Any such form of dimension ≥ 2 is isotropic, and all one-dimensional forms are equivalent; we conclude that there are precisely two equivalence classes of anisotropic representations, and $W(q^{\mathrm{H}}) = Z_2$.

$R(m_{\mathrm{II}}^{\mathbb{Z}}), m = 2^k, k \geq 2$: If $k > 2$, the image of multiplication by 2^{k-1} is always isotropic, and thus any anisotropic representation is induced from a representation of $4_{\mathrm{II}}^{\mathbb{Z}}$. In that case, the image of multiplication by 2 need not be isotropic, but in any case has an isotropic subspace of codimension 1. In other words, there are two kinds of anisotropic representations of $R(m_{\mathrm{II}}^{\mathbb{Z}})$, those with an element of order 4, and those without. The latter are representations of $R(2_{\mathrm{II}})$, and thus there are eight such representations, forming a subgroup Z_8. Given an anisotropic representation with an element of order 4, the direct sum with the representation $4^{\mathbb{Z}}$ has a quotient without elements of order 4; we therefore conclude that $W(R(m_{\mathrm{II}}^{\mathbb{Z}})) = W(R(4^{\mathbb{Z}})) = Z_2 \times Z_8$.

$R(m_{\mathrm{I}}^{\mathbb{Z}}), m = 2^k$: If $k \geq 3$, we again find that anisotropic representations are representations of $R(4_{\mathrm{II}}^{\mathbb{Z}})$, and thus $W(R(m_{\mathrm{I}}^{\mathbb{Z}})) = Z_2 \times Z_8$. If $k = 2$, anisotropic representations are representations of $R(2_{\mathrm{II}})$, and thus $W(R(4^{\mathbb{Z}})) = W(R(2_{\mathrm{II}}))$. Finally, if $k = 1$, we obtain a quotient of $W(R(2_{\mathrm{II}}))$ (taking each anisotropic representation of $R(2_{\mathrm{II}})$ to the corresponding representation of $R(2_{\mathrm{I}})$), and thus find $W(R(2_{\mathrm{I}})) = Z_2$.

$R(m^{\mathbb{Z}}), m = p^k, p$ odd: If $k > 1$, the image of multiplication by p^{k-1} is always an isotropic vector, and thus any anisotropic representation is induced from a representation for the case $k = 1$. Thus $W(R((p^k)^{\mathbb{Z}})) = W(R(p^{\mathbb{Z}})) = W(R(p^{\mathrm{E}}))$ is either $Z_2 \times Z_2$ or Z_4 as appropriate.

Theorem 4.6.14. *If (R, M, ψ, Φ) is a form k-algebra, then $W(R, M, \psi, \Phi)$ has a natural $W_\pi(Q_k(k))$-module structure given by the tensor product defined in Theorem 4.5.2.*

Proof. Let $\rho := (V, \rho_M, \rho_\Phi, \beta)$ be a k-representation of the form k-algebra R. Then by Theorem 3.3.2, ρ defines a unique nonsingular, $R \otimes R$-linear, $\mathrm{Hom}_k(M, U^0(k))$-valued Hermitian form β_ρ on V. For $(W, \beta') \in W_\pi(Q_k(k))$ the tensor product $(V \otimes_k W, \beta_\rho \otimes \beta')$ is defined as in Definition 4.6.2 using the fact that $\mathrm{Hom}_k(M, U^0(k))$ is a $Q_k(k)$-module. Hence $\beta_\rho \otimes \beta'$ is a nonsingular (since W is projective) $R \otimes R$-linear, $\mathrm{Hom}_k(M, U^0(k))$-valued Hermitian form on the R-module $V \otimes_k W$ and therefore (by Theorem 3.3.2) defines a k-representation of the form k-algebra R. This gives a $W_\pi(Q_k(k))$-module structure on the Witt group, since by Theorem 4.5.2, an isotropic self-dual code for ρ induces an isotropic self-dual code for the product. Similarly, an isotropic self-dual direct summand of W induces an isotropic self-dual code. $\qquad\square$

Remark. When k is a finite field, the Witt module of the form k-algebra $(k, Q_k(k))$ is isomorphic to the projective Witt ring of $Q_k(k)$, viewed as a module over itself.

Theorem 4.6.15. *Let (R, M, ψ, Φ) be a finite form ring. Then $W(R, M, \psi, \Phi)$ is a finite group.*

Proof. For a finite form ring (R, M, ψ, Φ), $\ker \lambda$ has a natural (linear) right R-module structure, which can be turned into a left R-module structure $\ker \lambda^J$ via

$$r\phi_0 := \phi_0[r^J].$$

Then a representation ρ induces a homomorphism $\rho^* : \ker \lambda^J \to V_\rho$; here $\phi_0 \in \ker \lambda$ is taken to the unique vector $\rho^*(\phi_0)$ such that

$$\rho_M(\psi(1))(v, \rho^*(\phi_0)) = \rho_\Phi(\phi_0)(v),$$

which exists since $\rho_M(\psi(1))$ is nonsingular and $\rho_\Phi(\phi_0)$ is linear.

Now, for an arbitrary finite form ring, we define an abelian group homomorphism

$$\kappa : W(R, M, \psi, \Phi) \to \mathrm{Hom}((M, \Phi) \otimes_{Q(R)} Q(\ker \lambda^J), F(\mathbb{Q}/\mathbb{Z}))$$

as follows:

$$\kappa(\rho)(m \otimes (v \otimes w))) = \rho_M(m)(\rho^*(v), \rho^*(w)) \qquad (4.6.1)$$
$$\kappa(\rho)(\phi \otimes [v]) = (\rho_\Phi(\phi)(\rho^*(v)), \rho_\Phi(\phi)(-\rho^*(v))). \qquad (4.6.2)$$

It is straightforward to verify that this gives a Quad-morphism and that κ is linear under direct sums and is trivial for Witt-null representations (since then $\rho^*(\ker \lambda^J)$ must be isotropic). As the codomain is finite, it will suffice to show that the kernel is also finite. Thus we may restrict our attention to representations for which $\ker \lambda$ induces an isotropic code; taking quotients, we may assume that λ is injective.

Suppose $\operatorname{rad} R$ is nontrivial. The radical of a finite ring is nilpotent, so for some minimal integer $n \geq 2$, $(\operatorname{rad} R)^n = \{0\}$. If we set $I = (\operatorname{rad} R)^{\lceil n/2 \rceil}$, then

$$M(I \otimes I) \cong I^J RI = R(\operatorname{rad} R)^{2\lceil n/2 \rceil} = 0 \,.$$

Since λ is injective, it follows also that $\Phi[I] = 0$ and $\Phi[I; R] \subset \{\psi(I)\}$. In particular, if ρ is a finite representation, for any vector $v \in V_\rho$, the code Iv is isotropic, and therefore an anisotropic representation is annihilated by I. It follows that the Witt groups of (R, M, ψ, Φ) and $(R, M, \psi, \Phi)/(I, \{\psi(I)\})$ must be isomorphic.

By induction (necessary since injectivity of λ is not preserved under quotients), we therefore reduce to the case that (R, M, ψ, Φ) is semisimple (that is, $\operatorname{rad} R = 0$, $\ker \lambda = 0$). The Witt group of a direct sum of form rings is clearly the direct sum of the corresponding Witt groups, so in fact it suffices to consider the simple case. Similarly, Morita equivalence preserves Witt groups and thus R can be taken to be either a finite field or a ring $k \times k$ with $(x, y)^J = (y, x)$. The computation of Witt groups in these cases is classical (see for instance W. Scharlau [475, §2.3] for the field case); in each case the resulting group is finite—see also Theorem 5.4.13. $\qquad\square$

Remark. The homomorphism κ will not be surjective, in general, since $\kappa(\rho)$ must respect the natural isomorphism

$$M \otimes_{R \otimes R} (\ker \lambda^J \otimes \ker \lambda^J) \cong \ker \lambda \otimes_R \ker \lambda^J \cong \ker \lambda \otimes_{Q(R)_\Phi} Q(\ker \lambda^J)_\Phi$$

given by

$$m \otimes (v \otimes w) \mapsto w[\psi^{-1}(m)^J] \otimes [v] \,.$$

Corollary 4.6.16. Let ρ be a finite representation of the form ring (R, M, ψ, Φ). There is a positive integer n_0 such that the representation $n_1 \rho + n_2 \bar{\rho}$ admits an isotropic self-dual code if and only if $n_1 - n_2$ is a multiple of n_0.

Proof. Take n_0 to be the order of ρ in the Witt group. $\qquad\square$

5

The Main Theorems

This chapter will introduce the Clifford-Weil groups and their invariants. The main results of this book (Theorems 5.5.5 and 5.5.7) will be established in §5.5. They show that under quite general conditions, the invariant ring of the Clifford-Weil group $\mathcal{C}(\rho)$ associated with a finite representation ρ of a form ring is spanned by the complete weight enumerators of self-dual isotropic codes of Type ρ (and arbitrary length).

The Clifford-Weil group will be defined in §5.3 as a central extension of the hyperbolic co-unitary group $\mathfrak{U}(R, \Phi)$ associated with a form ring (R, M, ψ, Φ). The hyperbolic co-unitary group is generated by the parabolic group $P(R, \Phi)$ introduced in §5.1, together with certain elements corresponding to generalized MacWilliams transforms. We will construct $\mathfrak{U}(R, \Phi)$ as a subgroup of a larger parabolic subgroup $P(\mathrm{Mat}_2(R), \Phi_2)$ (see Definition 5.2.4 in §5.2).

As a concrete example, consider Euclidean self-dual codes over a prime field $\mathbb{F}_p$, p odd, that contain the all-ones vector $\mathbf{1}$. These codes belong to the family p_1^{E} and their Type is given by the representation $\rho(p_1^{\mathrm{E}})$ defined in §2.3.2. For the genus-m weight enumerators of these codes, we will see in §7.4.1 that:

- the appropriate form ring is (R, M, ψ, Φ), where $R = \mathrm{Mat}_m(\mathbb{F}_p)$ and Φ is the set of all $\mathbb{F}_p$-valued pointed quadratic maps on $\mathbb{F}_p^m$,
- the parabolic group $P(R, \Phi) \cong \mathbb{F}_p^m \rtimes \mathrm{GL}_m(\mathbb{F}_p)$,
- the parabolic group $P(\mathrm{Mat}_2(R), \Phi_2) \cong \mathbb{F}_p^{2m} \rtimes \mathrm{GL}_{2m}(\mathbb{F}_p)$,
- the hyperbolic co-unitary group $\mathfrak{U}(R, \Phi) \cong \mathbb{F}_p^{2m} \rtimes \mathrm{Sp}_{2m}(\mathbb{F}_p)$, and
- the Clifford-Weil group $\mathcal{C}(\rho) \cong Z_a \times p_+^{1+2m} . \mathrm{Sp}_{2m}(\mathbb{F}_p)$, where $a = \gcd\{p+1, 4\}$.

In Chapter 7 we will give many other examples of Clifford-Weil groups and hyperbolic co-unitary groups associated with classical self-dual codes over fields.

5.1 Parabolic groups

Definition 5.1.1. Let Φ be an R-qmodule (see Definition 1.1.4). The associated *parabolic group* $P(R, \Phi)$ is defined to be the semidirect product $R^* \ltimes \Phi$; that is, $P(R, \Phi)$ consists of pairs (u, ϕ) where $u \in R$ is a unit, $\phi \in \Phi$, with product given by

$$(u, \phi)(u', \phi') = (uu', \phi[u'] + \phi').$$

Definition 5.1.2. $\rho := (V, \rho_M, \rho_\Phi, \beta)$ be a finite representation of the finite quadratic pair (M, Φ) over the finite ring R (see §1.6). Let $\mathbb{C}[V]$ denote the free complex vector space generated by V; that is, $\mathbb{C}[V]$ has a basis e_v $(v \in V)$ in one-to-one correspondence with V. We define an action of $P(R, \Phi)$ on $\mathbb{C}[V]$ as follows:

$$(u, \phi)e_v := \exp(2\pi i \rho_\Phi(\phi)(v)) \, e_{uv}, \quad \text{for } u \in R^*, \phi \in \Phi, v \in V.$$

This gives a unitary representation

$$P(R, \Phi) \to \mathrm{GL}(\mathbb{C}[V]), \; (u, \phi) \mapsto m_u d_\phi,$$

where $m_u(e_v) = e_{uv}$ and $d_\phi(e_v) = \exp(2\pi i \rho_\Phi(\phi)(v))e_v$, for $u \in R^*$, $\phi \in \Phi$, $v \in V$. The image of this representation is denoted by

$$P(\rho) := \langle m_u, d_\phi \mid u \in R^*, \phi \in \Phi \rangle \leq \mathrm{GL}(\mathbb{C}[V]).$$

This definition certainly applies to finite representations of form rings. However, we will need the greater freedom of quadratic pairs for Theorem 5.1.3 and Remark 5.1.5 below.

Given a code $C \subset V$, we define an element $\mu_C \in \mathbb{C}[V]$ by

$$\mu_C := \sum_{\{v \in V \mid Rv = C\}} e_v; \tag{5.1.1}$$

that is, μ_C is the formal sum of the vectors that generate C. In particular, $\mu_C = 0$ unless C has a single generator. If C is a linear code over a field, $\mu_C = 0$ unless C is one-dimensional. Note that a code $C = \langle v \rangle \leq V$ that is generated by a single vector v is isotropic (Definition 1.2.1) if and only if $\rho_\Phi(\phi)(v) = 0$ for all $\phi \in \Phi$. When C is isotropic, μ_C is fixed by $P(\rho)$: indeed,

$$m_u d_\phi \mu_C = \sum_{Rv=C} m_u d_\phi e_v = \sum_{Rv=C} e_{uv} = \sum_{Ru^{-1}w=C} e_w = \mu_C.$$

Theorem 5.1.3. *The vectors μ_C, where C ranges over isotropic, singly-generated codes, form a basis for the subspace of $\mathbb{C}[V]$ fixed by $P(\rho)$.*

Proof. A vector is invariant under $P(\rho)$ if and only if it is invariant under the groups $\langle d_\phi \mid \phi \in \Phi \rangle$ and $\langle m_u \mid u \in R^* \rangle$. The former subgroup acts diagonally, so it suffices to consider its action on the basis. We find that e_v is invariant under $\langle d_\phi \mid \phi \in \Phi \rangle$ precisely when v is isotropic, or equivalently when the code Rv is isotropic. It remains to determine the orbits of isotropic vectors under $\langle m_u \mid u \in R^* \rangle$. The following lemma completes the proof. □

Lemma 5.1.4. *Let R be a finite ring, and let V be a finite R-module. Define an equivalence relation on V by $v \sim w$ if and only if $v = uw$ for some unit $u \in R^*$. Then $v \sim w$ if and only if $Rv = Rw$.*

Proof. (Based on the proof of Proposition 5.1 in Wood [554].) "Only if": This is immediate. "If": Suppose $Rv = Rw$. Then there are $a, b \in R$ with $w = av$ and $v = bw$. Therefore $(1 - ba)v = 0$. Now $R = Rba + R(1 - ba)$. Since $Rba \subseteq Ra$ we also have $R = Ra + R(1 - ba)$. By Bass's theorem [342, Theorem 20.9], $a + R(1 - ba)$ contains a unit u, say $u = a + s(1 - ba)$. Then $uv = av + s(1 - ba)v = w$, as required.   □

Remark 5.1.5. Similarly, if we define the full weight enumerator of C (cf. Definition 2.1.3) to be

$$\text{fwe}(C) = \sum_{v \in C} e_v \in \mathbb{C}[V], \tag{5.1.2}$$

then

$$\text{fwe}(C) = \sum_{D \subset C} \mu_D \, ;$$

hence the elements $\text{fwe}(C)$, where C ranges over isotropic codes, again span the invariant space of $P(\rho)$. We will refer to the full weight enumerators $\text{fwe}(C)$ as *vector* invariants, to distinguish them from the complete weight enumerators that will appear in §5.7, which we call *polynomial* invariants.

5.2 Hyperbolic co-unitary groups

Definition 5.2.1. Let (R, M, ψ, Φ) be a form ring (Definition 1.7.1), and fix an element $f \in R$. We define the *co-unitary group* $U(f, R, \Phi)$ as follows:

$$U(f, R, \Phi) := \{(u, \phi) \in P(R, \Phi) \mid u^J f u - f = \psi^{-1}(\lambda(\phi))\}. \tag{5.2.1}$$

If $\mathcal{R}$ is the name of the form ring, we will sometimes write $U(f, \mathcal{R})$ instead of $U(f, R, \Phi)$. If (I, Γ) is an ideal in (R, M, ψ, Φ) (Definition 1.7.6), then we define

$$U(f, I, \Gamma) := \{(u, \phi) \in U(f, R, \Phi) \mid u \in 1 + I, \phi \in \Gamma\},$$

a subgroup of $U(f, R, \Phi)$.

Remark 5.2.2. The reason for calling $U(f, R, \Phi)$ a co-unitary group is that it is the dual of the obvious unitary group of a quadratic pair (M, Φ) associated with an element $f \in M$, given by

$$\{(u, \phi) \in P(R, \Phi) \mid f(u \otimes u) - f = \lambda(\phi)\} \tag{5.2.2}$$

(cf. Hahn and O'Meara [226, §5.3]). Since several varieties of unitary groups will appear in this chapter, it is worth saying that we will not mention the particular group in (5.2.2) again!

The following is straightforward:

Theorem 5.2.3. *For any form ring (R, M, ψ, Φ), any ideal (I, Γ), and any element $f \in R$, we have an exact sequence of groups:*

$$1 \to U(f, I, \Gamma) \to U(f, R, \Phi) \to U(f + I, R/I, \Phi/\Gamma).$$

In order to define the hyperbolic co-unitary group, we will use the form ring

$$\mathrm{Mat}_2(R, M, \psi, \Phi) := (\mathrm{Mat}_2(R), \mathrm{Mat}_2(M), \psi_2, \Phi_2), \qquad (5.2.3)$$

as defined in §1.10. Following our usual convention, this will also be written as $\mathrm{Mat}_2(R, \Phi)$. From (1.10.1),

$$\mathrm{Mat}_2(R, M, \psi, \Phi) := \mathrm{End}_R(R^2; \beta), \qquad (5.2.4)$$

where $\beta : R^2 \times R^2 \to M$, $\beta(x, y) := \psi(1)(x_1 \otimes y_1) + \psi(1)(x_2 \otimes y_2)$. Recall that the involution J_2 on $\mathrm{Mat}_2(R)$ is just transposition and elementwise application of the involution J. We have

$$\Phi_2 := \left\{ \begin{pmatrix} \phi_1 & m \\ & \phi_2 \end{pmatrix} \mid \phi_1, \phi_2 \in \Phi, m \in M \right\}. \qquad (5.2.5)$$

Imitating the formal matrix product $A^{J_2} \phi A$, we get an action of $\mathrm{Mat}_2(R)$ on Φ_2 by

$$\begin{pmatrix} \phi_1 & m \\ & \phi_2 \end{pmatrix} \left[\begin{pmatrix} a & b \\ c & d \end{pmatrix} \right] := \begin{pmatrix} \phi'_1 & m' \\ & \phi'_2 \end{pmatrix}, \qquad (5.2.6)$$

where

$$\phi'_1 := \phi_1[a] + \phi_2[c] + \{m(a \otimes c)\},$$
$$m' := \lambda(\phi_1)(a \otimes b) + m(a \otimes d) + \lambda(\phi_2)(c \otimes d) + \tau(m)(c \otimes b),$$
$$\phi'_2 := \phi_1[b] + \phi_2[d] + \{m(b \otimes d)\}.$$

One can check that this turns Φ_2 into a $\mathrm{Mat}_2(R)$-qmodule.

Similarly, imitating $A_1^{J_2} m A_2$ gives a twisted $\mathrm{Mat}_2(R)$-module structure on $\mathrm{Mat}_2(M)$ via

$$\begin{pmatrix} m_1 & m_2 \\ m_3 & m_4 \end{pmatrix} \cdot \left(\begin{pmatrix} a_1 & b_1 \\ c_1 & d_1 \end{pmatrix} \otimes \begin{pmatrix} a_2 & b_2 \\ c_2 & d_2 \end{pmatrix} \right) := \begin{pmatrix} m'_1 & m'_2 \\ m'_3 & m'_4 \end{pmatrix}, \qquad (5.2.7)$$

where

$$m'_1 = m_1(a_1 \otimes a_2) + m_2(a_1 \otimes c_2) + m_3(c_1 \otimes a_2) + m_4(c_1 \otimes c_2),$$
$$m'_2 = m_1(a_1 \otimes b_2) + m_2(a_1 \otimes d_2) + m_3(c_1 \otimes b_2) + m_4(c_1 \otimes d_2),$$
$$m'_3 = m_1(b_1 \otimes a_2) + m_2(b_1 \otimes c_2) + m_3(d_1 \otimes a_2) + m_4(d_1 \otimes c_2),$$
$$m'_4 = m_1(b_1 \otimes b_2) + m_2(b_1 \otimes d_2) + m_3(d_1 \otimes b_2) + m_4(d_1 \otimes d_2).$$

The mapping $\tau_2 : \mathrm{Mat}_2(M) \to \mathrm{Mat}_2(M)$ is defined to be the composition of transposition and applying τ componentwise:

$$\tau_2\left(\begin{pmatrix} m_1 & m_2 \\ m_3 & m_4 \end{pmatrix}\right) := \begin{pmatrix} \tau(m_1) & \tau(m_3) \\ \tau(m_2) & \tau(m_4) \end{pmatrix}. \tag{5.2.8}$$

$\{\!\!\{\ \}\!\!\}_2 : \mathrm{Mat}_2(M) \to \Phi_2$ is given by

$$\{\!\!\{ \begin{pmatrix} m_1 & m_2 \\ m_3 & m_4 \end{pmatrix} \}\!\!\}_2 := \begin{pmatrix} \{\!\!\{ m_1 \}\!\!\} & m_2 + \tau(m_3) \\ & \{\!\!\{ m_4 \}\!\!\} \end{pmatrix}, \tag{5.2.9}$$

and $\lambda_2 : \Phi_2 \to \mathrm{Mat}_2(M)$ is the obvious map

$$\lambda_2\left(\begin{pmatrix} \phi_1 & m \\ & \phi_2 \end{pmatrix}\right) := \begin{pmatrix} \lambda(\phi_1) & m \\ \tau(m) & \lambda(\phi_2) \end{pmatrix}. \tag{5.2.10}$$

The right module isomorphism $\psi_2 : \mathrm{Mat}_2(R)_{\mathrm{Mat}_2(R)} \to \mathrm{Mat}_2(M)_{1 \otimes \mathrm{Mat}_2(R)}$ is defined componentwise. One can now check by elementary calculations that $(\mathrm{Mat}_2(R), \mathrm{Mat}_2(M), \psi_2, \Phi_2)$ is indeed a form ring,

Using these definitions we find that

$$\epsilon_2 = \begin{pmatrix} \epsilon & 0 \\ 0 & \epsilon \end{pmatrix} \in \mathrm{Mat}_2(R),$$

and the mapping J_2 on $\mathrm{Mat}_2(R)$ is given by

$$\begin{pmatrix} a & b \\ c & d \end{pmatrix}^{J_2} = \begin{pmatrix} a^J & c^J \\ b^J & d^J \end{pmatrix}.$$

To obtain a nontrivial co-unitary group, we usually assume that $\psi(f) + \tau(\psi(f))$ is nonsingular, in the sense that $\psi(f) + \tau(\psi(f)) = \psi(u)$ for some unit $u \in R^*$. For example, we may take $f := \begin{pmatrix} 0 & 0 \\ 1 & 0 \end{pmatrix} \in \mathrm{Mat}_2(R)$ (with $u = \begin{pmatrix} 0 & \epsilon \\ 1 & 0 \end{pmatrix}$). This choice for f leads to the following important special case of the co-unitary group:

Definition 5.2.4. Let (R, Φ) be a form ring. The *hyperbolic co-unitary group* $\mathfrak{U}(R, \Phi)$ is defined to be

$$\mathfrak{U}(R, \Phi) := U\left(\begin{pmatrix} 0 & 0 \\ 1 & 0 \end{pmatrix}, \mathrm{Mat}_2(R), \Phi_2\right). \tag{5.2.11}$$

More explicitly, $\mathfrak{U}(R, \Phi)$ consists of the elements

$$\left(\begin{pmatrix} a & b \\ c & d \end{pmatrix}, \begin{pmatrix} \phi_1 & m \\ & \phi_2 \end{pmatrix}\right) \in \mathrm{Mat}_2(R) \times \Phi_2 \tag{5.2.12}$$

such that

$$\begin{pmatrix} c^J a & c^J b \\ d^J a - 1 & d^J b \end{pmatrix} = \psi_2^{-1}\begin{pmatrix} \lambda(\phi_1) & m \\ \tau(m) & \lambda(\phi_2) \end{pmatrix}. \tag{5.2.13}$$

Remark 5.2.5. To describe the isomorphism type of $\mathfrak{U}(R,\Phi)$, note that the map

$$\pi : \mathfrak{U}(R,\Phi) \to GL_2(R), \ (u,\phi) \mapsto u$$

defines a group homomorphism. The kernel of π is the set of elements $(1,\phi) \in \mathfrak{U}(R,\Phi)$, i.e.

$$\ker(\pi) = \left\{ \left(1, \begin{pmatrix} \phi_1 & 0 \\ & \phi_2 \end{pmatrix}\right) \mid \phi_1, \phi_2 \in \Phi, \ \lambda(\phi_1) = \lambda(\phi_2) = 0 \right\},$$

which is naturally isomorphic to $\ker(\lambda) \times \ker(\lambda)$. The image of π is the set of elements $\begin{pmatrix} a & b \\ c & d \end{pmatrix} \in \mathrm{Mat}_2(R)$ such that

$$\begin{pmatrix} a^J & c^J \\ b^J & d^J \end{pmatrix} \begin{pmatrix} 0 & 0 \\ 1 & 0 \end{pmatrix} \begin{pmatrix} a & b \\ c & d \end{pmatrix} - \begin{pmatrix} 0 & 0 \\ 1 & 0 \end{pmatrix} \in \psi_2^{-1}(\lambda_2(\Phi_2)). \tag{5.2.14}$$

In many important examples it is easy to describe the image of $\psi_2^{-1} \circ \lambda_2$ (typically this consists of all symmetric, skew-symmetric, even, or Hermitian elements), which enables us to establish an isomorphism between $\mathfrak{U}(R,\Phi)/\ker(\pi)$ and a subgroup of a classical group.

Remark 5.2.6. Since $\lambda_2(\Phi_2)$ consists of symmetric elements, (5.2.14) implies that elements $\begin{pmatrix} a & b \\ c & d \end{pmatrix} \in \pi(\mathfrak{U}(R,\Phi))$ satisfy

$$\begin{pmatrix} a^J & c^J \\ b^J & d^J \end{pmatrix} \begin{pmatrix} 0 & -\epsilon \\ 1 & 0 \end{pmatrix} \begin{pmatrix} a & b \\ c & d \end{pmatrix} = \begin{pmatrix} 0 & -\epsilon \\ 1 & 0 \end{pmatrix}, \tag{5.2.15}$$

which is equivalent to

$$\begin{pmatrix} 0 & 1 \\ -\epsilon^{-1} & 0 \end{pmatrix} \begin{pmatrix} a^J & c^J \\ b^J & d^J \end{pmatrix} \begin{pmatrix} 0 & -\epsilon \\ 1 & 0 \end{pmatrix} = \begin{pmatrix} a & b \\ c & d \end{pmatrix}^{-1}, \tag{5.2.16}$$

or equivalently

$$\begin{pmatrix} a & b \\ c & d \end{pmatrix} \begin{pmatrix} 0 & 1 \\ -\epsilon^{-1} & 0 \end{pmatrix} \begin{pmatrix} a^J & c^J \\ b^J & d^J \end{pmatrix} = \begin{pmatrix} 0 & 1 \\ -\epsilon^{-1} & 0 \end{pmatrix}. \tag{5.2.17}$$

Using the identities from Lemma 1.4.5 we find from (5.2.15) that

$$\begin{aligned}
c^J a &= \psi^{-1}(\tau(\psi(c^J a))) = a^J c^{J^2} \epsilon = a^J \epsilon c, \\
d^J b &= \psi^{-1}(\tau(\psi(d^J b))) = b^J d^{J^2} \epsilon = b^J \epsilon d, \\
d^J a - 1 &= \psi^{-1}(\tau(\psi(c^J b))) = b^J c^{J^2} \epsilon = b^J \epsilon c, \\
c^J b &= \psi^{-1}(\tau(\psi(d^J a - 1))) = (d^J a - 1)^J \epsilon = a^J \epsilon d - \epsilon.
\end{aligned} \tag{5.2.18}$$

and from (5.2.17) that

$$\begin{aligned}
ab^J &= b\epsilon^J a^J, \\
cd^J &= d\epsilon^J c^J, \\
ad^J &= 1 + b\epsilon^J c^J. \\
d\epsilon^J a^J &= \epsilon^J + cb^J.
\end{aligned} \tag{5.2.19}$$

Example 5.2.7. Let (M, Φ) be a quadratic pair over the ring R with associated structure maps $\{\} : M \to \Phi$ and $\lambda : \Phi \to M$. Then the hyperbolic co-unitary group of the triangular form ring $T(M, \Phi)$ (see §1.9) is isomorphic to the parabolic group

$$\mathfrak{U}(T(M, \Phi)) \cong P(\mathrm{Mat}_2(R, \Phi)) = \mathrm{GL}_2(R) \ltimes \Phi_2\,,$$

Proof. The hyperbolic co-unitary group $\mathfrak{U}(T(M, \Phi))$ is the set of pairs of matrices

$$\left(\begin{pmatrix} a_1 \ a_2 \ b_1 \ b_2 \\ 0 \ a_3 \ 0 \ b_3 \\ c_1 \ c_2 \ d_1 \ d_2 \\ 0 \ c_3 \ 0 \ d_3 \end{pmatrix}, \ (r_1, \phi_1) \begin{pmatrix} m_1 \ m_2 \\ \ \ m_3 \\ (r_2, \phi_2) \end{pmatrix} \right) \tag{5.2.20}$$

such that $a_1, a_3, b_1, b_3, c_1, c_3, d_1, d_3, m_1, m_3, r_1, r_2 \in R$, $a_2, b_2, c_2, d_2, m_2 \in M$, and $\phi_1, \phi_2 \in \Phi$ satisfy

$$\begin{array}{lll}
a_1 c_3 = r_1, & c_1 a_3 = r_1, & a_2(c_3 \otimes 1) + \tau(c_2)(1 \otimes a_3) = \lambda(\phi_1), \\
b_1 d_3 = r_2, & d_1 b_3 = r_2, & b_2(d_3 \otimes 1) + \tau(d_2)(1 \otimes b_3) = \lambda(\phi_2), \\
b_1 c_3 = m_1, & c_1 b_3 = m_3, & b_2(c_3 \otimes 1) + \tau(c_2)(1 \otimes b_3) = m_2, \\
a_1 d_3 - 1 = m_3, & d_1 a_3 - 1 = m_1, & a_2(d_3 \otimes 1) + \tau(d_2)(1 \otimes a_3) = \tau(m_2).
\end{array}$$

In particular, we have

$$\begin{pmatrix} a_1 \ c_1 \\ b_1 \ d_1 \end{pmatrix} \begin{pmatrix} 0 \ 1 \\ -1 \ 0 \end{pmatrix} \begin{pmatrix} a_3 \ b_3 \\ c_3 \ d_3 \end{pmatrix} = \begin{pmatrix} 0 \ 1 \\ -1 \ 0 \end{pmatrix},$$

and mapping the matrix (5.2.20) above to

$$\begin{pmatrix} a_3 \ b_3 \\ c_3 \ d_3 \end{pmatrix}$$

defines an epimorphism $\mathfrak{U}(T(M, \Phi)) \to \mathrm{GL}_2(R)$. The kernel K of this epimorphism consists of the matrices of the form (5.2.20) in $\mathfrak{U}(T(M, \Phi))$ that satisfy

$$a_3 = d_3 = 1 = a_1 = d_1, c_3 = b_3 = 0 = c_1 = b_1, m_1 = m_3 = r_1 = r_2 = 0$$

and

$$\tau(c_2) = \lambda(\phi_1), b_2 = \lambda(\phi_2), a_2 + \tau(d_2) = 0\,.$$

Hence the mapping $K \to \Phi_2$ that maps such a matrix to

$$\begin{pmatrix} \phi_1 \ a_2 \\ \ \ \phi_2 \end{pmatrix}$$

defines the desired isomorphism. To show that $\mathfrak{U}(T(M, \Phi))$ is isomorphic to the semidirect product, it remains to find a complement $\mathrm{GL}_2(R) \cong G \le \mathfrak{U}(T(M, \Phi))$ of K. Such a complement is given by the set of matrices of the form (5.2.20) in $\mathfrak{U}(T(M, \Phi))$ that satisfy

$$a_2 = b_2 = c_2 = d_2 = m_2 = \phi_1 = \phi_2 = 0\,. \qquad \square$$

In the applications when we deal with higher genus weight enumerators, it is convenient to have a name for the hyperbolic co-unitary group of a matrix ring of a form ring. Since the matrix ring construction is functorial, the hyperbolic co-unitary groups of different degrees are closely related.

Definition 5.2.8. Let (R, Φ) be a form ring. The *hyperbolic co-unitary group of degree* m, $\mathfrak{U}_m(R, \Phi)$, is defined to be the hyperbolic co-unitary group of the matrix form ring $\mathrm{Mat}_m(R, \Phi)$:

$$\mathfrak{U}_m(R, \Phi) = \mathfrak{U}(\mathrm{Mat}_m(R, \Phi)) = \mathfrak{U}(\mathrm{Mat}_m(R), \Phi_m). \tag{5.2.21}$$

Example. As already mentioned at the start of this chapter, if (R, Φ) is the form ring $(p)_1$ of Example 1.8.4, then $\ker(\lambda) = \langle \phi_0 \rangle$ yields $\mathfrak{U}_m(R, \Phi) \cong \mathbb{F}_p^{2m} \rtimes \mathrm{Sp}_{2m}(\mathbb{F}_p)$.

5.2.1 Generators for the hyperbolic co-unitary group

We now give some convenient generators for $\mathfrak{U}(R, \Phi)$, for which the action on a certain projective representation of $\mathfrak{U}(R, \Phi)$ (which yields the Clifford-Weil groups) can easily be described.

Remark. The parabolic group $P(R, \Phi)$ embeds naturally in $\mathfrak{U}(R, \Phi)$, by the map $d : P(R, \Phi) \to \mathfrak{U}(R, \Phi)$ defined by

$$d((u, \phi)) := \left(\begin{pmatrix} u^{-J} & u^{-J}\psi^{-1}(\lambda(\phi)) \\ 0 & u \end{pmatrix}, \begin{pmatrix} 0 & 0 \\ & \phi \end{pmatrix} \right). \tag{5.2.22}$$

Remark. Let $\iota \in R$ be a symmetric idempotent (defined in Remark 3.5.7), with associated elements $u_\iota \in \iota R\iota^J$ and $v_\iota \in \iota^J R\iota$ such that $\iota = u_\iota v_\iota$ and $\iota^J = v_\iota u_\iota$. Then the element

$$H_{\iota, u_\iota, v_\iota} := \left(\begin{pmatrix} 1 - \iota^J & v_\iota \\ -\epsilon^{-1}u_\iota^J & 1 - \iota \end{pmatrix}, \begin{pmatrix} 0 & \psi(-\epsilon\iota) \\ 0 & \end{pmatrix} \right) \in \mathfrak{U}(R, \Phi). \tag{5.2.23}$$

The only nontrivial verification is that $H_{\iota, u_\iota, v_\iota}$ is a unit, which follows from the fact that

$$H_{\iota, u_\iota, v_\iota}^{-1} = H_{\iota, -\epsilon^{-1}u_\iota^J, -v_\iota^J\epsilon} = \left(\begin{pmatrix} 1 - \iota^J & -v_\iota^J\epsilon \\ u_\iota^J & 1 - \iota \end{pmatrix}, \begin{pmatrix} 0 & \psi(-\epsilon\iota) \\ 0 & \end{pmatrix} \right).$$

Recall that a ring R is semiperfect (Lam [342, page 346]) if $R/\mathrm{rad}\,R$ is semisimple and idempotents in $R/\mathrm{rad}\,R$ lift to idempotents in R; in particular, all finite rings are semiperfect.

Theorem 5.2.9. *Suppose that R is semiperfect and let (R, Φ) be a form ring. Then $\mathfrak{U}(R, \Phi)$ is generated by $d(P(R, \Phi))$ together with the elements $H_{\iota, u_\iota, v_\iota}$ where ι runs through the conjugacy classes of symmetric idempotents in R.*

Proof. Let $\mathcal{G}$ be the subgroup generated by the specified elements. Fix an element

$$\left(\begin{pmatrix} \alpha & \beta \\ \gamma & \delta \end{pmatrix}, \begin{pmatrix} \phi_1 & \mu \\ & \phi_2 \end{pmatrix} \right) \in \mathfrak{U}(R, \Phi).$$

We will show that this can be reduced to the identity using elements of $\mathcal{G}$.

First, suppose δ is a unit. Then the element

$$\left(\begin{pmatrix} \delta^{-J} & \beta \\ 0 & \delta \end{pmatrix}, \begin{pmatrix} 0 & 0 \\ & \phi_2 \end{pmatrix} \right) \in d(P(R, \Phi)).$$

Left-multiplying by the inverse of this element, we obtain an element with $\delta = 1$, $\beta = 0$, $\phi_2 = 0$. Conjugating by $H_{1,1,1}$ then gives another element of $d(P(R, \Phi))$.

Thus $\mathcal{G}$ contains every element of $\mathfrak{U}(R, \Phi)$ for which δ is a unit. In particular, $\mathcal{G}$ contains the normal subgroup

$$U\left(\begin{pmatrix} 0 & 0 \\ 1 & 0 \end{pmatrix}, \mathrm{rad}(\mathrm{Mat}_2(R, \Phi)) \right).$$

(See Remark 1.7.8 for the definition of the radical of a form ring.) It will thus suffice to show that the quotient of $\mathcal{G}$ by this subgroup is isomorphic to

$$U\left(\begin{pmatrix} 0 & 0 \\ 1 & 0 \end{pmatrix} + \mathrm{rad}(\mathrm{Mat}_2(R, \Phi)), \mathrm{Mat}_2((R, \Phi)/\mathrm{rad}(R, \Phi)) \right).$$

So it suffices to prove the theorem when $\mathrm{rad}(R, \Phi) = 0$; indeed, every element of $P((R, \Phi)/\mathrm{rad}(R, \Phi))$ lifts to $P(R, \Phi)$, and similarly for the elements $H_{\iota, u_\iota, v_\iota}$, using the fact that idempotents lift and symmetric idempotents lift to symmetric idempotents. If ι is an idempotent in R and $\bar{\iota}R/\mathrm{rad}(R) \cong \bar{\iota}^J R/\mathrm{rad}(R)$, then also $\iota R \cong \iota^J R$, since the latter module is a projective right R-module and projective modules are isomorphic if and only if their factor modules modulo the radical are isomorphic (see for instance Reiner [457, Theorem 6.18]). Thus we may assume that R is semisimple and λ is injective. In particular, for $(u, \phi) \in \mathfrak{U}(R, \Phi)$, ϕ is uniquely determined by u, so can essentially be ignored.

Lemma 5.2.10. *If R is semisimple, then for all $r \in R$ there exist units u_1 and u_2 with $u_1 r u_2$ an idempotent.*

Proof. It clearly suffices to consider the case in which R is a matrix ring over a division ring, in which case the result is obvious. □

In particular, this is true for δ. Multiplying by

$$\begin{pmatrix} u_1^{-J} & 0 \\ 0 & u_1 \end{pmatrix} \text{ and } \begin{pmatrix} u_2^{-J} & 0 \\ 0 & u_2 \end{pmatrix},$$

we may thus arrange for δ to be an idempotent. We claim that the idempotent $\iota := 1 - \delta$ is symmetric.

More precisely, if we define

$$u_\iota = -\iota \epsilon^{-1} \gamma^J \iota^J \,,$$
$$v_\iota = \iota^J \beta \iota \,,$$

then, using (5.2.18),

$$
\begin{aligned}
u_\iota v_\iota &= -(1-\delta)\epsilon^{-1}\gamma^J(1-\delta^J)\beta(1-\delta) \\
&= -(1-\delta)\epsilon^{-1}\gamma^J\beta(1-\delta) \\
&= -(1-\delta)\epsilon^{-1}(\delta^J\alpha - 1)^J\epsilon(1-\delta) \\
&= (1-\delta) = \iota \,.
\end{aligned}
\tag{5.2.24}
$$

From (5.2.19) we get

$$
\begin{aligned}
v_\iota u_\iota &= -(1-\delta^J)\beta(1-\delta)\epsilon^{-1}\gamma^J(1-\delta^J) \\
&= -(1-\delta^J)\beta\epsilon^J\gamma^J(1-\delta^J) \\
&= -(1-\delta^J)(-1)(1-\delta^J) \\
&= (1-\delta^J) = \iota^J \,,
\end{aligned}
\tag{5.2.25}
$$

and thus ι is symmetric, as required.

If we now multiply on the right by

$$
\begin{pmatrix} 1 - \iota^J & v_\iota \\ -\epsilon^{-1}u_\iota^J & 1 - \iota \end{pmatrix}^{-1}
= \begin{pmatrix} \delta^J & -v_\iota^J\epsilon \\ u_\iota & \delta \end{pmatrix} \,,
$$

the resulting matrix $y := \begin{pmatrix} \star & \star \\ \star & \delta' \end{pmatrix}$ satisfies

$$
\begin{aligned}
\delta' &= -\gamma v_\iota^J\epsilon + \delta^2 \\
&= -\gamma(1-\delta^J)\beta^J\epsilon(1-\delta) + \delta \,.
\end{aligned}
$$

In particular

$$\delta'\delta = \delta$$

and

$$
\begin{aligned}
(1-\delta)\delta' &= -(1-\delta)\gamma(1-\delta^J)\beta^J\epsilon(1-\delta) \\
&= -(1-\delta)\gamma\beta^J\epsilon \\
&= -(1-\delta)(-\epsilon^J\epsilon) = (1-\delta) \,.
\end{aligned}
$$

Hence

$$\delta'(2-\delta') = \delta'(\delta + (1-\delta))(2-\delta') = 1 \,.$$

So δ' is a unit and therefore $y \in \mathcal{G}$. Thus $\mathcal{G} = \mathfrak{U}(R, \Phi)$, as required. This completes the proof of Theorem 5.2.9.   □

For a form ideal (I, Γ) in (R, Φ), we define

$$\mathfrak{U}(I, \Gamma) := U\left(\begin{pmatrix} 0 & 0 \\ 1 & 0 \end{pmatrix}, \mathrm{Mat}_2(I), \Gamma_2 \right),$$

where

$$\Gamma_2 := \left\{ \begin{pmatrix} \phi_1 & m \\ & \phi_2 \end{pmatrix} \mid \phi_1, \phi_2 \in \Gamma, m \in \psi(I) \subset M \right\}.$$

Then we have the following corollary.

Corollary 5.2.11. *For any form ring (R, M, ψ, Φ) where R is semiperfect, and any form ideal $(I, \Gamma) \subset (R, \Phi)$, there is an exact sequence*

$$1 \to \mathfrak{U}(I, \Gamma) \to \mathfrak{U}(R, \Phi) \to \mathfrak{U}(R/I, \Phi/\Gamma) \to 1.$$

Proof. It suffices to show that each of the given generators can be lifted from $\mathfrak{U}(R/I, \Phi/\Gamma)$ to $\mathfrak{U}(R, \Phi)$, which follows from the fact that any equivalence of idempotents in R/I lifts to an equivalence of idempotents in R (see e.g. Rowen [460, Proposition 2.7.27]); in particular, units lift. □

5.3 Clifford-Weil groups

Let (R, M, ψ, Φ), or simply (R, Φ), be a finite form ring. Since R is finite, R is semiperfect. Let $\rho := (V, \rho_M, \rho_\Phi, \beta)$ be a finite representation of (R, Φ) (cf. Definition 1.7.2). We wish to extend this representation of the parabolic subgroup $P(R, \Phi)$ on $\mathbb{C}[V]$ to a projective representation of $\mathfrak{U}(R, \Phi)$; by Theorem 5.2.9 it suffices to specify the action of the elements $H_{\iota, u_\iota, v_\iota}$, which we do by mapping $H_{\iota, u_\iota, v_\iota} \in \mathfrak{U}(R, \Phi)$ to $h_{\iota, u_\iota, v_\iota} \in \mathrm{GL}(\mathbb{C}[V])$, where

$$h_{\iota, u_\iota, v_\iota}(e_v) = |\iota V|^{-1/2} \sum_{w \in \iota V} \exp(2\pi i \beta(w, v_\iota v)) e_{(1-\iota)v+w}. \qquad (5.3.1)$$

This is a very general form of the MacWilliams transform.

Definition 5.3.1. Let ρ be a finite representation of a form ring (R, Φ). The subgroup

$$\mathcal{C}(\rho) := \langle m_u, d_\phi, h_{\iota, u_\iota, v_\iota} \mid u \in R^*, \phi \in \Phi, \iota = u_\iota v_\iota \text{ symmetric idempotent in } R \rangle$$

of $GL(\mathbb{C}[V])$ is called the *Clifford-Weil group* associated with the representation ρ.

Theorem 5.3.2. *Let (R, M, ψ, Φ) be a finite form ring and $\rho := (V, \rho_\Phi, \rho_M, \beta)$ a finite representation of (R, M, ψ, Φ). Then the mapping*

$$\mathfrak{U}(R, \Phi) \to \mathcal{C}(\rho)$$

defined on the generators by

$$d(u, \phi) \mapsto m_u d_\phi \quad for \ u \in R^*, \phi \in \Phi,$$
$$H_{\iota, u_\iota, v_\iota} \mapsto h_{\iota, u_\iota, v_\iota} \quad for \ symmetric \ idempotents \ \iota \in R,$$

gives a projective representation of $\mathfrak{U}(R, \Phi)$ *on* $\mathbb{C}[V]$ *— that is, a homomorphism from* $\mathfrak{U}(R, \Phi)$ *to* $\mathrm{PGL}(\mathbb{C}[V])$.

Proof. We first introduce a group $\mathcal{E}(V)$ which plays the role that the extraspecial group $E(m)$ played in the papers [96], [383] (see also §6.2). For an element $f \in R$ we define a (non-abelian) group structure $E(f, V)$ on $V \times \mathbb{Q}/\mathbb{Z}$ by

$$(x_1, q_1)(x_2, q_2) = (x_1 + x_2, \beta(x_1, f x_2) + q_1 + q_2), \qquad (5.3.2)$$

for $x_1, x_2 \in V$, $q_1, q_2 \in \mathbb{Q}/\mathbb{Z}$, and observe that

$$^{(u, \phi)}(x, q) := (ux, q + \rho_\Phi(\phi)(x)), \quad for \ u \in R^*, \phi \in \Phi, x \in V, q \in \mathbb{Q}/\mathbb{Z}, \quad (5.3.3)$$

defines a left action of $U(f, R, \Phi)$ on $E(f, V)$. In particular, $\mathfrak{U}(R, \Phi)$ acts naturally on the group

$$\mathcal{E}(V) := E\left(\begin{pmatrix} 0 & 0 \\ 1 & 0 \end{pmatrix}, V \times V \right). \qquad (5.3.4)$$

Products in $\mathcal{E}(V)$ are defined by extending β naturally to $V \times V$, so two elements of $(V \times V) \times \mathbb{Q}/\mathbb{Z}$ are multiplied by the rule

$$(z_1, x_1, q_1)(z_2, x_2, q_2) = (z_1 + z_2, x_1 + x_2, \beta(x_1, z_2) + q_1 + q_2).$$

With this multiplication, $\mathcal{E}(V)$ is a "Heisenberg group",

$$\mathcal{E}(V) \cong (\mathbb{Q}/\mathbb{Z}).(V \times V).$$

The group $\mathcal{E}(V)$ acts naturally on $\mathbb{C}[V]$ by

$$(z, x, q)e_y = \exp(2\pi i(q + \beta(y, z)))e_{y+x}, \quad for \ (z, x, q) \in \mathcal{E}(V), y \in V. \quad (5.3.5)$$

Furthermore, this representation is irreducible; in fact it is the unique irreducible representation of $\mathcal{E}(V)$ in which the central elements $(0, 0, q)$ act as scalar multiplication by $\exp(2\pi iq)$. The irreducibility of the representation can be seen easily from the fact that the subgroup

$$\langle ((z, 0), q) \mid z \in V, q \in \mathbb{Q}/\mathbb{Z} \rangle$$

acts diagonally with respect to the basis $(e_v \mid v \in V)$ with $|V|$ distinct characters (since β is nonsingular). Hence the subalgebra of $\mathrm{End}(\mathbb{C}[V])$ that commutes with this group consists of all diagonal matrices. This subgroup,

$$\langle ((0, x), 0) \mid x \in V \rangle,$$

acts as a transitive permutation group on $\{e_v \mid v \in V\}$. Hence the only diagonal matrices that commute with this action are the scalar matrices.

To complete the proof of the theorem, consider $\mathcal{E}(V)$ as a subgroup $\mathcal{E}(V) \leq GL(\mathbb{C}[V])$ via the above representation. Since the representation is irreducible, an element of the normalizer of $\mathcal{E}(V)$ in $U(\mathbb{C}[V])$ is determined modulo scalars by its action by conjugation on $\mathcal{E}(V)$.

In particular, the actions of m_u, d_ϕ and of the elements $h_{\iota,u_\iota,v_\iota}$ in $\mathcal{C}(\rho)$ by conjugation correspond to the standard action of the elements $d((u,0))$, $d((1,\phi))$ and $H_{\iota,u_\iota,v_\iota}$ of $\mathfrak{U}(R,\Phi)$ on $\mathcal{E}(V)$. Therefore the representation defined in the theorem gives a projective representation of $\mathfrak{U}(R,\Phi)$. $\qquad\square$

The above results provide an easy way to obtain generators for the Clifford-Weil groups, which we summarize in the following remark.

Remark 5.3.3. Let $\rho := (V, \rho_M, \rho_\Phi, \beta)$ be a finite representation of a form ring (R, M, ψ, Φ). Let $u_1, \ldots, u_r \in R^*$ generate the unit group R^*, let $\phi_1, \ldots, \phi_s$ be generators for the $\mathbb{Z}R^*$-module Φ, and let $\iota_1 = l_1 r_1, \ldots, \iota_t = l_t r_t$ (with $\iota_j^J = r_j l_j$) be representatives for the R^*-conjugacy classes of symmetric idempotents in R. Then the associated Clifford-Weil group $\mathcal{C}(\rho)$ is generated by the elements

$$m_{u_1}, \ldots, m_{u_r}, d_{\phi_1}, \ldots, d_{\phi_s}, h_{\iota_1, l_1, r_1}, \ldots, h_{\iota_t, l_t, r_t}, \qquad (5.3.6)$$

where $m_{u_j} = \rho((u_j, 0)) \in \rho(P(R, \Phi))$ (cf. Definition 5.1.2) is given by

$$m_{u_j} e_v = e_{u_j v}, \text{ for all } v \in V, j = 1, \ldots, r, \qquad (5.3.7)$$

$d_{\phi_i} = \rho((1, \phi_j)) \in \rho(P(R, \Phi))$ (cf. Definition 5.1.2) by

$$d_{\phi_j} e_v = \exp(2\pi i \rho_\Phi(\phi_j)(v)) e_v, \text{ for all } v \in V, j = 1, \ldots, s, \qquad (5.3.8)$$

and the partial MacWilliams transform h_{ι_j, l_j, r_j} is defined as above by

$$h_{\iota_j, l_j, r_j} e_v = |\iota_j V|^{-1/2} \sum_{w \in \iota_j V} \exp(2\pi i \beta(w, r_j v)) e_{(1 - \iota_j) v + w}. \qquad (5.3.9)$$

Note that the complex conjugate of any of these generators is also a generator, and thus $\overline{\mathcal{C}(\rho)} = \mathcal{C}(\rho)$.

To deal with higher genus weight enumerators, we introduce the Clifford-Weil groups $\mathcal{C}_m(\rho)$ of genus m, for all $m \in \mathbb{N}$. If either of the main theorems Theorem 5.5.5 or 5.5.7 applies to the representation ρ, then by Remark 2.1.10 the invariant ring of $\mathcal{C}_m(\rho)$ is spanned by the genus-m weight enumerators of self-dual codes of Type ρ.

Definition 5.3.4. Let $\rho = (V, \rho_M, \rho_\Phi, \beta)$ be a finite representation of a form ring (R, M, ψ, Φ) with underlying module V. By Morita theory, this corresponds to a unique finite representation $\rho \otimes R^m$ of the form ring $\mathrm{Mat}_m(R, M, \psi, \Phi)$ with underlying module $V \otimes R^m$. We define $\mathcal{C}_m(\rho) := \mathcal{C}(\rho \otimes R^m)$ to be the *associated Clifford-Weil group of genus m*.

Note that $\mathcal{C}_m(\rho)$ is the image of the corresponding projective representation of $\mathfrak{U}_m(R, \Phi)$. Generators for $\mathcal{C}_m(\rho)$ can be found by applying Remark 5.3.3 to the corresponding matrix form ring.

The following table lists the most important examples of Clifford-Weil groups. They are discussed in more detail in Chapters 6 and 7. The table emphasizes the simple and uniform structure of these groups. The correctness of the list follows from Remark 5.2.5. (Referring to the last line of the table, $\lambda(\Phi)$ was stated incorrectly in [385]: the correct $\lambda(\Phi)$ is the set of all *even* symmetric matrices, as defined in Definition 1.10.4.)

Table 5.1. The most important examples of Clifford-Weil groups $\mathcal{C}(\rho)$. Here R is a finite simple ring with involution and $\mathrm{Mat}_n(\mathbb{F}_q)$ is the ring of $n \times n$ matrices over the field $\mathbb{F}_q$, with transposition denoted by tr. For an appropriate center Z, $\mathcal{C}(\rho) = Z \cdot \mathcal{U}(R, \Phi) = Z \cdot (\ker(\lambda) \oplus \ker(\lambda)) \cdot \mathcal{G}(R, \Phi)$, where $\mathcal{G}(R, \Phi)$ is the classical group shown in the last column.

R	J	ϵ	$\mathcal{G}(R, \Phi)$
$\mathrm{Mat}_n(\mathbb{F}_q) \oplus \mathrm{Mat}_n(\mathbb{F}_q)$	$(r, s)^J = (s^{tr}, r^{tr})$	1	$\mathrm{GL}_{2n}(\mathbb{F}_q)$
$\mathrm{Mat}_n(\mathbb{F}_{q^2})$	$r^J = (r^q)^{tr}$	1	$U_{2n}(\mathbb{F}_{q^2})$
$\mathrm{Mat}_n(\mathbb{F}_q), \ q = p^m, \ p > 2$	$r^J = r^{tr}$	1	$\mathrm{Sp}_{2n}(\mathbb{F}_q)$
$\mathrm{Mat}_n(\mathbb{F}_q), \ q = p^m, \ p > 2$	$r^J = r^{tr}$	-1	$O_{2n}^+(\mathbb{F}_q)$
$\mathrm{Mat}_n(\mathbb{F}_q), \ q = p^m, \ p = 2$	$\psi^{-1}(\lambda(\Phi)) = \mathrm{Sym}_n(\mathbb{F}_q)$		$\mathrm{Sp}_{2n}(\mathbb{F}_q)$
$\mathrm{Mat}_n(\mathbb{F}_q), \ q = p^m, \ p = 2$	$\psi^{-1}(\lambda(\Phi)) = \mathrm{Ev}_n(\mathbb{F}_q)$		$O_{2n}^+(\mathbb{F}_q)$

For the classical examples of codes over prime fields, the Clifford-Weil groups are already discussed in [383]. The name of the real Clifford group $\mathcal{C}_m$ comes from the fact that the action of $\mathcal{C}_m$ on the unique invariant lattice, the *balanced Barnes-Wall lattice* (see §6.5) taken modulo the ideal $(\sqrt{2})$, coincides with the action of the full orthogonal group $O_{2m}^+(2) = \mathcal{C}_m/E(m)$ on the Clifford algebra of the quadratic form. To distinguish the generalizations $\mathcal{C}(\rho)$ of $\mathcal{C}_m$, and also to honor Weil's fundamental work [546] on generalized Gauss sums and the metaplectic representation, we call the groups $\mathcal{C}(\rho)$ Clifford-Weil groups. Weil in fact associates a group to every locally compact abelian group; in the case of a finite abelian group this is the Clifford-Weil group associated with a suitable finite representation of the form ring $R(m_{\mathrm{II},1}^{\mathbb{Z}})$ (cf. (2.3.54)), where m is twice the exponent of the group.

5.4 Scalar elements in $\mathcal{C}(\rho)$

We have defined the Clifford-Weil group $\mathcal{C}(\rho)$ to be a projective unitary representation of a hyperbolic co-unitary group. The scalar elements in $\mathcal{C}(\rho)$ can be obtained as words in the generators that map to 1 in the hyperbolic co-unitary group. Since the entries in the matrices of the generators generate a finite extension K of $\mathbb{Q}$, all scalar elements are roots of unity belonging to K. In particular, $\mathcal{C}(\rho)$ is a finite group; this also follows from the fact that it is a quotient of the "universal" Clifford-Weil group defined in Remark 5.4.8 below.

In the proofs of the main theorems and also for concrete applications, it is helpful to know some explicit elements in $\mathcal{C}(\rho)$ other than the generators. Recall that $\mathfrak{U}(R,\varPhi)$ contains a normal subgroup

$$\left\{\left(\begin{pmatrix} 1 & 0 \\ 0 & 1 \end{pmatrix}, \begin{pmatrix} \phi_1 & 0 \\ & \phi_2 \end{pmatrix}\right) \mid \phi_1, \phi_2 \in \ker(\lambda)\right\}.$$

For $\phi \in \varPhi$, the action of the element $(1,\phi) \in P(R,\varPhi)$ on $\mathbb{C}[V]$ is given by $e_v \mapsto \exp(2\pi i \rho_\varPhi(\phi)(v)) e_v$. The action of the element

$$T_\phi := \left(\begin{pmatrix} 1 & 0 \\ 0 & 1 \end{pmatrix}, \begin{pmatrix} \phi & 0 \\ & 0 \end{pmatrix}\right) = Hd(1,\phi)H^{-1} \in \mathfrak{U}(R,\varPhi), \tag{5.4.1}$$

where $H := H_{1,1,1}$, is given in the following lemma. For $\phi \in \ker(\lambda)$, there is an element $v_\phi \in V$ such that a certain image t_ϕ of T_ϕ acts as the addition of v_ϕ. Because $\ker(\lambda)$ is a linear R-module (see Remark 1.6.2), it is clear that $\rho_\varPhi(\ker(\lambda))$ consists of linear mappings. Since β is nonsingular, for each $\phi \in \ker(\lambda)$ there is a unique $v_\phi \in V$ such that

$$\rho_\varPhi(\phi)(w) = -\beta(v_\phi, w), \text{ for } w \in V. \tag{5.4.2}$$

Lemma 5.4.1. *Let $\rho = (V, \rho_M, \rho_\varPhi, \beta)$ be a finite representation of the form ring $(R, M, \psi, \varPhi)$ and let $h := h_{1,1,1} \in \mathcal{C}(\rho)$. For $\phi \in \ker(\lambda)$, let $v_\phi \in V$ be the unique element described above. Define*

$$t_\phi := hd_\phi h^{-1}.$$

Then

$$t_\phi(e_v) = e_{v+v_\phi}, \quad \text{for } v \in V. \tag{5.4.3}$$

Proof. Since d_ϕ multiplies the generators e_v by $\exp(2\pi i \rho_\varPhi(\phi)(v))$ for all $v \in V$, we have

$$t_\phi(e_v) = h(|V|^{-1/2} \sum_{w \in V} \exp(2\pi i(-\beta(v,w) + \rho_\varPhi(\phi)(w))) e_w)$$

$$= |V|^{-1} \sum_{x,w \in V} \exp(2\pi i(\beta(x-v,w) + \rho_\varPhi(\phi)(w))) e_x$$

$$= \sum_{x \in V} (|V|^{-1} \sum_{w \in V} \exp(2\pi i \beta(x - v - v_\phi, w))) e_x$$

$$= e_{v+v_\phi}.$$

$\square$

Corollary 5.4.2. *Let $\phi_1, \phi_2 \in \ker(\lambda)$. Then*

$$d_{\phi_2} t_{\phi_1} d_{\phi_2}^{-1} t_{\phi_1}^{-1} = \exp(2\pi i \rho_\Phi(\phi_2)(v_{\phi_1})) \, \mathrm{id} \,,$$

so the commutator acts as scalar multiplication by $\exp(2\pi i \rho_\Phi(\phi_2)(v_{\phi_1}))$, where v_{ϕ_1} is as in (5.4.2).

Proof. An elementary calculation using the fact that $\rho_\Phi(\phi_2)$ is a linear map.   □

If $\phi_2 \in \Phi$ is arbitrary, one has to multiply the commutator above by an element in $P(\rho)$ to get a scalar element:

Lemma 5.4.3. *Let $\phi_2 \in \Phi$, $\phi_1 \in \ker(\lambda)$. Then there is an element $\phi \in \Phi$ such that*

$$d_{\phi_2} t_{\phi_1} d_{\phi_2}^{-1} t_{\phi_1}^{-1} d_\phi = \exp(2\pi i \rho_\Phi(\phi_2)(v_{\phi_1})) \, \mathrm{id}$$

acts as scalar multiplication by $\exp(2\pi i \rho_\Phi(\phi_2)(v_{\phi_1}))$, where v_{ϕ_1} is as in Eq. (5.4.2).

Proof. Elementary calculations show that, for all $v \in V$,

$$d_{\phi_2} t_{\phi_1} d_{\phi_2}^{-1} t_{\phi_1}^{-1}(e_v) = \exp(2\pi i (\rho_\Phi(\phi_2)(v) - \rho_\Phi(\phi_2)(v + v_{\phi_1})) e_v \,.$$

Let $\phi := \phi_1[\psi^{-1}(\lambda(\phi_2))] \in \Phi$. Then, for $v \in V$,

$$\rho_\Phi(\phi)(v) = \rho_\Phi(\phi_1)(\psi^{-1}(\lambda(\phi_2))v) = \beta(v_{\phi_1}, \psi^{-1}(\lambda(\phi_2))v)$$
$$= \rho_\Phi(\phi_2)(v + v_{\phi_1})) - \rho_\Phi(\phi_2)(v) - \rho_\Phi(\phi_2)(v_{\phi_1}) \,,$$

and hence the element $d_\phi \in P(\rho)$ has the desired property.   □

Corollary 5.4.4. *Let $(V, \rho_M, \rho_\Phi, \beta)$ be a representation of the form ring (R, M, ψ, Φ) and define*

$$C_0 = \{w : w \in V \mid (v \mapsto \beta(w, v)) \in \rho_\Phi(\ker \lambda)\} \,. \tag{5.4.4}$$

Then, for every $w \in C_0$ and $\phi \in \Phi$, scalar multiplication by $\exp(2\pi i \rho_\Phi(\phi)(w))$ is in $\mathcal{C}(\rho)$. In particular, if there is a nonzero invariant of $\mathcal{C}(\rho)$, then C_0 is isotropic with respect to $\rho_\Phi(\Phi)$.

Other scalar elements arise from Gauss sums of nonsingular quadratic forms. If ι is a symmetric idempotent, an element $\phi \in \Phi[\iota]$ is *nonsingular* with respect to ι if multiplication by $\psi^{-1}(\lambda(\phi))$ gives an isomorphism from ιR to $\iota^J R$, or equivalently if the R-bilinear form

$$\beta_\phi : \iota R \otimes \iota R \to M, \; \beta_\phi(r, s) := \lambda(\phi)(r \otimes s)$$

is nonsingular. Such an element ϕ has two Clifford-Weil generators associated with it: the parabolic element d_ϕ, and the MacWilliams transform $h_{\iota, u_\iota, v_\iota}$, where $v_\iota = \psi^{-1}(\lambda(\phi))$, and $u_\iota \in \iota R \iota^J$ induces the inverse isomorphism, so $u_\iota v_\iota = \iota$, $v_\iota u_\iota = \iota^J$. Note that these elements satisfy the extra symmetries

$$v_\iota^J \epsilon = v_\iota \, , \;\; \epsilon^{-1} u_\iota^J = u_\iota \, ;$$

the first follows because $\psi(v_\iota)$ is invariant under τ, while the second follows by inversion.

Lemma 5.4.5. *Let $\iota = u_\iota v_\iota$ be a symmetric idempotent, and let ϕ be non-singular with respect to ι such that $\psi(v_\iota) = \lambda(\phi) \in \lambda(\Phi)$. In the hyperbolic co-unitary group, the element*

$$H_{\iota,u_\iota,v_\iota}\, d((1,\phi)) \;=\; \left(\begin{pmatrix} 1-\iota^J & v_\iota \\ -u_\iota & 1-2\iota \end{pmatrix}, \begin{pmatrix} 0 & -\psi(\epsilon\iota) \\ -\phi[-1] \end{pmatrix} \right)$$

has order 3.

Proof. We compute

$$(H_{\iota,u_\iota,v_\iota} d((1,\phi)))^2 = \left(\begin{pmatrix} 1-2\iota^J & -v_\iota \\ u_\iota & 1-\iota \end{pmatrix}, \begin{pmatrix} -\phi[u_\iota] & -\psi(\epsilon\iota) \\ & 0 \end{pmatrix} \right),$$

and thus

$$(H_{\iota,u_\iota,v_\iota} d((1,\phi)))^3 = \left(\begin{pmatrix} 1 & 0 \\ 0 & 1 \end{pmatrix}, \begin{pmatrix} 0 & 0 \\ & 0 \end{pmatrix} \right),$$

as required. $\square$

Remark. This is analogous to the element of order 3 given by Weil in [546, Eq. 9].

This leads to scalar elements in the Clifford-Weil groups, as follows.

Definition 5.4.6. Let ι be an idempotent, and let $\phi \in \Phi$ be nonsingular with respect to ι. For any finite representation ρ of (R, M, ψ, Φ), the *Gauss sum* $\gamma_\rho(\phi)$ is defined by

$$\gamma_\rho(\phi) := |\iota V|^{-1/2} \sum_{v \in \iota V} \exp(2\pi i \rho_\Phi(\phi)(v)) \,. \tag{5.4.5}$$

Theorem 5.4.7. *For all finite representations ρ, $\gamma_\rho(\phi)$ is a root of unity, and the corresponding scalar matrix is in $\mathcal{C}(\rho)$.*

Proof. Let $(V, \rho_M, \rho_\Phi, \beta)$ be a finite representation of (R, Φ) and let ι, ϕ be as in Definition 5.4.6. To shorten notation we set

$$\beta_\phi := \rho_M(\lambda(\phi)) : \iota V \times \iota V \to \mathbb{Q}/\mathbb{Z}, \;\; \beta_\phi(w,v) := \beta(w, \psi^{-1}(\lambda(\phi))v) \,.$$

Consider the element $(h_{\iota,u_\iota,v_\iota} d_\phi)^3 \in \mathcal{C}(\rho)$. That this is scalar multiplication by a root of unity follows from Lemma 5.4.5 and the fact that $\mathcal{C}(\rho)$ is finite. We first note that for $v \in \iota V$,

$$h_{\iota,u_\iota,v_\iota} d_\phi e_v = |\iota V|^{-1/2} \sum_{w \in \iota V} \exp(2\pi i(\beta_\phi(w,v) + \rho_\Phi(\phi)(v)))e_w \,,$$

so $\mathbb{C}[\iota V]$ is an invariant subspace for this transformation. Starting with e_0, we thus obtain:

$$h_{\iota,u_\iota,v_\iota} d_\phi e_0 = |\iota V|^{-1/2} \sum_{w \in \iota V} e_w \,,$$

$$(h_{\iota,u_\iota,v_\iota} d_\phi)^2 e_0 = |\iota V|^{-1} \sum_{v,w \in \iota V} \exp(2\pi i(\beta_\phi(v,w) + \rho_\Phi(\phi)(w)))e_v \,,$$

$$(h_{\iota,u_\iota,v_\iota} d_\phi)^3 e_0 = |\iota V|^{-3/2} \sum_{u,v,w \in \iota V} \exp(2\pi i(\rho_\Phi(\phi)(v+w) + \beta_\phi(u,v)))e_u \,.$$

Since $(h_{\iota,u_\iota,v_\iota} d_\phi)^3 \in \mathcal{C}(\rho)$ is a scalar element, we may put $u = 0$ in the last equation. Then $\beta_\phi(0,v) = 0$ for all $v \in V$. Since the sum over $w \in \iota V$ is constant for all $v \in \iota V$, we get

$$(h_{\iota,u_\iota,v_\iota} d_\phi)^3 e_0 = |\iota V|^{-1/2} \sum_{w \in \iota V} \exp(2\pi i \rho_\Phi(\phi)(w))e_0 \,,$$

as required. $\square$

Remark 5.4.8. More generally, any word w that evaluates to the identity in $\mathfrak{U}(R,\Phi)$ induces a scalar element in $\mathcal{C}(\rho)$ for every finite representation ρ of (R,Φ), and thus a map ν_w from the set of isomorphism classes of representations of (R,Φ) to the multiplicative group $\mathbb{C}^*$. Since the Clifford-Weil generators associated with a direct sum representation are the tensor products of the generators associated with the summands, ν_w is a semigroup map; since the Clifford-Weil groups of Witt-null representations have invariants, ν_w is trivial on the Witt-null subsemigroup.

In other words, ν_w induces a homomorphism from the Witt group $W(R,M,\psi,\Phi)$ (see Definition 4.6.11) to $\mathbb{C}^*$; since Witt groups of finite form rings are finite, we can replace $\mathbb{C}^*$ by the group of roots of unity, or equivalently by $\mathbb{Q}/\mathbb{Z}$. Since we also have $\nu_w \nu_{w'} = \nu_{ww'}$, this correspondence defines a central extension

$$\mathcal{C}(R,\Phi) := \mathfrak{U}(R,\Phi). \operatorname{Hom}(W(R,M,\psi,\Phi),\mathbb{Q}/\mathbb{Z}) \,, \tag{5.4.6}$$

which we call the *universal Clifford-Weil group*.

By the above construction, we have the following result.

Theorem 5.4.9. *For any representation ρ, the Clifford-Weil group $\mathcal{C}(\rho)$ is the image of a representation of the universal Clifford-Weil group $\mathcal{C}(R,\Phi)$ compatible with the scalar action of $\operatorname{Hom}(W(R,M,\psi,\Phi),\mathbb{Q}/\mathbb{Z})$ induced by ρ.*

Remark. As we remarked above, this gives another proof that Clifford-Weil groups are finite, as they are all quotients of the finite group $\mathcal{C}(R,\Phi)$.

Corollary 5.4.10. *For any idempotent ι and any element $\phi \in \Phi[\iota]$ that is nonsingular with respect to ι, the map $\rho \mapsto \gamma_\rho(\phi)$ induces a homomorphism from $W(R, M, \psi, \Phi)$ to the group of roots of unity.*

Remark 5.4.11. In particular, the Gauss sum is invariant under taking quotients, and thus one can use isotropic codes to simplify the computation of Gauss sums. In fact, even in anisotropic representations, one can sometimes use self-orthogonal codes to simplify Gauss sums. Suppose ϕ is nonsingular with respect to $\iota = 1$ (the case of general idempotents is analogous), and let C be a self-orthogonal code in the representation ρ. Then

$$
\begin{aligned}
\gamma_\rho(\phi) &= |V|^{-1/2} \sum_{v \in V} \exp(2\pi i \rho_\Phi(\phi)(v)) \\
&= |V|^{-1/2} \sum_{v \in V/C} \sum_{w \in C} \exp(2\pi i \rho_\Phi(\phi)(v + w)) \\
&= |V|^{-1/2} \sum_{v \in V/C} \exp(2\pi i \rho_\Phi(\phi)(v)) \\
&\qquad\qquad \sum_{w \in C} \exp(2\pi i [\rho_\Phi(\phi)(w) + \beta(\psi^{-1}(\lambda(\phi))\epsilon^{-1}v, w)]),
\end{aligned}
$$

where the sum over $v \in V/C$ is over an arbitrary system of coset representatives. Since the exponent of the inner sum is linear in w, the inner sum is 0 unless

$$
\rho_\Phi(\phi)(w) = \beta(-\psi^{-1}(\lambda(\phi))\epsilon^{-1}v, w)
$$

for all $w \in C$, or in other words unless

$$
v \in -\epsilon \psi^{-1}(\lambda(\phi))^{-1} S_\phi(C),
$$

when the inner sum is $|C|$. We thus find

$$
\gamma_\rho(\phi) = [C^\perp : C]^{-1/2} \sum_{v \in S_\phi(C)/C} \exp(2\pi i \rho_\Phi(\phi)(-\epsilon \psi^{-1}(\lambda(\phi))^{-1}v)).
$$

When C is isotropic, this simplifies to

$$
\gamma_\rho(\phi) = [C^\perp : C]^{-1/2} \sum_{v \in C^\perp/C} \exp(2\pi i \rho_\Phi(\phi)(v)) = \gamma_{\rho/C}(\phi),
$$

as expected.

Lemma 5.4.12. *Let ρ be a finite representation of the form ring (R, M, ψ, Φ), and suppose (I, Γ) is a form ideal such that*

$$
IV_\rho = 0, \quad \rho_\Phi(\Gamma) = 0.
$$

Then $\mathcal{C}(\rho)$ is naturally isomorphic to $\mathcal{C}(\rho')$, where ρ' is the corresponding representation of $(R/I, \Phi/\Gamma)$.

Proof. The generators of $\mathcal{C}(\rho')$ lift to generators of $\mathcal{C}(\rho)$, and different lifts have the same action on $\mathbb{C}[V]$. $\square$

Theorem 5.4.13. *Fix a finite form ring* (R, M, ψ, Φ). *A finite representation* ρ *is Witt-null if and only if the scalar subgroup of* $\mathcal{C}(\rho)$ *is trivial.*

Proof. If ρ is Witt-null, then $\mathcal{C}(\rho)$ has a nontrivial invariant, by Theorem 5.5.1 below, so must have a trivial scalar subgroup.

Conversely, assume $\mathcal{C}(\rho)$ has a trivial scalar subgroup. We proceed by induction on the sizes of R and Φ. Since passing to a quotient representation leaves the scalar subgroup of $\mathcal{C}(\rho)$ unchanged, we may assume that ρ is anisotropic; by Lemma 5.4.12, we may also assume that ρ is faithful (ρ_M and ρ_Φ are injective). Moreover, $\ker \lambda$ must be trivial, since otherwise Lemma 5.4.3, faithfulness, and anisotropy would force nontrivial scalar elements into $\mathcal{C}(\rho)$. But then the argument of Theorem 4.6.15 implies that (R, Φ) must be semisimple; as product form rings give rise to tensor product Clifford-Weil groups, it suffices to consider the case when (R, Φ) is simple. Let ι be a minimal symmetric idempotent in R. The faithful, anisotropic representations of (R, Φ) correspond via Morita equivalence to the faithful, anisotropic representations of the form ring $\iota(R, \Phi)$; moreover there is a natural (subject to the choice of isomorphism $\iota R \cong \iota^J R$) embedding of $\mathcal{C}(\iota(\rho))$ in $\mathcal{C}(\rho)$. It therefore suffices to consider the case when R is commutative. We have the following four possibilities for (R, Φ), up to rescaling:

1. $(\mathbb{F}_q \times \mathbb{F}_q, \mathbb{F}_q)$: $(x, y)^J \epsilon = (y, x)$, $\{\psi(x, y)\} = x + y$, $\lambda(x) = \psi(x, x)$,
2. $(\mathbb{F}_{q^2}, \mathbb{F}_q)$: $x^J \epsilon = x^q$, $\{\psi(x)\} = x + x^q$, $\lambda(x) = \psi(x)$,
3. $(\mathbb{F}_q, \mathbb{F}_q)$: $x^J \epsilon = x$, $\{\psi(x)\} = 2x$, $\lambda(x) = x$,
4. $(\mathbb{F}_q, 0)$: $x^J \epsilon = -x$, $\{\psi(x)\} = 0$.

Cases 1 and 4 have no faithful anisotropic representations (in case 1, $(0, 1)V$ is isotropic).

In case 2, a faithful anisotropic representation must be one-dimensional (since $(1, x)$ is isotropic if $x^{q+1} = -1$), so is in fact equivalent to the representation determined by $\rho_\Phi(1)(x) = \frac{1}{p} \mathrm{Tr}_{\mathbb{F}_q/\mathbb{F}_p}(x^{q+1})$. But then

$$\gamma_\rho(1) = |q|^{-1} \sum_{x \in \mathbb{F}_{q^2}} \exp(\frac{2\pi i}{p} \mathrm{Tr}_{\mathbb{F}_q/\mathbb{F}_p}(x^{q+1}))$$

$$= -1 + (1 + |q|^{-1}) \sum_{x \in \mathbb{F}_q} \exp(\frac{2\pi i}{p} \mathrm{Tr}_{\mathbb{F}_q/\mathbb{F}_p}(x))$$

$$= -1,$$

and thus $\mathcal{C}(\rho)$ has a nontrivial scalar element.

In case 3, q odd, all representations are isomorphic to a direct sum of some number of copies of the two anisotropic one-dimensional representations

$$\rho_\Phi(1)(x) = \frac{1}{p} \operatorname{Tr}_{\mathbb{F}_q/\mathbb{F}_p}(x^2),$$

$$\rho'_\Phi(1)(x) = \frac{1}{p} \operatorname{Tr}_{\mathbb{F}_q/\mathbb{F}_p}(\nu x^2),$$

where ν is a quadratic nonresidue; the only other anisotropic representation is $\rho + \bar{\rho}'$. We find that $\gamma_\rho(1) = -\gamma_{\rho'}(1) = -\gamma_\rho(\nu) = \gamma_{\rho'}(\nu)$, and thus for each anisotropic representation at least one of the two Gauss sums is nontrivial.

Finally, in case 3, q even, the only anisotropic representation (up to isomorphism) is the two-dimensional representation specified by

$$\rho_\Phi(1)((x,y)) = \frac{1}{2} \operatorname{Tr}_{\mathbb{F}_q/\mathbb{F}_2}(x^2 + xy + \omega y^2),$$

where ω is such that $\operatorname{Tr}_{\mathbb{F}_q/\mathbb{F}_2}(\omega) = 1$. We can write this as $\frac{1}{2}\operatorname{Tr}_{\mathbb{F}_q/\mathbb{F}_2}((x + \xi y)^{q+1})$, where $\xi \in \mathbb{F}_{q^2}$ has $\xi + \xi^q = 1$, at which point the computation from case 2 again gives a nontrivial Gauss sum.   □

Corollary 5.4.14. *For any finite representation ρ of a form ring (R, M, ψ, Φ), the scalar subgroup of $\mathcal{C}(\rho)$ is μ_m, where m is the order of ρ in the Witt group.*

Proof. If exponentiating by the positive integer m annihilates the scalar subgroup of $\mathcal{C}(\rho)$, then $\mathcal{C}(m\rho)$ has trivial scalar subgroup, so $m\rho$ is Witt-null.   □

Note that m here is the integer guaranteed by Corollary 4.6.16, having the property that an isotropic self-dual code of Type ρ and length N exists if and only if N is a multiple of m.

5.5 Clifford-Weil groups and full weight enumerators

Let ρ be a finite representation of the finite form ring (R, M, ψ, Φ). Recall that by Remark 5.1.5 the invariant space of $P(\rho)$ is spanned by the full weight enumerators of isotropic codes in ρ. In the light of this result, it is natural to ask if a similar result holds for the invariants of the Clifford-Weil group $\mathcal{C}(\rho)$. We can give only a partial answer.

Theorem 5.5.1. *For any self-dual isotropic code C, $\operatorname{fwe}(C)$ is invariant under $\mathcal{C}(\rho)$.*

Proof. Indeed, this is true for the parabolic subgroup of $\mathcal{C}(\rho)$, so it suffices to verify it for the generators of the form $h_{\iota, u_\iota, v_\iota}$. Moreover, by Theorem 3.5.9, ιC is self-dual; we can therefore restrict to the case $\iota = 1$, and then further assume $u_\iota = v_\iota = 1$. But then the statement follows from Example 2.2.6.   □

As before, we refer to the $\mathcal{C}(\rho)$-invariant elements in $\mathbb{C}[V]$ as vector invariants of $\mathcal{C}(\rho)$, and call the space consisting of the vector invariants the *invariant space* of $\mathcal{C}(\rho)$.

We believe the following to be true:

Conjecture 5.5.2. The Weight Enumerator Conjecture (1). *Let ρ be a finite representation of a finite form ring. Then the invariant space of $\mathcal{C}(\rho)$ is spanned by the full weight enumerators* fwe(C) *of isotropic self-dual codes C in ρ.*

Unfortunately, we can only prove this conjecture for certain (nevertheless very general) classes of representations.

Note that Theorems 5.4.13 and 5.5.1 together produce a weak version of the conjecture:

Theorem 5.5.3. *Let ρ be a finite representation of a finite form ring. An isotropic self-dual code of Type ρ and length N exists if and only if $\mathcal{C}(\rho)$ has an invariant of degree N.*

Proof. If $\mathcal{C}(\rho)$ has an invariant then the scalar subgroup must be trivial and then by Theorem 5.4.13 a code exists. Conversely, if a code exists, by Theorem 5.5.1 its full weight enumerator is an invariant.   □

Corollary 5.5.4. *Let ρ be a finite representation of the form ring (R, M, ψ, Φ). Then the Clifford group is a central extension of the hyperbolic co-unitary group,*

$$\mathcal{C}(\rho) \cong Z \, . \, \mathfrak{U}(R, \Phi) \, ,$$

where $Z \leq \mathbb{C}^*$ is a cyclic group of scalar matrices of order

$$
\begin{aligned}
|Z| &= \gcd\{ \text{ lengths of codes of Type } \rho \} \\
&= \min\{ \text{ lengths of codes of Type } \rho \} \\
&= \gcd\{ \text{ degrees of invariants of } \mathcal{C}(\rho) \} \\
&= \min\{ \text{ degrees of invariants of } \mathcal{C}(\rho) \} \\
&= \text{ order of } \rho \text{ in the Witt group of } (R, \Phi) \, . \quad (5.5.1)
\end{aligned}
$$

In this section we will continue to work with the full weight enumerators, specializing to complete weight enumerators in §5.7.

The first class where we can prove the weight enumerator conjecture arises from triangular form rings:

Theorem 5.5.5. *Suppose ρ is a finite representation of a finite triangular form ring $\tilde{R}$. Then the space of invariants of $\mathcal{C}(\rho)$ is spanned by the full weight enumerators* fwe(C), *where C ranges over self-dual, isotropic codes in ρ.*

Proof. Let (M, Φ) be a finite quadratic pair over the finite ring R, such that ρ is a representation of the corresponding triangular form ring $\tilde{R} = T(M, \Phi) := (T(M), T(M), T(\psi), T(\Phi))$. Then, by Lemma 1.9.2, there is a representation $\rho_0 = (V_0, \rho_M, \rho_\Phi, \beta)$ of (M, Φ) such that

$$\rho = T(\rho_0) = (\tilde{V}_0 \oplus V_0, T(\rho_M), T(\rho_\Phi), T(\beta)) \, ,$$

where $\tilde{V}_0 = \mathrm{Hom}_{\mathbb{Z}}(V_0, \mathbb{Q}/\mathbb{Z})$. Define a unitary map $h' : \mathbb{C}[V_0 \times \tilde{V}_0] \to \mathbb{C}[V_0 \times V_0]$ by

$$h' e_{v,\alpha} := |V_0|^{-1/2} \sum_{w \in V_0} \exp(2\pi i \alpha(w)) e_{v,w} \,.$$

If C is a self-dual, isotropic code in $V = V_0 \oplus \tilde{V}_0$, then by Lemma 1.9.3, $C = C_0 \times C_0^*$ for some isotropic code C_0 in V_0 and $C_0^* = C_0^{\perp} \leq \mathrm{Hom}_{\mathbb{Z}}(V_0, \mathbb{Q}/\mathbb{Z}) = \tilde{V}_0$, and we have

$$h' \, \mathrm{fwe}(C) = \mathrm{fwe}(C_0 \times C_0) \,.$$

Now $C_0 \times C_0$ is an arbitrary isotropic $\mathrm{Mat}_2(R)$-submodule in V_0^2 and the full weight enumerators of these codes span the ring of invariants of $P(\mathrm{Mat}_2(\rho_0))$ by Remark 5.1.5. Thus, to prove the result, it suffices to show that

$$h' \mathcal{C}(\rho) h'^{-1} = P(\mathrm{Mat}_2(\rho_0)) \,.$$

On the one hand, every generator on the left is readily verified to be in the parabolic group on the right. By Example 5.2.7

$$\mathfrak{U}(T(M, \Phi)) \cong P(\mathrm{Mat}_2(R), \Phi_2) \,,$$

giving the desired result since both groups are finite. $\qquad\square$

We can extend this somewhat:

Corollary 5.5.6. *Let R be a twisted ring. Suppose R contains an idempotent ι such that $\iota + \iota^J = 1$ and $\iota R \iota^J = 0$, i.e. R is isomorphic to a triangular twisted ring. Then for any form structure Φ on R and any finite representation ρ of (R, Φ), the invariant space of $\mathcal{C}(\rho)$ is spanned by the full weight enumerators $\mathrm{fwe}(C)$, where C ranges over self-dual, isotropic codes in ρ.*

Proof. If $\ker(\lambda) = 0$, we are done, since then (R, Φ) is isomorphic to a triangular form ring. Otherwise, define a subspace $C_0 \subset V$ by

$$C_0 = \{ w \in V \mid (v \mapsto \beta(w, v)) \in \rho(\ker \lambda) \} \,.$$

Then C_0 is contained in the dual of any isotropic code, so must in particular be contained in any self-dual, isotropic code. If C_0 is not isotropic, then no self-dual, isotropic code exists, but also $\mathcal{C}(\rho)$ has no invariants by Corollary 5.4.4. If C_0 is isotropic, then we can quotient out by C_0, at which point $\ker \lambda$ is contained in the annihilator of $C_0^{\perp}/C_0$; by induction, we can reduce to the triangular case. $\qquad\square$

The other situation where we can prove the conjecture is when R is a finite "quasi-chain" ring. Recall that a finite *chain ring* is a finite ring in which the left ideals form a chain under inclusion (cf. Honold and Landjev [267], McDonald [369]). We define a *quasi-chain ring* to be a product of matrix rings over finite chain rings; examples include matrix rings over finite fields and matrix rings over $\mathbb{Z}/m\mathbb{Z}$.

Theorem 5.5.7. *Let R be a finite quasi-chain ring. For any finite representation ρ of a form ring (R, M, ψ, Φ), the invariant space of $\mathcal{C}(\rho)$ is spanned by the full weight enumerators* fwe(C), *where C ranges over self-dual, isotropic codes in ρ.*

Proof. Suppose first that R contained a central symmetric idempotent ι other than 0 or 1. Then ι induces a factorization

$$\mathcal{C}(\rho) = \mathcal{C}(\iota\rho) \otimes \mathcal{C}((1 - \iota)\rho) .$$

Since the invariant space of a tensor product of groups is the tensor product of the respective invariant spaces, we can reduce to considering $\iota\rho$ and $(1 - \iota)\rho$. Thus we may assume that 0 and 1 are the only central symmetric idempotents in R. Similarly, if R contains a central idempotent $\iota \neq 0, 1$, then $\iota + \iota^J = 1$, since it is a central symmetric idempotent $\neq 0$, and $\iota\iota^J = 0$. Hence R is isomorphic to $R' \oplus R'$, where the involution J interchanges the two components. In particular R satisfies the hypotheses of Corollary 5.5.6, so again we are done.

We can therefore reduce to the case when $R = \mathrm{Mat}_m(R_0)$, where R_0 is a twisted chain ring. By Theorem 4.5.4, we then have

$$(R, \Phi) = \mathrm{Mat}_m(R_0, \Phi_0), \quad \rho = \rho_0 \otimes (R_0^m)^*, \quad V = V_0 \otimes (R_0^m)^* .$$

The $\mathrm{Mat}_m(R_0)$-submodules in V are of the form $C^m := C \otimes (R_0^m)^*$, where $C \leq V_0$ is an R_0-submodule. We observe that the invariants of the parabolic subgroup of $\mathcal{C}(\rho)$ are thus spanned by fwe(C^m), where C ranges over the isotropic codes in V_0.

We will need the following lemma about chain rings:

Lemma 5.5.8. *Let R_0 be a finite chain ring. For any finite left R_0-module M, the cyclic submodules of M form a rooted tree under inclusion.*

Proof. The root of the tree is of course provided by $\{0\}$. It clearly suffices to prove the lemma when M is cyclic. Let I be the ideal such that M is isomorphic to R_0/I. Then the submodules of M are in one-to-one correspondence with the ideals in R_0 containing I. Since these form a chain, the submodules of M form a chain, and thus the same is true for the cyclic submodules of M. □

Let ι be a primitive idempotent in R. Note that all such idempotents are equivalent. In particular ι and ι^J are equivalent, hence ι is a symmetric idempotent. After rescaling we may assume that $\iota^J = \iota$. Then $R_0 \cong \iota R\iota$ and $V_0 := \rho(\iota)V$ is a finite R_0-module.

Lemma 5.5.9. *With the notation above, let X_P denote the operation of averaging over the parabolic subgroup $P(\rho)$ of $\mathcal{C}(\rho)$. Then, for any isotropic code C in V_0,*

$$X_P \, h_{\iota,\iota,\iota} \, \mathrm{fwe}(C^m) \; = \; \sum_{\substack{C' \subset C^\perp \\ \exists x \in V_0 \,:\, C'=\langle C, x \rangle}} n_{C'} \, \mathrm{fwe}((C')^m) \,,$$

where $n_C < 1$ unless $C = C^\perp$.

Proof. It suffices to prove this when $C = 0$, since otherwise we can simply quotient out by C. Note that isotropic codes are self-orthogonal by definition. We first calculate that

$$h_{\iota,\iota,\iota} \, \mathrm{fwe}(0^m) = |V_0|^{-1/2} \, \mathrm{fwe}(V_0 \times 0^{m-1}) \,.$$

Averaging over the parabolic subgroup, we obtain

$$X_P h_{\iota,\iota,\iota} \, \mathrm{fwe}(0^m) = |V_0|^{-1/2} |R^*|^{-1} \sum_{v \in V_0} |\varPhi|^{-1} \sum_{\phi \in \varPhi} \exp(2\pi i \phi(v)) \sum_{u \in R^*} e_{u(v,0,\ldots,0)^{tr}}$$

$$= \sum_{\substack{\text{isotropic } v \,\in\, V_0}} |V_0|^{-1/2} |R^*|^{-1} \sum_{u \in R^*} e_{u(v,0,\ldots,0)^{tr}}$$

$$= \sum_{\substack{\text{isotropic } D \,\leq\, V \\ \exists x \in V_0 \,:\, D=\langle (x,0\ldots,0)^{tr} \rangle}} n'_D \mu_D \,, \tag{5.5.2}$$

where, for $D = \langle (v,0\ldots,0)^{tr} \rangle R$,

$$n'_D = |V_0|^{-1/2} |R^*|^{-1} |\, \mathrm{Stab}_{R^*}((v,0,\ldots,0)^{tr})| \,;$$

in particular $n'_0 = |V_0|^{-1/2}$. If $D \neq \{0\}$ then, since R_0 is a chain ring, D contains a unique maximal submodule D' that is generated by an element in V_0, and we find

$$\mu_D = \mathrm{fwe}(D) - \mathrm{fwe}(D') \,.$$

Substituting this in (5.5.2), we obtain

$$X_P h_{\iota,\iota,\iota} \, \mathrm{fwe}(0^m) = \sum_{\substack{\text{isotropic } C' \,\leq\, V \\ \exists x \in V_0 \,:\, C'=\langle (x,0,\ldots,0)^{tr} \rangle}} n_{C'} \, \mathrm{fwe}(C') \,,$$

where $n_{C'} = n'_{C'} - \sum_D n'_D$, and the sum is over cyclic modules D generated by an element of V_0 that are minimal among those containing C'. In particular, we find that $n_0 \leq n'_0 = |V_0|^{-1/2}$, which is smaller than 1 unless V_0 is trivial; i.e. unless 0 is self-dual. $\square$

We also need [383, Lemma 4.8]. Since the proof in [383] was somewhat imprecise, we give a different proof here.

Lemma 5.5.10. *Let V be a finite dimensional vector space, M a linear transformation on V, and $(P, \leq)$ a partially ordered set. Suppose there exists a spanning set v_p of V indexed by $p \in P$ on which M acts triangularly; that is,*

$$Mv_p = \sum_{q \geq p} c_{pq} v_q \,,$$

for suitable coefficients c_{pq}. Suppose furthermore that $c_{pp} = 1$ if and only if p is maximal in P. Then the fixed subspace of M in V is spanned by the elements v_p for p maximal.

Proof. Let $\leq'$ be a linear extension of $\leq$ such that the maximal elements of P are larger than all non-maximal elements of P. Let $\sum_p a_p v_p = 0$ be a linear relation among the v_p and let p_0 be the $\leq'$-minimal element with $a_{p_0} \neq 0$. Then this relation just expresses v_{p_0} in terms of the vectors v_q for $q >' p_0$. In such a case, we can clearly remove p_0 from P without affecting the form of the matrix M. The only time a maximal p_0 is removed is when we have a dependence between the v_p with p maximal. We may thus assume that the vectors v_p are actually linearly independent. But then the characteristic polynomial of M is $\prod_{p \in P}(\lambda - c_{pp})$. By assumption, the number of times 1 appears as a root is exactly the number of maximal elements of P; since we can exhibit that many linearly independent fixed vectors of M, they must form a basis of the fixed subspace. □

If a parabolic invariant p is invariant under $\mathcal{C}(\rho)$, then it must in particular satisfy

$$X_P \, h_{\iota, \iota, \iota} \, p \; = \; p \,,$$

so the theorem follows from the Lemma 5.5.9 and Lemma 5.5.10. This completes the proof of Theorem 5.5.7. □

Remark. Although Theorem 5.5.7 may be true for more general classes of rings than quasi-chain rings, the chain ring condition is necessary for our proof. If R_0 is not a chain ring, we again get an expression of the form

$$n_0 = n_0' + \sum_D c_D n_D'$$

via Möbius inversion, but the coefficients c_D need not be negative, and we may not obtain a useful bound on n_0. Thus in the above argument we need to assume that the Möbius function of the poset of cyclic modules over R_0 is negative off the diagonal: this is precisely the chain condition. If suitable bounds on the coefficients c_D compared to the size of n_D' could be obtained, it might be possible to enlarge the class of rings to which Theorem 5.5.7 applies. Note however that, even given such bounds, the above argument depends strongly on the fact that R is a product of rings each of which has a unique equivalence class of primitive idempotents. For example, the two rings mentioned in the footnote on page vi do not satisfy this latter condition.

5.6 Results from invariant theory

5.6.1 Molien series

Before proceeding further, we digress to give some general remarks about rings of invariants and Molien's theorem. For further information about this theorem, see the following references: [383], [454], [500], as well as Benson [33], Derksen and Kemper [144, Chap. 3], Molien [373], Smith [506], Stanley [512] and Sturmfels [516, p. 29].

The usual way to get detailed information about the structure of a ring of polynomial invariants starts by computing the corresponding Molien series. Let $\rho : G \to GL_n(\mathbb{C})$ be an n-dimensional representation of a finite group G. Then G acts on the polynomial ring

$$\mathbb{C}[x_1, \ldots, x_n] \cong \bigoplus_{N=0}^{\infty} \mathrm{Sym}_N(\mathbb{C}^n),$$

where $\mathrm{Sym}_N(\mathbb{C}^n)$, the N-th symmetric power of the G-module $\mathbb{C}^n$, is isomorphic to the space of homogeneous polynomials of degree N. Let

$$\mathrm{Inv}(G) = \mathbb{C}[x_1, \ldots, x_n]^G \cong \bigoplus_{N=0}^{\infty} \mathrm{Sym}_N(\mathbb{C}^n)^G$$

denote the *invariant ring*, that is, the ring of G-invariant polynomials, and let $\alpha(N) := \dim(\mathrm{Sym}_N(\mathbb{C}^n)^G)$ be the dimension of the vector space of homogeneous polynomial invariants of degree $N \geq 0$. Then Molien's theorem states that

$$\sum_{N=0}^{\infty} \alpha(N)t^N = \frac{1}{|G|} \sum_{g \in G} \frac{1}{\det(I_n - t\rho(g))}. \tag{5.6.1}$$

This generating function is called the *Molien series* for $\rho(G)$, and will be denoted by $\mathrm{MS}_G(t)$, or simply MS when the group is clear from the context.

Definition 5.6.1. A *good polynomial basis* for $\mathrm{Inv}(G)$ consists of homogeneous invariants $f_1, \ldots, f_\ell$ ($\ell \geq n$), where $f_1, \ldots, f_n$ are algebraically independent, such that either

$$\mathrm{Inv}(G) = \mathbb{C}[f_1, \ldots, f_n] \quad \text{if} \quad \ell = n, \tag{5.6.2}$$

or, if $\ell > n$,

$$\mathrm{Inv}(G) = \mathbb{C}[f_1, \ldots, f_n] \oplus f_{n+1}\mathbb{C}[f_1, \ldots, f_n] \oplus \cdots \oplus f_\ell\mathbb{C}[f_1, \ldots, f_n]. \tag{5.6.3}$$

In words, this says that any invariant of G can be written as a polynomial in $f_1, \ldots, f_n$ (if $\ell = n$), or as such a polynomial plus f_{n+1} times another such polynomial plus $\cdots$ plus f_ℓ times another such polynomial (if $\ell > n$). $f_1, \ldots, f_n$ are called *primary* or *basic* invariants and $1, f_{n+1}, \ldots, f_\ell$ (if $\ell > n$)

are the *secondary* invariants. Speaking loosely, (5.6.2) and (5.6.3) say that when describing an arbitrary invariant, $f_1, \ldots, f_n$ are "free" and can be used as often as needed, while $f_{n+1}, \ldots, f_\ell$ are "transients" and can each be used at most once. Equation (5.6.2) or (5.6.3) is the *Hironaka decomposition* of Inv(G).

Since the f_i are polynomials in n variables, if $\ell > n$ there must be algebraic dependencies (called *syzygies*) involving $f_{n+1}, \ldots, f_\ell$. If $\ell = n$ there are no syzygies, and Inv(G) is itself a polynomial ring. If $\ell > n$ there are $\binom{\ell-n+1}{2}$ syzygies expressing the products $f_i f_j$ ($n+1 \le i \le j \le \ell$) in terms of $f_1, \ldots, f_\ell$.

Note that the Molien series can be written down by inspection from the degrees of a good polynomial basis. Let $d_1 = \deg f_1, \ldots, d_\ell = \deg f_\ell$. Then

$$\mathrm{MS}_G(t) = \frac{1}{\prod_{i=1}^{n}(1 - t^{d_i})}, \quad \text{if} \quad \ell = n, \tag{5.6.4}$$

or

$$\mathrm{MS}_G(t) = \frac{1 + \sum_{j=n+1}^{\ell} t^{d_j}}{\prod_{i=1}^{n}(1 - t^{d_i})}, \quad \text{if} \quad \ell > n. \tag{5.6.5}$$

In this situation we will describe the invariant ring by writing a symbol like

$$\mathrm{Inv}_G = \frac{1}{f_1, f_2, \ldots, f_n}, \tag{5.6.6}$$

in the first case, or

$$\mathrm{Inv}_G = \frac{1, f_{n+1}, f_{n+2}, \ldots, f_\ell}{f_1, f_2, \ldots, f_n}, \tag{5.6.7}$$

in the second case.

One of the themes of this book is that weight enumerators of self-dual codes provide bases for these invariant rings. If a good polynomial basis for a ring can be found by taking f_i to be the weight enumerator of a code C_i, for $i = 1, \ldots, \ell$, we will indicate this by writing a symbol like

$$\mathrm{Inv}_G = \frac{1}{C_1, C_2, \ldots, C_n}, \tag{5.6.8}$$

in the first case, or

$$\mathrm{Inv}_G = \frac{1, C_{n+1}, C_{n+2}, \ldots, C_\ell}{C_1, C_2, \ldots, C_n}, \tag{5.6.9}$$

in the second case[1]. In the tables in the later chapters we will use a semicolon to separate the codes whose weight enumerators are algebraically independent, thus:

$$C_1, C_2, \ldots, C_n \, ; C_{n+1}, C_{n+2}, \ldots, C_\ell, \tag{5.6.10}$$

as in §8.2, for example.

---

[1] The "1" in the numerator of (5.6.9) is by convention, and refers to the "1" in the numerator of (5.6.7), not to a code!

Remark 5.6.2. A beautiful theorem of Chevalley [108] and Shephard and Todd [482] (see also Benson [33], Smith [506], Stanley [512]) states that the finite groups whose rings of invariants are polynomial rings are precisely the finite complex reflection groups. The complete list of the irreducible groups is given in [33], [482], [506]. Thus when the numerator of a Molien series is 1, the group is very often a reflection group, and more generally when the numerator is simple, the group is often closely related to a reflection group. This happens surprisingly often for Clifford-Weil groups, as shown by the (possibly incomplete) list given in Table 5.2. This behavior is the exception rather than the rule, as shown by Huffman and Sloane [283]. The first column in Table 5.2 gives the Type and the relevant weight enumerator, and the second column gives the order of the Clifford-Weil group. The third column specifies the associated reflection group. If the reflection group belongs to a real root system, we give its root system name (cf. Humphreys [287]). So A_n, B_n, D_n, G_2, H_3, F_4 are used to denote the real reflection groups isomorphic to S_{n+1}, $Z_2 \wr S_n$, $Z_2^{n-1} \rtimes S_n$, $Z_2 \times S_3$, $Z_2 \times A_5$, $2_+^{1+4}.(Z_3 \times Z_3).(Z_2 \times Z_2)$, respectively. The empty root system is denoted by $\emptyset$ and the "×" separates the different irreducible components. As usual we use Z_n to denote the cyclic group of order n in $GL_1(\mathbb{C})$ (instead of $G_3(n)$), and $I_2(8)$ is the dihedral group of order 16. For the other complex reflection groups we use the numbering introduced in [482] and denote the i-th Shephard-Todd group by X_i. If the Clifford-Weil group is a subgroup or supergroup of the reflection group, this is also indicated in the third column. The index refers to the index of the Clifford-Weil group in the complex reflection group; index $1/2$ means that the Clifford-Weil group contains the relevant complex reflection group as a subgroup of index 2. Reflection groups of order 2 have been omitted.

Remark 5.6.3. If R is a subring of $\mathbb{C}[x_1, \ldots, x_n] \cong \sum_{N=0}^{\infty} \mathrm{Sym}_N(\mathbb{C}^n)$, and $\alpha(N) := \dim(R \cap \mathrm{Sym}_N(\mathbb{C}^n))$, then the generating function $\sum_{N=0}^{\infty} \alpha(N) t^N$ is often called the Poincaré series for R. However, since such rings in this book will usually (although not always) arise as invariant rings for some group, we will always use the term Molien series for the generating function. (This is consistent with the usage elsewhere in the literature.) Similarly, we will use the symbols on the right-hand sides of (5.6.8) or (5.6.9) to describe the Hironaka decomposition of R, even if R is not actually the invariant ring of a group.

Remark 5.6.4. Many Molien series will be mentioned in this book, so to help keep track of them and to make it easy to see their coefficients we will give the appropriate identification number (e.g. [A008718]) for the corresponding entry in [504]. (The entry in [504] actually gives the coefficients of the "reduced" Molien series $\mathrm{MS}(t^{1/a})$, where a is the largest integer that still produces a power series in t.)

Table 5.2. Clifford-Weil groups that are closely related to complex reflection groups. The notation is explained in Remark 5.6.2.

Type (weight enum.)	Order	Group	See
2_I (hwe)	16	$I_2(8)$	(6.3.5)
2_I (genus 2)	2304	$F_4 = X_{28}$ (index $1/2$)	(6.3.8)
2_{II} (hwe)	192	X_9	(6.3.5)
2_{II} (genus 2)	92160	X_{31} (index $1/2$)	(6.4.5)
2^{lin} (cwe)	6	$A_2 \times \emptyset^2$	(7.2.1)
2_1^{lin} (cwe)	24	$A_3 \times \emptyset$	(7.2.4)
$2_{1,1'}^{lin}$ (cwe)	192	D_4	(7.2.7)
3 (hwe)	48	X_6	(7.4.3)
3_1 (cwe)	2592	X_{26} (index $1/2$)	(7.4.5)
4^E (cwe)	192	B_4 (index 2)	(7.6.19)
4_{II}^E (cwe)	3840	X_{29} (index 2)	(7.6.28)
4^H (hwe)	12	G_2	(7.3.22)
4_1^H (cwe)	576	$F_4 = X_{28}$ (index 2)	(7.3.5)
4_1^{H+} (cwe)	8	$\emptyset \times B_1 \times B_2$ (index 2)	(7.6.5)
4_1^{H+} (swe)	8	$\emptyset \times B_2$	(7.6.7)
4_S^{H+} (cwe)	6	$\emptyset^2 \times A_2$	(7.6.9)
4_S^{H+} (swe)	6	$\emptyset \times A_2$	(7.6.11)
4_{II}^{H+} (hwe)	12	G_2	(2.4.26)
$4_{II,1}^{H+}$ (cwe)	48	$A_3 \times \emptyset$ (index $1/2$) and $B_3 \times A_1$ (index 2)	(7.6.24)
$4_{II,1}^{H+}$ (swe)	48	B_3	(7.6.26)
$4^{\mathbb{Z}}$ (cwe)	64	$G(4,1,2) \times Z_4 \times \emptyset$ (index 2)	(8.2.1)
$4^{\mathbb{Z}}$ (swe)	32	$G(4,1,2) \times \emptyset$	(8.2.5)
$4_S^{\mathbb{Z}}$ (cwe)	192	$G(4,1,3) \times \emptyset$ (index 2)	(8.2.16)
$4_{II}^{\mathbb{Z}}$ (swe)	768	$G(4,1,3)$ (index $1/2$)	(8.2.19)
5 (swe)	120	$H_3 = X_{23}$	(7.4.19)
9^H (swe)	48	B_3	(5.8.1)
$GR(4,2)^H$ (swe)	48	$B_3 \times \emptyset$	(8.4.2)

5.6.2 Relative invariants

More generally, we may consider the relative invariants of a group. These do not form a ring, but rather are a module for the invariant ring of the group. Relative invariants are important for the study of maximal isotropic codes, as we will see in Chapter 10—whereas the weight-enumerators of self-dual isotropic codes are invariant under the Clifford-Weil group, those of maximal self-orthogonal isotropic codes are only relative invariants.

Definition 5.6.5. Let $G \leq \mathrm{GL}_n(\mathbb{C})$ be a group (assumed to be finite or reductive), let (S, s) be a *pointed* irreducible representation of G; that is, S is an irreducible representation and s is a nonzero vector in S. An (S, s)-*relative invariant* of G is a polynomial $p \in \mathbb{C}[x_1, \ldots, x_n]$ such that the representation of G spanned by the images of p is isomorphic to S in such a way that p is identified with s. Then we define $\mathrm{Inv}(G, S)_s$ to be the set of all such relative invariants, together with 0.

The properties of $\mathrm{Inv}(G, S)_s$ are summarized in the following remark.

Remark 5.6.6. (1) $\mathrm{Inv}(G, S)_s$ is a vector space, which inherits a grading from $\mathbb{C}[x_1, \ldots, x_n]$.

(2) If $1 = \langle 1 \rangle$ is the trivial representation, then $\mathrm{Inv}(G, 1)_1$ is the space of (ordinary) invariants of G.

(3) Given pointed irreducible representations (S, s), (S', s') such that $S \otimes S'$ is irreducible, there is a canonical injection

$$\mathrm{Inv}(G, S)_s \otimes \mathrm{Inv}(G, S')_{s'} \to \mathrm{Inv}(G, S \otimes S')_{s \otimes s'} .$$

If S' is 1-dimensional, this is an isomorphism. Hence $\mathrm{Inv}(G, S)_s$ is an $\mathrm{Inv}(G)$-module.

(4) The Molien series of $\mathrm{Inv}(G, S)_s$ is

$$\mathrm{MS}_{\mathrm{Inv}(G,S)_s}(t) = \frac{1}{|G|} \sum_{g \in G} \frac{\chi_S(g^{-1})}{\det(I_n - gt)} , \tag{5.6.11}$$

where χ_S is the character of S.

(5) $\mathrm{Inv}(G, S)_s$ is invariant under rescaling s.

One can also define the space $\mathrm{Inv}(G, S)$ of S-*relative invariants* to be the union of $\mathrm{Inv}(G, S)_s$ for all $s \in S$. We have the isomorphisms

$$\mathrm{Inv}(G, S)_s \cong \mathrm{Hom}_{\mathbb{C}G}(S, \mathbb{C}[x_1, \ldots, x_n]) ,$$
$$\mathrm{Inv}(G, S) \cong S \otimes_{\mathbb{C}} \mathrm{Inv}(G, S)_s . \tag{5.6.12}$$

The Molien series for $\mathrm{Inv}(G, S)$ is (cf. Benson [33, Theorem 2.5.3])

$$\mathrm{MS}_{\mathrm{Inv}(G,S)}(t) = \frac{\dim(S)}{|G|} \sum_{g \in G} \frac{\chi_S(g^{-1})}{\det(I_n - gt)} . \tag{5.6.13}$$

We may view $\mathrm{Inv}(G, S)_s$ as the space of polynomials that transform under G in the same way that s transforms in the G-module S. More precisely, since S is simple, $S = \langle sG \rangle$, and so S has a $\mathbb{C}$-basis of the form $(s, sg_2, \ldots, sg_d)$ with $g_2, \ldots g_d \in G$. Then, for all $g \in G$, the image sg is given by $\sum_{i=1}^{d} a_{gi} sg_i$ (where we put $g_1 := 1$). If $p \in \mathrm{Inv}(G, S)_s$ then, for all $g \in G$,

$$gp = p(xg) = \sum_{i=1}^{d} a_{gi} p(xg_i) . \tag{5.6.14}$$

5.6.3 Construction of invariants using differential operators

A convenient way to construct (relative) invariants is via invariant differential operators.

For a polynomial $p(x) \in \mathbb{C}[x_1, \ldots, x_n]$, let $p(\partial) := p(\partial_1, \ldots, \partial_n)$ be the corresponding differential operator, where $\partial_i = \frac{\partial}{\partial x_i}$ is the i-th partial derivative. Applying the chain rule we have the following:

Lemma 5.6.7. *Let $g \in \mathrm{GL}_n(\mathbb{C})$ and $p, q \in \mathbb{C}[x_1, \ldots, x_n]$. Then*

$$(p(\partial)q)(xg) = p(\partial g^{-tr})(q(xg)).$$

Proof. Since the equation is linear in p, it is enough to prove it in the case when p is a monomial $x_1^{n_1} \cdots x_n^{n_n}$. We argue by induction on the total degree $n_1 + \ldots + n_n$. Assume without loss of generality that $n_1 > 0$. Then $p = x_1 p_1$. Let $q_1 := \frac{\partial}{\partial x_1} q$. Then, by the chain rule,

$$q_1(xg) = ((\partial_1 g^{-tr})(q \circ g))(x).$$

By induction,

$$(p(\partial)q)(xg) = (p_1(\partial)(\frac{\partial}{\partial x_1} q))(xg) = (p_1(\partial)q_1)(xg)$$
$$= p_1(\partial g^{-tr})(q_1(xg)) = p(\partial g^{-tr})(q(xg)). \qquad (5.6.15)$$

$\square$

Let $G^* = \{g^{tr} \mid g \in G\} \leq \mathrm{GL}_n(\mathbb{C})$. The mapping $g \mapsto g^{-tr}$ is an isomorphism from G to G^*, via which any G-module also becomes a G^*-module. We have the following corollary:

Corollary 5.6.8. *Let $G \leq \mathrm{GL}_n(\mathbb{C})$ be a finite group and S an irreducible $\mathbb{C}G$-module. If $s \in S \setminus \{0\}$, $q \in \mathrm{Inv}(G)$ and $p \in \mathrm{Inv}(G^*, S)_s$ then*

$$p(\partial)(q) \in \mathrm{Inv}(G, S)_s.$$

In particular, if G is unitary, i.e. $g = \bar{g}^{tr}$ for all $g \in G$ (this applies to all the Clifford-Weil groups $G = \mathcal{C}(\rho)$), then $\overline{\mathrm{Inv}(G^{-tr}, S)} \cong \mathrm{Inv}(G, S^*)$, where S^* is the dual module of S. Hence, for any $p \in \mathrm{Inv}(G, S^*)$ and $q \in \mathrm{Inv}(G)$, the polynomial $\bar{p}(\partial)(q) \in \mathrm{Inv}(G, S)$, where $\bar{p}$ is obtained from p by applying complex conjugation to all the coefficients. This is a useful way to construct new (relative) invariants. If p is a weight enumerator of a code, then these coefficients are integers and $p = \bar{p}$.

5.6.4 Invariants and designs

Another reason for studying the Clifford-Weil groups and their invariants is that they can be used to construct spherical designs. This topic will re-appear in the next chapter, but since it involves the Molien series it is appropriate to discuss it here. For simplicity we restrict the discussion to the real case, that is, to subsets of the unit sphere in $\mathbb{R}^n$.

Definition 5.6.9. A set of M points $\mathcal{P} = \{P_1, \ldots, P_M\}$ on the unit sphere $\Omega_n := S^{n-1} := \{x = (x_1, \ldots, x_n) \in \mathbb{R}^n \mid x \cdot x = 1\}$ forms a *spherical s-design* if the identity

$$\int_{\Omega_n} f(x)d\mu(x) = \frac{1}{M} \sum_{i=1}^M f(P_i) \tag{5.6.16}$$

holds for all polynomials f of degree $\leq s$, where μ is the uniform measure on the sphere normalized to have total measure 1

Of course an s-design is also an s'-design for all $s' \leq s$. The largest s for which the points form an s-design is called the *strength* of the design.

It is known that if M is large enough then an s-design in Ω_n always exists (Seymour and Zaslavsky [481]). The problem is to find the smallest value of M for a given strength and dimension, or equivalently to find the largest strength s that can be achieved with M points in Ω_n.

There are several equivalent formulations.

Theorem 5.6.10. (See for example Delsarte, Goethals and Seidel [143], Reznick [458].) *The following are equivalent:*

(a) $P_1, \ldots, P_M$ *is a spherical s-design in* Ω_n.
(b) *We have*

$$\sum_{i=1}^M f(P_i) = 0, \tag{5.6.17}$$

for all harmonic polynomials f of degrees from 1 to s (f is harmonic if the Laplacian

$$\Delta f := \sum_{j=1}^n \frac{\partial^2 f}{\partial x_j^2}$$

vanishes).
(c) *We have*

$$\sum_{i=1}^M f(P_i^T) = \sum_{i=1}^M f(P_i), \tag{5.6.18}$$

for all polynomials f of degrees from 1 to s and all orthogonal transformations $T \in O(n, \mathbb{R})$.

(d) *The polynomial identities*

$$\frac{1}{M}\sum_{i=1}^{M}(P_i \cdot x)^{2s'} = \left(\prod_{j=0}^{s'-1}\frac{2j+1}{2j+n}\right)(x \cdot x)^{s'} \tag{5.6.19}$$

and

$$\frac{1}{M}\sum_{i=1}^{M}(P_i \cdot x)^{2s''+1} = 0 \tag{5.6.20}$$

hold, where s' and s'' are the integers defined by $\{2s', 2s''+1\} = \{s-1, s\}$.

Suppose now that ρ is a real n-dimensional representation of a finite group G, and let $\alpha'(N)$ be the dimension of the vector space of homogeneous harmonic polynomial invariants of degree $N \geq 0$. These numbers are given by the *harmonic Molien series*:

$$\sum_{N=0}^{\infty}\alpha'(N)t^N = \frac{1}{|G|}\sum_{g \in G}\frac{1-t^2}{\det(I_n - t\rho(g))}$$

$$= (1-t^2)\,\mathrm{MS}_G(t) \tag{5.6.21}$$

(Goethals and Seidel [192], [193], [194]).

Let $\mathcal{P} := \{P_1, \ldots, P_M\}$ be a union of m orbits under $\rho(G)$ (not necessarily all of the same size). There are $m(n-1)$ degrees of freedom in choosing $\mathcal{P}$. Then $\mathcal{P}$ is a spherical s-design if and only if the average of f over $\mathcal{P}$ is equal to the average of f over Ω_n, for all harmonic polynomials f of degrees 1 though s. This imposes a total of $e_s := \alpha'(1) + \cdots + \alpha'(s)$ conditions on $\mathcal{P}$. So if $m(n-1) \geq e_s$ (and we are lucky), we may hope that we can choose the orbits to form an s-design.

In particular, suppose that $\alpha'(1) = \cdots = \alpha'(s) = 0$ for some s. Then there are no conditions to be satisfied, and so every orbit under $\rho(G)$ is an s-design. This result seems to be due to Sobolev ([508], cf. [133]).

Furthermore, if $\alpha'(s+1) = 1$, and ψ is the unique (up to scalars) harmonic invariant of degree m, then the orbit of any real zero of ψ is at least an $(s+1)$-design. (A real zero must exist, since the average of ψ over the sphere is zero.)

For further information about spherical designs see Bachoc [22], Bajnok [25], [26]; Delsarte, Goethals and Seidel [143]; Goethals and Seidel [192], [193], [194]; Hardin and Sloane [259], [260]; Reznick [458]; Sloane, Hardin and Cara [505] and Sobolev [508].

5.7 Symmetrizations

We now return to the main theorems, and discuss what they say about complete weight enumerators.

There is a general method for obtaining all the invariants in a suitably symmetrized space. Let G, H be finite groups acting linearly on a $\mathbb{C}$-vector-space W. Assume that the actions of G and H commute. Let S_G (resp. S_H) be irreducible $\mathbb{C}G$- (resp. $\mathbb{C}H$)-modules and define $W(S_G)$ to be the S_G-homogeneous component of W, i.e. the span of all G-submodules of W that are isomorphic to S_G, and let $W(S_H)$ be the corresponding S_H-homogeneous component of W. Then $W(S_H)$ is a $\mathbb{C}G$-submodule of W, and the S_G-homogeneous component of $W(S_H)$ is

$$(W(S_H))(S_G) = W(S_G) \cap W(S_H) = (W(S_G))(S_H).$$

In other words, the S_H-homogeneous component of the S_G-homogeneous component equals the S_G-homogeneous component of the S_H-homogeneous component.

In our situation, let $\rho := (V, \rho_M, \rho_\Phi, \beta)$ be a finite representation of a form ring. To consider codes in $N\rho$, that is, codes of Type ρ and length N, we take $G = \mathcal{C}(\rho)$, $W = \mathbb{C}[\otimes^N V]$ and $H = S_N$. Then the fixed space of H is $\mathrm{Sym}_N(\mathbb{C}[V])$, the N-th symmetrization of $\mathbb{C}[V]$, which is the space of homogeneous complex polynomials of degree N in the variables $\{x_v, v \in V\}$. The fixed space for the action of $\mathcal{C}(\rho)$ on $\mathrm{Sym}_N(\mathbb{C}[V])$ is the space of homogeneous polynomial invariants of $\mathcal{C}(\rho)$ of degree N. The above remark says that this is equal to the subspace of vector invariants of $\mathcal{C}(\rho)$ of degree N that are fixed by S_N.

This shows that whenever the Weight Enumerator Conjecture 5.5.2 holds for the full weight enumerators and vector invariants of $\mathcal{C}(\rho)$, it also holds for the complete weight enumerators and polynomial invariants (and similarly for any other symmetrization that commutes with the action of $\mathcal{C}(\rho)$).

Theorem 5.7.1. *Let ρ be a Type such that, for all $N \geq 0$, the space of vector invariants of $\mathcal{C}(N\rho)$ is spanned by the full weight enumerators of codes of Type ρ and length N. Then the space of polynomial invariants of $\mathcal{C}(\rho)$ is spanned by the complete weight enumerators of self-dual codes of Type ρ.*

Proof. Fix a length N. The action of the symmetric group S_N commutes with the action of the Clifford-Weil group, and thus can be restricted to an action on the invariant space. Averaging the full weight enumerator fwe(C) over S_N gives the complete weight enumerator cwe(C) of C.   □

We can therefore restate Conjecture 5.5.2 in a weaker form, involving complete rather than full weight enumerators. (This is strictly weaker, since there is no way to 'anti-symmetrize'.)

Conjecture 5.7.2. The Weight Enumerator Conjecture (2). *Let ρ be a finite representation of a finite form ring. Then the space of homogeneous polynomial invariants of degree N of $\mathcal{C}(\rho)$ is spanned by the complete weight enumerators cwe(C) of isotropic self-dual codes C in $N\rho$.*

Corollary 5.7.3. *Conjecture 5.5.2 implies Conjecture 5.7.2.*

Since in practice we prefer to work with complete weight enumerators of codes, it is worth restating our main theorems in terms of complete weight enumerators:

Corollary 5.7.4. *Suppose ρ is a finite representation of a finite triangular form ring. Then the ring of polynomial invariants of $C(\rho)$ is spanned by the complete weight enumerators $\mathrm{cwe}(C)$, where C ranges over self-dual, isotropic codes of Type ρ.*

Corollary 5.7.5. *Let R be a quasi-chain ring. For any finite representation ρ of a form ring (R, M, ψ, Φ), the ring of polynomial invariants of $C(\rho)$ is spanned by the complete weight enumerators $\mathrm{cwe}(C)$, where C ranges over self-dual, isotropic codes of Type ρ.*

A product of complete weight enumerators of codes of Type ρ is again a complete weight enumerator of a code of the same Type:

$$\mathrm{cwe}(C)\,\mathrm{cwe}(C') = \mathrm{cwe}(C \perp C').$$

Therefore Corollaries 5.7.4 and 5.7.5 imply the following:

Corollary 5.7.6. *Let ρ be a representation of a form ring that satisfies the hypotheses of either Corollary 5.7.4 or 5.7.5. Then the space of homogeneous invariants of degree N of the associated genus-m Clifford-Weil group $\mathcal{C}_m(\rho)$ is spanned by the genus-m complete weight enumerators $\mathrm{cwe}_m(C)$, where C ranges over a set of representatives of the permutation-equivalence classes of self-dual, isotropic codes of Type ρ and length N. Moreover, if every length N code of Type ρ is generated by at most m elements, then these genus-m complete weight enumerators form a basis for the space of homogeneous invariants of degree N.*

Proof. It is clearly enough to show that the degree-m complete weight enumerators of pairwise permutation-inequivalent codes of length N and Type ρ are linearly independent, whenever these codes are generated by m elements. So let C be such a code, generated say by $c^{(1)}, \ldots, c^{(m)}$. Then the coefficient in $\mathrm{cwe}_m(C)$ of the monomial

$$X := \prod_{v \in V^m} x_v^{a_v(c^{(1)}, \ldots, c^{(m)})}$$

is nonzero. Any other code C' of Type ρ and length N for which the monomial X occurs in $\mathrm{cwe}_m(C')$ contains a code that is generated by $\sigma(c^{(1)}), \ldots, \sigma(c^{(m)})$, for some permutation $\sigma \in S_N$ of the coordinates. Hence

$$C' \supseteq \sigma(C) = \sigma(C^\perp) = \sigma(C)^\perp \supseteq C'^\perp = C',$$

so $C' = \sigma(C)$ and $\mathrm{cwe}_m(C') = \mathrm{cwe}_m(C)$. Since such monomials are linearly independent, this proves the corollary.   □

The following corollary provides information about the Molien series of the genus-m Clifford-Weil groups in terms of the numbers of inequivalent codes.

Corollary 5.7.7. *Assume that ρ satisfies the hypotheses of Corollary 5.7.6. Then the sequence of Molien series of $\mathcal{C}_m(\rho)$ converges monotonically as m increases:*

$$\lim_{m \to \infty} \mathrm{MS}_{\mathcal{C}_m(\rho)}(t) = \sum_{N=0}^{\infty} \nu_N t^N, \tag{5.7.1}$$

where ν_N is the number of permutation-equivalence classes of codes of Type ρ and length N.

By "converges monotonically" we mean that if $m \leq m'$ then the coefficient of t^N in $\mathrm{MS}_{\mathcal{C}_m(\rho)}(t)$ does not exceed the coefficient of t^N in $\mathrm{MS}_{\mathcal{C}_{m'}(\rho)}(t)$.

We may consider other symmetrizations. Let ρ be a finite representation of a form ring and let G be a subgroup of $\mathrm{Aut}(\rho)$. Recall from Definition 1.11.1 that the automorphism group $\mathrm{Aut}(\rho)$ of a Type $\rho = (V, \rho_M, \rho_\Phi, \beta)$ is the set of elements $g \in \mathrm{End}_R(V)$ such that

$$\beta(gv, gw) = \beta(v, w), \rho_\Phi(\phi)(gv) = \rho_\Phi(\phi)(v) \text{ for } v, w \in V, \phi \in \Phi.$$

Then the (permutation) action of G on $\mathbb{C}[V]$ commutes with the action of $\mathcal{C}(\rho)$, and we obtain an analogous classification of the invariants that are fixed under G. For instance, if H is a subgroup of the automorphism group of ρ, then

$$G = H \wr S_N$$

is a group of automorphisms of $N\rho$; this gives rise to an invariant theory classification of H-symmetrized weight enumerators.

Theorem 5.7.8. *Suppose there are no nonzero isotropic codes in ρ, and let H be the automorphism group of ρ. Then the commuting algebra of $\mathcal{C}(\rho)$ is spanned by H; in particular, the restriction of $\mathcal{C}(\rho)$ to the invariant space of H is irreducible.*

Proof. Indeed, we find that the commuting algebra is canonically isomorphic (as a vector space) to the invariant space of $\rho + \overline{\rho}$; it thus suffices to classify the self-dual, isotropic codes in that representation.

Let C be such a code. Associated with C are two codes $\pi_1(C)$ and $\pi_2(C)$, given by the projections onto the two factors of V in V^2. We have

$$\pi_1(C)^\perp \oplus \pi_2(C)^\perp \subset C \subset \pi_1(C) \oplus \pi_2(C).$$

Thus $\pi_1(C)^\perp$ and $\pi_2(C)^\perp$ are isotropic, so by assumption are trivial, and thus $\pi_1(C) = \pi_2(C) = V$. In particular, for any element $v \in V$, there exists a unique element $\eta_C(v) \in V$ such that $(v, \eta_C(v)) \in C$. We readily verify that η_C is an automorphism of ρ. $\qquad \square$

Lemma 5.7.9. *Let G be a finite group acting linearly on the $\mathbb{C}G$-module V and let W be a G-invariant subspace of $V = W \oplus W'$, where W' is also G-invariant. Let π and $\pi' = 1 - \pi$ be the corresponding G-invariant projections on W resp. W'. Then the invariant ring of G on W is the projection of the invariant ring of G on V onto the symmetric algebra $S(W)$ of W under the projection $S(\pi)$.*

Proof. Since π commutes with the action of G on V, also $S(\pi) = \pi \otimes \ldots \otimes \pi$ mod S_N commutes with the action of G on the N-th symmetric power $S^N(V)$. Therefore the image of a G-invariant element in $S(V)$ under $S(\pi)$ is a G-invariant element in $S(W)$. $\qquad\qquad\square$

For codes this reads as follows:

Corollary 5.7.10. *Assume that the Weight Enumerator Conjecture 5.7.2 holds for $\mathcal{C}(\rho)$. Let $W \leq \mathbb{C}[V]$ be a subspace that is invariant under $\mathcal{C}(\rho)$. Then the ring of polynomial invariants of $\mathcal{C}(\rho)$ (for the action on W) is spanned by the projections of complete weight enumerators of self-dual isotropic codes of Type ρ onto the symmetric algebra of W.*

In particular, using ρ-symmetrized weight enumerators (see Definition 2.1.5 and 2.1.8) instead of complete weight enumerators in Corollaries 5.7.6 and 5.7.7 allows us to use Molien series to count strict equivalence classes of codes. This is because a proof similar to that of Corollary 5.7.6 shows that, if C is a code of Type ρ generated by m elements, the genus-m ρ-symmetrized weight enumerator of C contains a certain monomial X, and X occurs in the ρ-symmetrized weight enumerator of another code C' of Type ρ if and only if C and C' are equivalent.

Corollary 5.7.11. *Let ρ be a representation of a form ring that satisfies the condition of either Corollary 5.7.4 or 5.7.5. Then the space of $\mathrm{Aut}(\rho)$-symmetrized homogeneous invariants of degree N of the associated genus-m Clifford-Weil group $\mathcal{C}_m(\rho)$ is spanned by the genus-m ρ-symmetrized weight enumerators $\mathrm{swe}_m^\rho(C)$, where C ranges over self-dual, isotropic codes of Type ρ and length N. Moreover, if every length N code of Type ρ is generated by at most m elements, then these genus-m ρ-symmetrized weight enumerators form a basis for the invariant space.*

If we replace the natural action of $\mathcal{C}_m(\rho)$ by its action $\mathcal{C}_m(\rho)^{\mathrm{Aut}(\rho)}$ on the $\mathrm{Aut}(\rho)$-invariant elements we obtain the following Corollary:

Corollary 5.7.12. *Assume that ρ satisfies the conditions of Corollary 5.7.11. Then the Molien series of $\mathcal{C}_m(\rho)^{\mathrm{Aut}(\rho)}$ converges monotonically to*

$$\lim_{m \to \infty} \mathrm{MS}_{\mathcal{C}_m(\rho)^{\mathrm{Aut}(\rho)}}(t) = \sum_{N=0}^{\infty} \nu_N t^N,$$

where ν_N is the number of strict equivalence classes of codes of Type ρ and length N.

Sometimes we are interested only in Hamming weight enumerators. As we will show in the next section, the appropriate symmetrization need not commute with the action of the Clifford-Weil group, and the ring spanned by the Hamming weight enumerators of codes of the given Type need not be an invariant ring for a finite group. However, we may obtain this ring by using invariant theory:

Definition 5.7.13. Let ρ be a finite representation of a form ring and let $\mathcal{C}(\rho)$ be the associated Clifford-Weil group. Let

$$\text{Inv}^{\text{Ham}}(\mathcal{C}(\rho)) := \{p(x,y,\dots,y) \in \mathbb{C}[x,y] \mid p(x_0, x_v : v \in V \setminus \{0\}) \in \text{Inv}(\mathcal{C}(\rho))\}$$

be the ring of polynomials that are obtained from the polynomials in $\text{Inv}(\mathcal{C}(\rho))$ by replacing the variable x_0 by x and all other variables by y.

Since Hamming weight enumerators are obtained by exactly this change of variables, we get:

Corollary 5.7.14. *Assume that ρ is as in Corollary 5.7.6. Then the ring spanned by the Hamming weight enumerators of self-dual isotropic codes of Type ρ is $\text{Inv}^{\text{Ham}}(\mathcal{C}(\rho))$.*

As we will see in the next section, the ring $\text{Inv}^{\text{Ham}}(\mathcal{C}(\rho))$ need not be the invariant ring of any finite group.

Although we will not study them in this book, for completeness we give a definition of formally self-dual codes using the language of Types.

Definition 5.7.15. A code C in a finite representation $N\rho$ of a form ring is *formally self-dual* if the complete weight enumerator is invariant under the associated Clifford-Weil group, i.e. if $\text{cwe}(C) \in \text{Inv}(\mathcal{C}(\rho))$. Often a weaker notion is more appropriate: a code C is "Hamming self-dual" if $\text{hwe}(C) \in \text{Inv}^{\text{Ham}}(\mathcal{C}(\rho))$.

5.8 Example: Hermitian codes over $\mathbb{F}_9$

We illustrate the results in this chapter by considering the case of Hermitian self-dual codes over $\mathbb{F}_9$. For a general discussion of higher genus weight enumerators of Hermitian codes over arbitrary fields $\mathbb{F}_{q^2}$ see §7.3.

As we saw in §2.3.3, the form ring (see (2.3.26)) is $(\mathbb{F}_9, \mathbb{F}_9, \text{id}, \mathbb{F}_9)$, and the representation ρ is given by

$$V := \mathbb{F}_9,$$

$$\rho_M(a)(v,w) := \frac{1}{3} \text{Tr}_{\mathbb{F}_9/\mathbb{F}_3} \, av\bar{w}, \text{ for } a \in \mathbb{F}_9, v, w \in V,$$

$$\rho_\Phi(a)(v) := \frac{1}{3} \text{Tr}_{\mathbb{F}_9/\mathbb{F}_3} \, av\bar{v}, \text{ for } a \in \mathbb{F}_9, v \in V,$$

where $^-$ is the nontrivial Galois automorphism of $\mathbb{F}_9$.

Let α be a primitive element of $\mathbb{F}_9$. Then with respect to the $\mathbb{C}$-basis

$$(0, 1, \alpha, \alpha^2, \alpha^3, \alpha^4, \alpha^5, \alpha^6, \alpha^7)$$

of $\mathbb{C}[V]$, the associated Clifford-Weil group $\mathcal{C}(\rho)$ is generated by the three matrices

$$\phi_1 := \mathrm{diag}(1, \zeta^2, \zeta, \zeta^2, \zeta, \zeta^2, \zeta, \zeta^2, \zeta),$$

$$\alpha := \begin{pmatrix} 1 & 0 & 0 & 0 & 0 & 0 & 0 & 0 & 0 \\ 0 & 0 & 0 & 0 & 0 & 0 & 0 & 0 & 1 \\ 0 & 1 & 0 & 0 & 0 & 0 & 0 & 0 & 0 \\ 0 & 0 & 1 & 0 & 0 & 0 & 0 & 0 & 0 \\ 0 & 0 & 0 & 1 & 0 & 0 & 0 & 0 & 0 \\ 0 & 0 & 0 & 0 & 1 & 0 & 0 & 0 & 0 \\ 0 & 0 & 0 & 0 & 0 & 1 & 0 & 0 & 0 \\ 0 & 0 & 0 & 0 & 0 & 0 & 1 & 0 & 0 \\ 0 & 0 & 0 & 0 & 0 & 0 & 0 & 1 & 0 \end{pmatrix}$$

and

$$h := \frac{1}{3} \begin{pmatrix} 1 & 1 & 1 & 1 & 1 & 1 & 1 & 1 & 1 \\ 1 & \zeta^2 & \zeta & 1 & \zeta & \zeta & \zeta^2 & 1 & \zeta^2 \\ 1 & \zeta & \zeta & \zeta^2 & 1 & \zeta^2 & \zeta^2 & \zeta & 1 \\ 1 & 1 & \zeta^2 & \zeta^2 & \zeta & 1 & \zeta & \zeta & \zeta^2 \\ 1 & \zeta & 1 & \zeta & \zeta & \zeta^2 & 1 & \zeta^2 & \zeta^2 \\ 1 & \zeta & \zeta^2 & 1 & \zeta^2 & \zeta^2 & \zeta & 1 & \zeta \\ 1 & \zeta^2 & \zeta^2 & \zeta & 1 & \zeta & \zeta & \zeta^2 & 1 \\ 1 & 1 & \zeta & \zeta & \zeta^2 & 1 & \zeta^2 & \zeta^2 & \zeta \\ 1 & \zeta^2 & 1 & \zeta^2 & \zeta^2 & \zeta & 1 & \zeta & \zeta \end{pmatrix}$$

(acting on $\mathbb{C}[V]$ on the left). This is a group of order 192. According to Theorem 5.5.7, the ring of polynomial invariants of $\mathcal{C}(\rho)$ is spanned by the complete weight enumerators of Hermitian self-dual codes over $\mathbb{F}_9$.

The Molien series for the nine-dimensional group $\mathcal{C}(\rho)$ in the present example will be given in (7.3.24) of §7.3, and shows that it would be hopeless to try to list all the codes whose complete weight enumerators generate the invariant ring. However, symmetrization leads to simpler rings.

The best-known symmetrization uses $G_1 := S_{\{0\}} \times S_{\mathbb{F}_9^*}$, so as to obtain information about the Hamming weight enumerators of these codes. This has the disadvantage that ϕ_1 does not act on the fixed space

$$W_1 := \langle e_0, \frac{1}{8} \sum_{i=0}^{7} e_{\alpha^i} \rangle_{\mathbb{C}} \leq \mathbb{C}[V]$$

of G_1. The action of α on W_1 is trivial, while h acts as $\bar{h} := \frac{1}{3} \begin{pmatrix} 1 & 1 \\ 8 & -1 \end{pmatrix}$. In fact, the subgroup of $\mathcal{C}(\rho)$ that commutes with G_1 acts on W_1 as

$$H_1 := \langle \bar{h}, -\mathrm{id} \rangle \cong Z_2 \times Z_2 \,.$$

The Molien series of H_1 is

$$\frac{1}{(1-t^2)^2} \,,$$

and the invariant ring of H_1 is spanned by

$$x^2 + 8y^2, \ y(x-y) \,,$$

as in [454, §7.8]. All self-dual codes of length 2 are of the form $\langle (1,a) \rangle_{\mathbb{F}_9}$ with $a \in \mathbb{F}_9$, $a\bar{a} = -1$, and have Hamming weight enumerator $x^2 + 8y^2$. In particular, the invariant ring of H_1 is *not* spanned by Hamming weight enumerators of self-dual codes.

In order to apply Corollary 5.7.5, we must choose a better symmetrization. We will symmetrize by the automorphism group of ρ, which is the unitary group of $\mathbb{F}_9$:

$$\mathrm{Aut}(\rho) = \{ g \in \mathbb{F}_9^* \mid g\bar{g} = 1 \} \,.$$

The orbits of $\mathrm{Aut}(\rho)$ on $V = \mathbb{F}_9$ are

$$T_0 := \{0\}, \ T_1 := \{a \in \mathbb{F}_9 \mid a\bar{a} = 1\} = (\mathbb{F}_9^*)^2, \ T_2 := \{a \in \mathbb{F}_9 \mid a\bar{a} = 2\} \,,$$

and the $\mathrm{Aut}(\rho)$-fixed space on $\mathbb{C}[V]$ is

$$W_2 := \left\langle e_0, \frac{1}{4} \sum_{i=0}^{3} e_{\alpha^{2i}}, \frac{1}{4} \sum_{i=0}^{3} e_{\alpha^{2i+1}} \right\rangle \,.$$

The Clifford-Weil group $\mathcal{C}(\rho)$ acts on W_2 by

$$\phi_1' := \mathrm{diag}(1, \zeta^2, \zeta), \ \alpha' := \begin{pmatrix} 1 & 0 & 0 \\ 0 & 0 & 1 \\ 0 & 1 & 0 \end{pmatrix}, \ h' := \frac{1}{3} \begin{pmatrix} 1 & 1 & 1 \\ 4 & 1 & -2 \\ 4 & -2 & 1 \end{pmatrix} \,.$$

These three matrices generate a group $H_2 = \mathcal{C}(\rho)^{\mathrm{Aut}(\rho)}$ of order 48, with Molien series[2]

$$\frac{1}{(1-t^2)(1-t^4)(1-t^6)} \qquad [A001399] \,. \tag{5.8.1}$$

The $\mathcal{C}(\rho)$-invariant projection of $S(\mathbb{C}[V])$ onto $S(W_2)$ is found by replacing every variable e_0 by x_0, $e_{\alpha^{2i}}$ by x_1 and $e_{\alpha^{2i+1}}$ by x_2 $(i = 0, \ldots, 3)$. The invariant ring of H_2 is spanned by the ρ-symmetrized weight enumerators of the codes with generator matrices

$$[\,1\ \alpha\,], \quad \begin{bmatrix} 1 & 1 & 1 & 0 \\ 0 & 1 & 2 & 1 \end{bmatrix}, \quad \begin{bmatrix} 1 & 1 & 1 & 1 & 1 & 1 \\ 1 & 1 & 1 & 0 & 0 & 0 \\ 0 & \alpha & 2\alpha & 0 & 1 & 2 \end{bmatrix} \,.$$

---

[2] The number A001399 in square brackets is the number of the of the corresponding sequence in [504] – see Remark 5.6.4.

The ρ-symmetrized weight enumerators of these codes are

$$q_2 := x_1^2 + 8x_2 x_3 ,$$
$$q_4 := x_1^4 + 16x_1 x_2^3 + 16x_1 x_3^3 + 48x_2^2 x_3^2 ,$$
$$q_6 := x_1^6 + 8x_1^3 x_2^3 + 8x_1^3 x_3^3 + 72x_1^2 x_2^2 x_3^2 + 144x_1 x_2^4 x_3$$
$$+ 144x_1 x_2 x_3^4 + 16x_2^6 + 320x_2^3 x_3^3 + 16x_3^6 , \qquad (5.8.2)$$

respectively. Their Hamming weight enumerators are

$$r_2 := x^2 + 8y^2 ,$$
$$r_4 := x^4 + 32xy^3 + 48y^4 ,$$
$$r_6 := x^6 + 16x^3 y^3 + 72x^2 y^4 + 288xy^5 + 352y^6 . \qquad (5.8.3)$$

The polynomials r_2, r_4 and r_6 generate the ring $\mathrm{Inv}^{\mathrm{Ham}}(\mathcal{C}(\rho))$ spanned by the Hamming weight enumerators of Hermitian self-dual codes. $\mathrm{Inv}^{\mathrm{Ham}}(\mathcal{C}(\rho))$ is a proper subring of the invariant ring $\mathrm{Inv}(H_1)$ described above. r_2 and r_4 are algebraically independent, as one easily sees by calculating the determinant of their Jacobi matrix. More precisely

$$\mathrm{Inv}^{\mathrm{Ham}}(\mathcal{C}(\rho)) = \mathbb{C}[r_2, r_4] \oplus r_6 \mathbb{C}[r_2, r_4] , \qquad (5.8.4)$$

with the syzygy

$$r_6^2 = \frac{3}{4} r_2^2 r_4 - \frac{3}{2} r_2^2 r_4^2 - \frac{1}{4} r_4^3 - r_2^3 r_6 + 3r_2 r_4 r_6 .$$

Note that $\mathrm{Inv}^{\mathrm{Ham}}(\mathcal{C}(\rho))$ is not the invariant ring of a finite group. For if it were, the order of the group would be $4 = \frac{2 \cdot 4}{2}$, the product of the degrees of the primary invariants divided by the number of secondary invariants. Since $\mathrm{Inv}^{\mathrm{Ham}}(\mathcal{C}(\rho))$ is contained in the invariant ring of H_1 and $|H_1| = 4$, one concludes that $\mathrm{Inv}^{\mathrm{Ham}}(\mathcal{C}(\rho)) = \mathrm{Inv}(H_1)$, a contradiction.

A better set of generators for the ring $\mathrm{Inv}^{\mathrm{Ham}}(\mathcal{C}(\rho))$ is

$$r_2, p_4 := x^2 y^2 - 2xy^3 + y^4, \ p_6 := x^3 y^3 - 3x^2 y^4 + 3xy^5 - y^6 , \qquad (5.8.5)$$

with $p_6^2 = p_4^3$.

6

Real and Complex Clifford Groups

This chapter summarizes some of the results of our earlier paper [383] and relates them to the new situation. Given the theory developed in the earlier chapters of this book, we can now give simpler proofs for some of the theorems in [383]. In particular, the main theorems follow from the general results in Chapter 5.

We hope that this concrete setting, one which will be familiar to at least some of our readers, will help in understanding the general theory.

The main subjects of our earlier paper [383] were certain real and complex 'Clifford groups', denoted respectively by $\mathcal{C}_m$ and $\mathcal{X}_m$, and their invariants. We start by giving some background information about the history of these groups.

6.1 Background

In 1959 Barnes and Wall [29] constructed a family of lattices in dimensions 2^m, $m = 0, 1, 2, \ldots$. They distinguished two geometrically similar lattices $L_m \subseteq L'_m$ in $\mathbb{Q}^{2^m}$, which are now called the *Barnes-Wall lattices*. These are important lattices, since they form an infinite family of fairly dense lattices for which many invariants, such as the minimal norm and even the minimal vectors, can be calculated explicitly. L_m is geometrically similar to L'_m, and for $m = 1, 2, 3$ are (geometrically similar to) the root lattices $\mathbb{Z}^2$, D_4 and E_8, and for $m = 4$ to the laminated lattice Λ_{16} [133]. The automorphism groups of these lattices (and many other things) were then studied in a series of papers by Barnes, Bolt, Room and Wall ([29], [57], [58], [59], [536]). Let $\mathcal{G}_m = \mathrm{Aut}(L_m) \cap \mathrm{Aut}(L'_m)$. It turns out that $\mathcal{G}_m = \mathrm{Aut}(L_m) = \mathrm{Aut}(L'_m)$ for $m \neq 3$. When $m = 3$ (and L_3 and L'_3 are two versions of E_8), $\mathcal{G}_3$ has index 270 in $\mathrm{Aut}(L_3)$. For all m, $\mathcal{G}_m$ is a subgroup of index 2 in a certain group $\mathcal{C}_m$ of structure $2^{1+2m}_+.O^+_{2m}(2)$. We follow Bolt et al. in calling $\mathcal{C}_m$ a Clifford group. This group and its complex analogue $\mathcal{X}_m$ are the subject of [383]. It will turn out that the Clifford group $\mathcal{C}_m = \mathcal{C}_m(\rho(2_{\mathrm{I}}))$ is the Clifford-Weil group of genus

m associated with the Type of binary self-dual codes and the complex Clifford group $\mathcal{X}_m = \mathcal{C}_m(\rho(2_{II}))$ is the Clifford-Weil group of genus m associated with the Type of doubly-even self-dual binary codes (see Theorem 6.2.1).

These groups have appeared in several different contexts in recent years. In 1972 Broué and Enguehard [82], [83] rediscovered the Barnes-Wall lattices and also determined their automorphism groups. The group-theoretic structure of the Clifford groups was investigated by Griess in 1973 [204] (see also [471], [476]). In 1995, Calderbank, Cameron, Kantor and Seidel [90] used the Clifford groups to construct orthogonal spreads and Kerdock sets, and asked "is it possible to say something about [their] Molien series, such as the minimal degree of an invariant?".

Around the same time, Runge [461], [462], [463], [464] (see also Duke [161], Oura [393]) investigated these groups in connection with Siegel modular forms. Among other things, Runge established the remarkable result that the space of homogeneous invariants for $\mathcal{C}_m$ of degree $2k$ is spanned by the complete weight enumerators of the codes $C \otimes_{\mathbb{F}_2} \mathbb{F}_{2^m}$, where C ranges over all binary self-dual (or Type 2_I) codes of length $2k$, and the space of homogeneous invariants for $\mathcal{X}_m$ of degree $8k$ is spanned by the complete weight enumerators of the codes $C \otimes_{\mathbb{F}_2} \mathbb{F}_{2^m}$, where C ranges over all binary doubly-even self-dual (or Type 2_{II}) codes of length $8k$. One of the results of [383] was a simpler proof of these two assertions, not involving Siegel modular forms.

Around 1996, the Clifford groups also appeared in two other superficially unrelated contexts: (i) the study of fault-tolerant quantum computation and the construction of quantum error-correcting codes (see Bennett, DiVincenzo, Smolin and Wootters [32], Calderbank, Rains, Shor and Sloane [95], [96], and Kitaev [310]); and (ii) in the construction of packings in Grassmannian spaces (see Calderbank, Hardin, Rains, Shor and Sloane [92], Conway, Hardin and Sloane [120], and Shor and Sloane [488]). The story of the astonishing coincidence (involving the group $\mathcal{C}_3$) that led to [92], [95], [96], and eventually to the present book is told in [96] and reproduced at the end of this section. Other recent references that mention these groups are Glasby [190], Kleidman and Liebeck [314] and Winter [548].

Independently, and slightly later, Sidelnikov [490], [491], [492], [493] (see also Kazarin [298]) came across the group $\mathcal{C}_m$ when studying spherical codes and designs. In particular, he showed that for $m \geq 3$ the lowest degree harmonic invariant of $\mathcal{C}_m$ has degree 8, and hence that the orbit under $\mathcal{C}_m$ of any point on a sphere in $\mathbb{R}^{2^m}$ is a spherical 7-design. (Venkov [534] had earlier shown that for $m \geq 3$ the minimal vectors of the Barnes-Wall lattices form 7-designs.)

In fact it is an immediate consequence of Runge's results that for $m \geq 3$ $\mathcal{C}_m$ has a unique harmonic invariant of degree 8 and no harmonic invariant of degree 10 (this follows from Corollary 6.2.4 below; see also Corollary 4.13 of [383]). The space of homogeneous invariants of degree 8 is spanned by the complete weight enumerators of $i_2^4 \otimes_{\mathbb{F}_2} \mathbb{F}_{2^m}$ and $e_8 \otimes_{\mathbb{F}_2} \mathbb{F}_{2^m}$, where i_2 is the repetition code of length 2 and e_8 is the $[8, 4, 4]$ Hamming code (see (2.4.2) and (2.4.8) in Chapter 2).

In the next section we will show how Runge's theorems now follow easily from our general theory.

Here is the story about C_3 from [96] [slightly edited].

> There is a remarkable story behind this paper. About two years ago one of the authors (P.W.S.) [P. W. Shor] was studying fault-tolerant quantum computation, and was led to investigate a certain group of 8×8 orthogonal matrices. P.W.S. asked another of us (N.J.A.S.) [N. J. A. Sloane] for the best method of computing the order of this group. N.J.A.S. replied by citing the computer algebra system MAGMA [68], [100], and gave as an illustration the MAGMA commands needed to specify a certain matrix group [the group C_3] that had recently arisen in connection with packings in Grassmannian spaces. This group was the symmetry group of a packing of 70 4-dimensional subspaces of $\mathbb{R}^8$ that had been discovered by computer search [120]. It too was an 8-dimensional group, of order 5160960. To our astonishment the two groups turned out to be identical [apart from some minor differences in signs] (not just isomorphic)! We then discovered that this group was a member of an infinite family of groups that played a central role in a joint paper [90] of another of the authors (A.R.C.) [A. R. Calderbank]. This is the family of real Clifford groups L_R [now called C_m], described in §2 [of [96]] (for $m = 3$, [C_3] has order 5160960).

> This coincidence led us to make connections which further advanced both areas of research (fault-tolerant quantum computing [486] and Grassmannian packings [488]).

> While these three authors were pursuing these investigations, the fourth author (E.M.R.) [E. M. Rains] happened to be present for a job interview and was able to make further contributions to the Grassmannian packing problem (see Calderbank, Hardin, Rains, Shor and Sloane [92]). As the latter involved packings of 2^k-dimensional subspaces in 2^n-dimensional space, it was natural to ask if the same techniques could be used for constructing quantum-error-correcting codes, which are also subspaces of 2^n-dimensional space. This question led directly to [95] and [96]. (Incidentally, he got the job.)

> A final postscript. (a) At the 1997 IEEE International Symposium on Information Theory, V. I. Sidelnikov presented a paper "On a finite group of matrices generating orbit codes on the Euclidean sphere" [491] (based on [490], [298]). It was no surprise to discover that— although Sidelnikov did not identify them in this way—these were the Clifford groups appearing in yet another guise. (b) We have also recently discovered that the complex Clifford groups L [now called X_m] described in §2 [of [96]] have also been studied by Duke [161],

Runge [461], [462], and Oura [393] in connection with multiple weight enumerators of codes and Siegel modular forms.

6.2 Runge's theorems

In [383] we defined the real Clifford group $\mathcal{C}_m$ to be the normalizer of an extraspecial 2-group $E(m) \cong 2^{1+2m}_+$ in the orthogonal group $O_{2^m}(\mathbb{R})$, and the complex Clifford group $\mathcal{X}_m$ to be the normalizer of the central product $E(m)\mathsf{Y}Z_4$ in the unitary group $U_{2^m}(\mathbb{Q}[\zeta_8])$. The extraspecial groups $E(m)$ are defined recursively, as follows:

$$E(1) := \left\langle \sigma_1 := \begin{pmatrix} 0 & 1 \\ 1 & 0 \end{pmatrix}, \sigma_2 := \begin{pmatrix} 1 & 0 \\ 0 & -1 \end{pmatrix} \right\rangle \cong D_8 \cong 2^{1+2}_+.$$

$E(1)$ is the automorphism group of the 2-dimensional square lattice, and is isomorphic to the dihedral group of order 8. Then $E(m) = \otimes^m E(1)$ is the m-fold tensor power of $E(1)$. Thus $E(m)$ is generated by the $2m$ tensor products

$$I_2 \otimes \cdots \otimes I_2 \otimes \sigma_1 \otimes I_2 \otimes \cdots \otimes I_2,$$

$$I_2 \otimes \cdots \otimes I_2 \otimes \sigma_2 \otimes I_2 \otimes \cdots \otimes I_2,$$

where in each of these tensor products there are a total of $m - 1$ identity matrices I_2. Then $\mathcal{C}_m \cong 2^{1+2m}_+.O^+_{2m}(2)$ and $\mathcal{X}_m \cong (2^{1+2m}_+\mathsf{Y}Z_8).\mathrm{Sp}_{2m}(\mathbb{F}_2) \cong (2^{1+2m}_+\mathsf{Y}Z_8).O_{2m+1}(\mathbb{F}_2)$. Using an appropriate set of generators, we showed in [383] that the genus-m complete weight enumerators of binary self-dual Type 2_I (resp. Type 2_II) codes are invariant under $\mathcal{C}_m$ (resp. $\mathcal{X}_m$).

We can now describe the groups $\mathcal{C}_m$ and $\mathcal{X}_m$ as the Clifford-Weil groups $\mathcal{C}_m(\rho(2_\mathrm{I}))$ and $\mathcal{C}_m(\rho(2_\mathrm{II}))$ associated with the representations $\rho(2_\mathrm{I})$ and $\rho(2_\mathrm{II})$ defined in §2.3.1. To simplify notation and for convenience of the reader we give the explicit definition of the matrix ring of the corresponding form rings (and their representations): Let $R := \mathrm{Mat}_m(\mathbb{F}_2)$ and let $V := \mathbb{F}_2^m$ be the natural left R-module. Let $M := R$ be the $R \otimes R$-module of all $\mathbb{F}_2$-bilinear forms from $V \times V$ to $\mathbb{F}_2$ and let $\psi := \mathrm{id}$. The involution on R is transposition. To define the quadratic forms, let $\phi_0 : V \to \frac{1}{4}\mathbb{Z}/\mathbb{Z}$ be defined by $\phi_0(v) := \frac{1}{4}v^{tr}v$ (mod $\mathbb{Z}$). Let $\Phi_\mathrm{II} \subset \mathrm{Quad}_0(V, \frac{1}{4}\mathbb{Z}/\mathbb{Z})$ be the R-qmodule generated by ϕ_0, and let $\Phi_\mathrm{I} = \{M\}$ be the R-qmodule generated by $2\phi_0$.

Finally, let $\rho_\mathrm{II} := (V, \rho_M, \rho_\Phi, \beta)$ be the finite representation of the form ring $(R, M, \psi, \Phi_\mathrm{II})$ defined by $\rho_M(\psi(1)) := \beta$, where $\beta(v, w) := \frac{1}{2}v^{tr}w$ and $\rho_\Phi(\phi_0) := \phi_0$; and let ρ_I be the restriction of ρ_II to the form ring $(R, M, \psi, \Phi_\mathrm{I})$. We leave it as an exercise for the reader to check that the form rings $(R, M, \psi, \Phi_\mathrm{I})$ and $(R, M, \psi, \Phi_\mathrm{II})$ are respectively isomorphic to the matrix rings $\mathrm{Mat}_m(R(2_\mathrm{I}))$ and $\mathrm{Mat}_m(R(2_\mathrm{II}))$ of the form rings $R(2_\mathrm{I})$ and $R(2_\mathrm{II})$ defined in §2.3.1.

Theorem 6.2.1. *The Clifford-Weil group* $\mathcal{C}(\rho_{\mathrm{I}}) = \mathcal{C}_m(\rho(2_{\mathrm{I}})) \subset O_{2^m}(\mathbb{R})$ *is conjugate to the real Clifford group* $\mathcal{C}_m$, *and the Clifford-Weil group* $\mathcal{C}(\rho_{\mathrm{II}}) = \mathcal{C}_m(\rho(2_{\mathrm{II}})) \subset U_{2^m}(\mathbb{C})$ *is conjugate to the complex Clifford group* $\mathcal{X}_m$.

Proof. For $a = \mathrm{I}$ or $a = \mathrm{II}$, let $N_a \trianglelefteq \mathcal{C}_m(\rho_a)$ be the full preimage of the kernel of $\pi : \mathfrak{U}_m(R, \Phi_a) \to GL_2(R)$. We claim that $N_{\mathrm{I}} \cong E(m)$ and $N_{\mathrm{II}} \cong E(m) \mathsf{Y} \langle \zeta_8 I_{2^m} \rangle$. Clearly $\ker(\pi) = \{(1, \begin{pmatrix} \phi_1 & 0 \\ & \phi_2 \end{pmatrix}) \mid \phi_1, \phi_2 \in \ker(\lambda)\} \cong \ker(\lambda) \times \ker(\lambda)$. Since $2\phi_0 : V \to \frac{1}{2}\mathbb{Z}/\mathbb{Z}$ is a nonzero linear mapping, $\ker(\lambda) \cong V^*$ is generated by $2\phi_0$ as an R-module. Fix an isomorphism $\ker(\lambda) \cong V^*$. Since β is non-degenerate, for each $\tau \in \ker(\lambda)$ there is a unique $w_\tau \in V$ such that $\tau(v) = \beta(w_\tau, v)$ for all $v \in V$. Then the element $(1, \begin{pmatrix} 0 & 0 \\ \tau & \end{pmatrix}) \in \ker(\pi)$ acts as the diagonal matrix $e_v \mapsto (-1)^{\tau(v)} e_v$ on $\mathbb{R}[V] \cong \mathbb{R}^{2^m}$. Conjugating by $h := H_{111}$, the element $(1, \begin{pmatrix} 0 & 0 \\ \tau & \end{pmatrix})^h = (1, \begin{pmatrix} \tau & 0 \\ 0 & \end{pmatrix}) \in \ker(\pi)$ maps e_v to e_{v+w_τ} (see Lemma 5.4.1). These elements generate $E(m)$. That the center of $\mathcal{C}(\rho_a)$ is $\{\pm 1\}$ if $a = \mathrm{I}$ (resp. $\langle \zeta_8 \rangle$ if $a = \mathrm{II}$) follows from the fact that binary self-dual codes exist for all even lengths and doubly-even binary self-dual codes exist if and only if the length is divisible by 8. Therefore N_a is a normal subgroup in $\mathcal{C}(\rho_a)$. Since $\mathcal{C}_m$ and $\mathcal{X}_m$ are defined as the normalizers of N_{I} and N_{II} respectively, it follows that $\mathcal{C}(\rho_{\mathrm{I}}) \leq \mathcal{C}_m$ and $\mathcal{C}(\rho_{\mathrm{II}}) \leq \mathcal{X}_m$.

To establish the other inclusion, note that ρ_a is a faithful representation, and hence $\mathcal{C}(\rho_a)/N_a \cong \pi(\mathfrak{U}_m(R, \Phi_a)) \leq GL_2(R)$. It therefore suffices to calculate the isomorphism type of the image. Using Remark 5.2.5, we have

$$\pi(\mathfrak{U}_m(R, \Phi_a)) =$$
$$\{A \in \mathrm{Mat}_{2m}(\mathbb{F}_2) \mid A^{tr} \begin{pmatrix} 0 & 0 \\ I_m & 0 \end{pmatrix} A - \begin{pmatrix} 0 & 0 \\ I_m & 0 \end{pmatrix} \in (\psi_a)_2^{-1}((\lambda_a)_2((\Phi_a)_2))\}.$$

If $a = \mathrm{I}$ then $\lambda_{\mathrm{I}}(\Phi) = \{0\}$ and hence $\lambda^{(m)}(\Phi^{(m)}) = \mathrm{Alt}_m(\mathbb{F}_2)$ is the set of symmetric matrices in $\mathrm{Mat}_m(\mathbb{F}_2)$ with all diagonal elements 0 (cf. Example 1.10.3). If $a = \mathrm{II}$ the image of $\psi_a^{-1} \circ \lambda_a$ consists of all symmetric matrices. Therefore

$$\pi(\mathfrak{U}_m(R, \Phi_{\mathrm{II}})) = \{A \in \mathrm{Mat}_{2m}(\mathbb{F}_2) \mid A^{tr} \begin{pmatrix} 0 & I_m \\ I_m & 0 \end{pmatrix} A = \begin{pmatrix} 0 & I_m \\ I_m & 0 \end{pmatrix}\}$$
$$\cong \mathrm{Sp}_{2m}(\mathbb{F}_2) \cong \mathcal{X}_m/(E(m) \mathsf{Y} Z_8)$$

and

$$\pi(\mathfrak{U}_m(R, \Phi_{\mathrm{I}})) =$$
$$\{\begin{pmatrix} a & b \\ c & d \end{pmatrix} \in \mathrm{Mat}_{2m}(\mathbb{F}_2) \mid b^{tr}c = d^{tr}a - I_m, \ c^{tr}a \in \mathrm{Alt}_m(\mathbb{F}_2), \ d^{tr}b \in \mathrm{Alt}_m(\mathbb{F}_2)\}$$
$$\cong O_{2m}^+(\mathbb{F}_2) \cong \mathcal{C}_m/E(m).$$

To see the isomorphism with the orthogonal group of plus type, we choose a basis $(b_1, \ldots, b_m, b'_1, \ldots, b'_m)$ such that both subspaces $\langle b_1, \ldots, b_m \rangle$ and $\langle b'_1, \ldots, b'_m \rangle$ are totally isotropic, and the corresponding bilinear form satisfies $(b_i, b'_j) = \delta_{ij}$ $(i, j = 1, \ldots, m)$. Referred to these coordinates, the quadratic form is $q((x, x')) = x^{tr} x'$, and hence $O_{2m}^+(\mathbb{F}_2)$ consists of the matrices $\begin{pmatrix} a & b \\ c & d \end{pmatrix} \in \mathrm{Mat}_{2m}(\mathbb{F}_2)$ such that

$$x^{tr} x' = (ax + bx')^{tr}(cx + dx') \text{ for all } x, x' \in \mathbb{F}_2^m,$$

i.e. such that

$$b^{tr}c + d^{tr}a = I_m, x^{tr}c^{tr}ax = x^{tr}d^{tr}bx = 0 \text{ for all } x \in \mathbb{F}_2^m.$$

Note that the condition $x^{tr}ax = 0$ for all $x \in \mathbb{F}_2^m$ is equivalent to $a \in \mathrm{Alt}_m(\mathbb{F}_2)$. For a more sophisticated discussion of the groups in these cases see §7.6.   □

Applying Corollary 5.7.10 we immediately obtain Runge's results.

Theorem 6.2.2. (Runge [464]; [383, Theorems 4.9, 6.2].)

(a) Let $N, m \geq 1$. The space of homogeneous polynomial invariants of degree N of $\mathcal{C}_m$ is spanned by the genus-m complete weight enumerators of Type 2_I codes of length N. This set is a basis when $m \geq N/2 - 1$.

(b) Let $N, m \geq 1$. The space of homogeneous polynomial invariants of degree N of $\mathcal{X}_m$ is spanned by the genus-m complete weight enumerators of Type 2_{II} codes of length N. This set is a basis when $m \geq N/2 - 1$.

Note the word "set" here: the same polynomial will appear many times, since we are not restricting ourselves to inequivalent codes. The linear independence of the genus-m complete weight enumerators of a set of representatives of equivalence classes of binary self-dual codes of length N with $N/2 - 1 \leq m$ follows from Corollary 5.7.6, together with the fact that any self-dual binary code of length N has a basis consisting of the all-ones vector $\mathbf{1}$ and $N/2 - 1$ other vectors.

Bolt et al. [57], [58], [59], [536] and Sidelnikov [490], [491], [492] also consider what may be called the p-Clifford group $\mathcal{C}_m^{(p)}$, obtained by replacing 2 in the definition of $\mathcal{C}_m$ by an odd prime p. As discussed in [383], this group can be defined as follows. The extraspecial p-group $E_p(m) \cong p_+^{1+2m}$ of exponent p is a subgroup of $U_{p^m}(\mathbb{C})$. To be precise, $E_p(1)$ is generated by the elements

$$\sigma_1 := v_x \mapsto v_{x+1} \quad \text{and} \quad \sigma_2 := v_x \mapsto \exp(2\pi i x/p)v_x, \ x \in \mathbb{Z}/p\mathbb{Z},$$

of $U_p(\mathbb{C})$, and $E_p(m)$ is the m-fold tensor power of $E_p(1)$. Then $\mathcal{C}_m^{(p)}$ is the normalizer of $E_p(m)$ in $U_{p^m}(\mathbb{Q}[\zeta_{ap}])$, where $a = \gcd\{p+1, 4\}$. This is a group of structure

$$\mathcal{C}_m^{(p)} \cong Z_a \times p_+^{1+2m}.\mathrm{Sp}_{2m}(\mathbb{F}_p)$$

and order

$$a\,p^{m^2+2m+1}\prod_{j=1}^{m}(p^{2j}-1)\,. \tag{6.2.1}$$

(cf. Winter [548]).

As in the case $p=2$, one can show that $\mathcal{C}_m^{(p)}$ is conjugate to the Clifford-Weil group $\mathcal{C}_m(\rho(p_1^{\mathrm{E}}))$ of Theorem 7.4.1 (for the special case $q=p$). We omit the detailed proof, which follows the lines of the proof of Theorem 6.2.1. Applying Corollary 5.7.10, we obtain Theorem 7.1 of [383]:

Theorem 6.2.3. *Let $N, m \geq 1$. The space of homogeneous polynomial invariants of degree N of $\mathcal{C}_m^{(p)}$ is spanned by the genus-m complete weight enumerators of self-dual codes over $\mathbb{F}_p$ of length N containing $\mathbf{1}$. This is a basis when $m \geq N/2 - 1$.*

As a special case of Corollary 5.7.7 we get the following nice interpretation of the Molien series of the three Clifford groups:

Corollary 6.2.4. *Let C_m denote the real Clifford group $\mathcal{C}_m$, the complex Clifford group $\mathcal{X}_m$ or the p-Clifford group $\mathcal{C}_m^{(p)}$. Then the Molien series of C_m converges monotonically as m increases:*

$$\lim_{m\to\infty}\mathrm{MS}_{C_m}(t)=\sum_{N=0}^{\infty}\nu_N t^N\,, \tag{6.2.2}$$

where ν_N is the number of equivalence classes of codes of length N and Types 2_{I}, 2_{II}, p_1^{E} respectively.

6.3 The real Clifford group $\mathcal{C}_m$

It follows from the discussion in the previous section that the real Clifford group $\mathcal{C}_m$ may be generated by the following elements of $O_{2^m}(\mathbb{R})$:

$$d_\phi : e_v \mapsto \exp(2\pi i\phi(v))\,e_v\,, \tag{6.3.1}$$

where ϕ is any $\frac{1}{2}\mathbb{Z}/\mathbb{Z}$-valued quadratic form on $V=\mathbb{F}_2^m$;

$$m_r : e_v \mapsto e_{rv}\,, \tag{6.3.2}$$

where r is any element of $\mathrm{GL}_m(\mathbb{F}_2)$; and the MacWilliams transformation corresponding to the symmetric idempotent $\mathrm{diag}(1,0,\dots,0)$, which is the matrix

$$h : e_{v_1,v_2,\dots,v_m} \mapsto \frac{1}{\sqrt{2}}(e_{0,v_2,\dots,v_m}+\exp(\pi i v_1)\,e_{1,v_2,\dots,v_m})\,. \tag{6.3.3}$$

As already mentioned, $\mathcal{C}_m$ has structure $2_+^{1+2m}.O_{2m}^+(2)$. Its order is

$$2^{m^2+m+2}(2^m - 1) \prod_{j=1}^{m-1} (4^j - 1). \tag{6.3.4}$$

We now illustrate Theorem 6.2.2 by looking at the genus-m weight enumerators of singly-even binary self-dual codes.

Example 6.3.1. 2_{I}: Singly-even binary self-dual codes (cont.)

(1) The case $m = 1$: $\mathcal{C}_1$ is generated by the matrices $d_\phi = \begin{pmatrix} 1 & 0 \\ 0 & -1 \end{pmatrix}$ corresponding to $\phi = \frac{1}{2}v^2$, and $h = h_2 = \frac{1}{\sqrt{2}}\begin{pmatrix} 1 & 1 \\ 1 & -1 \end{pmatrix}$. $\mathcal{C}_1$ is a dihedral group of order 16, with Molien series

$$\mathrm{MS} = \frac{1}{(1 - t^2)(1 - t^8)} \qquad [A008621]$$
$$= 1 + t^2 + t^4 + t^6 + 2t^8 + 2t^{10} + 2t^{12} + 2t^{14} + 3t^{16} + O\left(t^{18}\right). \tag{6.3.5}$$

The invariant ring is spanned by the weight enumerator $f_1 = x^2 + y^2$ of the code i_2 (see (2.4.2)) and the weight enumerator $f_2 = x^8 + 14x^4y^4 + y^8$ of the Hamming code e_8 (see (2.4.7)). It is nicer to replace f_2 by $f_2' = (f_1^4 - f_2)/4 = x^2y^2(x^2 - y^2)^2$, so we have, in the notation established in §5.6.1,

$$\mathrm{Inv} = \frac{1}{f_1,\, f_2'}. \tag{6.3.6}$$

or, in the alternative notation that just mentions the codes,

$$\mathrm{Inv} = \frac{1}{i_2,\, e_8}, \tag{6.3.7}$$

This is part of Gleason's original theorem (see Theorem 6.4.2).

(2) The case $m = 2$: $\mathcal{C}_2$ is generated by the matrices d_ϕ corresponding to the quadratic form $\phi = \begin{pmatrix} 1/2 & 0 \\ 0 & 0 \end{pmatrix}$, m_{r_1} and m_{r_2} corresponding to $r_1 = \begin{pmatrix} 0 & 1 \\ 1 & 0 \end{pmatrix}$ and $r_2 = \begin{pmatrix} 1 & 1 \\ 0 & 1 \end{pmatrix} \in \mathrm{GL}_2(\mathbb{F}_2)$, and $h = h_2 \otimes I_2$; namely

$$\begin{bmatrix} 1 & 0 & 0 & 0 \\ 0 & 1 & 0 & 0 \\ 0 & 0 & -1 & 0 \\ 0 & 0 & 0 & -1 \end{bmatrix}, \begin{bmatrix} 1 & 0 & 0 & 0 \\ 0 & 0 & 1 & 0 \\ 0 & 1 & 0 & 0 \\ 0 & 0 & 0 & 1 \end{bmatrix}, \begin{bmatrix} 1 & 0 & 0 & 0 \\ 0 & 0 & 0 & 1 \\ 0 & 0 & 1 & 0 \\ 0 & 1 & 0 & 0 \end{bmatrix}, \begin{bmatrix} t & t & 0 & 0 \\ t & -t & 0 & 0 \\ 0 & 0 & t & t \\ 0 & 0 & t & -t \end{bmatrix},$$

where $t = 1/\sqrt{2}$. (A smaller set of generators could easily be found.) $\mathcal{C}_2$ has order 2304 and Molien series

$$\mathrm{MS} = \frac{1 + t^{18}}{(1 - t^2)(1 - t^8)(1 - t^{12})(1 - t^{24})} \qquad [A008718]$$
$$= 1 + t^2 + t^4 + t^6 + 2t^8 + 2t^{10} + 3t^{12} + 3t^{14} + 4t^{16} + O\left(t^{18}\right). \tag{6.3.8}$$

This Molien series and its invariant ring were calculated in [359]. $\mathcal{C}_2$ has a subgroup of index 2, the reflection group $X_{28} = [3,4,3]$, No. 28 on the Shephard-Todd list [482]. X_{28} has a subgroup consisting of all 4×4 permutation matrices, so the invariants are symmetric functions of the four variables.

As a basis for the invariant ring we may take the genus-2 weight enumerators of the codes $i_2, e_8, d_{12}^+, g_{24}$ and $(d_{10}e_7f_1)^+$ where i_2 and e_8 are as above, d_{12}^+ and g_{24} are as in §2.4.1, and $(d_{10}e_7f_1)^+$ (in the standard gluing notation) will be given explicitly when we discuss gluing theory in §9.6. The latter code has weight enumerator

$$x^{18} + 17x^{14}y^4 + 51x^{12}y^6 + 187x^{10}y^8$$
$$+ 187x^8y^{10} + 51x^6y^{12} + 17x^4y^{14} + y^{18} . \qquad (6.3.9)$$

The codes d_{12}^+ and $(d_{10}e_7f_1)^+$ (or I_{18}) and their Hamming weight enumerators were given by Pless [414]. The genus-2 weight enumerators of these codes are as follows, using the symmetric function notation introduced in §2.4.1:

$$\text{cwe}_2(i_2) = \sigma_2 \qquad (6.3.10)$$

(as in (2.4.3)),

$$\text{cwe}_2(d_{12}^+) = \sigma_{12} + 15\,\sigma_{8,4} + 32\,\sigma_{6,6} + 120\,\sigma_{6,2,2,2} + 90\,\sigma_{4,4,4} + 480\,\sigma_{4,4,2,2} , \qquad (6.3.11)$$

$$\text{cwe}_2((d_{10}e_7f_1)^+) = \sigma_{18} + 17\,\sigma_{14,4} + 51\,\sigma_{12,6} + 102\,\sigma_{12,2,2,2} + 187\,\sigma_{10,8}$$
$$+ 170\,\sigma_{10,4,4} + 510\,\sigma_{10,4,2,2} + 357\,\sigma_{8,6,4} + 2652\,\sigma_{8,6,2,2}$$
$$+ 2040\,\sigma_{8,4,4,2} + 7140\,\sigma_{6,6,4,2} + 24990\,\sigma_{6,4,4,4} , \qquad (6.3.12)$$

while those for e_8 and g_{24} were given in (2.4.10) and (2.4.14). Then we have

$$\text{Inv} = \frac{1, \ (d_{10}e_7f_1)^+}{i_2, \ e_8, \ d_{12}^+, \ g_{24}} . \qquad (6.3.13)$$

Multiplying (6.3.8) by $1 - t^2$ we get the harmonic Molien series [A090176]

$$\frac{1 + t^{18}}{(1 - t^8)(1 - t^{12})(1 - t^{24})} = 1 + t^8 + t^{12} + t^{16} + t^{18} + t^{20} + 3\,t^{24} + \cdots . \qquad (6.3.14)$$

Note that (6.3.5) and (6.3.8) are consistent with Corollary 6.2.4, which says that the initial terms of the Molien series for $\mathcal{C}_m$ for $m \geq 1$ are

$$1 + t^2 + t^4 + t^6 + 2t^8 + 2t^{10} + O(t^{12}) , \qquad (6.3.15)$$

where the next term is $2t^{12}$ for $m = 1$ and $3t^{12}$ for $m > 1$. The coefficient of t^{12} is 3 for $m \geq 1$ because there are three Type 2_I codes of length 12, namely i_2^6, $e_8i_2^2$ and d_{12}^+, and their genus-2 weight enumerators are linearly independent. Since binary self-dual codes have been classified up through length 34 (see

Chapter 12), our main theorem would in principle allow us to calculate the coefficients of t^a in the Molien series of all groups $\mathcal{C}_m$ for $a \leq 34$. So far, we have carried out this calculation for $a \leq 26$ (cf. [381]). The result is that for $m \geq 2$ the Molien series of $\mathcal{C}_m$ is

$$1 + t^2 + t^4 + t^6 + 2t^8 + 2t^{10} + \sum_{N=12}^{\infty} a_N(m) t^N$$

where for $N \leq 26$ the coefficients $a_N(m)$ are given in Table 6.1. It is amusing to note that the group $\mathcal{C}_8$ mentioned in the last line of the table has order

$$23898798870542026677382030807969824768 0000 \,.$$

Table 6.1. Coefficients of t^{12} through t^{26} in Molien series for $\mathcal{C}_m$.

N	12	14	16	18	20	22	24	26	Sequence
$m = 2$	3	3	4	5	6	6	9	10	A008718
$m = 3$	3	4	6	7	10	12	18	22	A024186
$m = 4$	3	4	7	9	14	19	33	45	A110160
$m = 5$	3	4	7	9	16	23	46	74	A110868
$m = 6$	3	4	7	9	16	25	53	94	A110869
$m = 7$	3	4	7	9	16	25	55	102	A110876
$m \geq 8$	3	4	7	9	16	25	55	103	A110880

To illustrate the calculation, consider the coefficient of t^{14}. There are four Type 2_I codes of length 14, namely i_2^7, $e_8 i_2^3$, $d_{12}^+ i_2$ and e_7^{2+}. Their genus-m complete weight enumerators are linearly independent if $m \geq 3$, span a space of dimension 3 for $m = 2$ and a space of dimension 2 if $m = 1$. Therefore the coefficient of t^{14} in the Molien series of $\mathcal{C}_m$ is respectively 2 for $m = 1$, 3 for $m = 2$ and 4 for $m \geq 3$.

Similarly, the harmonic Molien series is

$$1 + t^8 + O(t^{12}) \,, \tag{6.3.16}$$

where the next terms are $0 \cdot t^{12} + 0 \cdot t^{14} + O(t^{16})$ for $m = 1$, $t^{12} + O(t^{16})$ for $m = 2$ and $t^{12} + t^{14} + O(t^{16})$ for $m \geq 3$.

Sidelnikov [491], [492] had shown that the lowest degree of a harmonic invariant of $\mathcal{C}_m$ is 8. We now have a stronger result.

Corollary 6.3.2. ([383, Cor. 4.13]) *For $m \geq 1$, the smallest degree of a harmonic invariant of $\mathcal{C}_m$ is 8, and there is a unique (up to scalar multiples) harmonic invariant of that degree. There are no harmonic invariants of degree 10. If $m = 1$, there are no harmonic invariants of degrees 12 or 14, while if $m \geq 2$ there is a unique harmonic invariant of degree 12.*

For $m = 2$, we may take the unique harmonic invariants of degrees 8 and 12 to be

$$h_8 := 10\,\mathrm{cwe}_2(e_8) - 7(\sigma_2)^4$$
$$= 3\,\sigma_8 - 28\,\sigma_{6,2} + 98\,\sigma_{4,4} - 84\,\sigma_{4,2,2} + 1512\,\sigma_{2,2,2,2} \qquad (6.3.17)$$

and

$$h_{12} := \sigma_{12} - 22\,\sigma_{10,2} + 15\,\sigma_{8,4} + 450\,\sigma_{8,2,2} + 76\,\sigma_{6,6}$$
$$- 780\,\sigma_{6,4,2} - 3240\,\sigma_{6,2,2,2} + 2700\,\sigma_{4,4,2,2} + 3450\,\sigma_{4,4,4}. \qquad (6.3.18)$$

It is easy to find real points $(x_{00}, x_{01}, x_{10}, x_{11}) \in S^3$ where both h_8 and h_{12} vanish. Since $\mathcal{C}_2$ has no harmonic invariants of degree 14, any orbit of such a point under $\mathcal{C}_2$ forms a spherical 15-design of size 2304.

(3) The case $m = 3$. We have a sentimental attachment to this group, and give the generators in full, since as remarked in §6.1 it was these matrices (or rather an equivalent set) which led to this book. $\mathcal{C}_3$ is generated by the matrices d_ϕ corresponding to the quadratic form $\phi = \mathrm{diag}\{1/2, 0, 0\}$, the matrices m_{r_1} and m_{r_2} corresponding to

$$r_1 = \begin{bmatrix} 1 & 0 & 1 \\ 0 & 1 & 0 \\ 0 & 0 & 1 \end{bmatrix}, \ r_2 = \begin{bmatrix} 0 & 1 & 0 \\ 0 & 0 & 1 \\ 1 & 0 & 0 \end{bmatrix} \in GL_4(\mathbb{F}_2),$$

and $h = h_2 \otimes I_4$;. These are the matrices

$$\begin{bmatrix} 1 & 0 & 0 & 0 & 0 & 0 & 0 & 0 \\ 0 & 1 & 0 & 0 & 0 & 0 & 0 & 0 \\ 0 & 0 & 1 & 0 & 0 & 0 & 0 & 0 \\ 0 & 0 & 0 & 1 & 0 & 0 & 0 & 0 \\ 0 & 0 & 0 & 0 & -1 & 0 & 0 & 0 \\ 0 & 0 & 0 & 0 & 0 & -1 & 0 & 0 \\ 0 & 0 & 0 & 0 & 0 & 0 & -1 & 0 \\ 0 & 0 & 0 & 0 & 0 & 0 & 0 & -1 \end{bmatrix} , \begin{bmatrix} 1 & 0 & 0 & 0 & 0 & 0 & 0 & 0 \\ 0 & 1 & 0 & 0 & 0 & 0 & 0 & 0 \\ 0 & 0 & 1 & 0 & 0 & 0 & 0 & 0 \\ 0 & 0 & 0 & 1 & 0 & 0 & 0 & 0 \\ 0 & 0 & 0 & 0 & 0 & 1 & 0 & 0 \\ 0 & 0 & 0 & 0 & 1 & 0 & 0 & 0 \\ 0 & 0 & 0 & 0 & 0 & 0 & 0 & 1 \\ 0 & 0 & 0 & 0 & 0 & 0 & 1 & 0 \end{bmatrix} ,$$

$$\begin{bmatrix} 1 & 0 & 0 & 0 & 0 & 0 & 0 & 0 \\ 0 & 0 & 1 & 0 & 0 & 0 & 0 & 0 \\ 0 & 0 & 0 & 0 & 1 & 0 & 0 & 0 \\ 0 & 0 & 0 & 0 & 0 & 0 & 1 & 0 \\ 0 & 1 & 0 & 0 & 0 & 0 & 0 & 0 \\ 0 & 0 & 0 & 1 & 0 & 0 & 0 & 0 \\ 0 & 0 & 0 & 0 & 0 & 1 & 0 & 0 \\ 0 & 0 & 0 & 0 & 0 & 0 & 0 & 1 \end{bmatrix} , \begin{bmatrix} t & t & 0 & 0 & 0 & 0 & 0 & 0 \\ t & -t & 0 & 0 & 0 & 0 & 0 & 0 \\ 0 & 0 & t & t & 0 & 0 & 0 & 0 \\ 0 & 0 & t & -t & 0 & 0 & 0 & 0 \\ 0 & 0 & 0 & 0 & t & t & 0 & 0 \\ 0 & 0 & 0 & 0 & t & -t & 0 & 0 \\ 0 & 0 & 0 & 0 & 0 & 0 & t & t \\ 0 & 0 & 0 & 0 & 0 & 0 & t & -t \end{bmatrix} ,$$

respectively, where again $t = 1/\sqrt{2}$. $\mathcal{C}_3$ has order $5160960 = 2^7\,8!$ (note that $O_6^+(\mathbb{F}_2) \cong S_8$). It has Molien series

$$\frac{\theta(t) + t^{154}\,\theta(t^{-1})}{(1 - t^2)(1 - t^{12})(1 - t^{14})(1 - t^{16})(1 - t^{24})^2(1 - t^{30})(1 - t^{40})} \qquad (6.3.19)$$

$$= 1 + t^2 + t^4 + t^6 + 2t^8 + 2t^{10} + 3t^{12} + 4t^{14} + 6t^{16} + 7t^{18} + 10t^{20} + 12t^{22} + O(t^{24}),$$

[A024186], where

$$\begin{aligned}
\theta(t) := &\; 1 + t^8 + t^{16} + t^{18} + 2t^{20} + t^{22} + 2t^{24} + 3t^{26} + 4t^{28} \\
&+ 2t^{30} + 5t^{32} + 4t^{34} + 7t^{36} + 6t^{38} + 7t^{40} + 8t^{42} \\
&+ 11t^{44} + 9t^{46} + 12t^{48} + 13t^{50} + 14t^{52} + 15t^{54} \\
&+ 17t^{56} + 17t^{58} + 20t^{60} + 19t^{62} + 20t^{64} + 20t^{66} \\
&+ 25t^{68} + 22t^{70} + 22t^{72} + 24t^{74} + 25t^{76}.
\end{aligned} \qquad (6.3.20)$$

It would clearly be hopeless to try to find codes whose weight enumerators would generate this ring.

(4) The case $m = 4$. Following Oura [393], we have computed the Molien series of $\mathcal{C}_4$. We find that the initial terms of $\mathrm{MS}(\mathcal{C}_4)$ are

$$\begin{aligned}
&1 + t^2 + t^4 + t^6 + 2t^8 + 2t^{10} + 3t^{12} + 4t^{14} + 7t^{16} + 9t^{18} + 14t^{20} \\
&+ 19t^{22} + 33t^{24} + 45t^{26} + 69t^{28} + 100t^{30} + 159t^{32} + 228t^{34} \\
&+ 355t^{36} + 526t^{38} + 815t^{40} + 1215t^{42} + 1861t^{44} + 2777t^{46} \\
&+ 4240t^{48} + 6318t^{50} + 9508t^{52} + 14107t^{54} + \cdots.
\end{aligned} \qquad (6.3.21)$$

For the full rational function see the entry [A110160] in [504]. It follows from that expression that the corresponding invariant ring has more than 2.10^{10} secondary invariants.

(5) For completeness, we mention that the Molien series for $E(1)$ is

$$\frac{1}{(1 - t^2)(1 - t^4)} \qquad [A008619], \qquad (6.3.22)$$

with basic invariants $x_0^2 + x_1^2$ and $x_0^2 x_1^2$. For arbitrary m the Molien series for $E(m)$ is

$$\frac{1}{2n^2} \left\{ \frac{1}{(1 - t)^n} + \frac{1}{(1 + t)^n} + \frac{n^2 + n - 2}{(1 - t^2)^{n/2}} + \frac{n^2 - n}{(1 + t^2)^{n/2}} \right\}, \qquad (6.3.23)$$

where $n = 2^m$.

6.4 The complex Clifford group $\mathcal{X}_m$

It also follows from the discussion in §6.2 that the complex Clifford group $\mathcal{X}_m$ may be generated by the elements d_ϕ, m_r and h (see (6.3.1), (6.3.2), (6.3.3)), regarded as elements of $U_{2^m}(\mathbb{C})$, except that now ϕ ranges over

$\frac{1}{4}\mathbb{Z}/\mathbb{Z}$-valued quadratic forms on V. The group $\mathcal{X}_m$ has structure $(2^{1+2m}_+ \mathsf{Y} Z_8)$. $\mathrm{Sp}_{2m}(\mathbb{F}_2)$ and order

$$2^{m^2+2m+2} \prod_{j=1}^{m}(4^j - 1). \tag{6.4.1}$$

We use this group to study the genus-m weight enumerators of doubly-even binary self-dual codes.

Example 6.4.1. 2_{II}: Doubly-even binary self-dual codes (cont.)

(1) The case $m = 1$: $\mathcal{X}_1$ is generated by h and the matrix $d_\phi = \begin{pmatrix} 1 & 0 \\ 0 & i \end{pmatrix}$ corresponding to $\phi = \frac{1}{4}v^2$. This is a unitary reflection group of order 192 (No. 9 on the Shephard-Todd list), with Molien series

$$\frac{1}{(1 - t^8)(1 - t^{24})} \qquad [A008620]. \tag{6.4.2}$$

The invariant ring is spanned by the weight enumerator $f_2 = x^8 + 14x^4y^4 + y^8$ of the Hamming code e_8 (see (2.4.7)), and the weight enumerator f_3 of the Golay code g_{24} (see (2.4.13)). It is nicer to replace f_3 by $f_3' = (f_2^3 - f_3)/42 = x^4y^4(x^4 - y^4)^4$, so we have

$$\mathrm{Inv} = \frac{1}{f_2,\ f_3'}. \tag{6.4.3}$$

or, in the alternative notation that just mentions the codes,

$$\mathrm{Inv} = \frac{1}{e_8,\ g_{24}}. \tag{6.4.4}$$

This is the other part of Gleason's theorem. The complete result is the following.

Theorem 6.4.2. (Gleason [191].) (a) *The weight enumerator of a binary self-dual code is a polynomial in the weight enumerators of the repetition code i_2 and the Hamming code e_8.*
(b) *The weight enumerator of a doubly-even binary self-dual code is a polynomial in the weight enumerators of the Hamming code e_8 and the Golay code g_{24}.*

(2) The case $m = 2$. References: Huffman [270], Duke [161]. $\mathcal{X}_2$ is generated by the matrices in (6.3.8) and the additional generator

$$d_\phi = \begin{bmatrix} 1 & 0 & 0 & 0 \\ 0 & 1 & 0 & 0 \\ 0 & 0 & i & 0 \\ 0 & 0 & 0 & i \end{bmatrix},$$

corresponding to the quadratic form $\phi = \mathrm{diag}\{1/4, 0, 0\}$. $\mathcal{X}_2$ has order 92160 and Molien series

$$\frac{1 + t^{32}}{(1 - t^8)(1 - t^{24})^2(1 - t^{40})} \qquad [A028288], \qquad (6.4.5)$$

$$\mathrm{Inv} = \frac{1, \; d_{32}^+}{e_8, \; g_{24}, \; d_{24}^+, \; d_{40}^+}. \qquad (6.4.6)$$

$\mathcal{X}_2$ has a reflection subgroup of index 2, No. 31 on the Shephard-Todd list.

(3) The case $m = 3$: $\mathcal{X}_3$ has order 743178240 and the Molien series [A039946] can be written as

$$\frac{\theta(t^8) + t^{352}\,\theta(t^{-8})}{(1 - t^8)(1 - t^{16})(1 - t^{24})^2(1 - t^{40})(1 - t^{56})(1 - t^{72})(1 - t^{120})}, \qquad (6.4.7)$$

where

$$\begin{aligned}
\theta(t) := \; & 1 + t^3 + 3t^4 + 3t^5 + 6t^6 + 8t^7 + 12t^8 + 18t^9 + 25t^{10} \\
& + 29t^{11} + 40t^{12} + 50t^{13} + 58t^{14} + 69t^{15} + 80t^{16} + 85t^{17} \\
& + 96t^{18} + 104t^{19} + 107t^{20} + 109t^{21} + 56t^{22}. \qquad (6.4.8)
\end{aligned}$$

Runge [462] gives the Molien series [A027633] for the commutator subgroup $\mathcal{H}_3 = \mathcal{X}_3'$, of index 2 in $\mathcal{X}_3$. The Molien series for $\mathcal{X}_3$ consists of the terms in the series for $\mathcal{H}_3$ that have exponents divisible by 4.

(4) The case $m = 4$: Oura [393] has computed the Molien series for $\mathcal{H}_4 = \mathcal{X}_4'$ [A027674], and that for $\mathcal{X}_4$ [A051354] can be obtained from it in the same way. Other related Molien series can be found in Bannai et al. [28], and of course in [504].

6.5 Barnes-Wall lattices

Let L_m and L'_m denote the Barnes-Wall lattices in $\mathbb{Q}^{2^m}$. We begin with a definition of these lattices that is essentially equivalent to that of Barnes and Wall [29], but then transform it into a much simpler definition, given originally in [383] and [384]. This also leads to a simple definition for the real Clifford group $\mathcal{C}_m$.

Let $b_0, \ldots, b_{2^m-1}$ be an orthonormal basis for $\mathbb{R}^{2^m}$, where the b_i are indexed by the elements of $V := \mathbb{F}_2^m$. For an affine subspace $U \subseteq V$, let $\chi_U \in \mathbb{Q}^{2^m}$ be the characteristic vector $\chi_U := \sum_{i=1}^{2^m} \epsilon_i b_i$, where $\epsilon_i = 1$ if i corresponds to an element of U, $\epsilon_i = 0$ otherwise. Then L_m (resp. L'_m) is spanned by the set of vectors

$$\{2^{\lfloor (m-d+\delta)/2 \rfloor} \chi_U \mid 0 \le d \le m, U \text{ is a } d\text{-dimensional affine subspace of } V\},$$

where $\delta = 1$ for L_m and $\delta = 0$ for L'_m.

The following are some of the properties of these lattices:

- L_m and L'_m are geometrically similar, i.e. differ only by a rotation and change of scale;
- L_m is a sublattice of index $2^{2^{m-1}}$ in L'_m;
- Let $\mathcal{G}_m = \mathrm{Aut}(L_m) \cap \mathrm{Aut}(L'_m)$. Then $\mathcal{G}_m = \mathrm{Aut}(L_m) = \mathrm{Aut}(L'_m)$ for $m \neq 3$. When $m = 3$, L_3 and L'_3 are two versions of the root lattice E_8, and $\mathcal{G}_3$ has index 270 in $\mathrm{Aut}(L_3)$. Note that there are two isomorphism classes of index 270 subgroups in $\mathrm{Aut}(L_3) = \mathrm{Weyl}(E_8) \cong 2.O_8^+(\mathbb{F}_2)$, related by triality. In $O_8^+(\mathbb{F}_2)$ the two groups are both isomorphic to $2^6 \rtimes \Omega_6^+(2) \cong \mathrm{Weyl}(D_8)'/\{\pm1\}$, but one lifts to a split extension, $\mathrm{Weyl}(D_8)'$, while the other lifts to a nonsplit extension, $\mathcal{G}_3$.
- For all m, $\mathcal{G}_m$ is a subgroup of index 2 in the Clifford group $\mathcal{C}_m$.

For other properties (minimal norm, density, kissing numbers, etc.), see [133].

A simpler construction can be given as follows. Extending scalars, let us define the $\mathbb{Z}[\sqrt{2}]$-lattice (cf. §9.1.6 below)

$$M_m := \sqrt{2}L'_m + L_m \,,$$

which we call the *balanced Barnes-Wall lattice*. Thus M_m is a free $\mathbb{Z}[\sqrt{2}]$-module equipped with a $\mathbb{Q}[\sqrt{2}]$-valued quadratic form.

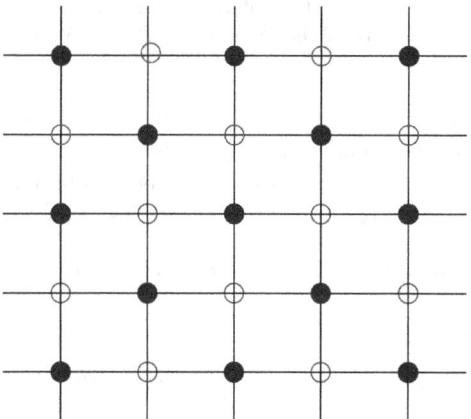

Fig. 6.1. The two Barnes-Wall lattices L_1 (solid circles) and L'_1 (solid or hollow circles) in two dimensions.

For example, for $m = 1$, L_1 and L'_1 are the two-dimensional lattices shown in Fig. 6.1. L_1 consists of the points marked with solid circles and L'_1 consists of the points marked with either solid or hollow circles. Both are geometrically similar to the square lattice $\mathbb{Z}^2$. We multiply the points of L'_1 by $\sqrt{2}$.

The eight minimal vectors of L_1 and $\sqrt{2}\,L_1'$ now have the same length and form the familiar configuration of points used in the 8-PSK signaling system (Fig. 6.2). The set of all $\mathbb{Z}[\sqrt{2}]$-integer combinations of these eight points gives a (dense) embedding of the balanced Barnes-Wall lattice M_1 into $\mathbb{R}^2$, where $\sqrt{2}$ is mapped to the real number $1.4142\ldots$. A second embedding is obtained similarly by multiplying L_1' by $-\sqrt{2}$. Putting both pictures together (as an orthogonal sum) yields a visualization of $M_1 \subseteq \mathbb{R}^4$. Note that we can recover L_1 from M_1 by taking just those vectors in M_1 whose components are integers.

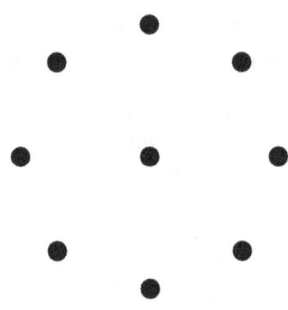

Fig. 6.2. The eight minimal vectors of L_1 and $\sqrt{2}\,L_1'$. The $\mathbb{Z}[\sqrt{2}]$ span of these points is a (dense) embedding of the "balanced" Barnes-Wall lattice M_1 in $\mathbb{R}^2$.

What makes the lattice M_m so attractive is that it is simply a tensor product of copies of M_1. The following result was established in [383]. (It is reminiscent of the $u|u+v$ construction for Reed-Muller codes, which is one of the other ways to obtain the Barnes-Wall lattices [133], [361].)

Theorem 6.5.1. *For all $m > 1$, the lattice M_m is a tensor product:*

$$M_m = M_{m-1} \otimes_{\mathbb{Z}[\sqrt{2}]} M_1$$
$$= M_1 \otimes_{\mathbb{Z}[\sqrt{2}]} M_1 \otimes_{\mathbb{Z}[\sqrt{2}]} \cdots \otimes_{\mathbb{Z}[\sqrt{2}]} M_1 \quad (\textit{with } m \textit{ factors}). \qquad (6.5.1)$$

Corollary 6.5.2. *For all $m \geq 1$, the automorphism group $\mathrm{Aut}(M_m)$ (the subgroup of the orthogonal group $O_{2^m}(\mathbb{R})$ that preserves M_m) is isomorphic to $\mathcal{C}_m$.*

We can recover L_m by taking the rational part of M_m; the purely irrational part is $\sqrt{2}\,L_m'$. Since this construction of the Barnes-Wall lattices deserves to be better known, we give some further details.

We need only two of the vectors in Fig. 6.2 to generate M_1, and we can take

$$G_1 = \begin{bmatrix} \sqrt{2} & 0 \\ 1 & 1 \end{bmatrix}$$

as a generator matrix. Then the m-fold tensor power of this matrix, $G_m = G_1^{\otimes m} = G_1 \otimes G_1 \otimes \cdots \otimes G_1$, is a generator matrix for M_m.

For example, $G_2 = G_1 \otimes G_1$ is

$$\begin{bmatrix} 2 & 0 & 0 & 0 \\ \sqrt{2} & \sqrt{2} & 0 & 0 \\ \sqrt{2} & 0 & \sqrt{2} & 0 \\ 1 & 1 & 1 & 1 \end{bmatrix}.$$

The rational part, L_2, is generated by

$$\begin{bmatrix} 2 & 0 & 0 & 0 \\ 2 & 2 & 0 & 0 \\ 2 & 0 & 2 & 0 \\ 1 & 1 & 1 & 1 \end{bmatrix}, \quad \text{or equivalently} \quad \begin{bmatrix} 2 & 0 & 0 & 0 \\ 0 & 2 & 0 & 0 \\ 0 & 0 & 2 & 0 \\ 1 & 1 & 1 & 1 \end{bmatrix}.$$

This lattice is geometrically similar to D_4 [133, Chap. 4, Eq. (90)]. The purely irrational part, $\sqrt{2}L_2'$, is generated by

$$\begin{bmatrix} 2\sqrt{2} & 0 & 0 & 0 \\ \sqrt{2} & \sqrt{2} & 0 & 0 \\ \sqrt{2} & 0 & \sqrt{2} & 0 \\ \sqrt{2} & \sqrt{2} & \sqrt{2} & \sqrt{2} \end{bmatrix}, \quad \text{or equivalently} \quad \sqrt{2}\begin{bmatrix} 2 & 0 & 0 & 0 \\ 1 & 1 & 0 & 0 \\ 1 & 0 & 1 & 0 \\ 1 & 0 & 0 & 1 \end{bmatrix},$$

which is another version of D_4 [133, Chap. 4, Eq. (86)].

We may avoid the use of coordinates and work directly with the lattices, provided we select an appropriate semilinear automorphism σ of M_m. The automorphism group $C_m = \text{Aut}(M_m)$ contains the normal subgroup $G_m = \text{Aut}(L_m) \cap \text{Aut}(L_m')$ of index 2. Let $\chi : C_m \to \{1, -1\}$ be the linear character with kernel G_m. Let $\sigma : M_m \to M_m$ be a group automorphism which is *semilinear*, i.e. satisfies $\sigma(av) = \bar{a}\sigma(v)$ for all $a \in \mathbb{Z}[\sqrt{2}]$ and all $v \in M_m$ (where $\bar{a}$ denotes the Galois conjugate of a), and such that $\sigma g = \chi(g)g\sigma$ for all $g \in C_m$. Since G_m acts absolutely irreducibly, such a σ is uniquely determined up to multiplication by elements in $\mathbb{Z}[\sqrt{2}]^*$. From our construction of M_m, we know that there is such an automorphism σ_0 with $\sigma_0^2 = 1$. For $u \in \{1, -1\}$ let

$$L_u(\sigma_0) := \{v \in M_m \mid \sigma_0(v) = uv\}.$$

Since $(a\sigma_0)^2 = a\bar{a}\sigma_0^2 = \text{Norm}(a)$, the automorphism σ_0 is uniquely determined up to a unit a of norm 1, hence $a = u_1 a_1^2$ for some $a_1 \in \mathbb{Z}[\sqrt{2}]^*$ and $u_1 \in \{1, -1\}$. For $u_2 := u_1 a_1 \bar{a_1} \in \{1, -1\}$, multiplication by a_1 induces an isometry between $L_u(\sigma_0)$ and $L_{uu_2}(a\sigma_0)$. Therefore $L_u(\sigma_0)$ is isometric to either L_m or $\sqrt{2}L_m'$.

Also, the usual procedure for constructing a $2n$-dimensional $\mathbb{Z}$-lattice from an n-dimensional $\mathbb{Z}[\sqrt{2}]$-lattice, by composing the bilinear form with the trace, constructs the Barnes-Wall lattices from their balanced analogues. This avoids the use of the automorphism group and the construction of the semilinear automorphism σ_0. We consider M_m as a 2^{m+1}-dimensional $\mathbb{Z}$-lattice, where the $\mathbb{Z}[\sqrt{2}]$-valued bilinear form (x,y) on M_m is replaced by the $\mathbb{Z}$-valued form $T_\alpha(x,y) := \text{Trace}(\alpha(x,y))$, α being any totally positive element of $\mathbb{Z}[\sqrt{2}]$. To obtain the Barnes-Wall lattices, it is enough to consider $\alpha = 1$ and $\alpha = \wp :=$ $2+\sqrt{2}$. If $(b_1,\ldots,b_{2^m})$ is a $\mathbb{Z}[\sqrt{2}]$-basis for M_m, then, since $(1,\sqrt{2})$ is a $\mathbb{Z}$-basis for $\mathbb{Z}[\sqrt{2}]$,

$$(b_1, \sqrt{2}b_1, b_2, \sqrt{2}b_2, \ldots, b_{2^m}, \sqrt{2}b_{2^m})$$

is a $\mathbb{Z}$-basis for M_m. If m is odd, the $\mathbb{Z}$-lattice (M_m, T_1) is similar to $L_m \perp L_m$ and $(M_m, T_\wp)$ is similar to L_{m+1}, while if m is even, (M_m, T_1) is similar to L_{m+1} and $(M_m, T_\wp)$ is similar to $L_m \perp L_m$.

For $m = 1$ this procedure yields the generator matrix

$$\tilde{G}_1 = \begin{bmatrix} 0 & 1 & 0 & 0 \\ 2 & 0 & 0 & 0 \\ 1 & 0 & 1 & 0 \\ 0 & 1 & 0 & 1 \end{bmatrix},$$

yielding the Gram matrices

$$\tilde{T}_1 = \tilde{G}_1 T_1 \tilde{G}_1^{tr} = 4 \begin{bmatrix} 1 & 0 & 0 & 1 \\ 0 & 2 & 1 & 0 \\ 0 & 1 & 1 & 0 \\ 1 & 0 & 0 & 2 \end{bmatrix} \text{ and } \tilde{T}_\wp = \tilde{G}_1 T_\wp \tilde{G}_1^{tr} = 4 \begin{bmatrix} 2 & 2 & 1 & 2 \\ 2 & 4 & 2 & 2 \\ 1 & 2 & 2 & 2 \\ 2 & 2 & 2 & 4 \end{bmatrix}$$

for $L_1 \perp L_1$ and L_2, respectively. Here T_1 and $T_\wp$ are block diagonal matrices with blocks $\begin{bmatrix} 2 & 0 \\ 0 & 4 \end{bmatrix}$ and $\begin{bmatrix} 4 & 4 \\ 4 & 8 \end{bmatrix}$ respectively.

6.6 Maximal finiteness in real case

The fact that the balanced Barnes-Wall lattice M_m has a simple tensor product construction allows us to show that for $m \geq 2$, the real Clifford group $\mathcal{C}_m$ is a maximal finite subgroup of $\text{GL}_{2^m}(\mathbb{R})$. The proof uses two lemmas.

Lemma 6.6.1. *If $m \geq 2$, then the $\mathbb{Z}$-span (denoted by $\overline{\mathbb{Z}[\mathcal{C}_m]}$) of the matrices in $\mathcal{C}_m$ acting on the 2^m-dimensional $\mathbb{Z}[\sqrt{2}]$-lattice M_m is $\mathbb{Z}[\sqrt{2}]^{2^m \times 2^m}$.*

Proof. We proceed by induction on m. Explicit calculations show that the lemma is <u>true for</u> $m = 2$ and $m = 3$. If $m \geq 4$ then $m - 2 \geq 2$, and by induction $\overline{\mathbb{Z}[\mathcal{C}_{m-2}]} = \mathbb{Z}[\sqrt{2}]^{2^{m-2} \times 2^{m-2}}$ and $\overline{\mathbb{Z}[\mathcal{C}_2]} = \mathbb{Z}[\sqrt{2}]^{4 \times 4}$. Since $M_m = M_2 \otimes_{\mathbb{Z}[\sqrt{2}]} M_{m-2}$, the automorphism group of M_m contains $\mathcal{C}_2 \otimes \mathcal{C}_{m-2}$. Hence

$$\mathbb{Z}[\sqrt{2}]^{2^m \times 2^m} \supseteq \overline{\mathbb{Z}[\mathcal{C}_m]} \supseteq \overline{\mathbb{Z}[\mathcal{C}_{m-2}]} \otimes_{\mathbb{Z}[\sqrt{2}]} \overline{\mathbb{Z}[\mathcal{C}_2]} = \mathbb{Z}[\sqrt{2}]^{2^m \times 2^m}. \qquad \square$$

The second lemma is [383, Lemma 5.5]:

Lemma 6.6.2. *Let $m \geq 2$. Let G be a finite group with $\mathcal{C}_m \leq G \leq GL(2^m, \mathbb{R})$ and let p be a prime. If p is odd, the maximal normal p-subgroup of G is trivial. The maximal normal 2-subgroup of G is either $E(m)$ if $G = \mathcal{C}_m$, or $Z(E(m)) = \langle -I_{2^m} \rangle$ if G properly contains $\mathcal{C}_m$.*

Theorem 6.6.3. *Let $m \geq 2$. Then the real Clifford group $\mathcal{C}_m$ is a maximal finite subgroup of $GL(2^m, \mathbb{R})$.*

Proof. (Sketch—see [383, Theorem 5.6] for more details.) Suppose, seeking a contradiction, that G is a finite subgroup of $GL(2^m, \mathbb{R})$ that properly contains $\mathcal{C}_m$.

1) All normal p-subgroups of G are central (by Lemma 6.6.2).

2) There is a totally real abelian number field K containing $\mathbb{Q}[\sqrt{2}]$ such that G is conjugate to a subgroup of $GL(2^m, K)$. Let K be a minimal such field and assume that $G \leq GL(2^m, K)$.

3) Let R be the ring of integers of K. Then G fixes an $R\mathcal{C}_m$-lattice. By Lemma 6.6.1 all $R\mathcal{C}_m$-lattices are of the form $I \otimes_{\mathbb{Z}[\sqrt{2}]} M_m$ for some fractional ideal I of R, so the group G fixes all $R\mathcal{C}_m$-lattices and hence also $R \otimes_{\mathbb{Z}[\sqrt{2}]} M_m$. So any choice of a $\mathbb{Z}[\sqrt{2}]$-basis for M_m gives rise to an embedding $G \hookrightarrow GL(2^m, R)$, by which we may regard G as a group of matrices. Without loss of generality we may assume that $G = \mathrm{Aut}(R \otimes_{\mathbb{Z}[\sqrt{2}]} M_m)$.

4) The Galois group $\Gamma := \mathrm{Gal}(K/\mathbb{Q}[\sqrt{2}])$ acts on G by acting componentwise on the matrices. Suppose $K \neq \mathbb{Q}[\sqrt{2}]$. It is enough to show that there is a nontrivial element $\sigma \in \Gamma$ that acts trivially on G, because then the matrices in G have their entries in the fixed field of σ, contradicting the minimality of K.

4o) Assume first that there is an odd prime p ramified in $K/\mathbb{Q}$, and let $\wp$ be a prime ideal in R that lies over p. Then p is also ramified in $K/\mathbb{Q}[\sqrt{2}]$ and therefore the action of the ramification group, the stabilizer in Γ of $\wp$, on $R/\wp$ is not faithful, hence the first inertia group

$$\Gamma_\wp := \{\, \sigma \in \mathrm{Gal}(K/\mathbb{Q}[\sqrt{2}]) \mid \sigma(x) \equiv x \pmod{\wp} \text{ for } x \in R \,\}$$

is nontrivial (see e.g. Fröhlich and Taylor [177, Corollary III.4.2]). Since $G_\wp := \{g \in G \mid g \equiv I_{2^m} \pmod{\wp}\}$ is a normal p-subgroup of G, $G_\wp = 1$ by Lemma 6.6.2. Therefore all the elements in $\Gamma_\wp$ act trivially on G, which is what we were seeking to prove.

4e) So 2 is the only ramified prime in K, which implies that $K = \mathbb{Q}[\zeta_{2^a} + \zeta_{2^a}^{-1}]$ for some $a \geq 3$, where $\zeta_t = \exp(2\pi i/t)$. If $a = 3$, then $K = \mathbb{Q}[\sqrt{2}]$, $G = \mathrm{Aut}(M_m) = \mathcal{C}_m$ and we are done. So assume $a > 3$ and let $\wp$ be the prime ideal of R over 2 (generated by $(1 - \zeta_{2^a})(1 - \zeta_{2^a}^{-1})$) and let $\sigma \in \Gamma$ be the Galois automorphism defined by $\sigma(\zeta_{2^a} + \zeta_{2^a}^{-1}) = \zeta_{2^a}^{2^{a-1}+1} + \zeta_{2^a}^{-2^{a-1}-1} = -(\zeta_{2^a} + \zeta_{2^a}^{-1})$. Then $\mathrm{id} = \sigma^2 \neq \sigma$ and

$$(\zeta_{2^a} + \zeta_{2^a}^{-1}) - \sigma(\zeta_{2^a} + \zeta_{2^a}^{-1}) = 2(\zeta_{2^a} + \zeta_{2^a}^{-1}) \in 2\wp.$$

Therefore $\sigma \in \Gamma_{2\wp}$. Since the subgroup $G_{2\wp} := \{g \in G \mid g \equiv I_{2^m} \pmod{2\wp}\}$ of G is trivial (cf. Bartels [30, Hilfssatz 1]) we conclude that σ acts trivially on G, and thus G is in fact defined over $\mathbb{Q}[\zeta_{2^{a-1}} + \zeta_{2^{a-1}}^{-1}]$, contradicting the minimality of K. $\square$

6.7 Maximal finiteness in complex case

There are analogues for the complex Clifford group $\mathcal{X}_m$ for most of the above results.

Let $\mathcal{M}_m := \mathbb{Z}[\zeta_8] \otimes_{\mathbb{Z}[\sqrt{2}]} M_m$ be the scalar extension of the balanced Barnes-Wall lattice M_m. As in the real case we obtain:

Proposition 6.7.1. *The subgroup of $U_{2^m}(\mathbb{Q}[\zeta_8])$ preserving $\mathcal{M}_m$ is precisely $\mathcal{X}_m$.*

For the analogue of Lemma 6.6.1, observe that the matrices in $\mathcal{X}_m$ generate a maximal order. Even for $m = 1$ the $\mathbb{Z}$-span of the matrices in $\mathcal{X}_1$ acting on $\mathcal{M}_1$ is the maximal order $\mathbb{Z}[\zeta_8]^{2\times2}$. Hence the induction argument used to prove Lemma 6.6.1 shows that $\overline{\mathbb{Z}[\mathcal{X}_m]} = \mathbb{Z}[\zeta_8]^{2^m \times 2^m}$. Therefore the analogue of Theorem 6.6.3 holds even for $m = 1$, with a similar proof which we omit here:

Theorem 6.7.2. *Let $m \geq 1$ and let G be a finite group such that $\mathcal{X}_m \leq G \leq U_{2^m}(\mathbb{C})$. Then there exists a root of unity ζ such that*

$$G = \langle \mathcal{X}_m, \zeta I_{2^m} \rangle.$$

In other words, $\mathcal{X}_m$ is a projective maximal finite subgroup of $U_{2^m}(\mathbb{C})$.

We can obtain a similar maximal finiteness result for the p-Clifford groups $\mathcal{C}_m^{(p)}$ for $p > 2$. Since these groups fix more than one lattice, the method of proof of Theorem 6.6.3 does not apply. However, using the classification of finite simple groups, we can show that $\mathcal{C}_m^{(p)}$ is projectively maximal finite in $U_{p^m}(\mathbb{C})$.

Theorem 6.7.3. *([383, Theorem 7.3]) Let G be a finite group such that $\mathcal{C}_m^{(p)} \leq G \leq U_{p^m}(\mathbb{C})$. Then there exists a root of unity ζ such that*

$$G = \langle \mathcal{C}_m^{(p)}, \zeta I_{p^m} \rangle.$$

6.8 Automorphism groups of weight enumerators

It follows from Gleason's Theorem (Theorem 6.4.2) and more generally from Corollary 5.7.10 that the weight enumerators of self-dual codes are fixed under the action of certain groups. In this section we show that, subject to certain natural restrictions, the converse sometimes holds: the full stabilizer of the appropriate weight enumerator *is* the Clifford group that we started with.

In [383, Corollary 5.7] we established these results by showing that the Clifford groups act irreducibly on the Lie algebra of the orthogonal group. We can now reprove this theorem in a much simpler way.

Theorem 6.8.1. *The real Clifford group C_m acts irreducibly on the Lie algebra of the orthogonal group; the complex Clifford group $\mathcal{X}_m$ and the p-Clifford groups $\mathcal{C}_m^{(p)}$ for $p > 2$ act irreducibly on the Lie algebra of the special unitary group.*

Proof. Let $\rho = \rho_m(2_I)$, $\rho_m(2_{II})$ or $\rho_m(p_I^E)$, corresponding to the three cases; let V be the underlying module for ρ, and let $W = \mathbb{C}[V]$. The Lie algebra in each case is a submodule of $\operatorname{End}(W) \cong W \otimes W^*$, irreducible under the action of the orthogonal or unitary group. It will thus suffice to show that the decomposition of $\operatorname{End}(W)$ into irreducible $\mathcal{C}(\rho)$-modules is the same as the decomposition into irreducible modules for the orthogonal or unitary group. Equivalently, by Schur's lemma, we thus wish to show that the space of $\mathcal{C}(\rho)$-invariants in $\operatorname{End}(W) \otimes \operatorname{End}(W^*) = W \otimes W \otimes W^* \otimes W^*$ is the same as the space of orthogonal or unitary invariants. By our main Theorem 5.5.7, this space is (up to canonical isomorphism) spanned by the genus-m full weight enumerators of self-dual codes in the representation $2\rho + 2\overline{\rho}$.

For the complex Clifford group and the p-Clifford groups for odd primes, we find that the only such codes are

$$\begin{bmatrix} 1 & 0 & 1 & 0 \\ 0 & 1 & 0 & 1 \end{bmatrix} \text{ and } \begin{bmatrix} 1 & 0 & 0 & 1 \\ 0 & 1 & 1 & 0 \end{bmatrix}. \tag{6.8.1}$$

Thus $\dim \mathcal{A} = 2$ and the action of $\mathcal{C}(\rho)$ on $\operatorname{End}(W)$ has two irreducible components; since multiples of the identity and traceless matrices form two such components for the action of the unitary group, we are done.

For the real Clifford group, the relevant codes are

$$\begin{bmatrix} 1 & 0 & 1 & 0 \\ 0 & 1 & 0 & 1 \end{bmatrix}, \begin{bmatrix} 1 & 0 & 0 & 1 \\ 0 & 1 & 1 & 0 \end{bmatrix} \text{ and } \begin{bmatrix} 1 & 1 & 0 & 0 \\ 0 & 0 & 1 & 1 \end{bmatrix}. \tag{6.8.2}$$

We thus have three irreducible components, i.e. multiples of the identity, traceless symmetric matrices, and antisymmetric matrices. $\square$

We have shown that any invariant of the real, complex, or p-Clifford group on the space $W^{\otimes 2} \otimes (W^*)^{\otimes 2}$ is invariant under the entire orthogonal or unitary group. For the real and complex Clifford groups, the same argument establishes the corresponding statement for $W^{\otimes 3} \otimes (W^*)^{\otimes 3}$; for instance, the invariants of $O(n)$ in $W^{\otimes 6}$ are spanned by invariants associated to matchings, and thus to codes equivalent to i_2^3. Most subgroups of classical groups do not have this property—see Guralnick and Tiep [225] (where Clifford groups are called "normalizers of groups of symplectic type").

We immediately conclude that:

Corollary 6.8.2. *If H is a closed subgroup of the orthogonal group and H contains the real Clifford group C_m, then either H is finite or $SO_{2m}(\mathbb{R}) \subset H$. Similarly, suppose either G is the complex Clifford group $\mathcal{X}_m$ and $p = 2$, or G is a p-Clifford group $\mathcal{C}_m^{(p)}$ for $p > 2$. If $H \supset G$ is a closed subgroup of the unitary group, then either $H/(\mathbb{C}^* \mathrm{id} \cap H)$ is finite or $SU_{p^m}(\mathbb{C}) \subset H$.*

Corollary 6.8.3. (a) *Let C be a Type 2_I code that is not generated by vectors of weight 2. Then the subgroup of $O_{2m}(\mathbb{R})$ fixing the genus-m weight enumerator of C is precisely C_m:*

$$C_m = \mathrm{Aut}_{O_{2m}(\mathbb{R})}(\mathrm{cwe}_m(C)).$$

(b) *Let C be a Type 2_{II} code of length N. Then the subgroup of $U_{2m}(\mathbb{C})$ fixing the genus-m weight enumerator of C is the group generated by $\mathcal{X}_m$ and scalar multiplication by complex N-th roots of unity:*

$$\mathrm{Aut}_{U_{2m}(\mathbb{C})}(\mathrm{cwe}_m(C)) = \langle \mathcal{X}_m, \zeta_N I_{2^m} \rangle.$$

(c) *Let p be an odd prime and let $C \leq \mathbb{F}_p^N$ be a Type p_I^E code. Then*

$$\mathrm{Aut}_{U_{p^m}(\mathbb{C})}(\mathrm{cwe}_m(C)) = \langle \mathcal{C}_m^{(p)}, \zeta_N I_{2^m} \rangle.$$

Proof. (a) If $m = 1$ this corollary can be proved directly (see [383, pp. 117-118]). Suppose $m > 1$. Let C be a Type 2_I code that is not equivalent to an orthogonal sum of copies of i_2, and let $H := \mathrm{Aut}_{O_{2m}(\mathbb{R})}(\mathrm{cwe}_m(C))$. Then H is a closed subgroup of $O_{2m}(\mathbb{R})$ containing C_m. Assume that $H \neq C_m$. Then H is infinite by Theorem 6.6.3 and hence H contains the group $SO_{2m}(\mathbb{R})$ by the corollary above. Since C is not generated by vectors of weight 2, $\mathrm{cwe}_m(C)$ is not invariant under $SO_{2m}(\mathbb{R})$, a contradiction. (b) and (c) follow similarly. □

To illustrate, suppose $m = 1$. Let C be a Type 2_I code of length N and let $W(x, y)$ be its Hamming weight enumerator. Let G be the subgroup of $O_2(\mathbb{R})$ that fixes $W(x, y)$. Provided C is not generated by vectors of weight 2, $G \cong C_1$, of order 16. If C is Type 2_{II}, the subgroup of the unitary group $U_2(\mathbb{C})$ that fixes $W(x, y)$ is (apart from its center, which of course must contain all complex N-th roots of unity) the familiar group $\mathcal{X}_1$ of order 192 arising in Gleason's theorem. We are not aware of any earlier proof (before [383], [384]) that the groups fixing $W(x, y)$ could never be larger than they have to be.

We end with a question for future research.

Research Problem 6.8.4. *To what extent can the results of Corollary 6.8.3 be extended to other Types?*

The reader will see that we used some quite strong properties: the maximal finiteness of the Clifford group, and the irreducibility of the action on the Lie algebra (although the latter condition could be weakened).

7

Classical Self-Dual Codes

All classical codes over finite fields, as well as the other Types of codes over finite fields defined in §2.3, are representations of "quasisimple" form rings. In this chapter we define these form rings, classify the finite examples, and explicitly describe the most important representations. We determine the associated Clifford-Weil groups and give the invariants and Molien series for complete weight enumerators of low genus. In most cases we also list codes whose weight enumerators provide a basis for the invariant rings. We note however that even in cases where we have been unable to find a complete set of such codes, we know from our main theorems that such codes certainly exist!

7.1 Quasisimple form rings

A *quasisimple* form ring is a form ring (R, M, ψ, Φ) such that all nontrivial ideals are of the form $(\{0\}, I)$, where $I \leq \ker(\lambda)$, or in other words the twisted ring (R, M, ψ) is simple (i.e. has no nontrivial ideals, cf. Remark 1.4.7). All the classical Types of self-dual codes over fields (cf. [454]) can be defined by representations of quasisimple form rings. As we will see below, for such form rings, the underlying ring R is a direct product of matrix rings over finite fields. In particular, R is a quasi-chain ring and so our main theorem (Theorem 5.5.7) applies to representations of these form rings.

As usual "quasisimple" is a generalization of the notion of simple. A simple form ring is a form ring with only the trivial ideals (cf. Definition 1.7.6), that is, a quasisimple form ring with $\ker(\lambda) = 0$. Simple form rings are also important for describing the structure of the hyperbolic co-unitary group $\mathfrak{U}(R, \Phi)$ over semiperfect rings R, since then this group has an epimorphic image $\mathfrak{U}((R, \Phi)/\operatorname{rad}(R, \Phi))$ which is a hyperbolic co-unitary group over the semisimple (i.e. direct sum of simple) form ring $(R, \Phi)/\operatorname{rad}(R, \Phi)$ (see Corollary 5.2.11). The kernel of this epimorphism can essentially be described by linear constraints.

Let (R, M, ψ, Φ) be a quasisimple form ring. In particular, (R, M, ψ) is a simple twisted ring, which means that the only ideals I in R with $I^J = I$ are R and $\{0\}$. Since J^2 is an inner automorphism of R, it follows that R is either $\mathrm{Mat}_g(D)$ or $\mathrm{Mat}_g(D) \oplus \mathrm{Mat}_g(D)^{op}$, for some division algebra D and some integer $g \geq 1$.

In the first case, let $d \mapsto \bar{d}$ be an involution of D that coincides with J on the center of D. Since J is an anti-automorphism of R, the composition of J and the canonical anti-automorphism $r \mapsto \bar{r}^{tr}$ is an automorphism of R, and hence by the Skolem-Noether theorem (see for example Jacobson [291, page 222]) is given by conjugation by a unit $u \in R^*$. Rescaling by u, we may assume that $r^J = \bar{r}^{tr}$ for all $r \in R$.

In the second case, the restriction of J to the center of R is given by $(r, s) \mapsto (s, r)$. The involution $(r, s) \mapsto (s, r)$ is an involution of R that coincides with J on the center of R. The composition of these two anti-automorphisms of R is an automorphism of R that fixes the two components of R. Again we can apply the Skolem-Noether theorem to find a unit $(u_1, u_2) \in R$ such that after rescaling by this unit the involution J satisfies $(r, s)^J = (s, r)$ for all $(r, s) \in R$.

In both cases the rescaling does not change the orthogonality of codes, and so the codes of the rescaled Type are exactly the same as the codes of the original Type.

If D is an infinite division algebra, Hahn and O'Meara [226] list the possible simple twisted rings. From this we obtain a classification of simple form rings (R, M, ψ, Φ), since in the simple case $\ker(\lambda) = \{0\}$, and hence the form structure Φ can be recovered from the twisted ring. However, it seems that the classification of infinite quasisimple form rings is an open problem.

We now restrict our discussion to the case where R is a finite ring. Then $D = k$ is a finite field of characteristic $p > 0$.

The following general lemma follows directly from the axioms for form rings—see §1.7.1:

Lemma 7.1.1. Let (R, M, ψ, Φ) be a form ring. Assume that there is a central element $x \in Z(R)$ such that $x + x^J = 1$. Then

$$\Phi = \ker(\lambda) \oplus \{\!\{M\}\!\}, \tag{7.1.1}$$

and $\lambda(\Phi) = \{m \in M \mid m = \tau(m)\}$ is the set of symmetric elements in M.

Proof. Let $S := \{m \in M \mid m = \tau(m)\}$ be the set of symmetric elements in M. Then $\lambda(\Phi) \subset S$. Let $x \in Z(R)$ satisfy $x + x^J = 1$. Then for $m \in S$, we have

$$\lambda\{\!\{m(1 \otimes x)\}\!\} = m(1 \otimes x) + \tau(m(1 \otimes x)) = m(1 \otimes (x + x^J)) = m.$$

so $\lambda : \Phi \to S$ is surjective and the mapping $S \to \Phi$ defined by $m \mapsto \{\!\{m(1 \otimes x)\}\!\}$ for all $m \in S$ is a right inverse of λ and therefore $\{\!\{M\}\!\}$ is a complement to $\ker(\lambda)$ in Φ.   □

Remark. In particular, if R is a quasisimple form k-algebra where k is a field of characteristic p, then for $p > 2$ the element $x = \frac{1}{2}$ satisfies $x + x^J = 1$. If $p = 2$ and the restriction of J to the center of R is nontrivial, then the existence of such an x is guaranteed by the surjectivity of the trace form $z \mapsto z + z^J$ from the center of R onto the fixed field of J. Hence if either $p > 2$ or the involution on the center is nontrivial, $\Phi = \ker(\lambda) \oplus \{\!\{M\}\!\}$.

For the Clifford groups associated with representations of quasisimple form rings we find by Corollary 5.5.4 that

$$C(\rho) \cong Z.\mathfrak{U}(R,\Phi) \cong Z.(\ker(\lambda) \times \ker(\lambda)).\mathcal{G}, \qquad (7.1.2)$$

where $\mathcal{G}$ is a classical group (depending on the twisted ring (R, M, ψ), see also Table 5.1) and $Z \leq \mathbb{C}^*$ is a cyclic group of scalar matrices with

$$|Z| = \gcd\{\text{ lengths of codes of Type } \rho\}. \qquad (7.1.3)$$

7.2 Split type

In this section we consider the ring $R := \mathrm{Mat}_g(k) \oplus \mathrm{Mat}_g(k)^{op}$ together with the involution J defined by $(r, s)^J := (s, r)$. Note that R is the triangular twisted ring $T(M)$ if M is the twisted $\mathrm{Mat}_g(k)$-module $\{0\}$. Since the involution J restricts to a nontrivial automorphism of the center of R, we may assume (by Hilbert's Theorem 90) that $\epsilon = 1$. Moreover, by the above remark, all form structures Φ on R satisfy

$$\Phi \cong \ker(\lambda) \oplus \{\!\{M\}\!\} \cong \ker(\lambda) \oplus \mathrm{Mat}_g(k),$$

where $\ker(\lambda)$ is a linear R-module. The resulting form rings (R, Φ) are said to be of *split type*. Since the twisted ring is triangular, the self-dual codes of a split Type are always of the form $C \times C^\perp$, where C is simply a linear code, possibly with some fixed codewords or dual vectors. Thus in particular our methods give information about *arbitrary* linear codes. We proceed to find the hyperbolic co-unitary group $\mathfrak{U}(R, \Phi)$. The kernel of the projection π (defined in Remark 5.2.5) onto the first component is $\ker(\lambda) \oplus \ker(\lambda)$. It remains to calculate the image of π, which is

$$\pi(\mathfrak{U}(R,\Phi)) = \{\left(\begin{pmatrix} A_1 & B_1 \\ C_1 & D_1 \end{pmatrix}, \begin{pmatrix} A_2 & B_2 \\ C_2 & D_2 \end{pmatrix} \right) \in \mathrm{Mat}_{2g}(k) \oplus \mathrm{Mat}_2(\mathrm{Mat}_g(k)^{op}) \mid$$

$$\left(\begin{pmatrix} A_2 & C_2 \\ B_2 & D_2 \end{pmatrix}, \begin{pmatrix} A_1 & C_1 \\ B_1 & D_1 \end{pmatrix} \right) \left(\begin{pmatrix} 0 & 0 \\ 1 & 0 \end{pmatrix}, \begin{pmatrix} 0 & 0 \\ 1 & 0 \end{pmatrix} \right) \left(\begin{pmatrix} A_1 & B_1 \\ C_1 & D_1 \end{pmatrix}, \begin{pmatrix} A_2 & B_2 \\ C_2 & D_2 \end{pmatrix} \right)$$

$$- \left(\begin{pmatrix} 0 & 0 \\ 1 & 0 \end{pmatrix}, \begin{pmatrix} 0 & 0 \\ 1 & 0 \end{pmatrix} \right) \in \psi^{-1}(\lambda_2(\Phi_2))\}.$$

Since the image of λ consists of all symmetric elements, we find that

$$\left(\begin{pmatrix} A_1 & B_1 \\ C_1 & D_1 \end{pmatrix}, \begin{pmatrix} A_2 & B_2 \\ C_2 & D_2 \end{pmatrix} \right) \in \pi(\mathfrak{U}(R, \Phi))$$

if and only if

$$\begin{pmatrix} A_2 & C_2 \\ B_2 & D_2 \end{pmatrix} \begin{pmatrix} 0 & 1 \\ -1 & 0 \end{pmatrix} \begin{pmatrix} A_1 & B_1 \\ C_1 & D_1 \end{pmatrix} = \begin{pmatrix} 0 & 1 \\ -1 & 0 \end{pmatrix}.$$

Therefore

$$\mathfrak{U}(R, \Phi) \cong (\ker(\lambda) \oplus \ker(\lambda)). \mathrm{GL}_{2g}(k).$$

Representations.
The finite representations $(V, \rho_M, \rho_\Phi, \beta)$ of the form ring (R, Φ) can be obtained by letting $g = 1$ and applying Morita theory. For an arbitrary finite left $\mathrm{Mat}_g(k)$-module W we define $W^* := \mathrm{Hom}(W, \mathbb{Q}/\mathbb{Z})$. Then W^* is a left $\mathrm{Mat}_g(k)^{op}$-module and $V = W \oplus W^*$ is a left R-module. The nonsingular bilinear form $\beta := \rho_M(\psi(1))$ is defined by $\beta((w_1, w_1'), (w_2, w_2')) := \frac{1}{p}\mathrm{Tr}(w_1'(w_2) + w_2'(w_1))$, where Tr denotes the trace from k into the prime field $\mathbb{F}_p \cong \mathbb{Z}/p\mathbb{Z}$. This gives the embedding ρ_M of M into $\mathrm{Bil}(V, \mathbb{Q}/\mathbb{Z})$. To define ρ_Φ it remains to specify the restriction $(\rho_\Phi)|_{\ker(\lambda)}$. Since $\ker(\lambda)$ is a linear R-module, we must specify a homomorphism $\ker(\lambda) \to V^*$.

7.2.1 q^{lin}: Linear codes over $\mathbb{F}_q$

In this section we describe the Type q^{lin} of all $\mathbb{F}_q$-linear codes in $\mathbb{F}_q^N$, together with the related Types q_1^{lin}, $q_{1'}^{\mathrm{lin}}$, and $q_{1,1'}^{\mathrm{lin}}$ of those for which the dual code, the code, and both the code and the dual contain $\mathbf{1}$, respectively. These Types were not discussed in §2.3. For any prime power $q = p^f$ we define three form rings of split type:

$$R(q^{\mathrm{lin}}) := (\mathbb{F}_q \oplus \mathbb{F}_q, M := \mathbb{F}_q \oplus \mathbb{F}_q, \mathrm{id}, \Phi := \{M\} \cong \mathbb{F}_q),$$
$$R(q_1^{\mathrm{lin}}) := (\mathbb{F}_q \oplus \mathbb{F}_q, M := \mathbb{F}_q \oplus \mathbb{F}_q, \mathrm{id}, \Phi := \{M\} \oplus \ker(\lambda) \cong \mathbb{F}_q \oplus \mathbb{F}_q),$$
$$R(q_{1,1'}^{\mathrm{lin}}) := (\mathbb{F}_q \oplus \mathbb{F}_q, M := \mathbb{F}_q \oplus \mathbb{F}_q, \mathrm{id}, \Phi := \{M\} \oplus \ker(\lambda) \cong \mathbb{F}_q \oplus \mathbb{F}_q \oplus \mathbb{F}_q).$$

The representation $\rho(q^{\mathrm{lin}}) := (V, \rho_M, \rho_\Phi, \beta)$ is as follows: $V := \mathbb{F}_q \oplus \mathbb{F}_q$, $\rho_M(a, b) := m_{a,b}$, with

$$m_{a,b}((x_1, y_1), (x_2, y_2)) := \frac{1}{p}\mathrm{Tr}(x_1 a y_2 + y_1 b x_2).$$

Then $\tau(a, b) = (b, a)$ and $\beta((x_1, y_1), (x_2, y_2)) = \frac{1}{p}\mathrm{Tr}(x_1 y_2 + y_1 x_2)$, where Tr denotes the trace from $\mathbb{F}_q$ to its prime field $\mathbb{F}_p = \mathbb{Z}/p\mathbb{Z}$. The map ρ_Φ sends $\{(a, b)\}$ to q_{a+b}, where $q_a((x_1, y_1)) := x_1 a y_1$. To define the representations $\rho(q_1^{\mathrm{lin}})$, $\rho(q_{1'}^{\mathrm{lin}})$ of $R(q_1^{\mathrm{lin}})$ and the representation $\rho(q_{1,1'}^{\mathrm{lin}})$ of $R(q_{1,1'}^{\mathrm{lin}})$ we need to specify the restriction of the respective representations ρ_Φ to the kernel of λ.

For $\rho(q_1^{\mathrm{lin}})$ and $\rho(q_{1'}^{\mathrm{lin}})$ we define $\rho_\Phi(\{\!\{(0,0)\}\!\}, b) := \varphi_b$ (respectively φ_b') by

$$\varphi_b((x_1, y_1)) := bx_1 \quad \text{and} \quad \varphi_b'((x_1, y_1)) := by_1\,,$$

for all $b \in \ker(\lambda) \cong \mathbb{F}_q$ and all $(x_1, y_1) \in V \cong \mathbb{F}_q \oplus \mathbb{F}_q$. For $\rho(q_{1,1'}^{\mathrm{lin}})$ the kernel of λ is isomorphic to $\mathbb{F}_q \oplus \mathbb{F}_q$. Here we put

$$\rho_\Phi(\{\!\{(0,0)\}\!\}, (a,b))(x_1, y_1) := ax_1 + by_1\,,$$

for all $(a,b) \in \ker(\lambda) \cong \mathbb{F}_q \oplus \mathbb{F}_q$ and all $(x_1, y_1) \in V \cong \mathbb{F}_q \oplus \mathbb{F}_q$.

Isotropic self-dual codes $C \leq V^N$ in the representation $\rho(q^{\mathrm{lin}})$ have the form $C = C_1 \oplus C_2$, where $C_1 \leq \mathbb{F}_q^N = (1,0)V^N$ and $C_2 \leq \mathbb{F}_q^N = (0,1)V^N$. We will denote typical codewords in C_1 and C_2 by $c_1 = (c_{11}, c_{12}, \ldots, c_{1N})$ and $c_2 = (c_{21}, c_{22}, \ldots, c_{2N})$ respectively. Since

$$\rho_M^N(m_{a,b})((c_1, c_2), (c_1', c_2')) = \sum_{i=1}^N (c_{1i}ac_{2i}' + c_{2i}bc_{1i}')\,,$$

it follows that C is self-orthogonal if $C_1 \perp C_2$, and C is self-dual if $C_1 = C_2^\perp$.

If $q = p$ is a prime, such codes were investigated by Bachoc [19] when constructing unimodular Hermitian lattices over imaginary quadratic fields K. If the prime p splits in the ring of integers O_K of K, then $O_K/pO_K \cong \mathbb{F}_p \oplus \mathbb{F}_p$ and complex conjugation gives an involution J on O_K/pO_K. Construction A over O_K applied to self-dual codes $C \leq (\mathbb{F}_p \oplus \mathbb{F}_p)^N$ then produces unimodular O_K-lattices of rank N. For $p = 2$ these codes were also considered by Betsumiya, Gulliver and Harada [41] and Betsumiya and Harada [45]. The idea of pairing a code with its dual is also used in the Calderbank-Shor-Steane construction of quantum codes (see Example 13.1.6). In our language, the Calderbank-Shor-Steane construction is simply the observation that Type $(q^2)^{\mathrm{H+}}$ is isomorphic to a sub-Type of q^{lin}.

The condition that C be isotropic does not impose any additional constraint if $\Phi = \{\!\{M\}\!\}$. For $\rho(q_1^{\mathrm{lin}})$ it implies $\rho_\Phi^N(\varphi_b)((c_1, c_2)) = \mathrm{Tr}(b \sum_{i=1}^N c_{1i}) = 0$ for all $b \in \mathbb{F}_q$, and so

$$\sum_{i=1}^N c_{1i} = 0 \quad \text{for all } (c_1, c_2) \in C\,.$$

Hence self-dual isotropic codes C in $N\rho(q_1^{\mathrm{lin}})$ are in one-to-one correspondence with $\mathbb{F}_q$-linear codes $C_1 \leq \mathbb{F}_q^N$ such that $C_1^\perp$ contains the all-ones vector. In particular, for $q = 2$ we get the Type 2_1^{lin} of even (i.e. singly-even) binary codes.

Similarly, for $\rho(q_{1'}^{\mathrm{lin}})$ we have the condition that for isotropic codes

$$\sum_{i=1}^N c_{2i} = 0 \quad \text{for all } (c_1, c_2) \in C\,.$$

Hence self-dual isotropic codes C in $N\rho(q_1^{\lin})$ are in one-to-one correspondence with $\mathbb{F}_q$-linear codes $C_1 \le \mathbb{F}_q^N$ containing the all-ones vector.

Finally, $\rho(q_{1,1'}^{\lin})$ combines both conditions, i.e. $\mathbf{1} \in C_1 \cap C_1^\perp$.

Clifford-Weil groups

Since linear codes exist for every length, the corresponding Clifford-Weil groups of genus g are

$$
\begin{aligned}
\mathcal{C}_g(\rho(q^{\lin})) &\cong & \mathfrak{U}_g(R(q^{\lin})) \cong \mathrm{GL}_{2g}(\mathbb{F}_q)\,, \\
\mathcal{C}_g(\rho(q_1^{\lin})) &\cong \mathcal{C}_g(\rho(q_{1'}^{\lin})) \cong \mathfrak{U}_g(R(q_1^{\lin})) \cong \mathbb{F}_q^{2g}.\,\mathrm{GL}_{2g}(\mathbb{F}_q)\,.
\end{aligned}
$$

Note that the groups $\mathcal{C}_g(\rho(q_1^{\lin}))$ and $\mathcal{C}_g(\rho(q_{1'}^{\lin}))$ are conjugate in $\mathrm{GL}_{q^{2g}}(\mathbb{C})$, and so their invariant rings are also isomorphic. We obtain $\mathrm{Inv}(\mathcal{C}_g(\rho(q_{1'}^{\lin})))$ from $\mathrm{Inv}(\mathcal{C}_g(\rho(q_1^{\lin})))$ by replacing $x_{a,b}$ by $x_{b,a}$. Also, if $C_1 \oplus C_2$ is a code of Type $\rho(q_1^{\lin})$, then $C_2 \oplus C_1$ is a code of Type $\rho(q_{1'}^{\lin})$, and conversely.

If $\mathbf{1} \in C_1 \cap C_1^\perp$, then the length of C_1 is divisible by p. Hence

$$
\mathcal{C}_g(\rho(q_{1,1'}^{\lin})) \cong Z_p.\mathfrak{U}_g(R(q_{1,1'}^{\lin})) \cong Z_p.\mathbb{F}_q^{4g}.\,\mathrm{GL}_{2g}(\mathbb{F}_q)\,.
$$

$\mathbb{F}_2$, Genus 1

The genus-1 Clifford-Weil group for the Type of linear binary codes is

$$
\mathcal{C}(\rho(2^{\lin})) \cong \mathfrak{U}(R(2^{\lin})) \cong S_3\,,
$$

of order 6 and with Molien series

$$
\mathrm{MS}_{\mathrm{cwe},2^{\lin}} = \frac{1}{(1-t)^2(1-t^2)(1-t^3)} \qquad [A000601]\,. \tag{7.2.1}
$$

The invariant ring is the polynomial ring in the complete weight enumerators p_N ($N = 1, 2, 3$) of $C_N \oplus \langle \mathbf{1} \rangle$, where $C_N = \langle \mathbf{1} \rangle^\perp \le \mathbb{F}_2^N$ is the even weight code of length N, and $f_0 := \mathrm{cwe}(0 \oplus \mathbb{F}_2) = x_{00} + x_{01}$:

$$
\mathrm{Inv}_{\mathrm{cwe},2^{\lin}} = \frac{1}{f_0, p_1, p_2, p_3}\,. \tag{7.2.2}
$$

For even linear binary codes, with respect to the basis $(e_{0,0}, e_{1,0}, e_{0,1}, e_{1,1})$ of $\mathbb{C}[V]$, the Clifford-Weil group $\mathcal{C}(\rho(2_1^{\lin}))$ is generated by the matrices $q := \rho(d(1, q_1)) = \mathrm{diag}(1, 1, 1, -1)$, $d := \rho(d(1, \varphi_1)) = \mathrm{diag}(1, -1, 1, -1)$ and

$$
h := h_{1,1,1} = \frac{1}{2}
\begin{bmatrix}
1 & 1 & 1 & 1 \\
1 & 1 & -1 & -1 \\
1 & -1 & 1 & -1 \\
1 & -1 & -1 & 1
\end{bmatrix}\,. \tag{7.2.3}
$$

Thus

$$\mathcal{C}(\rho(2_1^{\text{lin}})) \cong \mathfrak{U}(R(2_1^{\text{lin}})) \cong (Z_2 \times Z_2).S_3 \cong S_4 \,,$$

of order 24, with Molien series:

$$\text{MS}_{\text{cwe},2_1^{\text{lin}}} = \frac{1}{(1-t)(1-t^2)(1-t^3)(1-t^4)} \qquad [A001400]. \qquad (7.2.4)$$

The invariant ring is the polynomial ring in the complete weight enumerators p_N for $N = 1, \ldots, 4$:

$$\text{Inv}_{\text{cwe},2_1^{\text{lin}}} = \frac{1}{p_1, p_2, p_3, p_4} \,, \qquad (7.2.5)$$

with

$$p_1 = x_{0,0} + x_{0,1} \,,$$
$$p_2 = x_{0,0}^2 + x_{1,0}^2 + x_{0,1}^2 + x_{1,1}^2 \,,$$
$$p_3 = x_{0,0}^3 + 3x_{0,0}x_{1,0}^2 + 3x_{0,1}x_{1,1}^2 + x_{0,1}^3 \,,$$
$$p_4 = x_{0,0}^4 + x_{1,0}^4 + x_{0,1}^4 + x_{1,1}^4 + 6x_{0,0}^2 x_{1,0}^2 + 6x_{0,1}^2 x_{1,1}^2 \,. \qquad (7.2.6)$$

If we set $x_{0,0} = x_{0,1} = x$, $x_{1,0} = x_{1,1} = y$, these four polynomials collapse to two, and we conclude (no surprise) that the weight enumerator of an even binary code is a polynomial in x and y^2.

For $\rho(2_{1,1'}^{\text{lin}})$, the Type of even binary codes that contain the all-ones vector, we find

$$\mathcal{C}(\rho(2_{1,1'}^{\text{lin}})) \cong 2_+^{1+4}.S_3 \,,$$

of order 192, which is generated by $\mathcal{C}(\rho(2_1^{\text{lin}}))$ and the additional generator $d' := \rho(d(1, \varphi_1')) = \text{diag}(1, 1, -1, -1)$. Its Molien series is

$$\text{MS}_{\text{cwe},2_{1,1'}^{\text{lin}}} = \frac{1}{(1-t^2)(1-t^4)^2(1-t^6)} \qquad [A008763]. \qquad (7.2.7)$$

If q_4 denotes the complete weight enumerator of $\langle 1 \rangle \perp C_4$, then the invariant ring of $\mathcal{C}(\rho(2_{1,1'}^{\text{lin}}))$ is the polynomial ring

$$\text{Inv}_{\text{cwe},2_{1,1'}^{\text{lin}}} = \frac{1}{p_2, p_4, q_4, p_6} \,, \qquad (7.2.8)$$

Setting $x_{0,0} = x_{0,1} = x$, $x_{1,0} = x_{1,1} = y$, we find that the weight enumerator of an even code that contains the all-ones vector is a polynomial in $x^2 + y^2$ and $x^2 y^2$, or in other words is a symmetric polynomial in x^2 and y^2.

$\mathbb{F}_2$, Genus 2

For genus 2, $R_2 = \text{Mat}_2(\mathbb{F}_2) \oplus \text{Mat}_2(\mathbb{F}_2)$, $V_2 = \mathbb{F}_2^2 \oplus \mathbb{F}_2^2$, and the genus-2 Clifford-Weil group for linear binary codes is

$$\mathcal{C}_2(\rho(2^{\text{lin}})) = \mathfrak{U}_2(R(2^{\text{lin}})) \cong \text{GL}_4(\mathbb{F}_2)$$

of order 20160. Its Molien series is

$$\frac{\theta_0(t) + t^{65}\theta_0(t^{-1})}{\sigma_0(t)} \qquad [A104993]\,, \qquad (7.2.9)$$

where

$$\begin{aligned}
\theta_0(t) = {}& 1 + t + t^3 + 2\,t^5 + 9\,t^6 + 16\,t^7 + 33\,t^8 + 58\,t^9 + 93\,t^{10} \\
& + 154\,t^{11} + 254\,t^{12} + 418\,t^{13} + 682\,t^{14} + 1086\,t^{15} + 1665\,t^{16} \\
& + 2472\,t^{17} + 3554\,t^{18} + 4937\,t^{19} + 6672\,t^{20} + 8772\,t^{21} + 11210\,t^{22} \\
& + 13973\,t^{23} + 16967\,t^{24} + 20064\,t^{25} + 23154\,t^{26} + 26064\,t^{27} \\
& + 28668\,t^{28} + 30866\,t^{29} + 32557\,t^{30} + 33710\,t^{31} + 34287\,t^{32} \quad (7.2.10)
\end{aligned}$$

and

$$\sigma_0(t) = (1 - t^1)(1 - t^2)^2(1 - t^3)(1 - t^4)^4(1 - t^5)^2(1 - t^6)^3(1 - t^7)^2(1 - t^{15})\,.$$

The genus-2 Clifford-Weil group for even codes is isomorphic to

$$\mathcal{C}_2(\rho(2_1^{\mathrm{lin}})) \cong \mathfrak{U}_2(R(2_1^{\mathrm{lin}})) \cong \mathbb{F}_2^4 \rtimes \mathrm{GL}_4(\mathbb{F}_2)\,,$$

generated by $h \otimes I_4$, $q \otimes I_4$, $d \otimes I_4$ and the permutation matrices corresponding to the action of $R_2^* = \mathrm{GL}_2(\mathbb{F}_2) \times \mathrm{GL}_2(\mathbb{F}_2)$ on V_2. This has order 322560 and Molien series

$$\frac{\theta(t) + t^{92}\theta(t^{-1})}{\sigma(t)} \qquad [A092351]\,, \qquad (7.2.11)$$

where

$$\begin{aligned}
\theta(t) = {}& 1 + 2\,t^6 + 4\,t^7 + 6\,t^8 + 11\,t^9 + 19\,t^{10} + 25\,t^{11} + 44\,t^{12} + 64\,t^{13} \\
& + 107\,t^{14} + 164\,t^{15} + 269\,t^{16} + 399\,t^{17} + 619\,t^{18} + 907\,t^{19} \\
& + 1333\,t^{20} + 1896\,t^{21} + 2678\,t^{22} + 3686\,t^{23} + 5024\,t^{24} + 6713\,t^{25} \\
& + 8827\,t^{26} + 11420\,t^{27} + 14544\,t^{28} + 18216\,t^{29} + 22485\,t^{30} \\
& + 27334\,t^{31} + 32737\,t^{32} + 38672\,t^{33} + 45066\,t^{34} + 51810\,t^{35} \\
& + 58793\,t^{36} + 65888\,t^{37} + 72884\,t^{38} + 79660\,t^{39} + 85998\,t^{40} \\
& + 91735\,t^{41} + 96669\,t^{42} + 100708\,t^{43} + 103651\,t^{44} + 105488\,t^{45} \\
& + 53044\,t^{46} \qquad\qquad\qquad\qquad\qquad\qquad\qquad\qquad\qquad (7.2.12)
\end{aligned}$$

and

$$\begin{aligned}
\sigma(t) = {}& (1 - t^1)(1 - t^2)(1 - t^3)(1 - t^4)^2(1 - t^5)^2(1 - t^6)^2(1 - t^7) \\
& \times (1 - t^8)^3(1 - t^{12})(1 - t^{14})(1 - t^{15})\,. \qquad\qquad (7.2.13)
\end{aligned}$$

For $\rho(2_{1,1'}^{\mathrm{lin}})$ the genus-2 Clifford-Weil group is isomorphic to

$$\mathcal{C}_2(\rho(2_{1,1'}^{\mathrm{lin}})) \cong 2_+^{1+8} \rtimes \mathrm{GL}_4(\mathbb{F}_2)\,,$$

generated by $C_2(\rho(2_1^{\text{lin}}))$ and $d' \otimes I_4$. This has order 10321920 and Molien series

$$\frac{\theta_d(t^2) + t^{154}\theta_d(t^{-2})}{\sigma_d(t)} \qquad [A105319], \qquad (7.2.14)$$

where

$$\begin{aligned}
\theta_d(t) = {} & 1 + t + 3\,t^4 + 9\,t^5 + 24\,t^6 + 46\,t^7 + 117\,t^8 + 239\,t^9 + 541\,t^{10} \\
& + 1133\,t^{11} + 2370\,t^{12} + 4649\,t^{13} + 8923\,t^{14} + 16245\,t^{15} + 28601\,t^{16} \\
& + 48132\,t^{17} + 78194\,t^{18} + 121981\,t^{19} + 183920\,t^{20} + 267517\,t^{21} \\
& + 376916\,t^{22} + 514682\,t^{23} + 683056\,t^{24} + 881972\,t^{25} + 1110910\,t^{26} \\
& + 1366468\,t^{27} + 1644918\,t^{28} + 1940048\,t^{29} + 2245177\,t^{30} \\
& + 2551867\,t^{31} + 2851403\,t^{32} + 3133830\,t^{33} + 3389363\,t^{34} \\
& + 3608201\,t^{35} + 3781448\,t^{36} + 3901399\,t^{37} + 3962896\,t^{38}
\end{aligned}$$

$$(7.2.15)$$

and

$$\begin{aligned}
\sigma_d(t) = {} & (1 - t^2)(1 - t^4)^2(1 - t^6)^2(1 - t^8)^4(1 - t^{10})^2 \\
& \times (1 - t^{12})^2(1 - t^{14})(1 - t^{28})(1 - t^{30}).
\end{aligned} \qquad (7.2.16)$$

The Clifford-Weil groups of higher genus could be obtained in a similar way. We invite the reader to carry out similar calculations for codes over $\mathbb{F}_3$ and $\mathbb{F}_4$.

7.3 Hermitian type

We now return to the families of codes introduced in Chapter 2, beginning with the Hermitian codes in §2.3.3, and discuss their Clifford-Weil groups, Molien series and invariants. Whenever appropriate we give references to the sections of [454] where these codes are studied.

In this section the field k is $\mathbb{F}_q$, where $q = r^2$ is a square and $^- : k \to k$ is the map given by $a \mapsto a^r$, the Frobenius automorphism with fixed field $\mathbb{F}_r$. The twisted ring is (R, M, ψ), where $R = \text{Mat}_g(k)$, with the involution $a^J = \bar{a}^{tr}$ for all $a \in R$. As above we can choose $\epsilon = 1$, $M = R$, $\psi = id$ and $\tau = J$. Any form structures Φ must satisfy

$$\Phi = \{\!\{M\}\!\} \oplus \ker(\lambda).$$

Since $\psi^{-1}(\lambda(\Phi))$ is the set of symmetric elements in R (i.e. those $a \in R$ with $a^J = a$), we find that $\begin{pmatrix} A & B \\ C & D \end{pmatrix} \in \text{Mat}_{2g}(k)$ lies in $\pi(\mathfrak{U}(R, \Phi))$ if and only if

$$\begin{pmatrix} \overline{A}^{tr} & \overline{C}^{tr} \\ \overline{B}^{tr} & \overline{D}^{tr} \end{pmatrix} \begin{pmatrix} 0 & -1 \\ 1 & 0 \end{pmatrix} \begin{pmatrix} A & B \\ C & D \end{pmatrix} = \begin{pmatrix} 0 & -1 \\ 1 & 0 \end{pmatrix},$$

hence

$$\mathfrak{U}(R, \varPhi) \cong (\ker(\lambda) \oplus \ker(\lambda)).U_{2g}(k)$$

is an extension of a unitary group.

Representations

The finite representations $(V, \rho_M, \rho_\varPhi, \beta)$ of the form ring $(R, \varPhi)$ can be obtained by letting $g = 1$ and applying Morita theory. Let V be a left R-module. Then $V \cong (k^g)^N$ for some $N \geq 0$, and, up to rescaling and equivalence, we can choose the nonsingular bilinear form $\beta := \rho_M(\psi(1))$ to be defined by

$$\beta(v, v') := \frac{1}{p} \operatorname{Tr}(\sum_{i=1}^{gN} v_i \overline{v'_i}),$$

where Tr denotes the trace from k to the prime field $\mathbb{F}_p \cong \mathbb{Z}/p\mathbb{Z}$. This gives the embedding ρ_M of M into $\operatorname{Bil}(V, \mathbb{Q}/\mathbb{Z})$. To define $\rho_\varPhi$ it remains to specify the restriction $(\rho_\varPhi)_{|\ker(\lambda)}$. Since $\ker(\lambda)$ is a linear R-module, we must specify a homomorphism $\ker(\lambda) \to V^*$.

7.3.1 q^{H}: Hermitian self-dual codes over $\mathbb{F}_q$

We now consider the families defined in §2.3.3, namely Hermitian self-dual codes over $\mathbb{F}_q$ and Hermitian self-dual codes over $\mathbb{F}_q$ that contain the all-ones vector, where $q = r^2$ is a square. The corresponding representations are $\rho(q^{\mathrm{H}})$ and $\rho(q_1^{\mathrm{H}})$. The Galois automorphism $x \mapsto x^r$ of $\mathbb{F}_q$ over $\mathbb{F}_r$ is denoted by $^-$.

Clifford-Weil groups

Since the dimension of a Hermitian self-dual code is half its length, the length must be even. If $\alpha \in \mathbb{F}_q$ with $\alpha\overline{\alpha} = -1$, then $\langle (1, \alpha) \rangle$ is a self-dual code of Type $\rho(q^{\mathrm{H}})$ of length 2. We have thus established part (a) of the following proposition.

Proposition 7.3.1. *Let $q = p^{2f}$ and $g \in \mathbb{N}$.*

(a) $\mathcal{C}_g(\rho(q^{\mathrm{H}})) \cong Z_2.\mathfrak{U}_g(R(q^{\mathrm{H}})) \cong Z_2.\operatorname{GU}_{2g}(\mathbb{F}_q)$.
(b) For the codes that contain the all-ones vector,

$$\mathcal{C}_g(\rho(q_1^{\mathrm{H}})) \cong \begin{cases} (Z_2 \times p^{1+4gf}).\operatorname{GU}_{2g}(\mathbb{F}_q) & \text{if } p \text{ is odd}, \\ (2^{1+4gf}).\operatorname{GU}_{2g}(\mathbb{F}_q) & \text{if } p = 2. \end{cases}$$

Proof. It remains to prove (b). Clearly, if C is a Hermitian self-dual code containing $\mathbf{1}$, then C also contains $\alpha\mathbf{1}$, for all $\alpha \in \mathbb{F}_q$. Choosing $\alpha \in \mathbb{F}_q$ such that $\operatorname{Tr}_{\mathbb{F}_r/\mathbb{F}_p}(\alpha\overline{\alpha}) = 1$ implies that the length of C is divisible by p. Since the roots of unity in $\mathcal{C}_g(\rho(q_1^{\mathrm{H}}))$ belong to $\mathbb{Q}[\zeta_p]$, this proves (b). $\square$

The case $q = 4$

Reference: [454, pp. 226–228]. Let $R = V = \mathbb{F}_4$. With respect to the basis $(0, 1, \omega, \omega^2)$ of $\mathbb{C}[V]$, the Clifford-Weil group $\mathcal{C}(\rho(4^H))$ is generated by the three elements $h := h_{1,1,1}$ given in (7.2.3), $\phi_\omega := \mathrm{diag}\{1, -1, -1, -1\}$ and

$$r := m_\omega = \rho((\omega, 0)) = \begin{bmatrix} 1 & 0 & 0 & 0 \\ 0 & 0 & 1 & 0 \\ 0 & 0 & 0 & 1 \\ 0 & 1 & 0 & 0 \end{bmatrix}. \tag{7.3.1}$$

The group $\mathcal{C}(\rho(4_1^H))$ is generated by h, r, ϕ_ω and $\mathrm{diag}\{1, 1, -1, -1\}$.
    We find that

$$\mathcal{C}(\rho(4^H)) \cong D_{12} \times Z_3 \cong Z_2.\mathfrak{U}(R(4^H)), \tag{7.3.2}$$

of order $36 = 2^2 3^2$, and

$$\mathcal{C}(\rho(4_1^H)) \cong Z_2.\mathfrak{U}(R(4_1^H)) \cong 2_+^{1+4}.(Z_3 \times Z_3).2, \tag{7.3.3}$$

of order $576 = 2^6 3^2$. The latter is a subgroup of index 2 in the Weil group $W(F_4)$ of the root system F_4. The Molien series are

$$\mathrm{MS}_{\mathrm{cwe}, 4^H} = \frac{1 + 3t^6}{(1 - t^2)^2(1 - t^6)^2} \qquad [A092352], \tag{7.3.4}$$

and

$$\mathrm{MS}_{\mathrm{cwe}, 4_1^H} = \frac{1 + t^{12}}{(1 - t^2)(1 - t^6)(1 - t^8)(1 - t^{12})} \qquad [A028249], \tag{7.3.5}$$

respectively.
    To find the ring spanned by the Galois-invariant weight enumerators, we adjoin the additional generator

$$\sigma := \begin{bmatrix} 1 & 0 & 0 & 0 \\ 0 & 1 & 0 & 0 \\ 0 & 0 & 0 & 1 \\ 0 & 0 & 1 & 0 \end{bmatrix} \tag{7.3.6}$$

to obtain the following extensions of index 2:

$$\tilde{\mathcal{C}}(\rho(4^H)) := \langle \mathcal{C}(\rho(4^H)), \sigma \rangle \cong D_{12} \times S_3 \tag{7.3.7}$$

and

$$\tilde{\mathcal{C}}(\rho(4_1^H)) := \langle \mathcal{C}(\rho(4_1^H)), \sigma \rangle \cong W(F_4), \tag{7.3.8}$$

respectively, of orders 72 and 1152, with Molien series

$$\widetilde{\mathrm{MS}}_{\mathrm{cwe},\,4^{\mathrm{H}}} = \frac{1+t^6}{(1-t^2)^2(1-t^6)^2} \qquad [A092353] \qquad (7.3.9)$$

and

$$\widetilde{\mathrm{MS}}_{\mathrm{cwe},\,4_1^{\mathrm{H}}} = \frac{1}{(1-t^2)(1-t^6)(1-t^8)(1-t^{12})} \qquad [A008670] \qquad (7.3.10)$$

The invariant ring for the second of these two groups (the Weyl group $W(F_4)$, Shephard & Todd X_{28}) is

$$\widetilde{\mathrm{Inv}}_{\mathrm{cwe},\,4_1^{\mathrm{H}}} = \frac{1}{i_2,\, h_6,\, e_8 \otimes \mathbb{F}_4,\, (e_7 e_5)^+}, \qquad (7.3.11)$$

a polynomial ring in the complete weight enumerators of the four specified codes. For i_2, h_6 and $e_8 \otimes \mathbb{F}_4$ see (2.4.24), (2.4.26) and (2.4.10). Instead of the complete weight enumerator of the fourth code, $(e_7 e_5)^+$, we can use

$$f_{12} := (\sigma_{22} - 3x^2 y^2 - 3z^2 t^2)(\sigma_{22} - 3x^2 z^2 - 3y^2 t^2)(\sigma_{22} - 3x^2 t^2 - 3y^2 z^2), \quad (7.3.12)$$

where $\sigma_{22} = x^2 y^2 + x^2 z^2 + \cdots$ (6 terms). As usual we use the variables x, y, z, t instead of $x_0, x_1, x_\omega, x_{\omega^2}$ for the complete weight enumerators of codes over $\mathbb{F}_4$.

For the index-2 subgroup of order 576, i.e. without the assumption that C and $\bar{C}$ have the same complete weight enumerator, we have:

$$\mathrm{Inv}_{\mathrm{cwe},\,4_1^{\mathrm{H}}} = \frac{1,\, d_{12}'}{i_2,\, h_6,\, e_8 \otimes \mathbb{F}_4,\, (e_7 e_5)^+}, \qquad (7.3.13)$$

where d_{12}' is the code obtained from the usual binary self-dual code d_{12}^+ by multiplying the last four coordinates by ω. Instead of $\mathrm{cwe}(d_{12}')$ it is simpler to use

$$(x^2 - y^2)(x^2 - z^2)(x^2 - t^2)(y^2 - z^2)(y^2 - t^2)(z^2 - t^2). \qquad (7.3.14)$$

Next we consider the invariant rings for the groups $\mathcal{C}(\rho(4^{\mathrm{H}}))$ and $\tilde{\mathcal{C}}(\rho(4^{\mathrm{H}}))$, i.e. without the assumption that $\mathbf{1}$ is in the code. We will use the weight enumerators of the following codes: $i_2^{(a)} = \langle (1, \omega) \rangle$, and $h_6^{(a)}$, $h_6^{(b)}$, $h_6^{(c)}$, $h_6^{(d)}$, which are obtained from the hexacode h_6 (see §2.4.6) by multiplying the six coordinates by $(\omega, 1, 1, 1, 1, 1)$, $(\omega^2, 1, 1, 1, 1, 1)$, $(\omega, \omega, 1, 1, 1, 1)$ and $(\omega, \omega^2, 1, 1, 1, 1)$, respectively.

The invariant rings corresponding to the Molien series (7.3.4) and (7.3.5) are

$$\mathrm{Inv}_{\mathrm{cwe},\,4^{\mathrm{H}}} = \frac{1,\, h_6^{(a)},\, h_6^{(b)},\, h_6^{(c)}}{i_2,\, i_2^{(a)},\, h_6,\, h_6^{(d)}}, \qquad (7.3.15)$$

$$\widetilde{\mathrm{Inv}}_{\mathrm{cwe},\,4^{\mathrm{H}}} = \frac{1,\, h_6^{(a)} + h_6^{(b)}}{i_2,\, i_2^{(a)},\, h_6,\, h_6^{(d)}}, \qquad (7.3.16)$$

respectively, where $h_6^{(a)} + h_6^{(b)}$ is the direct sum code of length 12.

Now we consider the symmetrized weight enumerators, assuming $\mathbf{1}$ is in the code. The symmetrized weight enumerators of the codes mentioned in (7.3.11) (or equivalently (7.3.13)) span the ring spanned by all symmetrized weight enumerators of the codes of Type 4_1^H, which is

$$\frac{1, \ (e_7 e_5)^+}{i_2, \ h_6, \ e_8 \otimes \mathbb{F}_4} , \qquad (7.3.17)$$

or equivalently

$$\frac{1, \ \{(x^2 - z^2)(y^2 - z^2)\}^3}{x^2 + y^2 + 2z^2, \ \text{swe}(h_6), \ \{(x^2 - z^2)(y^2 - z^2)\}^2} \qquad (7.3.18)$$

(cf. (2.4.26)), with Molien series

$$\frac{1 + t^{12}}{(1 - t^2)(1 - t^6)(1 - t^8)} \qquad [A036410] . \qquad (7.3.19)$$

Remark. If we try to apply invariant theory directly to the symmetrized weight enumerators (for Type (4_1^H)), we are led to the group

$$G = \left\langle \frac{1}{2} \begin{pmatrix} 1 & 1 & 2 \\ 1 & 1 & -2 \\ 1 & -1 & 0 \end{pmatrix}, \ \begin{pmatrix} 0 & 1 & 0 \\ 1 & 0 & 0 \\ 0 & 0 & 1 \end{pmatrix}, \ \begin{pmatrix} 1 & & \\ & -1 & \\ & & -1 \end{pmatrix} \right\rangle ,$$

of order 48, the Weyl group of type B_3 (Shephard & Todd #2a), with Molien series

$$\frac{1}{(1 - t^2)(1 - t^4)(1 - t^6)} \qquad [A001399] . \qquad (7.3.20)$$

However, the G-invariant of degree 4 is

$$\delta_4 = (x^2 - z^2)(y^2 - z^2) ,$$

which cannot be obtained from the symmetrized weight enumerator of any self-dual code of length 4. So the ring of invariants of G is not the same as the ring (7.3.18) spanned by the symmetrized weight enumerators of codes of Type 4_1^H. In fact, the two rings have the same quotient field, so there is no group whose ring of invariants is (7.3.18).

The symmetrization needed to obtain the Hamming weight enumerators does not commute with all of $\mathcal{C}(\rho(4_1^H)))$, but it does commute with the subgroup $\mathcal{C}(\rho(4^H)))$. Now a code of Type 4^H can always be obtained from one of Type 4_1^H by multiplying certain columns by ω or ω^2, and so has the same Hamming weight enumerator as that code. Therefore the ring spanned by the Hamming weight enumerators of codes of Type 4_1^H is the same as that spanned by the Hamming weight enumerators of codes of Type 4^H . This is the invariant ring of the group

$$G = \left\langle \frac{1}{2} \begin{pmatrix} 1 & 3 \\ 1 & -1 \end{pmatrix}, \begin{pmatrix} 1 & 0 \\ 0 & -1 \end{pmatrix} \right\rangle, \tag{7.3.21}$$

of order 12, the Weyl group of type $G_2 \cong D_{12}$ (Shephard & Todd #2b), with

$$\mathrm{MS}_{\mathrm{hwe},\, 4^{\mathrm{H}}} = \mathrm{MS}_{\mathrm{hwe},\, 4_1^{\mathrm{H}}} = \frac{1}{(1-t^2)(1-t^6)} \qquad [A008620], \tag{7.3.22}$$

$$\mathrm{Inv}_{\mathrm{hwe},\, 4^{\mathrm{H}}} = \mathrm{Inv}_{\mathrm{hwe},\, 4_1^{\mathrm{H}}} = \frac{1}{i_2,\, h_6}, \tag{7.3.23}$$

(cf. [359]; [361, p. 621]). This might be regarded as a missing "fourth case" of Gleason's original theorem.

The case $q = 9$

For the field $\mathbb{F}_9$ we find

$$\mathcal{C}(\rho(9^{\mathrm{H}})) \cong Z_2.\mathfrak{U}(R(9^{\mathrm{H}})) \cong Z_2.\,\mathrm{GU}_2(\mathbb{F}_9),$$

of order 192, and

$$\mathcal{C}(\rho(9_1^{\mathrm{H}})) \cong Z_6.\mathfrak{U}(R(9_1^{\mathrm{H}})) \cong \pm 3_+^{1+4}.\,\mathrm{GU}_2(\mathbb{F}_9),$$

of order $46656 = 3^5.192$. The Molien series are

$$\mathrm{MS}_{\mathrm{cwe},\, 9^{\mathrm{H}}} = \frac{\theta_a(t)}{(1-t^2)^2(1-t^4)^2(1-t^6)^3(1-t^8)(1-t^{12})} \qquad [A092354], \tag{7.3.24}$$

where

$$\begin{aligned}
\theta_a(t) := {} & 1 + 3t^4 + 24t^6 + 74t^8 + 156t^{10} + 321t^{12} + 525t^{14} + 705t^{16} \\
& + 905t^{18} + 989t^{20} + 931t^{22} + 837t^{24} + 640t^{26} + 406t^{28} \\
& + 243t^{30} + 111t^{32} + 31t^{34} + 9t^{36} + t^{38},
\end{aligned}$$

and

$$\mathrm{MS}_{\mathrm{cwe},\, 9_1^{\mathrm{H}}} = \frac{\theta_b(t)}{(1-t^6)^2(1-t^{12})^3(1-t^{18})^3(1-t^{24})} \qquad [A092355], \quad (7.3.25)$$

where

$$\begin{aligned}
\theta_b(t) := {} & 1 + 24t^{12} + 192t^{18} + 958t^{24} + 3250t^{30} + 8190t^{36} + 15866t^{42} \\
& + 24729t^{48} + 31531t^{54} + 33133t^{60} + 28819t^{66} + 20586t^{72} \\
& + 11829t^{78} + 5304t^{84} + 1779t^{90} + 386t^{96} + 46t^{102} + t^{108}.
\end{aligned}$$

For the Hamming and symmetrized weight enumerators of these codes see §5.8.

7.4 Orthogonal (or Euclidean) type, p odd

Now $R = \mathrm{Mat}_g(k)$ with involution J defined by $r^J = r^{tr}$ for all $r \in R$, $\epsilon = 1$ and the characteristic of k is an odd prime p. Since $\epsilon = 1$, the element $\psi(1)$ is symmetric. Also $\Phi = \ker(\lambda) \oplus \{\!\{M\}\!\}$ and $\psi^{-1}(\lambda(\Phi))$ is the set of symmetric elements in R, i.e. the set of elements in R that are fixed under the involution J. The condition for $\begin{pmatrix} A & B \\ C & D \end{pmatrix} \in \mathrm{Mat}_{2g}(k)$ to be in $\pi(\mathfrak{U}(R, \Phi))$ is now

$$\begin{pmatrix} A^{tr} & C^{tr} \\ B^{tr} & D^{tr} \end{pmatrix} \begin{pmatrix} 0 & -1 \\ 1 & 0 \end{pmatrix} \begin{pmatrix} A & B \\ C & D \end{pmatrix} = \begin{pmatrix} 0 & -1 \\ 1 & 0 \end{pmatrix},$$

hence

$$\mathfrak{U}(R, \Phi) \cong (\ker(\lambda) \oplus \ker(\lambda)). \mathrm{Sp}_{2g}(k)$$

is an extension of a symplectic group.

Representations

To define a representation $(V, \rho_M, \rho_\Phi, \beta)$ it is enough to specify a symmetric nonsingular form β on the R-module V and an R-homomorphism $\ker(\lambda) \to V^*$. The R-modules V are of the form $V = (k^g)^N$, and up to rescaling and equivalence β can be chosen to be

$$\beta(v, v') := \sum_{i=1}^{gN} v_i v'_i.$$

7.4.1 q^{E} (odd): Euclidean self-dual codes over $\mathbb{F}_q$

Here we consider the families defined in §2.3.2, namely Euclidean self-dual codes over $\mathbb{F}_q$ and Euclidean self-dual codes over $\mathbb{F}_q$ that contain $\mathbf{1}$, where q is odd. The corresponding representations $\rho(q^{\mathrm{E}})$ and $\rho(q_1^{\mathrm{E}})$ were given in §2.3.2.

Clifford-Weil groups (q odd)

Theorem 7.4.1. (Cf. [454, Thm. 22], [383, §7].) *Let* $q := p^f$ *for an odd prime* p *and let* $a := \gcd\{q+1, 4\}$. *Then*

(a) *for Euclidean self-dual codes over* $\mathbb{F}_q$:

$$\mathcal{C}_g(\rho(q^{\mathrm{E}})) \cong Z_a \times \mathrm{Sp}_{2g}(\mathbb{F}_q),$$

(b) *for Euclidean self-dual codes over* $\mathbb{F}_q$ *that contain* $\mathbf{1}$:

$$\mathcal{C}_g(\rho(q_1^{\mathrm{E}})) \cong Z_a \times p_+^{1+2gf}. \mathrm{Sp}_{2g}(\mathbb{F}_q).$$

Proof. The structure of the hyperbolic co-unitary group follows from the arguments given above. Clearly, if C is a Euclidean code with $1 \in C \subset C^{\perp}$, then the length of C is divisible by p. Also, if $C = C^{\perp} \le \mathbb{F}_q^N$ is self-dual, then N must be even, since $\dim(C) = \frac{N}{2}$. If $q \equiv 1 \pmod 4$, then there is an element $\alpha \in \mathbb{F}_q$ such that $\alpha^2 = -1$, and the length 2 code $\langle (1, \alpha) \rangle \le \mathbb{F}_q^2$ is self-dual over $\mathbb{F}_q$. If $q \equiv 3 \pmod 4$ then there is a Euclidean self-dual $\mathbb{F}_q$-code of length 4 (see Remark 2.4.18).

The entries of the matrices that generate $\mathcal{C}_g(\rho(q^{\mathrm{E}}))$ and $\mathcal{C}_g(\rho(q_1^{\mathrm{E}}))$ belong to $\mathbb{Q}[\sqrt{p}, \zeta_p]$. Therefore the scalar elements in these Clifford-Weil groups are roots of unity in this field. So it suffices to construct $i\,\mathrm{id} \in \mathcal{C}_g(\rho(q^{\mathrm{E}}))$ and in $\mathcal{C}_g(\rho(q_1^{\mathrm{E}}))$ if $q \equiv 3 \pmod 4$. This will be done with the help of Theorem 5.4.7. Let ι be a symmetric primitive idempotent in $\mathrm{Mat}_g(\mathbb{F}_q)$, let $h := h_{\iota,\iota,\iota}$ and let $\phi := \rho_\phi(\{\psi(u\iota)\})$, where $u \in \mathbb{F}_q$ is an element of trace 1. Let $\zeta_p := \exp(2\pi i/p)$ be a primitive p-th root of unity. Then, by Theorem 5.4.7,

$$(h d_\phi)^3 = q^{-1/2} \sum_{x \in \mathbb{F}_q} \zeta_p^{\mathrm{Tr}(ux^2)} \,\mathrm{id} = \gamma\,\mathrm{id} \quad (\text{say}).$$

We claim that $\gamma^2 = -1$. The equation $x^2 + y^2 = z$, $z \in \mathbb{F}_q$, has a unique solution $(x, y) = (0, 0) \in \mathbb{F}_q \times \mathbb{F}_q$ if $z = 0$ and $\frac{q^2 - 1}{q - 1} = q + 1$ solutions $(x, y) \in \mathbb{F}_q \times \mathbb{F}_q$ if $z \ne 0$, because the two-dimensional quadratic form $x^2 + y^2$ is the unique anisotropic form of dimension 2 over $\mathbb{F}_q$ and so is the norm form of $\mathbb{F}_{q^2}$ over $\mathbb{F}_q$. Therefore

$$\gamma^2 = q^{-1} \sum_{x,y \in \mathbb{F}_q} \zeta_p^{\mathrm{Tr}(u(x^2+y^2))} = q^{-1} + q^{-1}(q+1) \sum_{z \in \mathbb{F}_q^*} \zeta_p^{\mathrm{Tr}(uz)}.$$

Since the map $z \mapsto \mathrm{Tr}(uz)$, $\mathbb{F}_q \to \mathbb{F}_p$ is surjective, $\sum_{z \in \mathbb{F}_q} \zeta_p^{\mathrm{Tr}(uz)} = 0$. Hence

$$\gamma^2 = q^{-1} + q^{-1}(q+1)(-1) = q^{-1}(1 - q - 1) = -1.$$

To finish the proof for case (b), we need to distinguish between the two possible extraspecial p-groups. The two possibilities are distinguished by the presence of elements of order p^2 in the group of minus type. It thus suffices to show that $O_p(\mathcal{C}_g(\rho(q_1^{\mathrm{E}})))$ has exponent p. By a theorem of P. Hall (see Huppert [288, Satz 13.10]) it is enough to show that $Z(O_p(\mathcal{C}_g(\rho(q_1^{\mathrm{E}}))))$ is the unique nontrivial characteristic subgroup of $O_p(\mathcal{C}_g(\rho(q_1^{\mathrm{E}})))$. See also Winter [548] for a more detailed investigation of the automorphism groups of the extraspecial groups. Let H be a characteristic subgroup of $O_p(G)$ and assume without loss of generality that H contains $Z := Z(O_p(\mathcal{C}_g(\rho(q_1^{\mathrm{E}}))))$. Then H/Z is an invariant subspace of $O_p(\mathcal{C}_g(\rho(q_1^{\mathrm{E}})))/Z$ under the action of $\mathrm{Sp}_{2g}(\mathbb{F}_{p^f}) \le \mathrm{Sp}_{2gf}(\mathbb{F}_p)$. Since this action is irreducible, $H/Z = O_p(\mathcal{C}_g(\rho(q_1^{\mathrm{E}})))/Z$ and hence $H = O_p(\mathcal{C}_g(\rho(q_1^{\mathrm{E}})))$ or $H = Z$. $\qquad\square$

Corollary 7.4.2. *If $q \equiv 3 \pmod 4$ then there is a Euclidean self-dual code $C \le \mathbb{F}_q^N$ over $\mathbb{F}_q$ if and only if the length N is divisible by 4, and there is a*

Euclidean self-dual code containing **1** *if and only if the length N is divisible by $4p$, where $p = \mathrm{char}(\mathbb{F}_q)$.*

If $q \equiv 1 \pmod 4$ then there is a Euclidean self-dual code $C \le \mathbb{F}_q^N$ over $\mathbb{F}_q$ if and only if the length N is even, and there is a Euclidean self-dual code containing **1** *if and only if the length N is divisible by $2p$, where $p = \mathrm{char}(\mathbb{F}_q)$.*

Remark. The groups in Theorem 7.4.1 were first studied in the present context by Gleason [191].

The case $q = 3$

Reference: [454, §7.3, pp. 225–226]. $k = \mathbb{F}_3$:

(a)

$$C(\rho(3)) = \left\langle \frac{1}{\sqrt{3}} \begin{pmatrix} 1 & 1 & 1 \\ 1 & \omega & \omega^2 \\ 1 & \omega^2 & \omega \end{pmatrix}, \begin{pmatrix} 1 & 0 & 0 \\ 0 & 0 & 1 \\ 0 & 1 & 0 \end{pmatrix}, \ \mathrm{diag}(1, \omega, \omega) \right\rangle,$$

of order 96 and structure $\cong Z_4 \times \mathrm{SL}_2(3)$, where $\omega = \exp(2\pi i/3)$;

$$\mathrm{MS}_{\mathrm{cwe},\,3} = \frac{1 + 4t^{12} + t^{24}}{(1 - t^4)(1 - t^{12})^2} \qquad [A092076], \tag{7.4.1}$$

$$\mathrm{Inv}_{\mathrm{cwe},\,3} = \frac{f^{(0)}, f^{(1)}, f^{(2)}, f^{(3)}, f^{(4)}, f^{(5)}}{q_4, q_6^2, s^{12}}, \tag{7.4.2}$$

where (we give explicit polynomials here, rather than codes and use the variables x, y, z instead of x_0, x_1, x_2) $r = y + z$, $s = y - z$, $q_4 = x(x^3 + r^3)$, $q_6 = 8x^6 - 20x^3 r^3 - r^6$, $f^{(0)} = 1$, $f^{(1)} = s^2 r_4 q_6$, $r_4 = r(8x^3 - r^3)$, $f^{(2)} = s^4 r_4^2$, $f^{(3)} = s^6 q_6$, $f^{(4)} = s^8 r_4$, $f^{(5)} = s^{10} r_4^2 q_6$.

For the Hamming weight enumerators, the group becomes the reflection group $3[6]2$ (Coxeter [135, p. 176], Shephard and Todd X_6), of order 48, generated by

$$\frac{1}{\sqrt{3}} \begin{pmatrix} 1 & 2 \\ 1 & -1 \end{pmatrix}, \begin{pmatrix} 1 & 0 \\ 0 & \omega \end{pmatrix},$$

$$\mathrm{MS}_{\mathrm{hwe},\,3} = \frac{1}{(1 - t^4)(1 - t^{12})} \qquad [A008620]. \tag{7.4.3}$$

$$\mathrm{Inv}_{\mathrm{hwe},\,3} = \frac{1}{x^4 + 8xy^3, \ y^3(x^3 - y^3)^3}, \tag{7.4.4}$$

corresponding to the codes t_4, g_{12} of §2.4.5. This result is also due to Gleason [191] (see also [34], [359]).

(b)

$$C(\rho(3_1)) \cong Z_4 \times 3_+^{1+2} . \mathrm{SL}_2(3),$$

of order 2592, generated by $C(\rho_a)$ and $\mathrm{diag}(1, \omega, \omega^2)$;

$$\text{MS}_{\text{cwe}, 3_1} = \frac{1 + t^{24}}{(1 - t^{12})^2(1 - t^{36})} \qquad [A007980], \qquad (7.4.5)$$

$$\text{Inv}_{\text{cwe}, 3_1} = \frac{1, XQ(\mathbb{F}_3, 23)}{e_3^{4+}, g_{12}, S(36)}, \qquad (7.4.6)$$

where e_3^{4+} is the other indecomposable code of length 12 and Type $\rho(3_1)$ (see Table 12.9), $XQ(\mathbb{F}_3, p)$ denotes an extended quadratic residue code of length $p + 1$ (see Theorem 2.4.9), and $S(n)$ denotes a Pless double-circulant code of length n (see §11.3.2).

Though the appropriate symmetrization does not commute with the action of $\mathcal{C}(\rho(3_1))$, we can calculate the ring spanned by the Hamming weight enumerators of these codes, just by setting $y = z$ in the complete weight enumerators. $\text{Inv}^{\text{Ham}}(\mathcal{C}(\rho(3_1)))$ is generated by the Hamming weight enumerators of g_{12} and e_3^{4+}, and hence is equal to the polynomial ring in $\text{hwe}(g_{12}) = x^{12} + 264x^6y^6 + 440x^3y^9 + 24y^{12}$ and $\frac{1}{8}(\text{hwe}(g_{12}) - \text{hwe}(e_3^{4+})) = y^3(y^3 - x^3)^3$:

$$\text{MS}_{\text{hwe}, 3_1} = \frac{1}{(1 - t^{12})^2} \qquad [A000027], \qquad (7.4.7)$$

$$\text{Inv}_{\text{hwe}, 3_1} = \frac{1}{g_{12}, e_3^{4+}}. \qquad (7.4.8)$$

The case $q = 3$, genus 2

(a)

$$\mathcal{C}_2(\rho(3)) \cong Z_4 \times \text{Sp}_4(\mathbb{F}_3), \qquad (7.4.9)$$

of order 207360;

$$\text{MS}_{\text{cwe}_2, 3}(t) = \frac{\theta(t)}{(1 - t^8)(1 - t^{12})^4(1 - t^{20})^2(1 - t^{36})^2} \qquad [A092069], \qquad (7.4.10)$$

where

$$\begin{aligned}
\theta(t) := {} & 1 + t^4 + 6t^{12} + 30t^{16} + 57t^{20} + 207t^{24} + 565t^{28} + 1000t^{32} \\
& + 2031t^{36} + 3880t^{40} + 5804t^{44} + 8696t^{48} + 12991t^{52} \\
& + 16595t^{56} + 20527t^{60} + 25965t^{64} + 29418t^{68} + 31536t^{72} \\
& + 34772t^{76} + 35273t^{80} + 33093t^{84} + 31969t^{88} + 29068t^{92} \\
& + 23862t^{96} + 20052t^{100} + 16217t^{104} + 11369t^{108} + 7996t^{112} \\
& + 5554t^{116} + 3097t^{120} + 1642t^{124} + 930t^{128} + 350t^{132} \\
& + 104t^{136} + 51t^{140} + 9t^{144} + t^{148} + t^{152}.
\end{aligned}$$

(b)

$$\mathcal{C}_2(\rho(3_1)) \cong Z_4 \times 3_+^{1+4}. \text{Sp}_4(\mathbb{F}_3), \qquad (7.4.11)$$

of order 50388480;

$$\mathrm{MS\,cwe}_2,\, 3_1 \;=\; \frac{\theta(t^{12})}{(1-t^{12})^2(1-t^{24})^2(1-t^{36})^3(1-t^{60})^2} \qquad [A092070]\,,$$

$$(7.4.12)$$

where

$$\theta(t) := 1 + 8t^2 + 60t^3 + 292t^4 + 1090t^5 + 3127t^6 + 7116t^7$$
$$+ 13411t^8 + 21536t^9 + 29963t^{10} + 36631t^{11} + 39638t^{12}$$
$$+ 37973t^{13} + 32135t^{14} + 23906t^{15} + 15462t^{16} + 8507t^{17}$$
$$+ 3858t^{18} + 1369t^{19} + 342t^{20} + 52t^{21} + 3t^{22}\,.$$

The case $q = 9$

$k = \mathbb{F}_9$: **(a)**

$$\mathcal{C}(\rho(9)) \cong \pm\,\mathrm{Sp}_2(\mathbb{F}_9)\,, \qquad\qquad (7.4.13)$$

of order 1440;

$$\mathrm{MS}_{\mathrm{cwe},\,9} \;=\; \frac{\theta(t)}{(1-t^2)(1-t^4)(1-t^6)^4(1-t^8)(1-t^{10})^2} \qquad [A092071]\,,$$

$$(7.4.14)$$

where

$$\theta(t) := 1 + 2t^6 + 17t^8 + 36t^{10} + 89t^{12} + 167t^{14} + 278t^{16} + 428t^{18} + 590t^{20}$$
$$+ 704t^{22} + 760t^{24} + 745t^{26} + 643t^{28} + 504t^{30} + 365t^{32}$$
$$+ 223t^{34} + 118t^{36} + 56t^{38} + 23t^{40} + 6t^{42} + 4t^{44} + t^{46}\,.$$

(b)

$$\mathcal{C}(\rho(9_1)) \cong \pm 3^{1+4}.\,\mathrm{Sp}_2(\mathbb{F}_9)\,, \qquad\qquad (7.4.15)$$

of order 349920;

$$\mathrm{MS}_{\mathrm{cwe},\,9_1} \;=\; \frac{\theta(t)}{(1-t^6)(1-t^{12})^2(1-t^{18})^3(1-t^{24})(1-t^{30})^2} \qquad [A092072]\,,$$

$$(7.4.16)$$

where

$$\theta(t) := 1 + 4t^{12} + 32t^{18} + 154t^{24} + 602t^{30} + 1820t^{36} + 4383t^{42}$$
$$+ 8857t^{48} + 15425t^{54} + 23464t^{60} + 31635t^{66}$$
$$+ 38191t^{72} + 41354t^{78} + 40262t^{84} + 35271t^{90}$$
$$+ 27662t^{96} + 19295t^{102} + 11885t^{108} + 6373t^{114}$$
$$+ 2885t^{120} + 1079t^{126} + 323t^{132} + 68t^{138} + 12t^{144} + 3t^{150}\,.$$

The case $q = 5$

References: [347], [454, §7.9]. $k = \mathbb{F}_5$: **(a)**

$$C(\rho(5)) \cong \pm \mathrm{Sp}_2(\mathbb{F}_5), \tag{7.4.17}$$

of order 240;

$$\mathrm{MS}_{\mathrm{cwe},\,5} = \frac{\theta(t)}{(1-t^4)(1-t^6)^2(1-t^{10})^2} \qquad [A028344], \tag{7.4.18}$$

where

$$\theta(t) = 1 + t^2 + t^8 + 4t^{10} + 10t^{12} + 13t^{14} + 10t^{16} + 5t^{18} + 6t^{20} + 5t^{22} + 3t^{24} + t^{26}.$$

For the symmetrized (or Lee) weight enumerator, the group becomes the reflection group $[3,5]$, a three-dimensional representation of the icosahedral group (Shephard and Todd X_{23}), of order 120, generated by

$$\begin{pmatrix} 1 & 2 & 2 \\ 1 & \xi + \xi^4 & \xi^2 + \xi^3 \\ 1 & \xi^2 + \xi^3 & \xi + \xi^4 \end{pmatrix}, \mathrm{diag}\{1, \xi, \xi^4\}, \begin{pmatrix} 1 & 0 & 0 \\ 0 & 0 & 1 \\ 0 & 1 & 0 \end{pmatrix},$$

$$\mathrm{MS}_{\mathrm{swe},\,5} = \frac{1}{(1-t^2)(1-t^6)(1-t^{10})} \qquad [A008672]. \tag{7.4.19}$$

$$\mathrm{Inv}_{\mathrm{swe},\,5} = \frac{1}{\alpha,\,\beta,\,\gamma} = \frac{1}{c_2,\,c_6,\,e_{10}^+} \tag{7.4.20}$$

where

$$\alpha = x^2 + 4yz$$
$$\beta = x^4 yz - x^2 y^2 z^2 - x(y^5 + z^5) + 2y^3 z^3$$
$$\gamma = 5x^6 y^2 z^2 - 4x^5(y^5 + z^5) - 10x^4 y^3 z^3 + 10x^3(y^6 z + yz^6) + 5x^2 y^4 z^4$$
$$\quad - 10x(y^7 z^2 + y^2 z^7) + 6y^5 z^5 + y^{10} + z^{10}.$$

As codes we may take

$$c_2 := [12], \quad c_6 := [(100)(133)], \tag{7.4.21}$$

and either of the following

$$d_5^{2+} = [(01234)(00000), (00000)(01234), 1111111111]$$
$$e_{10}^+ = [(00014)(00023), 1111111111], \tag{7.4.22}$$

for the code of length 10. We will meet these codes again in §9.1.6.

For the Hamming weight enumerators (Patterson [405]) we set $y = z$ in the symmetrized weight enumerators and obtain:

$$\text{MS}_{\text{hwe}, 5} = \frac{1 + t^{10} + t^{20}}{(1 - t^2)(1 - t^6)} \quad [A097950]. \tag{7.4.23}$$

$$\text{Inv}_{\text{hwe}, 5} = \frac{1, \ \overline{\gamma}, \ \overline{\gamma}^2}{\overline{\alpha}, \ \overline{\beta}} = \frac{1, \ e_{10}^+, \ (e_{10}^+)^2}{c_2, \ c_6}$$

where

$$\overline{\alpha} = x^2 + 4y^2,$$
$$\overline{\beta} = y^2(x - y)^2(x^2 + 2xy + 2y^2),$$
$$\overline{\gamma} = y^4(x - y)^4(5x^2 + 12xy + 8y^2).$$

(b)

$$\mathcal{C}(\rho(5_1)) \cong \pm 5_+^{1+2} \, \text{Sp}_2(\mathbb{F}_5), \tag{7.4.24}$$

of order 30000;

$$\text{MS}_{\text{cwe}, 5_1} = \frac{\theta(t)}{(1 - t^{10})(1 - t^{20})^2(1 - t^{30})^2} \quad [A028345], \tag{7.4.25}$$

where

$$\theta(t) = 1 + 3t^{20} + 13t^{30} + 18t^{40} + 28t^{50} + 34t^{60} + 17t^{70} + 4t^{80} + 2t^{90}.$$

The degree 10 invariant is the cwe of either of the codes of length 10 given in (7.4.22).

7.5 Symplectic type, p odd

Now $R = \text{Mat}_g(k)$, $r^J = r^{tr}$ for $r \in R$, $\epsilon = -1$ and $\text{char}(k) = p$ is odd. Since $\epsilon = -1$, we have $\tau(\psi(1)) = \psi(-1) = -\psi(1)$, so $\psi(1)$ is skew symmetric. Then $\Phi = \ker(\lambda) \oplus \{M\}$ and $\psi^{-1}(\lambda(\Phi))$ is the set of skew symmetric elements in R, i.e. the set of $r \in R$ such that $r^J = -r$. The condition for $\begin{pmatrix} A & B \\ C & D \end{pmatrix} \in \text{Mat}_{2g}(k)$ to be in $\pi(\mathfrak{U}(R, \Phi))$ is

$$\begin{pmatrix} A^{tr} & C^{tr} \\ B^{tr} & D^{tr} \end{pmatrix} \begin{pmatrix} 0 & 1 \\ 1 & 0 \end{pmatrix} \begin{pmatrix} A & B \\ C & D \end{pmatrix} = \begin{pmatrix} 0 & 1 \\ 1 & 0 \end{pmatrix},$$

hence

$$\mathfrak{U}(Mat_g(k), \Phi) \cong (\ker(\lambda) \oplus \ker(\lambda)).O_{2g}^+(k)$$

is an extension of an orthogonal group of plus type.

Representations

To define a representation $(V, \rho_M, \rho_\Phi, \beta)$ it is enough to specify a skew-symmetric nonsingular form β on the R-module V and an R-homomorphism

$\ker(\lambda) \to V^*$. The R-modules V are of the form $V = (k^g)^m$. For a skew-symmetric nonsingular form β on V to exist, $m = 2N$ must be even, and then up to rescaling and equivalence β can be chosen to be

$$\beta(v, v') := \sum_{j=1}^{N} \left(\sum_{i=1+2g(j-1)}^{g+2g(j-1)} v_i v'_{2gj-i} - v_{2gj-i} v'_i \right).$$

7.5.1 q^{H+} (odd): Hermitian $\mathbb{F}_r$-linear codes over $\mathbb{F}_q$, $q = r^2$

The Type of these codes (see the end of §2.3.4) is given by the representation $\rho(q^{H+}) := (V, \rho_M, \rho_\Phi, \beta)$ of the form ring $R(q^{H+}) := (\mathbb{F}_r, \mathbb{F}_r, \mathrm{id}, \{0\})$, where $V = K \cong \mathbb{F}_q$ is an extension of $R = \mathbb{F}_r = k$ of degree 2. Let $\bar{}$ denote the nontrivial Galois automorphism of K over k and choose $\alpha \in K$ such that

$$\alpha = -\bar{\alpha}.$$

We define

$$\beta : V \times V \to \frac{1}{p}\mathbb{Z}/\mathbb{Z} \text{ by } \beta(x, y) := \frac{1}{p} \mathrm{Tr}(x\alpha\bar{y}) \text{ for all } x, y \in V,$$

where Tr denotes the trace from K to $\mathbb{F}_p \cong \mathbb{Z}/p\mathbb{Z}$ and ρ_M by $\rho_M(\psi(1)) := \beta$. Note that $\beta(x, ay) = -\beta(y, ax)$ for all $a \in k$, so $\rho_M(M)$ consists of skew symmetric forms. Hence $\{\rho_M(M)\} = \{0\}$, $\Phi = \ker(\lambda)$ and $\lambda = 0$. If we choose $\Phi := \{\varphi[a] \mid a \in k\} \cong k$ and

$$\rho_\Phi(\varphi)(x) := \frac{1}{p} \mathrm{Tr}(x\alpha),$$

then the codes of length N and Type $\rho(q_1^{H+}) := (K, \rho_M, \rho_\Phi, \beta)$ are K/k-Hermitian self-dual (with respect to the non-standard form β) k-linear codes in K^N that contain $\mathbf{1}$. If we choose $\Phi := \{0\}$ then the codes of length N and Type $\rho(q^{H+}) := (K, \rho_M, 0, \beta)$ are precisely the K/k-Hermitian self-dual k-linear codes in K^N.

Clifford-Weil groups (genus g)

Since the code $(1) \le K$ is a self-dual code of length 1 both of Type $\rho(q_1^{H+})$ and Type $\rho(q^{H+})$, the Clifford-Weil groups are respectively

$$\mathcal{C}_g(\rho(q^{H+})) \cong \mathfrak{U}(R, \Phi) \cong O_{2g}^+(\mathbb{F}_r),$$

$$\mathcal{C}_g(\rho(q_1^{H+})) \cong \mathfrak{U}(R, \Phi) \cong \mathbb{F}_q^{2g}.O_{2g}^+(\mathbb{F}_r).$$

The case $q = 9$, genus 1

For $k = \mathbb{F}_3$, $K = \mathbb{F}_9$: **(a)**

$$\mathcal{C}(\rho(q^{\mathrm{H}+})) \cong Z_2 \times Z_2 \cong O_2^+(\mathbb{F}_3),$$

of order 4 (there are no scalars);

$$\mathrm{MS}_{\mathrm{cwe},\,q^{\mathrm{H}+}} = \frac{1 + 2t^2 + 4t^3 + t^4}{(1-t)^4(1-t^2)^5} \quad [A092091]. \tag{7.5.1}$$

(b)

$$\mathcal{C}(\rho(q_1^{\mathrm{H}+})) \cong S_3 \times S_3 \cong (Z_3 \times Z_3).O_2^+(\mathbb{F}_3), \tag{7.5.2}$$

of order 36,

$$\mathrm{MS}_{\mathrm{cwe},\,q_1^{\mathrm{H}+}} = \frac{\theta(t)}{(1-t^2)^4(1-t^3)^4(1-t^6)} \quad [A052365], \tag{7.5.3}$$

where

$$\begin{aligned}
\theta(t) := {}&1 + t + 2t^3 + 10t^4 + 17t^5 + 19t^6 + 20t^7 + 29t^8 + 37t^9 \\
&+ 34t^{10} + 23t^{11} + 12t^{12} + 7t^{13} + 3t^{14} + t^{15}.
\end{aligned}$$

7.6 Characteristic 2, orthogonal and symplectic types

Now let $p = 2$. Since $1 = -1$ in characteristic 2, it does not make sense to use the value of ϵ to distinguish the orthogonal and symplectic cases. There are two possibilities:

$$\psi^{-1}(\lambda(\Phi)) = \{r \in R \mid r^J = r\} \quad \text{(orthogonal type)}$$

or

$$\psi^{-1}(\lambda(\Phi)) = \{0\} \quad \text{(symplectic type)} .$$

In both situations we get the quadratic pair (M, Φ) as the module for a certain "universal quadratic ring" $W(R)$ which is a quotient of $Q(R)$.

By Morita theory, it suffices to consider the case $g = 1$, hence $R = k$ is a finite field of characteristic 2, $r^J = r$ for all $r \in R$ and $\epsilon = 1$. For any form ring (R, M, ψ, Φ) over R, the pair (M, Φ) is a quadratic pair over R which satisfies $m(1 \otimes r) = m(r \otimes 1)$ for all $r \in R, m \in M$, since the involution J is trivial. Therefore (M, Φ) is a module over

$$W(R) := Q(R)/\langle (1 \otimes r) - (r \otimes 1) \mid r \in R\rangle .$$

Recall that $Q(R)$ is the quadratic ring $(M_{Q(R)}, \Phi_{Q(R)})$, where $M_{Q(R)} = R \otimes_{\mathbb{Z}} R$ and $\Phi_{Q(R)}$ is given by

$$\langle [r] \mid r \in R \rangle / \langle [x+y+z]-[x+y]-[x+z]-[y+z]+[x]+[y]+[z], [0] \mid x, y, z \in R \rangle \,.$$

Therefore

$$M_{W(R)} = M_{Q(R)} / \langle (1 \otimes r) - (r \otimes 1) \mid r \in R \rangle \cong R$$

and

$$\Phi_{W(R)} = \Phi_{Q(R)} \cong W_4(R) \,,$$

where

$$W_4(R) := \{ (a^2, 2b) \mid a, b \in R \} \,.$$

Here '$2b$' is a formal symbol, and does not mean $b + b \in R$ (which is of course 0 since $\operatorname{char}(R) = 2$). $W_4(R)$ is the ring of *Witt vectors* over R, and has multiplication and addition defined by

$$(a^2, 2b)((a')^2, 2b') = ((aa')^2, 2(a^2 b' + b(a')^2)) \,,$$

$$(a^2, 2b) + ((a')^2, 2b') = ((a+a')^2, 2(b + b' - aa')) \,.$$

The isomorphism $\Phi_{W(R)} \to W_4(R)$ is defined by $[r] \mapsto (r^2, 0)$ for all $r \in R$. Note that $[r, 1] \mapsto (0, 2r)$ under this isomorphism. Also

$$W_4(\mathbb{F}_{2^f}) \cong \mathbb{Z}_2[\zeta_{2^f - 1}]/(4) \,.$$

Let (M, Φ) be a quadratic pair over R. Then M is an R-module, Φ is a $W_4(R)$-module such that the conditions for a $Q(R)$-module on page 120 are satisfied, i.e. $\tau : M \to M$ is R-linear and, for all $\phi \in \Phi$, $m \in M$, $r, r' \in R$, we have

$$\lambda(\phi[r]) = \lambda(\phi)(r \otimes r) \,, \{m\}[r] = \{m(r \otimes r)\} \,, \text{ and } \phi[r, r'] = \{\lambda(\phi)(r \otimes r')\} \,.$$

If (R, M, ψ, Φ) is a form ring, we may assume without loss of generality that $M = R$ and $\psi = \mathrm{id}$. The form structure Φ is a $W_4(R)$-module such that $\lambda(\Phi) \leq M$ and $\ker(\lambda)$ is a linear R-module. This leaves the following possibilities for (Φ, λ):

(A) $\lambda = 0$, $\Phi = \ker(\lambda) = R^t$ for some t.
(B) $\Phi = R \oplus \ker(\lambda)$ and λ is surjective.
(C) $\Phi = W_4(R) \oplus R^t$, λ is surjective and $\ker(\lambda) = 2W_4(R) \oplus R^t \cong R^{t+1}$ for some t.

The hyperbolic co-unitary groups can be calculated in the same way as for odd primes. To get the Clifford-Weil groups in the general situation we now assume that $R = \operatorname{Mat}_g(k)$ for some $g \in \mathbb{N}$. The condition for $\begin{pmatrix} a & b \\ c & d \end{pmatrix} \in \operatorname{Mat}_2(R)$ to be in $\pi(\mathfrak{U}(R, \Phi))$ is

$$\begin{pmatrix} c^{tr} a & c^{tr} b \\ d^{tr} a - 1 & d^{tr} b \end{pmatrix} = \begin{pmatrix} \psi^{-1}(\lambda(\phi_1)) & \psi^{-1}(m) \\ \psi^{-1}(\tau(m)) & \psi^{-1}(\lambda(\phi_2)) \end{pmatrix} \,,$$

for some $m \in M, \phi_1, \phi_2 \in \Phi$. In the orthogonal cases (B) and (C), when $\psi^{-1}(\lambda(\Phi))$ is the set of all symmetric elements in $R = \mathrm{Mat}_g(k)$, we find

$$\mathfrak{U}(R, \Phi) = (\ker(\lambda) \oplus \ker(\lambda)) . \mathrm{Sp}_{2g}(k) .$$

In the symplectic case (A), $\psi^{-1}(\lambda(\Phi)) = \mathrm{Ev}_g(k)$ (see Definition 1.10.4) and

$$\mathfrak{U}(R, \Phi) = (\ker(\lambda) \oplus \ker(\lambda)) . O_{2g}^+(k) .$$

We now give examples of Types of codes over fields of characteristic 2 for each of these three form structures.

7.6.1 $q^{\mathrm{H}+}$ (even): Hermitian $\mathbb{F}_r$-linear codes over $\mathbb{F}_q$, $q = r^2$

Here we consider certain of the families defined in §2.3.4, namely Hermitian self-dual $\mathbb{F}_r$-linear codes over $\mathbb{F}_q$, $q = r^2$, q even, and the same but containing the all-ones vector $\mathbf{1}$. The corresponding representations are $\rho(q^{\mathrm{H}+})$ and $\rho(q_1^{\mathrm{H}+})$. These are our first examples for case (A). Let $K := \mathbb{F}_q$ and $k := \mathbb{F}_r = R$.

Clifford-Weil groups (genus g)

Since $\langle (1) \rangle_k \leq K$ is a self-dual code of Types $\rho(q^{\mathrm{H}+})$ and $\rho(q_1^{\mathrm{H}+})$, such self-dual codes exist for all lengths N. Therefore the genus-g Clifford-Weil groups are

$$\mathcal{C}_g(\rho(q^{\mathrm{H}+})) \cong \mathfrak{U}_g(R(q^{\mathrm{H}+})) \cong O_{2g}^+(\mathbb{F}_r) ,$$

$$\mathcal{C}_g(\rho(q_1^{\mathrm{H}+})) \cong \mathfrak{U}_g(R(q_1^{\mathrm{H}+})) \cong \mathbb{F}_r^{2g} \rtimes O_{2g}^+(\mathbb{F}_r) .$$

The case $q = 4$, genus 1

Reference: [454, §7.6]. $k = \mathbb{F}_2$: **(a)** Additive self-dual codes over $\mathbb{F}_4$ using trace-Hermitian inner product.

$$\mathcal{C}(\rho(4^{\mathrm{H}+})) = \langle h \rangle \cong Z_2 \cong O_2^+(\mathbb{F}_2) ,$$

where h is given in (7.2.3) (there are no scalars):

$$\mathrm{MS}_{\mathrm{cwe}, 4^{\mathrm{H}+}} = \frac{1}{(1-t)^3(1-t^2)} \qquad [A002623] , \tag{7.6.1}$$

$$\mathrm{Inv}_{\mathrm{cwe}, 4^{\mathrm{H}+}} = \frac{1}{i_1, i_1', i_1'', i_2} ; \tag{7.6.2}$$

$$\mathrm{MS}_{\mathrm{hwe}, 4^{\mathrm{H}+}} = \frac{1}{(1-t)(1-t^2)} \qquad [A008619] , \tag{7.6.3}$$

$$\text{Inv}_{\text{hwe}, 4^{\text{H+}}} = \frac{1}{i_1, i_2}. \tag{7.6.4}$$

(b) Same, but containing **1**.

$$\mathcal{C}(\rho(4_1^{\text{H+}})) = \langle h, \text{diag}(1, 1, -1, -1) \rangle \cong D_8 \cong (Z_2 \times Z_2).O_2^+(\mathbb{F}_2),$$

of order 8;

$$\text{MS}_{\text{cwe}, 4_1^{\text{H+}}} = \frac{1 + t^3}{(1 - t)(1 - t^2)^2(1 - t^4)} \qquad [A005232], \tag{7.6.5}$$

$$\text{Inv}_{\text{cwe}, 4_1^{\text{H+}}} = \frac{1, c_3}{i_1, i_2, i_2', c_4}; \tag{7.6.6}$$

$$\text{MS}_{\text{swe}, 4_1^{\text{H+}}} = \frac{1}{(1 - t)(1 - t^2)(1 - t^4)} \qquad [A008642], \tag{7.6.7}$$

$$\text{Inv}_{\text{cwe}, 4_1^{\text{H+}}} = \frac{1}{i_1, i_2, c_4}; \tag{7.6.8}$$

with hwe as in (7.6.3), (7.6.4).

(c) Type $4_{\text{S}}^{\text{H+}}$: Same, but with **1** in the shadow. The number of weight N vectors in the shadow of a code of Type $4^{\text{H+}}$ is given by $\text{hwe}_C(3/2, 1/2)$ and is thus positive. It follows that every code of Type $4^{\text{H+}}$ is equivalent to a code for which the shadow contains the all-ones vector. This Type also corresponds to the case of real self-dual quantum codes under the additive construction discussed in §13.2.

$$\mathcal{C}(\rho(4_{\text{S}}^{\text{H+}})) = \langle h, \text{diag}(1, -1, 1, 1) \rangle \cong S_3,$$

of order 6;

$$\text{MS}_{\text{cwe}, 4_{\text{S}}^{\text{H+}}} = \frac{1}{(1 - t)^2(1 - t^2)(1 - t^3)} \qquad [A000601], \tag{7.6.9}$$

$$\text{Inv}_{\text{cwe}, 4_1^{\text{H+}}} = \frac{1}{i_1', i_1'', i_2, c_3'}; \tag{7.6.10}$$

$$\text{MS}_{\text{swe}, 4_{\text{S}}^{\text{H+}}} = \frac{1}{(1 - t)(1 - t^2)(1 - t^3)} \qquad [A001399], \tag{7.6.11}$$

$$\text{Inv}_{\text{cwe}, 4_{\text{S}}^{\text{H+}}} = \frac{1}{i_1', i_2, c_3'}. \tag{7.6.12}$$

The codes mentioned above are the following:

$$i_1 = [1], \quad \text{cwe} = \text{swe} = \text{hwe} = x + y,$$
$$i_1' = [\omega], \quad \text{cwe} = \text{swe} = x + z,$$
$$i_1'' = [\overline{\omega}], \quad \text{cwe} = x + t, \text{swe} = x + z,$$
$$i_2 = [11, \omega\omega], \quad \text{cwe} = x^2 + y^2 + z^2 + t^2,$$

$$\text{swe} = x^2 + y^2 + 2z^2, \text{hwe} = x^2 + 3y^2 ,$$

$$i_2' = [11, \omega\overline{\omega}], \quad \text{cwe} = x^2 + y^2 + 2zt,$$

$$\text{swe} = x^2 + y^2 + 2z^2, \text{hwe} = x^2 + 3y^2 ,$$

$$i_2'' = [1\omega, \omega\overline{\omega}], \quad \text{cwe} = x^2 + yz + zt + yt,$$

$$\text{swe} = x^2 + 2yz + z^2, \text{hwe} = x^2 + 3y^2 ,$$

$$i_2''' = [1\omega, \omega 1], \quad \text{cwe} = x^2 + t^2 + 2yz,$$

$$\text{swe} = x^2 + z^2 + 2yz, \text{hwe} = x^2 + 3y^2 ,$$

$$c_3 = [111, \omega\omega 0, \omega 0\omega], \quad \text{cwe} = x^3 + y^3 + 3xz^2 + 3yt^2,$$

$$\text{swe} = x^3 + y^3 + 3xz^2 + 3yz^2, \text{hwe} = x^3 + 3xy^2 + 4y^3 ,$$

$$c_3' = [\overline{\omega\omega\omega}, 110, 101], \quad \text{cwe} = x^3 + t^3 + 3xy^2 + 3z^2t,$$

$$\text{swe} = x^3 + 3xy^2 + 4z^3, \text{hwe} = x^3 + 3xy^2 + 4y^3 ,$$

$$c_4 = [1111, \omega(\omega 00)] ,$$

$$c_4' = [\omega\omega\omega\omega, 1(100)]$$

The case $q = 4$, genus 2

(a) $C_2(\rho(4^{\text{H}+})) \cong O_4^+(\mathbb{F}_2) \cong (S_3 \times S_3).2$ of order 72, generated by $I_4 \otimes h$ and the permutation matrices corresponding to the action of $GL_2(\mathbb{F}_2)$ on $\mathbb{F}_4^2$;

$$\text{MS} = \frac{\theta_0(t) + t^{30}\theta_0(t^{-1})}{(1-t)^3(1-t^2)^5(1-t^3)^3(1-t^4)^3(1-t^6)^2} \qquad [A092201], \quad (7.6.13)$$

where

$$\theta_0 := 1 + 10t^3 + 25t^4 + 53t^5 + 121t^6 + 247t^7 + 471t^8 + 803t^9$$
$$+ 1201t^{10} + 1674t^{11} + 2182t^{12} + 2616t^{13} + 2911t^{14} + 1509t^{15} .$$

(b) $C_2(\rho(4_1^{\text{H}+})) \cong 2^4.O_4^+(\mathbb{F}_2) \cong 2^4.(S_3 \times S_3).2$ of order 1152, generated by $C_2(\rho(4^{\text{H}+}))$ and $I_4 \otimes \text{diag}(1, 1, -1, -1)$;

$$\text{MS} = \frac{\theta_1(t) + t^{56}\theta_1(t^{-1})}{(1-t)(1-t^2)^2(1-t^3)^3(1-t^4)^6(1-t^6)(1-t^8)^2(1-t^{12})} , \quad (7.6.14)$$

[A092203] , where

$$\theta_1 := 1 + t^3 + 5t^4 + 18t^5 + 45t^6 + 88t^7 + 196t^8 + 394t^9 + 804t^{10}$$
$$+ 1512t^{11} + 2702t^{12} + 4529t^{13} + 7218t^{14} + 11019t^{15} + 16064t^{16}$$
$$+ 22609t^{17} + 30555t^{18} + 39889t^{19} + 50303t^{20} + 61476t^{21}$$
$$+ 72888t^{22} + 84047t^{23} + 94299t^{24} + 102995t^{25} + 109674t^{26}$$
$$+ 113791t^{27} + 57614t^{28} .$$

The case $q = 16$

For $k = \mathbb{F}_4$: **(a)**

$$\mathcal{C}(\rho(16^{\mathrm{H}+})) \cong S_3 \cong O_2^+(\mathbb{F}_4),$$

of order 6 (there are no scalars);

$$\mathrm{MS}_{\mathrm{cwe},\,16^{\mathrm{H}+}} = \frac{\theta(t) + t^{16}\theta(t^{-1})}{(1-t)^5(1-t^2)^6(1-t^3)^5} \qquad [A092496]\,, \qquad (7.6.15)$$

where

$$\theta(t) := 1 + 10t^2 + 40t^3 + 90t^4 + 180t^5 + 340t^6 + 420t^7 + 215t^8\,.$$

(b)

$$\mathcal{C}(\rho(16_1^{\mathrm{H}+})) \cong Z_2^4.S_3\,,$$

of order 96,

$$\mathrm{MS}_{\mathrm{cwe},\,16_1^{\mathrm{H}+}} = \frac{\theta(t) + t^{30}\theta(t^{-1})}{(1-t)(1-t^2)^4(1-t^3)^7(1-t^4)^4} \qquad [A092497]\,, \qquad (7.6.16)$$

where

$$\theta(t) := 1 + 4t^3 + 34t^4 + 88t^5 + 237t^6 + 516t^7 + 1161t^8 + 2176t^9$$
$$+ 3726t^{10} + 5478t^{11} + 7524t^{12} + 9296t^{13} + 10805t^{14} + 5610t^{15}\,.$$

7.6.2 q^{E} (even): Euclidean self-dual $\mathbb{F}_q$-linear codes

Here we consider one of the families defined in §2.3.2, namely Euclidean self-dual $\mathbb{F}_q$-linear codes when $q = 2^f$ is a power of 2. The corresponding representation is $\rho(q^{\mathrm{E}})$. This is also an example for case (A).

Clifford-Weil groups (genus g)

Theorem 7.6.1. *For $g \in \mathbb{N}$ the associated Clifford-Weil group of genus g is*

$$\mathcal{C}_g(\rho(q^{\mathrm{E}})) \cong 2_+^{1+2fg}.O_{2g}^+(\mathbb{F}_{2^f})\,.$$

Proof. This follows from our general description of $\mathfrak{U}_g(R(q^{\mathrm{E}}))$ together with the fact that the length of any such code has to be even (of course there is a code of length 2). Note that $\mathcal{C}_g(\rho(q^{\mathrm{E}}))$ has a real representation of degree 2^{fg}. Therefore the largest normal 2-subgroup of this Clifford-Weil group is an extraspecial group of plus type, isomorphic to the central product of fg copies of the dihedral group D_8. $\qquad\square$

The case $q = 2$

The genus-g Clifford-Weil groups are the real Clifford groups $\mathcal{C}_g = \mathcal{C}_g(\rho(2_\mathrm{I}))$ of [383]. Their invariant theory has already been discussed in Chapter 6.

For genus 1 there is another connection with invariant theory which we do not completely understand in our current context. Namely, if C is a code of Type 2_I then its shadow (cf §1.12) is a union of two cosets of its doubly-even subcode C_0. The difference between the weight enumerators of these two cosets turns out to be a relative invariant for the *complex* Clifford group $\mathcal{X}_1$ of order 192, with character depending "linearly" on N. (Conway and Sloane [130].) There is thus a ring structure on such invariants. The corresponding Molien series and generators are as follows:

$$\mathrm{MS} = \frac{1 + t^{18}}{(1 - t^8)(1 - t^{12})} \quad [A008647], \tag{7.6.17}$$

$$\mathrm{Inv} = \frac{1,\ xy(x^8 - y^8)(x^8 - 34x^4y^4 + y^8)}{x^8 + 14x^4y^4 + y^8,\ x^2y^2(x^4 - y^4)^2}. \tag{7.6.18}$$

The codes are[1]

$$\frac{1, (d_{10}e_7f_1)^+}{e_8,\ d_{12}^+}.$$

However, since a difference of weight enumerators always has constant term 0, we do not obtain the full ring of invariants. This same ring, with the same constant term 0 condition, also appears in the theory of harmonic weight enumerators (cf. Bachoc [20]). There is presumably a connection.

Research Problem 7.6.2. *Explain the above observations using the machinery developed in this book. Investigate also the fact that if C is a maximal isotropic ternary code of odd length, then a similar phenomenon occurs if we consider the difference between the weight enumerators of the two nonzero cosets of C in $C^\perp$.*

The case $q = 4$

Reference: [454, §7.5]. The Clifford-Weil group is $\mathcal{C}(\rho(4^\mathrm{E})) \cong 2_+^{1+4}.S_3$, of order 192, generated by $\mathrm{diag}(1, 1, -1, -1)$, r (see (7.3.1)) and h (see (7.2.3)):

$$\mathrm{MS}_{\mathrm{cwe}, 4^\mathrm{E}} = \frac{1 + t^{16}}{(1 - t^2)(1 - t^4)(1 - t^6)(1 - t^8)} \quad [A008769], \tag{7.6.19}$$

$$\mathrm{Inv}_{\mathrm{cwe}, 4^\mathrm{E}} = \frac{1,\ BC_{16}}{i_2,\ RS_4,\ c_6,\ e_8 \otimes \mathbb{F}_4}. \tag{7.6.20}$$

Adjoining the Galois automorphism (7.3.6), we obtain the group $\tilde{\mathcal{C}}(\rho(4^\mathrm{E}))$ of order 384 (the Weyl group of type B_4, Shephard-Todd #2a):

---

[1] These codes were incorrectly specified in [454, Eq. (100)].

$$\widetilde{\mathrm{MS}}_{\mathrm{cwe},\,4\mathrm{E}} \;=\; \frac{1}{(1-t^2)(1-t^4)(1-t^6)(1-t^8)} \qquad [A001400]\,,$$

$$\widetilde{\mathrm{Inv}}_{\mathrm{cwe},\,4\mathrm{E}} \;=\; \frac{1}{i_2,\,RS_4,\,c_6,\,e_8\otimes\mathbb{F}_4}\,.$$

The code c_6 mentioned above is the $[6,3,3]_4$ code with generator matrix

$$\begin{bmatrix} 1 & 1 & 1 & 1 & 1 & 1 \\ 0 & 0 & 0 & 1 & \omega & \overline{\omega} \\ 1 & \omega & \overline{\omega} & 0 & 0 & 0 \end{bmatrix}, \qquad (7.6.21)$$

and $\mathrm{cwe}(c_6) = x^6 + \cdots (4 \text{ such terms}) + 6x^3 yzt + \cdots (4 \text{ terms}) + 9x^2 y^2 z^2 + \cdots (4 \text{ terms})$, $\mathrm{hwe}(c_6) = x^6 + 6x^3 y^3 + 27x^2 y^4 + 18xy^5 + 12y^6$. The code BC_{16} mentioned in (7.6.20) was found by Betsumiya and Choie [37]. This is a $[16,8,4]_4$ code with generator matrix $[I\,A]$, where A is an 8×8 circulant matrix with first row $\omega,1,\omega,1,\omega,1,\overline{\omega},1$.

Neither the Hamming nor symmetrized weight enumerators can be obtained directly from invariant theory, but must be found by collapsing the complete weight enumerators. It suffices to take Galois-invariant polynomials. Then we have:

$$\mathrm{MS}_{\mathrm{swe},\,4\mathrm{E}} \;=\; \frac{1+t^8+t^{16}}{(1-t^2)(1-t^4)(1-t^6)} \qquad [A028309]\,,$$

$$\mathrm{Inv}_{\mathrm{swe},\,4\mathrm{E}} \;=\; \frac{1,\,e_8\otimes\mathbb{F}_4,\,(e_8\otimes\mathbb{F}_4)^2}{i_2,\,c_4,\,c_6}\,;$$

$$\mathrm{MS}_{\mathrm{hwe},\,4\mathrm{E}} \;=\; \frac{1+t^6}{(1-t^2)(1-t^4)} \qquad [A028310]\,,$$

$$\mathrm{Inv}_{\mathrm{hwe},\,4\mathrm{E}} \;=\; \frac{1,\,c_6}{i_2,\,c_4}\,.$$

7.6.3 $q_{\mathrm{II}}^{\mathrm{H}+}$ (even): Even Trace-Hermitian $\mathbb{F}_r$-linear codes, $q = r^2$

Here we consider another of the families defined in §2.3.4, namely even trace-Hermitian codes over $\mathbb{F}_q$ which are linear over $\mathbb{F}_r$, where $q = r^2$ is even, and the same but containing the all-ones vector $\mathbf{1}$. The corresponding representations are $\rho(q_{\mathrm{II}}^{\mathrm{H}+})$ and $\rho(q_{\mathrm{II},1}^{\mathrm{H}+})$. These are our first examples for case (B).

Clifford-Weil groups (genus g)

For both Types $q_{\mathrm{II}}^{\mathrm{H}+}$ and $q_{\mathrm{II},1}^{\mathrm{H}+}$, λ is surjective and the involution J on $R = \mathbb{F}_r$ is trivial. Therefore

$$\mathfrak{U}_g(R,\Phi) = \ker(\lambda)^{2g}.O_{2g}^+(\mathbb{F}_r)\,.$$

The lengths of codes of Type $q_{\mathrm{II}}^{\mathrm{H+}}$ or $q_{\mathrm{II},1}^{\mathrm{H+}}$ are even. On the other hand, the natural representations of the Clifford-Weil groups are real (and hence only involve square roots of unity). Therefore the extraspecial normal 2-subgroup of $\mathcal{C}_g(\rho(q_{\mathrm{II},1}^{\mathrm{H+}}))$ is of +-type and the genus-g Clifford-Weil groups are

$$\mathcal{C}_g(\rho(q_{\mathrm{II}}^{\mathrm{H+}})) \cong Z_2.\mathfrak{U}_g(R(q_{\mathrm{II}}^{\mathrm{H+}})) \cong 2.O_{2g}^+(\mathbb{F}_r),$$

$$\mathcal{C}_g(\rho(q_{\mathrm{II},1}^{\mathrm{H+}})) \cong Z_2.\mathfrak{U}_g(R(q_{\mathrm{II},1}^{\mathrm{H+}})) \cong Z_2.\mathbb{F}_r^{2g}.O_{2g}^+(\mathbb{F}_r) \cong 2_+^{1+2gf}.O_{2g}^+(\mathbb{F}_{2^f}),$$

where $r = 2^f$.

The case $q = 4$, genus 1

Reference: [454, §7.7]. These are even additive trace-Hermitian self-dual codes over $\mathbb{F}_4$. The Clifford-Weil groups are generated by the groups in §7.6.1 together with the matrix $\phi_0 := \mathrm{diag}(1, -1, -1, -1)$:

$$\mathcal{C}(\rho(4_{\mathrm{II}}^{\mathrm{H+}})) = \langle \mathcal{C}(\rho(4^{\mathrm{H+}})), \phi_0 \rangle \cong S_3 \times Z_2,$$

of order 12;

$$\mathrm{MS}_{\mathrm{cwe},\, 4_{\mathrm{II}}^{\mathrm{H+}}} = \frac{1 + t^2 + 2t^4}{(1 - t^2)^3(1 - t^6)} \qquad [A092498],$$

$$\mathrm{Inv}_{\mathrm{cwe},\, 4_{\mathrm{II}}^{\mathrm{H+}}} = \frac{1,\, i_2''',\, c_4,\, c_4'}{i_2,\, i_2',\, i_2'',\, h_6},$$

where the codes are as in §7.6.1 and h_6 denotes the hexacode defined in (2.4.26).

$$\mathrm{MS}_{\mathrm{hwe},\, 4_{\mathrm{II}}^{\mathrm{H+}}} = \mathrm{MS}_{\mathrm{hwe},\, 4_{\mathrm{II},1}^{\mathrm{H+}}} = \frac{1}{(1 - t^2)(1 - t^6)} \qquad [A008620], \qquad (7.6.22)$$

$$\mathrm{Inv}_{\mathrm{hwe},\, 4_{\mathrm{II}}^{\mathrm{H+}}} = \frac{1}{i_2,\, h_6}. \qquad (7.6.23)$$

If the code contains **1** then:

$$\mathcal{C}(\rho(4_{\mathrm{II},1}^{\mathrm{H+}})) = \langle \mathcal{C}(\rho(4_1^{\mathrm{H+}})), \phi_0 \rangle \cong S_4 \times Z_2,$$

of order 48;

$$\mathrm{MS}_{\mathrm{cwe},\, 4_{\mathrm{II},1}^{\mathrm{H+}}} = \frac{1 + t^4}{(1 - t^2)^2(1 - t^4)(1 - t^6)} \qquad [A014126], \qquad (7.6.24)$$

$$\mathrm{Inv}_{\mathrm{cwe},\, 4_{\mathrm{II},1}^{\mathrm{H+}}} = \frac{1,\, ABCD}{D^2,\, A^2 + B^2 + C^2,\, A^4 + B^4 + C^4,\, A^6 + B^6 + D^6}$$

$$= \frac{1,\, c_4}{i_2,\, i_2',\, c_4',\, h_6}, \qquad (7.6.25)$$

where

$$A := x + y, \ B := x - y, \ C := z + t, \ D := z - t.$$

For the symmetrized weight enumerators we find

$$\mathrm{MS}_{\mathrm{swe}, \, 4_{\mathrm{II},1}^{\mathrm{H}+}} = \frac{1}{(1 - t^2)(1 - t^4)(1 - t^6)} \qquad [A001399] \qquad (7.6.26)$$

and

$$\mathrm{Inv}_{\mathrm{swe}, \, 4_{\mathrm{II},1}^{\mathrm{H}+}} = \frac{1}{\text{symmetric polynomials in } A^2, \, B^2, \, C^2}$$

$$= \frac{1}{i_2, \, c_4', \, h_6}. \qquad (7.6.27)$$

Most of the codes are as in §7.6.1; as usual h_6 is the hexacode.

7.6.4 $q_{\mathrm{II}}^{\mathrm{E}}$ (even): Generalized Doubly-even Euclidean Self-dual codes over $\mathbb{F}_q$

These codes are also considered by Nebe, Quebbemann, Rains and Sloane [382]. The Type is given by the representation $\rho(q_{\mathrm{II}}^{\mathrm{E}})$ of §2.3.2. However, we will describe this representation again, to show that this is an example for case (C).

Let $q = 2^f$, $k = \mathbb{F}_q$, $V = \mathbb{F}_q$, $\beta(x, y) := \frac{1}{2} \mathrm{Tr}(xy)$. Define $\phi : V \to W_4(\mathbb{F}_2) \cong \frac{1}{4}\mathbb{Z}/\mathbb{Z}$ by

$$\phi(x) := \mathrm{Tr}((x^2, 0)) = \sum_{i=0}^{f-1} (x^2, 0)^{2^i} = \left(\left(\sum_{i=0}^{f-1} x^{2^i} \right)^2, 2 \sum_{i<j} x^{2^i} x^{2^j} \right),$$

where Tr denotes the trace from $W_4(k)$ to $W_4(\mathbb{F}_2) \cong \mathbb{Z}/4\mathbb{Z}$. Let $C \leq V^N$ be a self-dual isotropic code. Then C is self-dual with respect to the usual Euclidean inner product on V^N. Being isotropic means that

$$\phi(c) = \mathrm{Tr}\left(\sum_{j=1}^{N} (c_j^2, 0) \right) = \mathrm{Tr}\left(\left(\sum_{j=1}^{N} c_j \right)^2, 2 \sum_{i<j} c_i c_j \right) = 0,$$

for all $c = (c_1, \ldots, c_N) \in C$. Therefore the codes of Type ρ are exactly the generalized doubly-even codes.

Clifford-Weil groups (genus g)

Theorem 7.6.3. *Let* $\mathcal{C}_g(\rho(q_{\mathrm{II}}^{\mathrm{E}}))$ *be the Clifford-Weil group of genus g corresponding to the representation* $\rho(q_{\mathrm{II}}^{\mathrm{E}})$ *above. Then*

$$\mathcal{C}_g(\rho(q_{\mathrm{II}}^{\mathrm{E}})) \cong Z.(\mathbb{F}_q^g \oplus \mathbb{F}_q^g). \mathrm{Sp}_{2g}(\mathbb{F}_q) \cong Z.Z_2^{2gf}. \mathrm{Sp}_{2g}(\mathbb{F}_{2^f}),$$

where $Z \cong \mathbb{Z}/4\mathbb{Z}$ *if* $f = \log_2(q)$ *is even and* $Z \cong \mathbb{Z}/8\mathbb{Z}$ *if* $f = \log_2(q)$ *is odd.*

Proof. By our general theory $C_g(\rho(q_{II}^E))$ has an epimorphic image

$$\mathfrak{U}_g(R(q_{II}^E)) \cong (\mathbb{F}_q^g \oplus \mathbb{F}_q^g).\,\mathrm{Sp}_{2g}(\mathbb{F}_q)\,.$$

The kernel Z of this epimorphism is a cyclic group consisting of scalar matrices. Since the invariant ring of $C_g(\rho(q_{II}^E))$ is spanned by weight enumerators of self-dual isotropic codes C, the order of Z is the greatest common divisor of the lengths of these self-dual isotropic codes. Hence the theorem follows from Remark 2.4.7 if one notes that any self-dual isotropic code contains **1** and hence has length a multiple of 4.   □

The case $k = \mathbb{F}_2$, arbitrary genus

The codes are the usual doubly-even self-dual binary codes. The Clifford-Weil groups are the complex Clifford groups of [383] and their invariant theory is described in Chapter 6.

The case $k = \mathbb{F}_4$, genus 1

If $k = \mathbb{F}_4$, then

$$G := \mathcal{C}(\rho(4_{II}^E)) \cong (Z_4 \mathsf{Y} D_8 \mathsf{Y} D_8).\,\mathrm{Alt}_5\,,$$

of order 3840, a subgroup of index 2 of the complex reflection group X_{29} (Shephard-Todd X_{29}). Then

$$\mathrm{MS}_{X_{29}} \;=\; \frac{1}{(1-t^4)(1-t^8)(1-t^{12})(1-t^{20})} \qquad [A008669]\,,$$

$$\mathrm{Inv}_{X_{29}} \;=\; \frac{1}{XQ(\mathbb{F}_4,3),\,XQ(\mathbb{F}_4,7),\,XQ(\mathbb{F}_4,11),\,XQ(\mathbb{F}_4,19)}\,,$$

where $XQ(\mathbb{F}_4,p)$ is the extended quadratic residue code of length $p+1$ over $\mathbb{F}_4$ (see Theorem 2.4.2).

$$\mathrm{MS}_{\mathrm{cwe},\,4_{II}^E} \;=\; \frac{1+t^{40}}{(1-t^4)(1-t^8)(1-t^{12})(1-t^{20})} \qquad [A020702]\,. \qquad (7.6.28)$$

The elements in $X_{29} \setminus G$ act as the Frobenius automorphism $a \mapsto \bar{a}$ on the variables of the complete weight enumerators of these codes. That the weight enumerator p_C is invariant under X_{29}, means that $p_C = \overline{p_C}$. To get the full invariant ring for G, it remains to find a self-dual even code BC_{40} (say) of length 40 over $\mathbb{F}_4$, for which the complete weight enumerator is not invariant under the Frobenius automorphism. Such a code was also found by Betsumiya and Choie [37]. This is the code BC_{40} with generator matrix $[I\,A]$, where A is a 20×20 circulant matrix with first row

$$\omega\,\omega\,0\,0\,\omega\,\omega\,\omega\,\bar{\omega}\,1\,1\,\omega\,\omega\,\omega\,1\,\omega\,1\,0\,0\,0\,0\,. \qquad (7.6.29)$$

So we have:

$$\mathrm{Inv}_{\mathrm{cwe},\,4_{II}^E} \;=\; \frac{1,\,BC_{40}}{XQ(\mathbb{F}_4,3),\,XQ(\mathbb{F}_4,7),\,XQ(\mathbb{F}_4,11),\,XQ(\mathbb{F}_4,19)}\,.$$

The case $k = \mathbb{F}_8$

The Molien series of $\mathcal{C}(\rho(8_{\mathrm{II}}^{\mathrm{E}}))$ is

$$\mathrm{MS}_{\mathrm{cwe},\,8_{\mathrm{II}}^{\mathrm{E}}} = \theta(t)/\sigma(t) \qquad [A069247]\,,$$

where

$$\begin{aligned}
\theta(t) := 1 &+ 5t^{16} + 77t^{24} + 300t^{32} + 908t^{40} + 2139t^{48} + 3808t^{56} + 5864t^{64} \\
&+ 8257t^{72} + 10456t^{80} + 12504t^{88} + 14294t^{96} + 15115t^{104} \\
&+ 15115t^{112} + 14294t^{120} + 12504t^{128} + 10456t^{136} + 8257t^{144} \\
&+ 5864t^{152} + 3808t^{160} + 2139t^{168} + 908t^{176} + 300t^{184} \\
&+ 77t^{192} + 5t^{200} + t^{216}
\end{aligned}$$

and

$$\sigma(t) := (1 - t^8)^2 (1 - t^{16})^2 (1 - t^{24})^2 (1 - t^{56})(1 - t^{72})\,.$$

8

Further Examples of Self-Dual Codes

This chapter describes some families of self-dual codes that cannot be obtained from representations of quasisimple form rings: codes over $\mathbb{Z}/m\mathbb{Z}$ (§8.1), then the special cases of codes over $\mathbb{Z}/4\mathbb{Z}$ (§8.2) and $\mathbb{Z}/8\mathbb{Z}$ (§8.3), codes over more general Galois rings (§8.4), and codes over $\mathbb{F}_{q^2} + \mathbb{F}_{q^2} u$ with $u^2 = 0$ (§8.5).

8.1 $m^{\mathbb{Z}}$: Codes over $\mathbb{Z}/m\mathbb{Z}$

We begin with codes of Type $m^{\mathbb{Z}}$, that is, self-dual $\mathbb{Z}/m\mathbb{Z}$-linear codes. The representation $\rho(m^{\mathbb{Z}})$ was given in §2.3.5.

Let $m = \prod_{i=1}^{s} p_i^{a_i}$ be the prime decomposition of m, where the p_i are distinct primes. By the Chinese Remainder Theorem

$$\mathbb{Z}/m\mathbb{Z} \cong \mathbb{Z}/p_1^{a_1}\mathbb{Z} \oplus \ldots \oplus \mathbb{Z}/p_s^{a_s}\mathbb{Z}.$$

Since the involution J is trivial, the projections onto the primary components of $\mathbb{Z}/m\mathbb{Z}$ are symmetric idempotents in the form ring $R(m^{\mathbb{Z}})$, that decompose the form ring into a direct sum of form rings

$$R(m^{\mathbb{Z}}) \cong R((p_1^{a_1})^{\mathbb{Z}}) \oplus \ldots \oplus R((p_s^{a_s})^{\mathbb{Z}}).$$

Also the representation is just the direct sum of the representations of the primary components, and hence the associated Clifford-Weil group is the tensor product of the Clifford-Weil groups of the components. So it is enough to treat the case when

$$m = p^a$$

is a prime power. We first discuss the case when p is odd.

Theorem 8.1.1. *Let $m = p^a$ where p is odd. Let $n := 1$ if a is even, $n := 2$ if a is odd and $p \equiv 1 \pmod{4}$, and $n := 4$ if a is odd and $p \equiv 3 \pmod{4}$. Then*

$$\mathcal{C}_g(\rho(m^{\mathbb{Z}})) \cong Z_n . \mathrm{Sp}_{2g}(\mathbb{Z}/m\mathbb{Z}).$$

If in addition the code contains the all-ones vector then

$$C_g(\rho(m_1^{\mathbb{Z}})) \cong (Z_{mn}).(Z_m^{2g}).\mathrm{Sp}_{2g}(\mathbb{Z}/m\mathbb{Z}).$$

Proof. We first treat Type $m^{\mathbb{Z}}$. Since p is odd, λ is bijective and we find

$$\begin{aligned}
\mathfrak{U}_g(\rho(m^{\mathbb{Z}})) = \{ & \begin{pmatrix} A & B \\ C & D \end{pmatrix} \in (\mathbb{Z}/m\mathbb{Z})^{2g \times 2g} \mid \\
& D^{tr}A - C^{tr}B = I_g, C^{tr}A = A^{tr}C, D^{tr}B = B^{tr}D \} \\
= & \mathrm{Sp}_{2g}(\mathbb{Z}/m\mathbb{Z}).
\end{aligned}$$

It remains to determine the scalar elements in $C_g(\rho(m^{\mathbb{Z}}))$. If $a = 2b$ is even then (p^b) is an isotropic self-dual code of Type $\rho(m^{\mathbb{Z}})$ and length 1. Hence $n = 1$ in this case. If m is not a square, then the length N of any self-dual code has to be even (since $|C||C^{\perp}| = |V^N| = m^N$). On the other hand, if $p \equiv 1$ (mod 4), then -1 is a p-adic square, hence there is an element $i \in \mathbb{Z}/m\mathbb{Z}$ with $i^2 = -1$, and $(1, i)$ generates a self-dual isotropic code of length 2. Hence $n = 2$ in this case. It remains to consider the case $p \equiv 3 \bmod 4$, a odd. Then the Gauss sum (see Theorem 5.4.7)

$$\gamma := \frac{1}{\sqrt{m}} \sum_{x=0}^{m-1} \exp(2\pi i x^2/m)$$

is a scalar in $C(\rho(m^{\mathbb{Z}}))$. By Dirichlet's formula (see for instance Terras [520, Theorem 8.3]) $\gamma = i$ (since $m \equiv 3 \bmod 4$) is a primitive fourth root of unity. By Corollary 2.4.19 there is always a self-dual isotropic code of length 4, so $n = 4$ in this last case.

For Type $m_1^{\mathbb{Z}}$, the lengths of the codes are divisible by m, because $\mathbf{1}$ is isotropic. Therefore mn divides the order of the center of $C(\rho(m_1^{\mathbb{Z}}))$. On the other hand, the entries of the matrices in $C(\rho(m_1^{\mathbb{Z}}))$ belong to $\mathbb{Q}[\zeta_m, \sqrt{m}]$. All roots of unity in this field are nm-th roots of unity. □

Corollary 8.1.2. *Let m, n be as in Theorem 8.1.1. There is a self-dual code $C \leq (\mathbb{Z}/m\mathbb{Z})^N$ (respectively a self-dual code containing $\mathbf{1}$) if and only if the length N is divisible by n (respectively nm).*

The corollary above can also be deduced from the following general observation:

Proposition 8.1.3. *Let ρ be a Type. Assume that there is a self-dual isotropic code of Type ρ and length N and that $\mathbf{1}$ is isotropic. Then there is a self-dual isotropic code of Type ρ and length N that contains $\mathbf{1}$.*

Proof. By Lemma 4.6.7, an isotropic code (of Type ρ and length N) can be extended to an isotropic self-dual code if and only if there exists an isotropic self-dual code of that length. Thus the lengths N for which there exists an isotropic self-dual code containing the vector $\mathbf{1}$ are precisely those lengths N for which (i) there exists an isotropic self-dual code of Type ρ and length N, and (ii) $\langle \mathbf{1} \rangle$ is isotropic. □

We now treat the case $p = 2$. Let $m = 2^a$ be a power of 2. Recall from Definition 1.10.4 that a symmetric matrix in $(\mathbb{Z}/m\mathbb{Z})^{g \times g}$ is called *even* if its diagonal entries belong to $2\mathbb{Z}/m\mathbb{Z}$. By analogy with the theta-group in the theory of Siegel modular forms (cf. §9.1.5 or Freitag [176, Appendix 1]), we define

$$\Theta_{2g}(\mathbb{Z}/m\mathbb{Z}) := \left\{ \begin{pmatrix} A & B \\ C & D \end{pmatrix} \in \mathrm{Sp}_{2g}(\mathbb{Z}/m\mathbb{Z}) \mid \right. \tag{8.1.1}$$
$$\left. C^{tr}A = A^{tr}C \text{ and } D^{tr}B = B^{tr}D \text{ are both even} \right\}.$$

Then we have:

Theorem 8.1.4. *Let $m = 2^a$. Let $n := 1$ if a is even, $n := 2$ if a is odd. Then*

$$\mathcal{C}_g(\rho(m^{\mathbb{Z}})) \cong Z_n . Z_2^{2g} . \Theta_{2g}(\mathbb{Z}/m\mathbb{Z}),$$

which is isomorphic to $2_+^{1+2g} . \Theta_{2g}(\mathbb{Z}/m\mathbb{Z})$ if $a = 1$ and to $(Z_n \times Z_2^{2g}) . \Theta_{2g}(\mathbb{Z}/m\mathbb{Z})$ if $a > 1$.

Proof. For the form ring $R(m^{\mathbb{Z}})$ we find that $\ker(\lambda) = \langle \frac{m}{2} \rangle \cong \mathbb{Z}/2\mathbb{Z}$ and that the image of λ in $M \cong \mathbb{Z}/m\mathbb{Z}$ is $2\mathbb{Z}/m\mathbb{Z}$. Therefore $\mathfrak{U}_g(\rho(m^{\mathbb{Z}})) = Z_2^{2g} . U$, where $U = \Theta_{2g}(\mathbb{Z}/m\mathbb{Z})$ is defined above. It remains to determine the group Z of scalar elements in $\mathcal{C}_g(\rho(m^{\mathbb{Z}}))$. As before, if m is a square then $(\sqrt{m})$ is an isotropic self-dual code of Type $\rho(m^{\mathbb{Z}})$ and length 1, and if m is not a square then the length of a self-dual code is even. If $m = 2^{2b+1}$ then the code with generator matrix

$$\begin{bmatrix} 2^b & 2^b \\ 0 & 2^{b+1} \end{bmatrix}$$

is a self-dual code of length 2. Hence $Z \cong Z_n$ is as stated in the Theorem. The largest normal 2-subgroup of $\mathcal{C}_g(\rho(m^{\mathbb{Z}}))$ is generated by the elements $\rho(d(1, \phi))$ and $\rho(t_\phi)$ with $\phi \in \ker(\lambda)$. The two subgroups $\langle \rho(d(1, \phi)) \mid \phi \in \ker(\lambda) \rangle$ and $\langle \rho(t_\phi) \mid \phi \in \ker(\lambda) \rangle$ are both elementary abelian 2-subgroups of $\mathcal{C}_g(\rho(m^{\mathbb{Z}}))$, Using the commutator relations in Corollary 5.4.2, we find that $O_2(\mathcal{C}_g(\rho(m^{\mathbb{Z}})))$ is extraspecial of $+$ type if $a = 1$ and elementary abelian in all other cases. $\qquad\square$

Theorem 8.1.5. *Let m be a power of 2. If the code also contains $\mathbf{1}$ then we have*

$$\mathcal{C}_g(\rho(m_1^{\mathbb{Z}})) \cong Z_m . Z_m^{2g} . \Theta_{2g}(\mathbb{Z}/m\mathbb{Z}).$$

Proof. This follows from Theorem 8.1.4 together with Proposition 8.1.3, once we note that now $\ker(\lambda) = \langle \varphi \rangle \cong \mathbb{Z}/m\mathbb{Z}$. $\qquad\square$

If m is even we may make the additional assumption that the Euclidean norm of every codeword is divisible by $2m$. This is the representation $\rho(m_{\mathrm{II}}^{\mathbb{Z}})$. As above it suffices to treat the case where $m = 2^a$ is a power of 2.

Theorem 8.1.6. *Let* $m = 2^a$. *Then*

$$\mathcal{C}_g(\rho(m_{\mathrm{II}}^{\mathbb{Z}})) \cong Z_8.Z_2^{2g}.\mathrm{Sp}_{2g}(\mathbb{Z}/m\mathbb{Z}).$$

Proof. Now λ is surjective and $\ker(\lambda) \cong \mathbb{Z}/2\mathbb{Z}$. Therefore $\mathfrak{U}_g(\rho(m_{\mathrm{II}}^{\mathbb{Z}})) = Z_2^{2g}.\mathrm{Sp}_{2g}(\mathbb{Z}/m\mathbb{Z})$. It remains to determine the scalars in $\mathcal{C}_g(\rho(m^{\mathbb{Z}}))$. By Theorem 2.4.15(c) there is always a self-dual isotropic code of length 8, since the octacode $\mathcal{O}_8$ is a lift of the Hamming code e_8.

The Gauss sum

$$\gamma := \gamma(\phi_0) = \frac{1}{\sqrt{m}} \sum_{v=0}^{m-1} \exp(\pi i v^2/m),$$

corresponding to the nonsingular element $\phi_0 \in \Phi$ with $\phi_0(x) = \frac{1}{2m}x^2$ for all $x \in \mathbb{Z}/m\mathbb{Z}$, is a primitive eighth root of unity. For $m = 2$ and 4 this is easily checked by explicit calculations. For larger m, the code generated by 2 is isotropic and thus the invariance of Gauss sums under quotient representations (cf. Corollary 5.4.10 and the following remark) reduces the computation to that for $m/4$. It follows that

$$\gamma = \frac{1+i}{\sqrt{2}}$$

is a primitive eighth root of unity.   □

Assuming in addition that the code contains **1** we get the following theorem, which is an immediate consequence of the above results and Proposition 8.1.3.

Theorem 8.1.7. *Let* m *be a power of* 2 *and let* $n := \max(2m, 8)$. *Then*

$$\mathcal{C}_g(\rho(m_{\mathrm{II},1}^{\mathbb{Z}})) \cong Z_n.Z_m^{2g}.\mathrm{Sp}_{2g}(\mathbb{Z}/m\mathbb{Z}).$$

8.2 $4^{\mathbb{Z}}$: Self-dual codes over $\mathbb{Z}/4\mathbb{Z}$

For general references for these codes see §2.4.9. Other references will be indicated in the individual subsections.

8.2.1 $4^{\mathbb{Z}}$: Type I self-dual codes over $\mathbb{Z}/4\mathbb{Z}$

Reference: Klemm [315].

The Clifford-Weil group $\mathcal{C}(\rho(4^{\mathbb{Z}}))$ is generated by $\phi := \mathrm{diag}(1, i, 1, i)$,

$$h := \frac{1}{2}\begin{pmatrix} 1 & 1 & 1 & 1 \\ 1 & i & -1 & -i \\ 1 & -1 & 1 & -1 \\ 1 & -i & -1 & i \end{pmatrix} \quad \text{and} \quad m_3 := \begin{pmatrix} 1 & 0 & 0 & 0 \\ 0 & 0 & 0 & 1 \\ 0 & 0 & 1 & 0 \\ 0 & 1 & 0 & 0 \end{pmatrix},$$

and has order 64;

$$\text{MS}_{\text{cwe},\,4^{\mathbb{Z}}} = \frac{1+t^{10}}{(1-t)(1-t^4)^2(1-t^8)} \qquad [A092531]\,, \tag{8.2.1}$$

$$\text{Inv}_{\text{cwe},\,4^{\mathbb{Z}}} = \frac{1,\,(BCD)^2(B^4-C^4)}{A,\,B^4+C^4,\,D^4,\,B^4C^4}\,, \tag{8.2.2}$$

where (as in Eq. (2.4.39))

$$A := x_0 + x_2,\ B := x_1 + x_3,\ C := x_0 - x_2,\ D := x_1 - x_3\,. \tag{8.2.3}$$

In terms of codes,

$$\text{Inv}_{\text{cwe},\,4^{\mathbb{Z}}} = \frac{1,\,\mathcal{J}_{10}}{i_1,\,\mathcal{D}_4^{\oplus},\,\mathcal{D}_4^{\oplus'},\,\mathcal{O}_8}\,. \tag{8.2.4}$$

The codes mentioned here and in the following sections were defined in §2.4.9.

Symmetrizing by $(\mathbb{Z}/4\mathbb{Z})^*$ (i.e. setting $x_1 = x_3$ in the complete weight enumerator), we get the ring spanned by the symmetrized weight enumerators of codes of Type $4^{\mathbb{Z}}$, with Molien series

$$\text{MS}_{\text{swe},\,4^{\mathbb{Z}}} = \frac{1}{(1-t)(1-t^4)(1-t^8)} \qquad [A092532]\,, \tag{8.2.5}$$

$$\text{Inv}_{\text{swe},\,4^{\mathbb{Z}}} = \frac{1}{A,\,B^4+C^4,\,B^4C^4}$$

$$= \frac{1}{i_1,\,\mathcal{D}_4^{\oplus},\,\mathcal{O}_8}\,. \tag{8.2.6}$$

For the Hamming weight enumerators, we have

$$\text{MS}_{\text{hwe},\,4^{\mathbb{Z}}} = \frac{1+t^8}{(1-t)(1-t^4)} \qquad [A092533]\,, \tag{8.2.7}$$

$$\text{Inv}_{\text{hwe},\,4^{\mathbb{Z}}} = \frac{1,\,y^4(x-y)^4}{x+y,\,y(x-y)(x^2+xy+2y^2)}$$

$$= \frac{1,\,\mathcal{O}_8}{i_1,\,\mathcal{D}_4^{\oplus}}\,. \tag{8.2.8}$$

8.2.2 $4_{\mathrm{I}}^{\mathbb{Z}}$: Type I self-dual codes over $\mathbb{Z}/4\mathbb{Z}$ containing 1

These are codes of Type $4^{\mathbb{Z}}$ that contain the all-ones vector. Then

$$\mathcal{C}(4_{\mathrm{I}}^{\mathbb{Z}}) = \langle h\,,m_3\,,\phi\,,\varphi := \text{diag}(1,i,-1,-i)\rangle\,,$$

of order 1024;

$$\mathrm{MS}_{\mathrm{cwe},\,4_1^Z} \;=\; \frac{(1+t^{12})(1+t^{16})}{(1-t^4)(1-t^8)^2(1-t^{16})} \qquad [A004657]\,, \qquad (8.2.9)$$

$$\mathrm{Inv}_{\mathrm{cwe},\,4_1^Z} = \frac{(1,\; A^{12}+B^{12}+C^{12}+D^{12}) \times (1,\; f_{16})}{A^4+B^4+C^4+D^4,\, A^8+B^8+C^8+D^8,\, f_8,\, A^4B^4C^4D^4}\,, \qquad (8.2.10)$$

where

$$f_8 := A^4D^4 + B^4C^4\,,$$
$$f_{16} := (ABCD)^2(A^4B^4 + C^4D^4 - A^4C^4 - B^4D^4)\,,$$

and A, B, C, D are as above. In terms of codes,

$$\mathrm{Inv}_{\mathrm{cwe},\,4_1^Z} \;=\; \frac{1,\, \mathcal{K}_{12},\, \mathcal{J}_{16},\, \mathcal{K}_{12}+\mathcal{J}_{16}}{\mathcal{K}_4,\, \mathcal{K}_8,\, \mathcal{O}_8,\, \mathcal{K}_{16}}\,. \qquad (8.2.11)$$

Symmetrizing by $(\mathbb{Z}/4\mathbb{Z})^*$ (i.e. setting $x_1 = x_3$ in the complete weight enumerator), we get the ring spanned by the symmetrized weight enumerators of codes of Type 4_1^Z with Molien series

$$\mathrm{MS}_{\mathrm{swe},\,4_1^Z} \;=\; \frac{1+t^{12}}{(1-t^4)(1-t^8)^2} \qquad [A004652]\,, \qquad (8.2.12)$$

$$\mathrm{Inv}_{\mathrm{swe},\,4_1^Z} \;=\; \frac{1,\; A^4B^4C^4}{A^4+B^4+C^4,\; A^8B^8C^8,\; B^4C^4}$$
$$=\; \frac{1,\, \mathcal{K}_{12}}{\mathcal{K}_4,\, \mathcal{K}_8,\, \mathcal{O}_8}\,. \qquad (8.2.13)$$

Since this symmetrization does not commute with the generator φ of $\mathcal{C}(4_1^Z)$, this is not the invariant ring of the corresponding symmetrized group. This ring may also be described as $R_0 \oplus B^4C^4R_0 \oplus B^8C^8R_0$, where R_0 is the ring of symmetric polynomials in A^4, B^4, C^4.

For the Hamming weight enumerators we have

$$\mathrm{MS}_{\mathrm{hwe},\,4_1^Z} \;=\; \frac{(1+t^8)(1+t^{12})}{(1-t^4)(1-t^8)} \qquad [A092535]\,, \qquad (8.2.14)$$

$$\mathrm{Inv}_{\mathrm{hwe},\,4_1^Z} \;=\; \frac{(1,\; y^2(x^2+3y^2)(x^2-y^2)^2) \times (1,\; y^4(x^2-y^2)^4)}{(x^2+3y^2)^2,\; y^4(x-y)^4}$$
$$=\; \frac{1,\, \mathcal{K}_8,\, \mathcal{K}_{12},\, \mathcal{K}_8+\mathcal{K}_{12}}{\mathcal{K}_4,\, \mathcal{O}_8}\,. \qquad (8.2.15)$$

8.2.3 $4_{\mathrm{S}}^{\mathbb{Z}}$: Type I Self-dual Codes over $\mathbb{Z}/4\mathbb{Z}$ with 1 in the shadow.

$$\mathcal{C}(4_{\mathrm{S}}^{\mathbb{Z}}) = \langle h\ , m_3\ , \phi\ , s := \mathrm{diag}(1, \eta, 1, -\eta) \rangle\ ,$$

where $\eta = \exp(2\pi i/8)$, of order 192;

$$\mathrm{MS}_{\mathrm{cwe},\, 4_{\mathrm{S}}^{\mathbb{Z}}} = \frac{1 + t^{18}}{(1-t)(1-t^4)(1-t^8)(1-t^{12})} \qquad [A092508]\,, \qquad (8.2.16)$$

$$\mathrm{Inv}_{\mathrm{cwe},\, 4_{\mathrm{S}}^{\mathbb{Z}}} = \frac{1,\ B^2C^2D^2(B^4 - C^4)(B^4 + D^4)(C^4 + D^4)}{A, B^4 + C^4 - D^4, B^8 + C^8 + D^8, B^{12} + C^{12} - D^{12}}. \qquad (8.2.17)$$

The symmetrized weight enumerators and the Hamming weight enumerators of the codes of Type $4_{\mathrm{S}}^{\mathbb{Z}}$ span the same rings as those of Type $4^{\mathbb{Z}}$, that is, (8.2.6) and (8.2.8).

8.2.4 $4_{\mathrm{II}}^{\mathbb{Z}}$: Type II self-dual codes over $\mathbb{Z}/4\mathbb{Z}$

$$\mathcal{C}(\rho(4_{\mathrm{II}}^{\mathbb{Z}})) = \langle h\ , r\ , \phi_0 := \mathrm{diag}(1, \eta, -1, \eta) \rangle\ ,$$

where $\eta = \exp(2\pi i/8)$, of order 1536;

$$\mathrm{MS}_{\mathrm{cwe},\, 4_{\mathrm{II}}^{\mathbb{Z}}} = \frac{1 + t^8 + 2t^{16} + 2t^{24} + t^{32} + t^{40}}{(1 - t^8)^3(1 - t^{24})} \qquad [A051462]\,. \qquad (8.2.18)$$

The invariants are polynomials in A^2, B^2, C^2, D^2 (which reduces the group to order 96). As primary invariants we may take $A^8 + B^8 + C^8, D^8, A^4B^4 + A^4C^4 - B^4C^4$ and $A^{24} + B^{24} + C^{24}$. We omit the secondary invariants. For the symmetrized weight enumerators we have

$$\mathrm{MS}_{\mathrm{swe},\, 4_{\mathrm{II}}^{\mathbb{Z}}} = \frac{1 + t^{16}}{(1 - t^8)^2(1 - t^{24})} \qquad [A007980]\,, \qquad (8.2.19)$$

$$\mathrm{Inv}_{\mathrm{swe},\, 4_{\mathrm{II}}^{\mathbb{Z}}} = \frac{1,\ f_{16}}{f_8,\ f_8',\ f_{24}}\,, \qquad (8.2.20)$$

where now

$$f_8 := x^8 + 28x^6z^2 + 70x^4z^4 + 28x^2z^6 + z^8 + 128y^8\,,$$
$$f_{16} := \{x^2z^2(x^2 + z^2)^2 - 4y^8\}\{(x^4 + 6x^2z^2 + z^4)^2 - 64y^8\}\,,$$
$$f_{24} := y^8(x^2 - z^2)^8\,,$$
$$f_8' := \{xz(x^2 + z^2) - 2y^4\}^2\,.$$

In terms of codes,

$$\mathrm{Inv}_{\mathrm{swe},\, 4_{\mathrm{II}}^{\mathbb{Z}}} = \frac{1,\ \mathcal{C}_{16}}{\mathcal{K}_8,\ \mathcal{O}_8,\ \mathcal{K}_{24}}. \qquad (8.2.21)$$

8.2.5 $4_{\text{II},1}^{\mathbb{Z}}$: Type II self-dual codes over $\mathbb{Z}/4\mathbb{Z}$ containing 1

References: Bonnecaze, Solé, Bachoc and Mourrain [66],   Calderbank and Sloane [99], Harada, Solé and Gaborit [257].

It follows from Theorem 2.4.14 that any Type $4_{\text{II}}^{\mathbb{Z}}$ code contains a word of the form $(\pm 1)^N$. It is therefore only a mild restriction to require that $\mathbf{1}$ is in the code, which greatly enlarges the group. We have:

$$\mathcal{C}(\rho(4_{\text{II},1}^{\mathbb{Z}})) = \langle h \, , m_3 \, , \phi_0 \, , \varphi \rangle \, ,$$

of order 6144;

$$\text{MS}_{\text{cwe}, 4_{\text{II},1}^{\mathbb{Z}}} = \frac{(1+t^{16})(1+t^{32})}{(1-t^8)^2(1-t^{16})(1-t^{24})} \qquad [A007979] \, , \qquad (8.2.22)$$

$$\text{Inv}_{\text{cwe}, 4_{\text{II},1}^{\mathbb{Z}}} = \frac{(1, \, f_{16}) \times (1, \, f_{32})}{A^8 + B^8 + C^8 + D^8, f_8, A^{16} + \cdots + D^{16}, A^{24} + \cdots + D^{24}} \, , \qquad (8.2.23)$$

where now

$$f_8 := A^4C^4 + C^4D^4 + D^4B^4 + B^4A^4 - A^4D^4 - B^4C^4 \, ,$$
$$f_{16} := (ABCD)^4 \, ,$$
$$f_{32} := (ABCD)^2(A^4 + C^4)(C^4 + D^4)(D^4 + B^4)(B^4 + A^4)$$
$$\times (A^4 - D^4)(B^4 - C^4) \, .$$

In terms of codes,

$$\text{Inv}_{\text{cwe}, 4_{\text{II},1}^{\mathbb{Z}}} = \frac{1, \, C_{16}, \, C_{32}, \, C_{16} + C_{32}}{\mathcal{K}_8, \, \mathcal{O}_8, \, \mathcal{K}_{16}, \, \mathcal{K}_{24}} \, . \qquad (8.2.24)$$

C_{32} was not mentioned in Chapter 2, for the good reason that we do not have a presentable code with the desired weight enumerator—although a random code of Type $4_{\text{II},1}^{\mathbb{Z}}$ and length 32 works. (We know from Theorem 5.5.7 that C_{32} exists.)

The symmetrized weight enumerators are the same as for $4_{\text{II}}^{\mathbb{Z}}$.

Calderbank and Sloane [99] also consider the further condition that the Lee weights be divisible by 4. This is readily verified to be a non-quadratic condition, and thus is outside the scope of our theory. Luckily, any code satisfying this condition has a linear image under the Gray map (2.4.35), which makes it uninteresting.

8.3 $8^{\mathbb{Z}}$: Self-dual codes over $\mathbb{Z}/8\mathbb{Z}$

For $\mathbb{Z}/8\mathbb{Z}$ the orders of the Clifford-Weil groups are:

$$|\mathcal{C}(\rho(8_{\text{I}}^{\mathbb{Z}}))| = 2^{10}, \; |\mathcal{C}(\rho(8_{\text{I}}^{\mathbb{Z}}))| = 2^{16}, \; |\mathcal{C}(\rho(8_{\text{II}}^{\mathbb{Z}}))| = 2^{12}3, \; |\mathcal{C}(\rho(8_{\text{II},1}^{\mathbb{Z}}))| = 2^{17}3 \, ;$$

$$\mathrm{MS}_{\mathrm{cwe},\,8_{\mathrm{I}}^{\mathbb{Z}}} = \frac{\theta_a(t^2)}{(1-t^2)(1-t^8)^5(1-t^{16})^2} \qquad [A092544]\,, \qquad (8.3.1)$$

where

$$\begin{aligned}
\theta_a(t) := {}& 1 + 4t^3 + 13t^4 + 31t^5 + 66t^6 + 123t^7 + 212t^8 + 346t^9 \\
& + 495t^{10} + 694t^{11} + 904t^{12} + 1097t^{13} + 1313t^{14} + 1421t^{15} \\
& + 1473t^{16} + 1484t^{17} + 1390t^{18} + 1133t^{20} + 913t^{21} + 700t^{22} \\
& + 501t^{23} + 330t^{24} + 218t^{25} + 131t^{26} + 62t^{27} + 30t^{28} + 7t^{29} \\
& + t^{30} + 3t^{31}\,;
\end{aligned}$$

$$\mathrm{MS}_{\mathrm{cwe},\,8_{\mathrm{I}}^{\mathbb{Z}}} = \frac{\theta_b(t^8)}{(1-t^8)^3(1-t^{16})^3(1-t^{32})^2} \qquad [A092545]\,, \qquad (8.3.2)$$

where

$$\begin{aligned}
\theta_b(t) := {}& 1 + 35t^2 + 237t^3 + 943t^4 + 2250t^5 + 4089t^6 + 5659t^7 \\
& + 6323t^8 + 5680t^9 + 4057t^{10} + 2311t^{11} + 909t^{12} + 246t^{13} \\
& + 27t^{14} + t^{15}\,;
\end{aligned}$$

$$\mathrm{MS}_{\mathrm{cwe},\,8_{\mathrm{II}}^{\mathbb{Z}}} = \frac{\theta_c(t^8)}{(1-t^8)^4(1-t^{16})(1-t^{24})^3} \qquad [A092546]\,, \qquad (8.3.3)$$

where

$$\begin{aligned}
\theta_c(t) := {}& 1 + 4t + 158t^2 + 1134t^3 + 3964t^4 + 9015t^5 + 14318t^6 + 16531t^7 \\
& + 14322t^8 + 9016t^9 + 3978t^{10} + 1129t^{11} + 155t^{12} + 3t^{13}\,;
\end{aligned}$$

$$\mathrm{MS}_{\mathrm{cwe},\,8_{\mathrm{II},\mathbf{1}}^{\mathbb{Z}}} = \frac{\theta_d(t^{16})}{(1-t^{16})^3(1-t^{32})^2(1-t^{48})^3} \qquad [A092547]\,, \qquad (8.3.4)$$

where

$$\begin{aligned}
\theta_d(t) := {}& 1 + 10t + 635t^2 + 6481t^3 + 30054t^4 + 85114t^5 + 166002t^6 \\
& + 235709t^7 + 254210t^8 + 205865t^9 + 123812t^{10} + 53334t^{11} \\
& + 15059t^{12} + 2247t^{13} + 115t^{14}\,.
\end{aligned}$$

8.4 Codes over more general Galois rings

In this section we will treat the Types of codes over Galois rings defined in §2.3.5. Since these are generalizations of codes over finite fields, the associated Clifford-Weil groups are analogues of those over fields. On the other hand, the Galois rings $m^{\mathbb{Z}}$ are also special cases, so the reader may compare the results in this section with those in the previous three sections and Chapter 7.

8.4.1 $\mathrm{GR}(p^e, f)^{\mathrm{E}}$: Euclidean self-dual $\mathrm{GR}(p^e, f)$-linear codes.

Codes of length N over the Galois ring $\mathrm{GR}(p^e, f)$ are also codes of length fN over $\mathbb{Z}/p^e\mathbb{Z} = \mathrm{GR}(p^e, 1)$. It is therefore helpful to know the associated $\mathbb{Z}_p$-bilinear forms on the f-dimensional $\mathbb{Z}_p$-lattice

$$O(p, f) := \mathbb{Z}_p[\zeta_{p^f - 1}].$$

Lemma 8.4.1. *There is a unit $a \in O(p, f)$ such that the lattice $O(p, f)$ has an orthonormal $\mathbb{Z}_p$-basis with respect to the bilinear form $B_a : (x, y) \mapsto \mathrm{Tr}(axy)$, where Tr denotes the trace from $O(p, f)$ to $\mathbb{Z}_p$.*

Proof. Since $O(p, f)$ is an unramified extension of $\mathbb{Z}_p$, the discriminant of the trace-bilinear form B_1 is a unit. The discriminant of B_a is equal to that of B_1 multiplied by the norm of a. Since the norm is surjective on units, there is some element $a \in O(p, f)$ such that the discriminant of B_a is 1. By Cassels [103, Corollary on p. 116] this implies that B_a is equivalent to the standard form when p is odd. For $p = 2$ we note in addition that squaring is a Galois automorphism of $\mathbb{F}_{2^f} = O(2, f)/(2)$. In particular, $\mathbb{F}_{2^f} = \mathbb{F}_{2^f}^2$. If $B_a(x, x) = \mathrm{Tr}(ax^2) \in 2\mathbb{Z}_2$ for all $x \in O(2, f)$, then reduction modulo 2 implies that a is in the radical of the trace form from $\mathbb{F}_{2^f}$ to $\mathbb{F}_2$. Since this form is nondegenerate, $a \in 2O(2, f)$ is not a unit. Therefore the lattice $(O(2, f), B_a)$ is an odd unimodular lattice of discriminant 1 and hence is equivalent to the standard lattice (see [103, §8.5]). □

Theorem 8.4.2. *Let $p > 2$ be an odd prime. Let $n := 1$ if e is even, $n := 2$ if e is odd and $p^f \equiv 1 \pmod 4$, and $n := 4$ if e is odd and $p^f \equiv 3 \pmod 4$. Then*

$$\mathcal{C}_g(\rho(\mathrm{GR}(p^e, f)^{\mathrm{E}})) \cong Z_n. \mathrm{Sp}_{2g}(\mathrm{GR}(p^e, f)).$$

If in addition the code contains $p^s 1$ for some $0 \leq s < e$, then setting $a := \max\{e - 2s, 0\}$ we get

$$\mathcal{C}_g(\rho(\mathrm{GR}(p^e, f)_{p^s}^E)) \cong (Z_{p^a n}).(Z_{p^{e-s}}^{2gf}). \mathrm{Sp}_{2g}(\mathrm{GR}(p^e, f)).$$

Proof. The proof is similar to that of Theorem 8.1.1. Again, λ is surjective because the characteristic is odd and the hyperbolic co-unitary group is isomorphic to the symplectic group. To determine the scalar elements in the general case, note that for even e the code $(p^{e/2})$ is of Type $\mathrm{GR}(p^e, f)^{\mathrm{E}}$. If e is odd, then the length of any self-dual code must be even, since $\mathrm{GR}(p^e, f)$ has an odd number (namely e) of composition factors. If $p^f \equiv 1 \pmod 4$, there is a p-adic integer $a \in \mathbb{Z}_p[\zeta_{p^f-1}]$ with $a^2 = -1$. If a also denotes the image of a in $\mathrm{GR}(p^e, f) = \mathbb{Z}_p[\zeta_{p^f-1}]/(p^e)$ then $(1, a)$ is a self-dual code of length 2. If $p^f \equiv 3 \pmod 4$, then f is odd and $p \equiv 3 \pmod 4$. A length N code of Type $\rho(\mathrm{GR}(p^e, f)^{\mathrm{E}})$ is also a code of length Nf and Type $\rho((p^e)^{\mathbb{Z}})$, so by Theorem 8.1.1 Nf and hence N is a multiple of 4. For the case $\mathrm{GR}(p^e, f)_{p^s}^E$, note that $p^s 1$ is isotropic if and only if the length is divisible by p^{e-2s}. As in the case $m_1^{\mathbb{Z}}$ (cf. Proposition 8.1.3), the theorem now follows from Lemma 4.6.7. □

Now assume $p = 2$. In this case, λ is not surjective. To describe the hyperbolic co-unitary groups, we need to generalize the notion of Θ-group given in (8.1.1). Recall that a symmetric matrix $S \in \mathrm{Mat}_n(\mathrm{GR}(2^e, f))$ is called even if the diagonal entries of S lie in $2\,\mathrm{GR}(2^e, f)$ (cf. Definition 1.10.4). Then

$$\Theta_{2g}(\mathrm{GR}(2^e, f)) := \{\begin{pmatrix} A & B \\ C & D \end{pmatrix} \in \mathrm{Sp}_{2g}(\mathrm{GR}(2^e, f)) \mid \tag{8.4.1}$$
$$C^{tr}A = A^{tr}C \text{ and } D^{tr}B = B^{tr}D \text{ are both even } \}.$$

A proof completely analogous to that of Theorem 8.1.4 now establishes:

Theorem 8.4.3. *Let $n = 0$ if e is even or $n = 1$ if e is odd.*

(a) Then
$$\mathcal{C}_g(\rho(\mathrm{GR}(2^e, f)^{\mathrm{E}})) \cong Z_{2^n}.Z_2^{2gf}.\Theta_{2g}(\mathrm{GR}(2^e, f)),$$

which is isomorphic to $2_+^{1+2fg}.\Theta_{2g}(\mathrm{GR}(2^e, f)) = 2_+^{1+2fg}.O_{2g}^+(\mathbb{F}_{2^f})$ if $e = 1$ and to $(Z_{2^n} \times (Z_2^{2fg})).\Theta_{2g}(\mathrm{GR}(2^e, f)_{2^s}^{\mathrm{E}})$ if $e > 1$.
(b) If in addition we assume that the codes contains $2^s 1$ we have

$$\mathcal{C}_g(\rho(\mathrm{GR}(2^e, f)_{2^s}^{\mathrm{E}})) \cong Z_{2^a}.(Z_{2^{e-s}})^{2gf}.\Theta_{2g}(\mathrm{GR}(2^e, f)),$$

where $a = \max\{e - 2s, n\}$.

For the generalized doubly-even codes we have:

Theorem 8.4.4. *Let $n = 2$ if f is even or $n = 3$ if f is odd. Then*

$$\mathcal{C}_g(\rho(\mathrm{GR}(2^e, f)_{\mathrm{II}}^{\mathrm{E}})) \cong Z_{2^n}.Z_2^{2fg}.\mathrm{Sp}_{2g}(\mathrm{GR}(2^e, f))$$

and

$$\mathcal{C}_g(\rho(\mathrm{GR}(2^e, f)_{\mathrm{II},2^s}^{\mathrm{E}})) \cong Z_{2^a}.Z_{2^{e-s}}^{2fg}.\mathrm{Sp}_{2g}(\mathrm{GR}(2^e, f)),$$

where $a = \max\{e - 2s, n\}$.

Proof. The structure of the hyperbolic co-unitary group follows from our general theory. Also the second statement follows from the first, using Proposition 8.1.3. It remains to determine the scalar elements in $\mathcal{C}_g(\rho(\mathrm{GR}(2^e, f)_{\mathrm{II}}^{\mathrm{E}}))$. By Theorem 2.4.15(c) there is always a self-dual isotropic code of length 8, since the octacode $\mathcal{O}_8$ is a lift of the Hamming code e_8. The code $RS_4(\mathrm{GR}(2^e, 2))$ over $\mathrm{GR}(2^e, 2)$ with generator matrix

$$\begin{bmatrix} 1 & \omega & \omega^2 & 0 \\ 1 & 1 & 1 & 2\omega + 1 \end{bmatrix},$$

where $\omega \in \mathrm{GR}(2^e, 2)$ denotes a primitive cube root of unity, is a self-dual isotropic lift of the extended quadratic residue code $XQ(\mathbb{F}_4, 3) = RS_4$. If f is even, then $\mathrm{GR}(2^e, f)$ contains $\mathrm{GR}(2^e, 2)$ and we obtain a self-dual isotropic code of length 4 by extension of scalars. Therefore the scalar elements of

$\mathcal{C}_g(\rho(\mathrm{GR}(2^e, f)_{\mathrm{II}}^{\mathrm{E}}))$ lie in the group of 8th (resp. 4th) roots of unity if f is odd (resp. even).

By Lemma 8.4.1, any length N code of Type $\rho(\mathrm{GR}(2^e, f)_{\mathrm{II}}^{\mathrm{E}})$ is also a length fN code of Type $\rho((2^e)_{\mathrm{II}}^{\mathbb{Z}})$. Remark 2.4.7 shows that Nf is divisible by 8. Therefore N is also a multiple of 8 if f is odd.

It suffices to show that there is no self-dual isotropic code of Type $\rho(\mathrm{GR}(2^e, f)_{\mathrm{II}}^{\mathrm{E}})$ and length 2. Let C be such a code. Then C has a generator matrix of the form

$$\begin{bmatrix} 2^a & x \\ 0 & 2^c \end{bmatrix},$$

where we may assume that $a \leq c$ after interchanging the columns. Since $|C| = 2^{ef}$ we have $a + c = e$. If $\alpha \in \mathrm{GR}(2^e, f)$ satisfies $\mathrm{Tr}(\alpha^2) = 1$, then $\mathrm{Tr}((2^c\alpha)^2) = 2^{2c}$ and hence $2c \geq e + 1$. Since $\mathrm{Tr}(2^c xy)$ is divisible by 2^e for all $y \in \mathrm{GR}(2^e, f)$, the nondegeneracy of Tr implies that

$$x = 2^{e-c}x' \in 2^{e-c}\,\mathrm{GR}(2^e, f).$$

The isotropy of the vectors $(2^a y, 2^a x'y)$ for all $y \in \mathrm{GR}(2^e, f)$ implies that $1 + (x')^2 \in 4\,\mathrm{GR}(2^e, f)$. In particular, $x' \notin 2\,\mathrm{GR}(2^e, f)$ is a unit. Hence $(1 + x')^2 = 1 + (x')^2 - 2x'$ generates the ideal $2\,\mathrm{GR}(2^e, f)$ and gives a zero-divisor $(1 + x') \in \mathrm{GR}(2^e, f)/(2\,\mathrm{GR}(2^e, f)) \cong \mathbb{F}_{2^f}$, a contradiction.   □

8.4.2 $\mathrm{GR}(p^e, f)^{\mathrm{H}}$: Hermitian self-dual $\mathrm{GR}(p^e, f)$-linear codes.

Let $f = 2l$ be even and define the Galois automorphism $^-$ of $\mathrm{GR}(p^e, f)$ over $\mathrm{GR}(p^e, l)$ by

$$\overline{\sum_{i=0}^{e-1} p^i \zeta_i} := \sum_{i=0}^{e-1} p^i \zeta_i^{(p^l)},$$

where $\zeta_i \in T := \{0, 1, \zeta, \ldots, \zeta^{p^f - 2}\}$ and $\zeta \in \mathrm{GR}(p^e, f)$ denotes a primitive $(p^f - 1)$st root of unity.

Since $\psi^{-1}(\lambda(\varPhi)) = \mathrm{GR}(p^e, l)$ is the fixed space of the Galois automorphism $^-$ we find that $\begin{pmatrix} A & B \\ C & D \end{pmatrix} \in \mathrm{Mat}_{2g}(\mathrm{GR}(p^e, 2l))$ lies in $\pi(\mathfrak{U}_g(R(\mathrm{GR}(p^e, 2l)^{\mathrm{H}})))$ if and only if

$$\begin{pmatrix} \overline{A}^{tr} & \overline{C}^{tr} \\ \overline{B}^{tr} & \overline{D}^{tr} \end{pmatrix} \begin{pmatrix} 0 & -1 \\ 1 & 0 \end{pmatrix} \begin{pmatrix} A & B \\ C & D \end{pmatrix} = \begin{pmatrix} 0 & -1 \\ 1 & 0 \end{pmatrix},$$

hence

$$\mathfrak{U}_g(\mathrm{GR}(p^e, 2l)^{\mathrm{H}}) \cong (\ker(\lambda) \oplus \ker(\lambda)).U_{2g}(\mathrm{GR}(p^e, 2l))$$

is an extension of the full unitary group.

If e is even, then the code $[p^{e/2}]$ is a Hermitian self-dual code of length 1. For odd e, such a code has even dimension and, by the surjectivity of the norm, there is a code of length 2 (generated by any $(1, a)$ with $a \in \mathrm{GR}(p^e, 2l)$ satisfying $a\bar{a} = -1$). Therefore the Clifford-Weil groups are as follows:

Proposition 8.4.5. *Let $a = 1$ if e is even or $a = 2$ if e is odd. Then*

(a) $\mathcal{C}_g(\rho(\mathrm{GR}(p^e, 2l)^{\mathrm{H}})) \cong Z_a.U_{2g}(\mathrm{GR}(p^e, 2l))$.

(b) *For the codes that contain $p^s\mathbf{1}$, let $b = \max\{e-2s, 0\}$ and let $c = \mathrm{lcm}(a, p^b)$.*
Then

$$\mathcal{C}_g(\rho(\mathrm{GR}(p^e, 2l)^{\mathrm{H}}_{p^s})) \cong Z_c.(Z^{2gf}_{p^{e-s}}).U_{2g}(\mathrm{GR}(p^e, 2l)).$$

8.4.3 GR$(p^e, 2l)^{\mathrm{H}+}$: Trace-Hermitian self-dual GR(p^e, l)-linear codes.

For the form ring $R(\mathrm{GR}(p^e, 2l)^{\mathrm{H}+})$ the qmodule Φ is $\{0\}$. Therefore $\lambda_2(\Phi_2)$ is the set of all skew-Hermitian matrices. Using the formula in Remark 5.2.5 we find that the hyperbolic co-unitary group

$$\mathfrak{U}_g(\mathrm{GR}(p^e, 2l)^{\mathrm{H}+}) \cong O^+_{2g}(\mathrm{GR}(p^e, l))$$

is isomorphic to the orthogonal group of the quadratic form $\sum^g_{i=1} x_i x_{2g+1-i}$ on $\mathrm{GR}(p^e, l)^{2g}$, and

$$\mathfrak{U}_g(\mathrm{GR}(p^e, 2l)^{\mathrm{H}+}_{p^s}) \cong (Z^{2lg}_{p^{e-s}}).O^+_{2g}(\mathrm{GR}(p^e, l)).$$

Since the $\mathrm{GR}(p^e, l)$-linear code with generator matrix $[1]$ is a self-dual isotropic code of Type $\rho(\mathrm{GR}(p^e, 2l)^{\mathrm{H}+}_1)$, self-dual isotropic codes exist for all lengths, and the corresponding Clifford-Weil groups are isomorphic to the hyperbolic co-unitary groups above:

Proposition 8.4.6. $\mathcal{C}_g(\rho(\mathrm{GR}(p^e, 2l)^{\mathrm{H}+})) \cong O^+_{2g}(\mathrm{GR}(p^e, l))$,
$\mathcal{C}_g(\rho(\mathrm{GR}(p^e, 2l)^{\mathrm{H}+}_{p^s})) \cong (Z^{2lg}_{p^{e-s}}).O^+_{2g}(\mathrm{GR}(p^e, l))$.

8.4.4 Clifford-Weil groups for GR$(4, 2)$.

The smallest Galois ring which is neither a field nor a ring of the form $\mathbb{Z}/m\mathbb{Z}$ is $\mathrm{GR}(4, 2)$, a ring with 16 elements. So already the Clifford-Weil groups of genus 1 are subgroups of $\mathrm{GL}_{16}(\mathbb{C})$. If $\omega \in \mathrm{GR}(4, 2)$ denotes a primitive cube root of unity then the alphabet $V = \mathrm{GR}(4, 2)$ is $\{a + b\omega \mid a, b \in \mathbb{Z}/4\mathbb{Z}\}$.

For Euclidean self-dual codes we find

$$\mathcal{C}(\rho(\mathrm{GR}(4, 2)^{\mathrm{E}})) \cong Z^4_2.\Theta_2(\mathrm{GR}(4, 2)),$$

of order $2^{11}3$, with Molien series

$$\mathrm{MS}_{\mathrm{cwe}}(\mathrm{GR}(4, 2)^{\mathrm{E}}) = \frac{f_1(t) + t^{72}f_1(t^{-1})}{(1 - t)(1 - t^3)(1 - t^4)^5(1 - t^6)^4(1 - t^8)^5} \quad [A099720],$$

where

$$f_1 := 1 + t^3 + t^4 + 4t^6 + 15t^7 + 75t^8 + 101t^9 + 274t^{10} + 555t^{11} + 1232t^{12}$$
$$+ 2187t^{13} + 4122t^{14} + 7245t^{15} + 12514t^{16} + 20155t^{17} + 31998t^{18}$$
$$+ 48747t^{19} + 72408t^{20} + 103925t^{21} + 144878t^{22} + 197573t^{23}$$
$$+ 261102t^{24} + 338155t^{25} + 425254t^{26} + 524445t^{27} + 629888t^{28}$$
$$+ 740973t^{29} + 851614t^{30} + 957155t^{31} + 1053412t^{32} + 1133809t^{33}$$
$$+ 1196064t^{34} + 1233375t^{35} + 623575t^{36}.$$

If in addition we assume that the code contains the all-ones vector, then

$$\mathcal{C}(\rho(\mathrm{GR}(4,2)_{\mathrm{I}}^{\mathrm{E}})) \cong Z_4.Z_4^4.\Theta_2(\mathrm{GR}(4,2)),$$

of order $2^{17}3$, with Molien series

$$\mathrm{MS}_{\mathrm{cwe}}(\mathrm{GR}(4,2)_{\mathrm{I}}^{\mathrm{E}}) = \frac{f_2(t^4) + t^{156}f_2(t^{-4})}{(1-t^4)^2(1-t^8)^5(1-t^{12})^5(1-t^{16})^4} \qquad [A099748],$$

where

$$f_2 := 1 + 13t^2 + 180t^3 + 2815t^4 + 26097t^5 + 178909t^6 + 914207t^7$$
$$+ 3741734t^8 + 12638167t^9 + 36300248t^{10} + 90284593t^{11}$$
$$+ 197531731t^{12} + 384302984t^{13} + 670899605t^{14} + 1057839714t^{15}$$
$$+ 1514398168t^{16} + 1975782327t^{17} + 2355691682t^{18} + 2571102793t^{19}.$$

For the generalized doubly-even codes we get

$$\mathcal{C}(\rho(\mathrm{GR}(4,2)_{\mathrm{II}}^{\mathrm{E}})) \cong Z_4.Z_2^4.\mathrm{Sp}_2(\mathrm{GR}(4,2)),$$

of order $2^{14}15$, with Molien series

$$\mathrm{MS}_{\mathrm{cwe}}(\mathrm{GR}(4,2)_{\mathrm{II}}^{\mathrm{E}}) = \frac{f_3(t^4) + t^{156}f_3(t^{-4})}{(1-t^4)^3(1-t^8)^5(1-t^{12})^5(1-t^{20})^3} \qquad [A099750],$$

where

$$f_3(t) := 1 + 11t^2 + 283t^3 + 4055t^4 + 37722t^5 + 243578t^6$$
$$+ 1179852t^7 + 4535052t^8 + 14380814t^9 + 38708195t^{10}$$
$$+ 90379766t^{11} + 186147868t^{12} + 342605290t^{13} + 569177435t^{14}$$
$$+ 860160090t^{15} + 1189401593t^{16} + 1511365669t^{17}$$
$$+ 1770220838t^{18} + 1914917488t^{19}.$$

Since the automorphism group here is nontrivial, we may symmetrize by $\{\pm 1\} = \mathrm{Aut}(\rho(\mathrm{GR}(4,2)_{\mathrm{II}}^{\mathrm{E}}))$ to obtain the Molien series for the symmetrized weight enumerators:

$$\mathrm{MS}_{\mathrm{swe}}(\mathrm{GR}(4,2)_{\mathrm{II}}^{\mathrm{E}}) = \frac{f_3'(t^4)}{(1-t^4)^2(1-t^8)^3(1-t^{12})^3(1-t^{20})^2} \qquad [A099752],$$

where

$$f_3'(t) := 1 + 2t^2 + 15t^3 + 51t^4 + 170t^5 + 500t^6 + 1136t^7 + 2126t^8 + 3439t^9$$
$$+ 4822t^{10} + 5908t^{11} + 6473t^{12} + 6325t^{13} + 5437t^{14} + 4124t^{15}$$
$$+ 2764t^{16} + 1596t^{17} + 764t^{18} + 305t^{19} + 95t^{20} + 20t^{21} + 5t^{22} + 2t^{23}.$$

For doubly-even codes that contain the all-ones vector, we find

$$\mathcal{C}(\rho(\mathrm{GR}(4,2)_{\mathrm{II},1}^{\mathrm{E}})) \cong Z_8.Z_4^4.\mathrm{Sp}_2(\mathrm{GR}(4,2)),$$

of order $2^{19}15$, with Molien series

$$\mathrm{MS}_{\mathrm{cwe}}(\mathrm{GR}(4,2)_{\mathrm{II},1}^{\mathrm{E}}) = \frac{f_4(t^8) + t^{320} f_4(t^{-8})}{(t^8-1)^4(t^{16}-1)^4(t^{24}-1)^5(t^{40}-1)^3} \qquad [A099595],$$

where

$$f_4(t) := 1 + 2t + 348t^2 + 24480t^3 + 663188t^4 + 9100890t^5$$
$$+ 77529608t^6 + 463193328t^7 + 2097567377t^8 + 7588564500t^9$$
$$+ 22768177018t^{10} + 58238217746t^{11} + 129682589691t^{12}$$
$$+ 255463425888t^{13} + 450744907318t^{14} + 719144650774t^{15}$$
$$+ 1044999703594t^{16} + 1390473277756t^{17} + 1700743431532t^{18}$$
$$+ 1917273923644t^{19} + 1995079651434/2t^{20}.$$

We next consider Hermitian self-dual codes over $\mathrm{GR}(4,2)$.
For Type $\rho(\mathrm{GR}(4,2)^{\mathrm{H}})$ we find

$$\mathcal{C}(\rho(\mathrm{GR}(4,2)^{\mathrm{H}})) \cong U_2(\mathrm{GR}(4,2)),$$

of order $2^5 3^2$, with Molien series

$$\mathrm{MS}_{\mathrm{cwe}}(\mathrm{GR}(4,2)^{\mathrm{H}}) = \frac{g_1(t) + t^{59} g_1(t^{-1})}{(1-t)(1-t^2)^4(1-t^4)^6(1-t^6)^3(1-t^{12})^2} \qquad [A099757],$$

where

$$g_1 := 1 + 16t^4 + 18t^5 + 143t^6 + 170t^7 + 696t^8 + 1073t^9$$
$$+ 2786t^{10} + 4420t^{11} + 8881t^{12} + 13782t^{13} + 23018t^{14} + 33979t^{15}$$
$$+ 50086t^{16} + 69748t^{17} + 94127t^{18} + 123476t^{19} + 155400t^{20}$$
$$+ 192999t^{21} + 229352t^{22} + 270342t^{23} + 306276t^{24} + 343736t^{25}$$
$$+ 372818t^{26} + 399845t^{27} + 415884t^{28} + 425872t^{29}.$$

The automorphism group of $\rho(\mathrm{GR}(4,2)^{\mathrm{H}})$ is isomorphic to Z_6 and is generated by $-\omega$. It has four orbits on $\mathrm{GR}(4,2)$, represented by $0, 2, 1, 1 + 2\omega$, of lengths $1, 3, 6, 6$ respectively. The group

$$U_2(\mathrm{GR}(4,2))/Z_6 \cong Z_2 \times S_4$$

of order 48 acts on the symmetrized space with Molien series

$$\mathrm{MS}_{\mathrm{swe}}(\mathrm{GR}(4,2)^{\mathrm{H}}) = \frac{1}{(1-t)(1-t^2)(1-t^4)(1-t^6)} \qquad [A099770], \quad (8.4.2)$$

and invariant ring spanned by the symmetrized weight enumerators of the codes with generator matrix $[2]$, $[1, 2+\omega]$, $\begin{bmatrix} 1\,1\,1\,1 \\ 2\,2\,0\,0 \\ 0\,2\,2\,0 \end{bmatrix}$ and

$$\begin{bmatrix} 1\,1+\omega\,0 & 2\omega\,1 & 1+\omega \\ 0 & 2\,0 & 2\omega\,0 & 2 \\ 0 & 0\,1\,2+3\omega\,\omega\,2\omega+3 \\ 0 & 0\,0 & 0\,2\,2\omega+2 \end{bmatrix}.$$

(The length 6 code was found by a random search.)

Note that this is another instance where the invariant ring of a symmetrized Clifford-Weil group is a polynomial ring. If we continue to symmetrize (of course these symmetrizations do not commute with the action of $\mathcal{C}(\rho(\mathrm{GR}(4,2)^{\mathrm{H}}))$, so the rings spanned by the corresponding symmetrized weight enumerators are not necessarily invariant rings of groups) we also get polynomial rings for the Lee- and Hamming-weight enumerators:

$$\mathrm{Inv}^{\mathrm{Lee}}(\mathcal{C}(\rho(\mathrm{GR}(4,2)^{\mathrm{H}}))) = \frac{1}{x+3y, xy+y^2-2z^2, x^2z^2-2xyz^2+y^2z^2}$$

(where x, y, z correspond to $\{0\}, 2R, R^*$ respectively) and

$$\mathrm{Inv}^{\mathrm{Ham}}(\mathcal{C}(\rho(\mathrm{GR}(4,2)^{\mathrm{H}}))) = \frac{1}{x+3y, xy-y^2}.$$

We now also assume that the code contains the all-ones vector. Then

$$\mathcal{C}(\rho(\mathrm{GR}(4,2)^{\mathrm{H}}_1)) \cong Z_4.Z_4^4.U_2(\mathrm{GR}(4,2)),$$

of order $2^{15}3^2$, with Molien series

$$\mathrm{MS}_{\mathrm{cwe}}(\mathrm{GR}(4,2)^{\mathrm{H}}_1) = \frac{g_2(t^4)+t^{148}g_2(t^{-4})}{(1-t^4)^2(1-t^8)^9(1-t^{12})^3(1-t^{24})^2} \qquad [A100023],$$

where

$$\begin{aligned}
g_2(t) := {}& 1 + 21t^2 + 348t^3 + 4167t^4 + 36071t^5 + 229176t^6 \\
& + 1098673t^7 + 4151088t^8 + 12776903t^9 + 32923056t^{10} \\
& + 72635374t^{11} + 139845777t^{12} + 238772588t^{13} + 366395941t^{14} \\
& + 510689826t^{15} + 651563740t^{16} + 764743937t^{17} + 828011969t^{18}.
\end{aligned}$$

In this case there is no nice symmetrization, since $\mathrm{Aut}(\rho(\mathrm{GR}(4,2)_1^{\mathrm{H}})) = \{1\}$.

For the trace-Hermitian additive codes, the Clifford-Weil groups are quite small:

$$\mathcal{C}(\rho(\mathrm{GR}(4,2)^{\mathrm{H}+})) \cong O_2^+(\mathrm{GR}(4,1)) \cong O_2^+(\mathbb{Z}/4\mathbb{Z}) \cong \mathbb{Z}_2 \times \mathbb{Z}_2 \,,$$

of order 2^2, with Molien series

$$\mathrm{MS}_{\mathrm{cwe}}(\mathrm{GR}(4,2)^{\mathrm{H}+})$$
$$= \frac{1 + 9t^2 + 27t^3 + 27t^4 + 27t^5 + 27t^6 + 9t^7 + t^9}{(1-t)^7(1-t^2)^9} \qquad [A100024]\,.$$

Symmetrizing by the automorphism group of $\rho(\mathrm{GR}(4,2)^{\mathrm{H}+})$,

$$\mathrm{Aut}(\rho(\mathrm{GR}(4,2)^{\mathrm{H}+})) = \mathrm{Aut}(\rho(\mathrm{GR}(4,2)^{\mathrm{H}})) \cong \mathbb{Z}_6 \,,$$

we find that the Molien series for the symmetrization of $\mathcal{C}(\rho(\mathrm{GR}(4,2)^{\mathrm{H}+}))$ is

$$\mathrm{MS}_{\mathrm{swe}}(\mathrm{GR}(4,2)^{\mathrm{H}+}) = \frac{1}{(1-t)^3(1-t^2)} \qquad [A002623]\,,$$

for which the invariant ring is spanned by the symmetrized weight enumerators of the $\mathbb{Z}/4\mathbb{Z}$-linear codes generated by $[1]$, $[1+2\omega]$, $\begin{bmatrix} 2 \\ 2\omega \end{bmatrix}$ and $\begin{bmatrix} 1 & 2+\omega \\ \omega & 2\omega + \omega^2 \end{bmatrix}$.
Note that the last two codes are linear over $\mathrm{GR}(4,2)$. The symmetrized weight enumerators of these codes are $x_0 + 2x_1 + x_2$, $x_0 + 2x_{1+2\omega} + x_2$, $x_0 + 3x_2$ and $x_0^2 + 3x_2^2 + 12x_1 x_{1+2\omega}$.

If in addition we assume that the code contains the all-ones vector, then

$$\mathcal{C}(\rho(\mathrm{GR}(4,2)_1^{\mathrm{H}+})) \cong Z_4^2 . O_2^+(\mathbb{Z}/4\mathbb{Z}) \,,$$

of order 2^6, with Molien series

$$\mathrm{MS}_{\mathrm{cwe}}(\mathrm{GR}(4,2)_1^{\mathrm{H}+}) = \frac{h(t) + t^{41} h(t^{-1})}{(1-t)(1-t^2)^6(1-t^4)^7(1-t^8)^2} \qquad [A100025]\,,$$

where

$$h(t) := 1 + 14t^3 + 43t^4 + 115t^5 + 334t^6 + 808t^7 + 1752t^8 + 3557t^9$$
$$+ 6448t^{10} + 10800t^{11} + 17020t^{12} + 24972t^{13} + 34704t^{14} + 45824t^{15}$$
$$+ 57298t^{16} + 68464t^{17} + 78120t^{18} + 85092t^{19} + 88922t^{20}\,.$$

8.5 Self-dual codes over $\mathbb{F}_{q^2} + \mathbb{F}_{q^2} u$

In this section we study self-dual codes over the ring $R = \mathbb{F}_{q^2} + \mathbb{F}_{q^2} u$, for $q = p^f$, where $u^2 = 0$ and $ua = a^q u$ for all $a \in \mathbb{F}_{q^2}$. These are certainly "non-classical" codes.

We have $R \cong \mathcal{M}/p\mathcal{M}$, where $\mathcal{M}$ is the maximal order in the quaternion division algebra over the unramified extension of degree f of the p-adic numbers. The most important case is $q = 2$. In this special case, self-dual codes have been studied by Bachoc [19] in connection with the construction of interesting modular lattices, and Gaborit [178] has found a mass formula. These codes have also been investigated by Betsumiya, Ling and Nemenzo [47].

To define self-dual codes, we use the R-valued Hermitian form $R^N \times R^N \to R$ given by $(x, y) := \sum_{i=1}^{N} x_i \overline{y_i}$, where $\bar{} : R \to R$ is the involution $\overline{a + bu} := a^q - bu$. Then

$$(a' + b'u)\overline{(a + bu)} = a'a^q + (ab' - ba')u,$$

for all $a, b, a', b' \in \mathbb{F}_{q^2}$. A code $C \leq R^N$ is self-dual if $C = C^\perp := \{v \in R^N \mid (v, c) = 0 \text{ for all } c \in C\}$.

To express this self-duality in our language of Types, we need a representation of a form ring.

Let $m_0 : R \times R \to \frac{1}{p}\mathbb{Z}/\mathbb{Z}$ be the bilinear form defined by

$$m_0(a + bu, a' + b'u) := \frac{1}{p}\operatorname{Tr}(ab' - a'b),$$

where Tr denotes the trace from $\mathbb{F}_{q^2}$ to $\mathbb{F}_p \cong \mathbb{Z}/p\mathbb{Z}$. Let $M := m_0(1 \otimes R)$, where the right action of $R \otimes R$ on M is left multiplication on the arguments. Let $\psi : R_R \to M_{1 \otimes R}$ be the R-module isomorphism defined by $\psi(1) := m_0$. Then

$$\psi(r + su)(a + bu, a' + b'u) = m_0(1 \otimes (r + su))(a + bu, a' + b'u)$$
$$= m_0((a + bu), (r + su)(a' + b'u)) = \frac{1}{p}\operatorname{Tr}(r(ab' - ba') + saa'^q),$$

for $r + su, a + bu, a' + b'u \in R$. Define $\tau : M \to M$ by $\tau(m)(x, y) := m(y, x)$. The involution $J : R \to R$ induced by ψ and τ is given by $(r + su)^J = r - s^q u$, and $\epsilon = -1$ (since $m_0 = \psi(1)$ is skew-symmetric). Then (R, M, ψ) is a twisted R-module.

Let $\Phi := \{\!\{M\}\!\}$, where $\{\!\{\}\!\} : M \to \Phi$ is the obvious diagonal evaluation

$$\{\!\{m_0(1 \otimes (r + su))\}\!\}(a + bu) := m_0(1 \otimes (r + su))(a + bu, a + bu) = \operatorname{Tr}(saa^q).$$

Note that $\operatorname{Tr}(saa^q) = \operatorname{Tr}_{\mathbb{F}_q/\mathbb{F}_p}((s + s^q)(aa^q))$ for all $s, a \in \mathbb{F}_{q^2}$, and $\Phi \cong \mathbb{F}_q$ (as abelian groups) via the isomorphism $\{\!\{\psi(r + su)\}\!\} \mapsto s + s^q$. For $s \in \mathbb{F}_q$ we therefore may define $\phi_s \in \Phi$ by

$$\phi_s(a) = \frac{1}{p}\operatorname{Tr}_{\mathbb{F}_q/F_p}(saa^q) \text{ for } a \in \mathbb{F}_{q^2}.$$

Since $\{\!\{\}\!\}$ is surjective, this defines a unique mapping $\lambda : \Phi \to M$, satisfying

$$\lambda\{\!\{m\}\!\} = m + \tau(m), \text{ with } \lambda(\phi_s) = \psi(su),$$

for all $m \in M$, $s \in \mathbb{F}_q$. In particular, λ is injective. This defines a form ring (R, M, ψ, Φ). The identity is a finite representation ρ of this form ring, and the self-dual codes $C \leq R^N$ above are precisely the codes of Type ρ.

The hyperbolic co-unitary group $\mathfrak{U}(R, \Phi)$ contains a normal subgroup $\mathcal{N} = \mathfrak{U}((u), \Phi)$ for which the quotient is a subgroup of $\mathfrak{U}(\mathbb{F}_{q^2}, \{0\})$. Note that $\lambda(\Phi) = \psi(\mathbb{F}_q u)$ and λ is injective. Therefore the projection onto the first component $\pi : \mathfrak{U}(R, \Phi) \to \mathrm{GL}_2(R)$ is injective. In fact, since $R = \mathbb{F}_{q^2} + \mathbb{F}_{q^2} u$, $\mathfrak{U}(R, \Phi)$ has a subgroup $H \cong O_2^+(\mathbb{F}_{q^2})$ consisting of the elements

$$\left\{\left(\begin{pmatrix} a & b \\ c & d \end{pmatrix}, \begin{pmatrix} 0 & \psi(cb) \\ & 0 \end{pmatrix}\right) \mid a, b, c, d \in \mathbb{F}_{q^2}, ca = db = 0, cb + da = 1\right\}.$$

This subgroup H is isomorphic to $\mathfrak{U}(\mathbb{F}_{q^2}, \{0\})$, and is a complement to the normal subgroup $\mathcal{N} = \mathfrak{U}((u), \Phi)$ given by

$$\left\{\left(\begin{pmatrix} 1 + au & bu \\ cu & 1 + a^q u \end{pmatrix}, \begin{pmatrix} \phi_{-c} & 0 \\ & \phi_b \end{pmatrix}\right) \mid c, b \in \mathbb{F}_q, a \in \mathbb{F}_{q^2}\right\},$$

which is isomorphic to $\mathbb{F}_{q^2} \oplus \mathbb{F}_q \oplus \mathbb{F}_q$. Therefore

$$\mathfrak{U}(R, \Phi) \cong (\mathbb{F}_{q^2} \oplus \mathbb{F}_q \oplus \mathbb{F}_q) \rtimes O_2^+(\mathbb{F}_{q^2}).$$

Example 8.5.1. Let $q := 2$, $R = \mathbb{F}_4 + \mathbb{F}_4 u$. Then the hyperbolic co-unitary group $\mathfrak{U}(R, \Phi)$ is generated by

$$g_1 := \left(\begin{pmatrix} \omega & 0 \\ 0 & \omega^2 \end{pmatrix}, 0\right),$$

$$g_2 := \left(\begin{pmatrix} 1 + u & 0 \\ 0 & 1 + u \end{pmatrix}, 0\right),$$

$$g_3 := \left(\begin{pmatrix} 1 & u \\ 0 & 1 \end{pmatrix}, \begin{pmatrix} 0 & 0 \\ & \phi_1 \end{pmatrix}\right),$$

$$h := \left(\begin{pmatrix} 0 & 1 \\ 1 & 0 \end{pmatrix}, \begin{pmatrix} 0 & 1 \\ & 0 \end{pmatrix}\right),$$

where g_1, g_2, g_3 generate the parabolic group $P(R, \Phi) \leq \mathfrak{U}(R, \Phi)$. Since λ is injective, $\mathfrak{U}(R, \Phi)$ is isomorphic to its image under the projection of the first component. This image contains a normal subgroup $\mathcal{N} \cong \mathbb{F}_4 + \mathbb{F}_2 + \mathbb{F}_2$ generated by

$$\begin{pmatrix} 1 + \omega u & 0 \\ 0 & 1 + \omega^2 u \end{pmatrix}, \begin{pmatrix} 1 + u & 0 \\ 0 & 1 + u \end{pmatrix}, \begin{pmatrix} 1 & u \\ 0 & 1 \end{pmatrix}, \begin{pmatrix} 1 & 0 \\ u & 1 \end{pmatrix},$$

and the quotient group is isomorphic to the S_3 generated by the matrices

$$\begin{pmatrix} \omega & 0 \\ 0 & \omega^2 \end{pmatrix}, \begin{pmatrix} 0 & 1 \\ 1 & 0 \end{pmatrix}.$$

Hence

$$\mathfrak{U}(R,\Phi) \cong (Z_2^2 \times Z_2^2) \rtimes S_3,$$

where S_3 acts faithfully on one copy of Z_2^2 and with kernel Z_3 on the other copy.

The Molien series of $\mathcal{C}(\rho)$ is

$$\frac{\theta(t) + t^{45}\theta(t^{-1})}{(1 - t^2)^5(1 - t^3)(1 - t^4)^6(1 - t^6)^4} \qquad [A092548], \qquad (8.5.1)$$

where

$$\begin{aligned}
\theta(t) := {}& 1 + t + 4t^2 + 3t^3 + 53t^4 + 104t^5 + 458t^6 + 858t^7 + 2474t^8 + 4839t^9 \\
& + 10667t^{10} + 19018t^{11} + 34193t^{12} + 55481t^{13} + 86078t^{14} \\
& + 125990t^{15} + 173466t^{16} + 230402t^{17} + 287430t^{18} + 346462t^{19} \\
& + 393648t^{20} + 431930t^{21} + 450648t^{22} .
\end{aligned}$$

Various interesting symmetrizations are possible:

(a) Right multiplication by $\mathbb{F}_4^* = \langle \omega \rangle$ has six orbits on $\mathbb{F}_4 + \mathbb{F}_4 u$, represented by $0, 1, u, 1 + u, \omega + u, \omega^2 + u$. The corresponding symmetrization yields a matrix group $\mathcal{C}^{(6)}(\rho)$ of degree 6 isomorphic to $\mathcal{C}(\rho)$, with Molien series

$$\frac{1 + t^5 + t^8 + t^{13}}{(1 - t)(1 - t^2)^2(1 - t^4)^2(1 - t^6)} \qquad (8.5.2)$$

[A092549] and invariant ring

$$\mathrm{Inv}_{\mathrm{swe}, \rho} = \mathrm{Inv}(\mathcal{C}^{(6)}(\rho)) = \frac{1, C_5, C_8, C_5 \oplus C_8}{C_1, V_2, C_2, V_4, C_4, V_6}. \qquad (8.5.3)$$

Codes for which the symmetrized weight enumerators generate the invariant ring have been found by Annika Günther in Aachen, as follows. For even length N the code

$$V_N := [\mathbf{1}, (0, u, 0, \dots, 0, u), (0, 0, u, 0, \dots, 0, u), \dots, (0, \dots, 0, u, u)]$$

(where the words in the brackets generate the code as a left-module over R) is a self-dual isotropic code of Type ρ. For the other codes we have

$$\begin{aligned}
& C_1 := [(u)], C_2 := [(1 + u, \omega + u)], \\
& C_4 := [(1, \omega, \omega + u, \omega + \omega u), (0, u, 0, u), (0, 0, u, u)], \\
& C_5 := [(1, 1, 1, 1, 0), (0, 1, \omega, \omega^2, 1), (u, u, 0, 0, u)].
\end{aligned}$$

The code C_8 of length 8 was found by a random search and has generator matrix $[I_4 \; A]$ where

$$A = \begin{bmatrix}
\omega u & 1 & \omega + \omega u & \omega^2 \\
\omega u & 1 + u & \omega^2 + \omega u & \omega + u \\
\omega u & 1 + \omega u & 1 & 1 \\
1 + \omega u & \omega u & 0 & 0
\end{bmatrix}.$$

(b) The unit group R^* has three orbits on R, namely $\{0\}, R^*, uR^*$. Symmetrizing by R^* gives a matrix group

$$\mathcal{C}^{(3)}(\rho) = \langle \mathrm{diag}(1, 1, -1), \frac{1}{4} \begin{pmatrix} 1 & 3 & 12 \\ 1 & 3 & -4 \\ 1 & -1 & 0 \end{pmatrix} \rangle \cong D_8 ,$$

for which the Molien series $1/((1 - t)(1 - t^2)(1 - t^4))$ and invariant ring are given by Bachoc [19, Theorem 4.4]. If the variables X, Y, Z correspond to the three orbits, then

$$\mathrm{Inv} = \frac{1}{X + 3Z, 2Y^2 - Z(X + Z), Y^2(X - Z)^2} ;$$

in terms of codes

$$\mathrm{Inv} = \frac{1}{C_1, V_2, V_4} .$$

The symmetrized weight enumerators are

$$X + 3Z, \quad X^2 + 12Y^2 + 3Z^2, \quad X^4 + 18X^2Z^2 + 24XZ^3 + 21Z^4 + 192Y^4 .$$

9

Lattices

In this chapter we attempt (although with only partial success) to fit self-dual lattices into our framework. We will show that the hyperbolic co-unitary groups of the main theorems of Chapter 5 are in a strong sense exactly the analogues for codes of the groups $\mathrm{Sp}_{2n}(\mathbb{Z})$ and the theta-groups $\Gamma_{\vartheta,n}$ that arise when studying unimodular lattices. In particular, (a) the groups $\mathrm{Sp}_{2n}(\mathbb{R})$, $U_{n,n}(\mathbb{R})$ and $\mathrm{SO}^*_{2n}(\mathbb{R})$ *are* the hyperbolic co-unitary groups associated with form $\mathbb{R}$-algebras, and (b) the gluing theory construction gives rise to finite representations of form orders for which the corresponding co-unitary groups are quotients of discrete subgroups of $\mathrm{Sp}_{2n}(\mathbb{R})$, etc. Thus for example the hyperbolic co-unitary group associated with ternary self-dual codes is the quotient $\mathrm{SL}_2(\mathbb{Z})/\Gamma(3)$; similarly for any odd prime p. Again, for an odd prime p, the hyperbolic co-unitary group associated with self-dual codes over $\mathbb{F}_p$ that contain the all-ones vector is a quotient of the group $(\mathbb{Z} \times \mathbb{Z}).\mathrm{SL}_2(\mathbb{Z})$ appearing in the theory of Jacobi forms.

Unfortunately, we have so far been unable to construct a general theory that includes all possible cases, and what follows is more intended to be suggestive than definitive. Our dream of obtaining a common framework that simultaneously includes weight enumerators and elliptic, Siegel and Jacobi modular forms has not yet been completely realized.

Nevertheless, the machinery developed here will be used in the following chapter to obtain new results about maximally integral lattices (in §9.7). We also discuss an application to gluing theory for lattices. Since we have only briefly alluded to gluing theory so far, we also include a discussion of gluing theory for codes in §9.6. This section, which can be read independently of the lattice-theoretic sections, will be used in the following three chapters.

There are several difficulties when we come to deal with lattices. One of the first steps in the proof of our main theorems for codes was Theorem 5.1.3, which gave an invariant-theory interpretation for the space of weight enumerators of isotropic codes. For lattices, in contrast, even the analogue of this fact appears to be nontrivial. The difficulties here are two-fold. First, if we allow all isotropic lattices, i.e. even integral lattices, the resulting ring of theta series is only automorphic for the upper triangular subgroup of $SL_2(\mathbb{Z})$. Since this subgroup is not co-compact in $SL_2(\mathbb{R})$, its ring of automorphic forms is not finitely generated (indeed, there are infinitely many linearly independent forms of weight 1 in that ring). Thus we need to restrict our attention to some subset of integral lattices. The second difficulty arises because the most natural way to do that is to insist that the lattice have level dividing some integer $\ell \geq 1$. But in fact taking $\ell > 1$ seems to make things worse. For example, for $\ell \gg 1$ the corresponding modular curve is no longer rational and in fact has arbitrarily large genus. This is in sharp contrast to the code case, where any ring generated by finitely many weight enumerators has an associated variety which is unirational.

This points out the fact that for codes there is only one way to relax self-duality, whereas for lattices we have such a notion for every positive integer ℓ, and even this can be refined by the notion of genus.

Even if we restrict our attention to self-dual lattices there are multiple notions of self-duality. In particular, for every $s \geq 1$, we have the notion of an *s-modular* lattice as defined by Quebbemann [439], [440] (see §9.1.2 below). Problems arise here since for s-modular lattices there may be more than one similarity transformation between the lattice and its dual. This is a serious obstacle to characterizing the theta series of s-modular lattices and defining analogues of the Clifford-Weil groups.

Another source of difficulty is that sometimes the ring spanned by the theta series of the lattices is strictly smaller than the ring of associated modular forms. For instance, it is a classical theorem that the theta series of an even unimodular lattice is a modular form for the group $SL_2(\mathbb{Z})$. On the other hand, the ring of *all* such modular forms has the structure $\mathbb{C}[E_4, E_6]$ where E_4 and E_6 are Eisenstein series of weights 4 and 6, respectively. Since a lattice of dimension N gives rise to a modular form of weight $N/2$ and even unimodular lattices only exist in dimensions that are multiples of 8, we cannot obtain the Eisenstein series E_6 as a linear combination of theta series. However, if we restrict our attention to valid weights, i.e. weights that are multiples of 4, we do indeed obtain the full space of modular forms.

A similar problem would have arisen in the case of Type II binary codes (codes of Type $\rho(2_{II})$ in our new language) had we replaced the Clifford-Weil group by the index 2 subgroup with center Z_4 (instead of Z_8). So we could in principle fix the problem by replacing $SL_2(\mathbb{Z})$ by a central extension $Z_4 \mathsf{Y} SL_2(\mathbb{Z})$, where the Z_4 acts with a character, to force the ring of invariant modular forms to be the subring of the ring of modular forms of $SL_2(\mathbb{Z})$ consisting of the forms of weights divisible by 4. But even this is not quite the true analogue of the Clifford-Weil group for Type II binary codes. Indeed, in the latter case, although the codes only exist in dimensions that are multiples of 8, the associated Clifford-Weil group acts in all dimensions. Thus the "correct" group for even unimodular lattices should really be a central extension $Z_8 . PSL_2(\mathbb{Z})$ together with a canonical multiplier system of weight $1/2$. It is however unclear which central extension to use, let alone how to define the associated multiplier system.

Furthermore, even for the full Siegel modular group $\mathrm{Sp}_{2g}(\mathbb{Z})$ and in dimensions that are multiples of 8, there are cases where the space of Siegel modular forms is not spanned by the higher-order theta series of even unimodular lattices. That is, where the algebra spanned by the higher-order theta series is not integrally closed in its field of fractions. Kohnen and Salvati Manni [332] have shown that there is a modular form of genus 20 and weight 16 that is not in the space spanned by the Siegel theta series of 32-dimensional even unimodular lattices.

Yet another serious difficulty is shown by the following example. Because of this, the main theorems cannot possibly hold in full generality for lattices. Let p be an odd prime and consider a fixed genus of lattices Λ of level p (cf. §9.1.2) and dimension $N \equiv 0 \pmod 4$, $N > 4$. The theta series of Λ is a modular form for $\Gamma_0(p)$ of weight $N/2$ and trivial character. $\Gamma_0(p)$ has two cusps, at ∞ and 0. At ∞ a theta series necessarily takes the value 1 (the number of norm 0 vectors in Λ) and at 0 by the Jacobi identity the value is $(\det \Lambda)^{-1/2}$. On the other hand, the Eisenstein series $E_{N/2}(z)$ and $E_{N/2}(pz)$ are both modular forms for $\Gamma_0(p)$ with trivial character and their cuspidal values are linearly independent. It follows that there is a linear combination of the Eisenstein series that is not a linear combination of theta series. (Incidentally, the linear combination of the Eisenstein series which takes the correct values at the cusps is equal to the average

$$\frac{\sum_\Lambda \frac{1}{\mathrm{Aut}(\Lambda)} \Theta_\Lambda}{\sum_\Lambda \frac{1}{\mathrm{Aut}(\Lambda)}}$$

of the theta series over the genus.)

We begin in §9.1 by defining lattices, various kinds of theta series that can be associated with them, and related notions such as modular lattices, shadows, etc. We then show how this theory fits into our framework of form rings. The topics mentioned in §9.1 have been the subject of a large number of books and papers, and our discussion must necessarily be very incomplete.

In §9.2 we consider form $\mathbb{R}$-algebras. In particular, over the reals there is a natural additional condition to impose, corresponding to positive-definiteness of quadratic forms, and the corresponding form $\mathbb{R}$-algebras can in fact be completely classified. Section 9.3 associates to any positive definite form $\mathbb{R}$-algebra an analogue of the upper half-plane from modular form theory, and §9.4 discusses form orders (certain discrete subrings of form $\mathbb{R}$-algebras) and the corresponding notion of isotropic self-dual lattices.

Section 9.5 applies the above machinery to the important special case of unimodular lattices, and §§9.6, 9.7 discuss gluing theory for codes and lattices respectively. In particular, in §9.7 we associate to any isotropic but not self-dual lattice a *finite* representation of a form order, and thus a corresponding Type of code.

9.1 Lattices and theta series

9.1.1 Preliminary definitions

The main objects of interest in the theory of lattices are lattices in Euclidean space, and unless specified otherwise that is what the word "lattice" will mean in this chapter.

More formally, a *lattice* Λ in a finite-dimensional real vector space V is a discrete and co-compact subgroup of V (cf. [133][1], Cassels [103], Ebeling [164], Kitaoka [312], Martinet [368], Milnor and Husemoller [371], W. Scharlau [475], Serre [480]).

We maintain the convention that we used for codes: vectors are row vectors. If $u \in V$, $v \in V$, $u \cdot v := uv^{tr}$ will denote the standard inner product, and $\mathrm{Norm}(v) := v \cdot v$ is the *norm* of v. If $v_1, v_2, \ldots, v_N$ is a basis for Λ, the matrix M with rows $v_1, v_2, \ldots, v_N$ is a *generator matrix* for Λ, $B := MM^{tr}$ is its *Gram matrix* and $\det \Lambda := \det B$. For a lattice Λ in V, the *dual lattice* is

$$\Lambda^{\#} := \{u \in V \mid u \cdot v \in \mathbb{Z} \text{ for } v \in \Lambda\}.$$

A lattice Λ in $\mathbb{R}^N$ considered as an abstract group is isomorphic to $\mathbb{Z}^N$, and any basis for Λ is a basis for $\mathbb{R}^N$. Conversely any basis for $\mathbb{R}^N$ generates a lattice. We will occasionally discuss lattices without explicit reference to an ambient vector space and simply note that Λ always embeds into $\Lambda \otimes_{\mathbb{Z}} \mathbb{R}$ as a lattice.

Of particular interest are *integral* lattices, which have the property that $u \cdot v \in \mathbb{Z}$ for all $u, v \in \Lambda$, and there is a refined notion of *even* lattice, for which $u \cdot u \in 2\mathbb{Z}$ for all $u \in \Lambda$. These are analogous to the notions of even and doubly-even binary codes (cf. §2.3).

The analogue for lattices of the weight enumerator of a code is the *theta series*

$$\Theta_\Lambda(\tau) := \sum_{v \in \Lambda} e(\,^1\!/_2 v \cdot v)\,, \tag{9.1.1}$$

where $e(\tau) := \exp(2\pi i \tau)$ and $\tau \in \mathbb{C}$, $\Im(\tau) > 0$. Thus $e : \mathbb{C}/\mathbb{Z} \to \mathbb{C}^*$ is a group homomorphism. Similarly, for $u \in V$,

$$\Theta_{\Lambda+u}(\tau) := \sum_{v \in \Lambda+u} e(\,^1\!/_2 v \cdot v)\,.$$

It is traditional[2] to express theta series in terms of the variable $q = e(\tau) = e^{2\pi i \tau}$, when Θ_Λ becomes simply the generating function for the norms (divided

---

[1] It would not be appropriate here to define every term from lattice theory that we will need: we refer to these references, especially [133], for any undefined terms.

[2] It is also traditional (see for instance [133]) to take $q = e^{\pi i \tau}$, especially when working with non-even lattices, as this avoids fractional exponents. We will use $q = e^{2\pi i \tau}$ throughout.

by 2) of the vectors in Λ. Introducing the variable τ turns a formal power series with fractional exponents into a holomorphic function on the upper half-plane ($\Im(\tau) > 0$), which allows one to use the theory of complex functions. A more important reason for using the τ variable is that it simplifies the expression for the relevant group action, even more than using the homogeneous form for the weight enumerators of codes simplifies the MacWilliams identities (as in (2.2.5), for instance).

To be precise, for any N-dimensional lattice Λ, integral or not, we have the *Jacobi identity* (see for instance Ebeling [164], Serre [480])

$$\Theta_{\Lambda^\#}(\tau) \;=\; (\det \Lambda)^{1/2} \, (i/\tau)^{N/2} \, \Theta_\Lambda(-1/\tau) \,.$$

The proof uses the continuous version of the Poisson summation formula, and otherwise is essentially the same as the proof of the MacWilliams identity (cf. Theorem 2.2.4). For an integral lattice we also have

$$\Theta_\Lambda(\tau + 2) \;=\; \Theta_\Lambda(\tau) \,,$$

and for an even lattice

$$\Theta_\Lambda(\tau + 1) \;=\; \Theta_\Lambda(\tau) \,.$$

Thus the theta series of an even unimodular lattice $\Lambda = \Lambda^\#$ is acted on by the group generated by

$$\tau \mapsto \tau + 1, \; \tau \mapsto -\frac{1}{\tau} \,,$$

which is $\mathrm{PSL}_2(\mathbb{Z})$. For the general element $\begin{pmatrix} a & b \\ c & d \end{pmatrix}$, $ad - bc = 1$, of $\mathrm{SL}_2(\mathbb{Z})$,

$$\Theta_\Lambda\!\left(\frac{a\tau + b}{c\tau + d}\right) \;=\; (c\tau + d)^{N/2} \, \Theta_\Lambda(\tau) \,,$$

for an even unimodular lattice Λ.

More generally, if Λ is an even lattice of *even-level* ℓ (that is, if $\sqrt{\ell}\,\Lambda^\#$ is also an even lattice, see §9.1.2), then we have

$$\Theta_\Lambda\!\left(\frac{a\tau + b}{c\tau + d}\right) \;=\; \chi_\Lambda(d)(c\tau + d)^{N/2} \, \Theta_\Lambda(\tau) \,,$$

where $\chi_\Lambda(d)$ is the Legendre symbol

$$\chi_\Lambda(d) \;:=\; \left(\frac{(-1)^{N/2} \det(\Lambda)}{d}\right) \;\in\; \{\pm 1\} \,,$$

for all $\begin{pmatrix} a & b \\ c & d \end{pmatrix} \in \mathrm{SL}_2(\mathbb{Z})$ such that c is a multiple of ℓ. Hence the subgroup

$$\Gamma_0(\ell) \;:=\; \{\begin{pmatrix} a & b \\ c & d \end{pmatrix} \subseteq \mathrm{SL}_2(\mathbb{Z}) \mid \ell \text{ divides } c\} \leq \mathrm{SL}_2(\mathbb{Z})$$

acts on $\Theta_\Lambda(\tau)$ with character χ_Λ (cf. Ebeling [164, Theorem 3.2]).

Example 9.1.1. Let

$$\theta_2(\tau) := \sum_{m=-\infty}^{\infty} q^{1/2(m+1/2)^2} = 2q^{1/8} + 2q^{9/8} + 2q^{25/8} + \cdots , \qquad (9.1.2)$$

$$\theta_3(\tau) := \sum_{m=-\infty}^{\infty} q^{m^2/2} = 1 + 2q^{1/2} + 2q^2 + 2q^{9/2} + \cdots , \qquad (9.1.3)$$

$$\theta_4(\tau) := \sum_{m=-\infty}^{\infty} (-q)^{m^2/2} = 1 - 2q^{1/2} + 2q^2 - 2q^{9/2} + \cdots . \qquad (9.1.4)$$

The first nontrivial unimodular lattice is $\mathbb{Z}$ itself, for which $\Theta_{\mathbb{Z}}(\tau) = \theta_3(\tau)$. The first nontrivial even unimodular lattice is the 8-dimensional root lattice E_8, with

$$\Theta_{E_8}(\tau) = {}^{1}/{}_{2} \left\{ \theta_2(\tau)^8 + \theta_3(\tau)^8 + \theta_4(\tau)^8 \right\}$$
$$= 1 + 240q + 2160q^2 + 6720q^3 + \cdots \qquad (9.1.5)$$

in which the coefficient of q^m is 240 times the sum of the cubes of the divisors of m. The 24-dimensional *Leech lattice* Λ_{24} has theta series

$$\Theta_{\Lambda_{24}}(\tau) = \Theta_{E_8}(\tau)^3 - 720\Delta_{24}(\tau)$$
$$= 1 + 196560q^2 + 16773120q^3 + 398034000q^4 + \cdots \qquad (9.1.6)$$

where

$$\Delta_{24}(\tau) = q \prod_{m=1}^{\infty} (1 - q^m)^{24}$$
$$= q - 24q^2 + 252q^3 - \cdots \qquad (9.1.7)$$

is the Ramanujan cusp form (see [133, Chap. 4] for many other ways of writing these theta series).

The following result is the analogue for lattices of Gleason's theorem 6.4.2.[3]

Theorem 9.1.2. (a) *The theta series of a unimodular lattice is a polynomial in the theta series of the lattices $\mathbb{Z}$ and E_8.*
(b) *The theta series of an even unimodular lattice is a polynomial in the theta series of the E_8 lattice and the Leech lattice Λ_{24}.*

In Part (a), $\Theta_{E_8}(\tau)$ may be replaced by

---

[3] Although this result is certainly "classical", we do not know where it first appeared in print.

$$\Delta_8(\tau) = (\Theta_{\mathbb{Z}}(\tau)^8 - \Theta_{E_8}(\tau))/16$$

$$= q^{1/2} \prod_{m=1}^{\infty} ((1 - q^{(m-1)/2})(1 - q^{2m}))^8$$

$$= q^{1/2} - 8q + 28q^{3/2} - 64q^2 + \cdots \tag{9.1.8}$$

and in Part (b), $\Theta_{\Lambda_{24}}(\tau)$ may be replaced by $\Delta_{24}(\tau)$. The Poincaré series for these two rings coincide with the Molien series (6.3.5) and (6.4.2) occurring in Gleason's theorem.

9.1.2 Modular lattices and Atkin-Lehner involutions

Most of the material in this and the following subsection is taken from the paper [453].

We define (not necessarily even) strongly modular lattices, generalizing Quebbemann's original notion (see [439] and [440]), and the corresponding modular groups (see (9.1.12) and (9.1.13)) under which the theta series of strongly modular lattices are invariant.

The definition of a strongly modular lattice can be given in a quite general context. Let Λ be a lattice in $\mathbb{R}^N$ such that the inner product $u \cdot v \in \mathbb{Q}$ for all $u, v \in \Lambda$, and let Π be a (possibly infinite) set of rational primes. The Π-*dual* $\Lambda^{\#\Pi}$ of Λ consists of the vectors $v \in \Lambda \otimes \mathbb{Q}$ such that $v \cdot \Lambda \subseteq \mathbb{Z}_p$ for $p \in \Pi$ and $v \cdot \Lambda^{\#} \subseteq \mathbb{Z}_p$ for $p \notin \Pi$.

In particular, with Ω the set of all rational primes,

$$\Lambda^{\#\emptyset} = \Lambda, \; \Lambda^{\#\Omega} = \Lambda^{\#}, \; (\Lambda^{\#\dot{\Pi}})^{\#\Pi} = \Lambda,$$

and more generally

$$(\Lambda^{\#\Pi_1})^{\#\Pi_2} = \Lambda^{\#(\Pi_1 \triangle \Pi_2)},$$

where $\triangle$ denotes a symmetric difference. We will also need the notation $\overline{\Pi} := \Omega \setminus \Pi$, and when there is no possibility of confusion we abbreviate $\Pi = \{p\}$ to p. Furthermore,

$$\Lambda^{\#\Pi} \otimes \mathbb{Z}_p = \begin{cases} \Lambda^{\#} \otimes \mathbb{Z}_p, \; p \in \Pi, \\ \Lambda \otimes \mathbb{Z}_p, \; p \notin \Pi. \end{cases}$$

We also define

$$\det_{\Pi}(\Lambda) := (\det \Lambda / \det \Lambda^{\#\Pi})^{1/2}$$
$$= [\Lambda^{\#\Pi} : \Lambda \cap \Lambda^{\#\Pi}] / [\Lambda : \Lambda \cap \Lambda^{\#\Pi}].$$

If $\det \Lambda$ is written as rs where the prime factors of r belong to Π and $\gcd(s, p) = 1$ for all $p \in \Pi$, then $\det_{\Pi}(\Lambda) = r$.

Suppose now that Λ is integral. The *level* of Λ is the smallest number ℓ' such that $\sqrt{\ell'} \, \Lambda^{\#}$ is integral. If Λ is even, the *even-level* of Λ is the smallest number

ℓ such that $\sqrt{\ell}\,\Lambda^{\#}$ is even. The Π-levels ℓ'_{Π} and ℓ_{Π} are defined analogously, replacing $\Lambda^{\#}$ by $\Lambda^{\#_{\Pi}}$.

Quebbemann [440] associates certain Gauss sums with Λ. We do the same, but in a slightly more explicit fashion.

We will use the oddity and p-excess of an integral lattice Λ, as in [133, Chap. 15]. These terms are defined as follows. The discussion is simplified if we enlarge the set of rational primes to include -1. By convention, $\mathbb{Z}_{-1} = \mathbb{Q}_{-1} = \mathbb{R}$. Any rational or p-adic integer A can be written uniquely in the form $A = p^{\alpha}a$, where a is prime to p if $p \neq -1$ or a is positive and $\alpha \in \{0,1\}$ if $p = -1$. Then $p(A) = p^{\alpha}$ is called the p-part of A, and $p'(A) = a$ is the p'-part. A p-adic antisquare is a number of the form $p^{\text{odd}}u_{-}$ if $p \geq 3$, where u_{-} is any p-adic unit that is a quadratic nonresidue mod p, or a number of the form $2^{\text{odd}}u_{\pm 3}$ if $p = 2$, where u_i is any 2-adic unit that is congruent to i (mod 8), for $i = \pm 1, \pm 3$. There are no (-1)-adic antisquares.

The oddity and p-excess are rational invariants, so we may assume the Gram matrix for Λ has been diagonalized and is

$$\text{diag}\{p^{\alpha}a,\ p^{\beta}b,\ p^{\gamma}c, \ldots\}.$$

The p-signature of Λ is then

$$p^{\alpha} + p^{\beta} + p^{\gamma} + \cdots + 4m \quad (p \neq 2),$$
$$a + b + c + \cdots + 4m \quad (p = 2),$$

where m is the number of p-adic antisquares among $p^{\alpha}a,\ p^{\beta}b,\ p^{\gamma}c, \ldots$. Thus the (-1)-signature of Λ is just its ordinary signature, which equals the dimension for positive definite lattices. For $p \neq -1$ the p-signature is only to be regarded as defined modulo 8. The 2-signature, which often behaves specially, is called the oddity of Λ. The p-excess of Λ is defined to be

$$p\text{-signature} - \text{dimension} \quad (p \neq 2),$$
$$\text{dimension} - p\text{-signature} \quad (p = 2).$$

These quantities are related by the "product formula", which in this notation becomes the oddity formula

$$\dim \Lambda + \sum_{p \geq 3} p\text{-excess}(\Lambda) \equiv \text{oddity}(\Lambda) \pmod 8.$$

Let

$$\gamma_2(\Lambda) := \xi^{\text{oddity}(\Lambda)}, \quad \gamma_p(\Lambda) := \xi^{-p\text{-excess}(\Lambda)} \quad (p \geq 3),$$

where $\xi = e^{\pi i/4}$, and define

$$\gamma_{\Pi}(\Lambda) := \prod_{p \in \Pi} \gamma_p(\Lambda).$$

The oddity formula now becomes

$$\gamma_\Omega(\Lambda) = \xi^{\dim \Lambda}.$$

The following lemma shows that $\gamma_{\Pi}(\Lambda)$ agrees with Quebbemann's Gauss sum.

Lemma 9.1.3. *For an even lattice* Λ,

$$\gamma_{\Pi}(\Lambda) = (\det{}_{\Pi} \Lambda)^{-1/2} \sum_{v \in \Lambda^{\#\Pi}/\Lambda} e(1/2 \, v \cdot v).$$

(Compare Definition 5.4.6.)

It is classical (cf. Miyake [372]) that if Λ is a lattice of even-level ℓ, then its theta series Θ_Λ is a modular form for $\Gamma_0(\ell)$ with respect to an appropriate character. Kitaoka [312] describes how a somewhat larger subset of $SL_2(\mathbb{Z})$ acts on Θ_Λ, up to an unspecified constant. Quebbemann [440] has determined this constant, but only for one representative from each coset of $\Gamma_0(\ell)$. The following is a more explicit theorem which includes their results. Here $\Pi(m)$ denotes the set of primes dividing m, and $\left(\frac{m}{n}\right)$ denotes the Kronecker-Jacobi symbol (cf. Cohen [119, p. 28]).

Theorem 9.1.4. *Let* Λ *be an even lattice of even-level* ℓ, *and let* $S = \begin{pmatrix} a & b \\ c & d \end{pmatrix}$ *be any element of* $SL_2(\mathbb{Z})$ *such that cd is a multiple of* ℓ. *Then*

$$\Theta_\Lambda\left(\frac{az+b}{cz+d}\right) = (\det{}_{\Pi(d)} \Lambda)^{-1/2} \chi_{c,d}(\Lambda) \left(\sqrt{cz+d}\right)^{\dim \Lambda} \Theta_{\Lambda^{\#\Pi(d)}}(z), \quad (9.1.9)$$

where in both cases the square root is that with positive real part, and $\chi_{c,d}(\Lambda)$ is equal to

$$\gamma_{\Pi(d)}(\Lambda)^{-1} \left(\frac{c}{\det_{\Pi(d)} \Lambda}\right) \left(\frac{d}{\det_{\Pi(c)} \Lambda}\right)$$

multiplied either by

$$\left(\frac{d}{|c|}\right)^{\dim \Lambda} \left(\frac{\left(\frac{-1}{c}\right)}{\det \Lambda}\right) \xi^{-(c-1)\dim \Lambda}$$

if c is odd, or by

$$\left(\frac{c}{d}\right)^{\dim \Lambda} \left(\frac{\left(\frac{-1}{d}\right)}{\det \Lambda}\right) \xi^{(d-1)\dim \Lambda} \qquad (9.1.10)$$

if c is even.

See [453] for the proof. For a sketch of a different proof see Theorem 9.1.19 below.

Remark 9.1.5. (1) There is an apparent inconsistency in (9.1.9). Since

$$\Theta_{\Lambda^{\#\Pi(d)}}\left(z + \frac{\ell}{\gcd(c,\ell)}\right) = \Theta_{\Lambda^{\#\Pi(d)}}(z),$$

$\chi_{c,d}(\Lambda)$ must be periodic in d of period $c\ell/\gcd(c,\ell)$. For c odd or $c \equiv 0$ (mod 8) this is manifestly true, but otherwise (9.1.10) appears to have the wrong period. For instance, for $c \equiv 4$ (mod 8),

$$\chi_{c,d+(c\ell/\gcd(c,\ell))}(\Lambda) = (-1)^{\lambda+\dim \Lambda} \chi_{c,d}(\Lambda),$$

where $\lambda = \log_2(\det_2 \Lambda)$. However, since $\ell | cd$, it follows that in the 2-adic Jordan decomposition of Λ the forms of levels 1 and 4 are both Type II and so have even dimension. This implies that $\lambda \equiv \dim \Lambda$ (mod 2).

Similarly, for $c \equiv \pm 2$ (mod 8), the correct period is restored by the identities

$$\lambda \equiv \dim \Lambda \equiv 0 \pmod 2, \quad \left(\frac{-1}{\det \Lambda}\right) = (-1)^{(\dim \Lambda)/2}.$$

If both c and d are odd (so Λ has odd even-level), then similar reasoning allows us to simplify $\chi_{c,d}(\Lambda)$ to

$$\chi_{c,d}(\Lambda) = \gamma_{\Pi(d)}(\Lambda)^{-1}\left(\frac{c}{\det_{\Pi(d)} \Lambda}\right)\left(\frac{d}{\det_{\Pi(c)} \Lambda}\right).$$

(2) When ℓ divides c, the usual formula (cf. Miyake [372, Theorem 4.9.3]) for the action of $\Gamma_0(\ell)$ on the theta series of lattices of even-level ℓ can be recovered with the help of the identity

$$\xi^{(t-1)} = \epsilon_t\left(\frac{-2}{t}\right) = \epsilon_t^{-1}\left(\frac{2}{t}\right),$$

for odd t, where $\epsilon_t = 1$ if $t \equiv 1$ (mod 4) and $\epsilon_t = i$ if $t \equiv 3$ (mod 4).

If Λ is any integral lattice of level ℓ, $\sqrt{2}\Lambda$ is an even lattice of even-level dividing 4ℓ. We can apply Theorem 9.1.4 to obtain:

Corollary 9.1.6. *Let Λ be an integral lattice of level ℓ, and let $S = \begin{pmatrix} a & b \\ c & d \end{pmatrix}$ be any element of $SL_2(\mathbb{Z})$ such that cd is a multiple of 2ℓ. Then (9.1.9) holds if either d is odd and b is even, or c is odd and a is even.*

A *modularity* σ of an integral lattice Λ is a similarity mapping $\Lambda^{\#\Pi}$ to Λ for some set of primes Π. We say that σ has[4] *scale* s (or is an s-*modularity*) if σ multiplies norms by s. Given σ, Π can be taken to be the set of primes dividing s. A 1-modularity is just an automorphism of Λ.

---

[4] The term "level" is often used here instead of "scale", but "level" has too many meanings already.

Corollary 9.1.7. *Suppose Λ has even-level ℓ and admits an s-modularity. Then for any matrix*

$$W_s := s^{-1/2} \begin{pmatrix} sa & b \\ sc & d \end{pmatrix} \tag{9.1.11}$$

of determinant 1, with d a multiple of s and sc a multiple of ℓ, we have

$$\Theta_\Lambda\left(\frac{sa\tau + b}{sc\tau + d}\right) = \chi_{c,d}(\Lambda)\,(s^{1/2}c\tau + s^{-1/2}d)^{(\dim \Lambda)/2}\,\Theta_\Lambda(\tau).$$

The matrix W_s in (9.1.11) is called an *Atkin-Lehner involution* [17] of scale s. The next result combines known properties of these involutions with a slight generalization of a result from [379] on modularities.

Theorem 9.1.8. *If W_{s_1} and W_{s_2} are Atkin-Lehner involutions then $W_{s_1}W_{s_2}$ is an Atkin-Lehner involution of scale $s_1 s_2 / \gcd(s_1, s_2)^2$. Moreover, W_s^{-1} is an Atkin-Lehner involution of scale s. If σ_1 is an s_1-modularity and σ_2 is an s_2-modularity then $\sigma_1\sigma_2/\gcd(s_1, s_2)$ is a modularity of scale $s_1 s_2 / \gcd(s_1, s_2)^2$. Moreover, if σ is an s-modularity then so is $s\sigma^{-1}$.*

It follows from Theorem 9.1.8 that the number of distinct scales of modularities of a lattice is a power of 2, and indeed the scales have a natural elementary abelian 2-group structure. Moreover, the total number of modularities is equal to the number of scales of modularity times $|\operatorname{Aut}\Lambda|$.

We will say that an integral lattice Λ is $\{s_1, s_2, \ldots\}$-*modular* if it has modularities of scales $s_1, s_2, \ldots$. Two special cases warrant a shorthand notation. (a) Λ is ℓ-*modular* if its level divides ℓ and Λ is $\{1, \ell\}$-modular. (b) Λ is *strongly ℓ-modular* if its level divides ℓ and Λ is $\{s : s\|\ell\}$-modular, where $a\|b$ means $a|b$ and $\gcd(a, b/a) = 1$.

Corollary 9.1.7 states that if Λ is an even $\{s_1, s_2, \ldots\}$-modular lattice of even-level ℓ, then its theta series is an automorphic form for the group $\Gamma_0(\ell)^{+\{s_1, s_2, \ldots\}}$, i.e. the group generated by $\Gamma_0(\ell)$ together with all its Atkin-Lehner involutions of levels $s_1, s_2, \ldots$. For ease in discussing strongly modular lattices we abbreviate $\Gamma_0(\ell)^{+\{s:s\|\ell\}}$ to $\Gamma_0(\ell)^+$.

If Γ is any group of 2×2 matrices, $^1\!/_2\,\Gamma$ will denote the group

$$\left\{ \begin{pmatrix} a & 2b \\ c/2 & d \end{pmatrix} : \begin{pmatrix} a & b \\ c & d \end{pmatrix} \in \Gamma \right\}.$$

Corollary 9.1.6 implies that if Λ is an $\{s_1, s_2, \ldots\}$-modular lattice of level ℓ, ℓ odd, then its theta series is an automorphic form for the group

$$^1\!/_2\,\Gamma_0(4\ell)^{+\{4, s_1, s_2, \ldots\}}. \tag{9.1.12}$$

The initial 4 arises because $\sqrt{2}\Lambda$ has an obvious 4-modularity. As a special case, the theta series of any lattice of odd level is an automorphic form for $^1\!/_2\,\Gamma_0(4\ell)^{+\{4\}}$ (a subgroup of $\Gamma_0(\ell)$). If ℓ is even, (9.1.12) must be replaced by

$$^1\!/_2\,\Gamma_0(4\ell)^{+\{4e_1, 4e_2, \ldots, d_1, d_2, \ldots\}}, \tag{9.1.13}$$

where $e_1, e_2, \ldots$ are the even s_i's and $d_1, d_2, \ldots$ are the odd s_i's.

9.1.3 Shadows

Shadows of codes were defined in §1.12. There is an analogous notion for lattices.

Let Λ be an integral lattice, or more generally a 2-*integral* lattice (i.e. $u \cdot v \in \mathbb{Z}_2$ for all $u, v \in \Lambda$), and set $\Lambda_0 = \{u \in \Lambda : u \cdot u \in 2\mathbb{Z}_2\}$. If Λ is even, $\Lambda = \Lambda_0$; otherwise Λ_0 is a sublattice of index 2. Λ_0 is called the *even sublattice* of Λ.

Following Conway and Sloane [129], [130], we define the *shadow* $S(\Lambda)$ to be $\Lambda^{\#}$ if Λ is even, $(\Lambda_0)^{\#} \setminus \Lambda^{\#}$ if Λ is odd. Equivalently,

$$S(\Lambda) := \{v \in \Lambda \otimes \mathbb{Q} \; : \; 2u \cdot v \equiv u \cdot u \; (\mathrm{mod} \; 2\mathbb{Z}_2) \quad \text{for all} \quad u \in \Lambda\} \, .$$

Hence the shadow $S(\Lambda)$ consists of the characteristic vectors of Λ, multiplied by 2. The theta series of $S(\Lambda)$ is given by

$$\Theta_{S(\Lambda)}(z) = (\det \Lambda)^{1/2} \left(\frac{\xi}{\sqrt{z}} \right)^{\dim \Lambda} \Theta_\Lambda \left(1 - \frac{1}{z} \right) . \tag{9.1.14}$$

Example 9.1.9. For example, $\mathbb{Z}^N$ has theta series $\theta_3(\tau)^N$, and its shadow $\mathbb{Z}^N + (1/2, 1/2, \ldots, 1/2)$ has theta series $\theta_2(\tau)^N$.

We also define the *Π-shadow* $S_\Pi(\Lambda)$ as follows: if $2 \in \Pi$,

$$S_\Pi(\Lambda) := S(\Lambda^{\#_\Pi}) \, ,$$

and if $2 \notin \Pi$,

$$S_\Pi(\Lambda) := \sqrt{\ell_2'} \, S(\sqrt{\ell_2'} \, \Lambda^{\#_\Pi}) \, ,$$

where ℓ_2' is the 2-*level* of Λ (the smallest number ℓ_2' such that $\sqrt{\ell_2'} \Lambda^{\#}$ is 2-integral). The Π-shadow is a coset of the Π-dual $\Lambda^{\#_\Pi}$, and in fact $v \pm w \in \Lambda^{\#_\Pi}$ for $v, w \in S_\Pi(\Lambda)$. In particular, $S_\Omega(\Lambda) = S(\Lambda)$ is a coset of $\Lambda^{\#}$, and $S_\emptyset(\Lambda)$ is a coset of Λ. The theta series of $S_\Pi(\Lambda)$ may be computed from Corollary 9.1.6 and (9.1.14).

It is clear from the definition of $S(\Lambda)$ that any two vectors in the same coset of Λ in $S(\Lambda)$ have the same norm modulo $2\mathbb{Z}_2$. If Λ has odd determinant it is possible to say more.

Theorem 9.1.10. *Let Λ be a 2-integral lattice of odd determinant and let Π be a set of primes. Then every vector in $S_\Pi(\Lambda)$ has norm $\equiv$ (oddity Λ)/4 (mod $2\mathbb{Z}_2$).*

Remark 9.1.11. (1) For unimodular lattices, Theorem 9.1.10 together with the product formula implies that for $v \in S(\Lambda)$, $v \cdot v \equiv (\dim \Lambda)/4$ (mod 2), a result that has been rediscovered several times (see van der Blij [54], Braun [81], Conway and Sloane [129], Milnor and Husemoller [371] and Serre [479]). (2) Reference [453] gives three proofs of Theorem 9.1.10. The third proof can be used to extend Lemma 9.1.3 to integral lattices, since it shows that

$$\gamma_2(\Lambda) = (\det_2 \Lambda)^{-1/2} \sum_{v \in S_2(\Lambda)/\Lambda} e(1/2\, v \cdot v)$$

(compare Remark 5.4.11).

9.1.4 Jacobi forms

In the case of codes we were often able to force particular vectors into our codes. For lattices this gives rise to the notion of a *Jacobi form* (cf. Eichler and Zagier [166]), defined as follows: for a fixed vector $v_0 \in \Lambda \otimes \mathbb{Q}$,

$$\Theta_{\Lambda,v_0}(\tau, z) := \sum_{v \in \Lambda} e(1/2\,\tau v \cdot v + v \cdot v_0 z),$$

for $\Im(\tau) > 0, \Im(z) > 0$. Note that this incorporates the notion of the theta series of a coset of a lattice, since

$$\Theta_{v_0 + \Lambda}(\tau) = e(1/2\,\tau v_0 \cdot v_0)\,\Theta_{\Lambda,v_0}(\tau, \tau)\,.$$

A similar comment applies to the Jacobi-Siegel theta series defined below.

The action of $SL_2(\mathbb{R})$ on τ extends to the following action of the group $\mathbb{R}^2 \rtimes SL_2(\mathbb{R})$ on τ and z:

$$\begin{pmatrix} a & b \\ c & d \end{pmatrix} \in SL_2(\mathbb{R}) : (\tau, z) \mapsto (\frac{a\tau + b}{c\tau + d}, \frac{z}{c\tau + d}),$$

$$(s, t) \in \mathbb{R}^2 : (\tau, z) \mapsto (\tau, z + s\tau + t)\,.$$

If Λ is an even unimodular lattice and $v_0 \in \Lambda$, the Jacobi theta series satisfies a transformation law under the subgroup $\mathbb{Z}^2 \rtimes SL_2(\mathbb{Z})$.

Theorem 9.1.12. *Let Λ be an even unimodular lattice of dimension N, and fix $v_0 \in \Lambda$. Then for any pair of elements $\begin{pmatrix} a & b \\ c & d \end{pmatrix} \in SL_2(\mathbb{Z})$, $(s, t) \in \mathbb{Z}^2$,*

$$\Theta_{\Lambda,v_0}(\frac{a\tau + b}{c\tau + d}, \frac{z}{c\tau + d}) =$$
$$(c\tau + d)^{N/2} e(v_0 \cdot v_0(1/2\,\tau s^2 + zs - z^2 \frac{c}{c\tau + d}))\,\Theta_{\Lambda,v_0}(\tau, z + s\tau + t)\,.$$

9.1.5 Siegel theta series

The analogue for lattices of the higher genus weight enumerator is the Siegel theta series (cf. Siegel [494]; Freitag [176], Klingen [317]). Let g be an integer ≥ 1. For binary codes, a genus-g weight enumerator is effectively parameterized by the sizes of the intersections among g-tuples of codewords. For lattices, the analogue is the Gram matrix of a g-tuple of lattice vectors.

This requires us to replace the complex variable τ, $\Im(\tau) > 0$, in (9.1.1) by a matrix variable T in the *"Siegel half-plane"* $\mathfrak{S}_g$, which consists of complex $g \times g$ symmetric matrices $T = X + iY$ with Y positive definite. $\mathfrak{S}_g$ is acted on by the symplectic group[5] $\mathrm{Sp}_{2g}(\mathbb{R}) :=$

$$\left\{ \begin{pmatrix} A & B \\ C & D \end{pmatrix} \in \mathrm{Mat}_{2g}(\mathbb{R}) \mid AB^{tr} \text{ and } CD^{tr} \text{ symmetric}, AD^{tr} - BC^{tr} = I \right\},$$

via $T \mapsto (AT + B)(CT + D)^{-1}$.

Then the *Siegel theta series* of genus g for an N-dimensional lattice Λ is

$$\Theta_\Lambda^{(g)}(T) := \sum_{v \in \Lambda^g} e(\tfrac{1}{2} \mathrm{Tr}(Tvv^{tr})),$$

where $v \in \Lambda^g$ is to be interpreted as a $g \times N$ matrix. If Λ is integral, then $\Theta_\Lambda^{(g)}(T)$ has a Fourier expansion

$$\Theta_\Lambda^{(g)}(T) = \sum_M a(M) e(\tfrac{1}{2} \mathrm{Tr}(TM)),$$

where M runs through positive semi-definite integral symmetric $g \times g$ matrices (cf. Freitag [176, p. 44]).

Siegel theta series of genus 1 are the ordinary theta series of §9.1.1. For genus $g \geq 2$ they are much harder to write down.

Example 9.1.13. The genus-2 theta series for the square lattice $\mathbb{Z}^2$ is

$$\Theta_{\mathbb{Z}^2}^{(2)}(T) = \sum_v e(\tfrac{1}{2} \mathrm{Tr}(Tvv^{tr})),$$

where the sum is over all matrices $v \in \mathrm{Mat}_2(\mathbb{Z})$. Its Fourier expansion is

$$\Theta_{\mathbb{Z}^2}^{(2)}(T) = \sum_M a(M) e(\tfrac{1}{2} \mathrm{Tr}(TM)),$$

where now the sum is over all positive semi-definite integral symmetric matrices $M \in \mathrm{Mat}_2(\mathbb{Z})$. Here $a(M)$ is the number of ways to choose a pair of vectors in $\mathbb{Z}^2$ with Gram matrix M, so we have

$$a\left(\begin{pmatrix} 0 & 0 \\ 0 & 0 \end{pmatrix} \right) = 1, \ a\left(\begin{pmatrix} 1 & 0 \\ 0 & 0 \end{pmatrix} \right) = 4, \ a\left(\begin{pmatrix} 1 & 0 \\ 0 & 1 \end{pmatrix} \right) = 8, \ a\left(\begin{pmatrix} 2 & 1 \\ 1 & 2 \end{pmatrix} \right) = 0,$$

etc. Note that $a(M) = a(U^{tr} M U)$ for all $U \in \mathrm{GL}_2(\mathbb{Z})$.

Again, these theta series transform nicely when Λ is an even unimodular lattice. The Siegel theta series of genus g for an even unimodular lattice of dimension N is a weight $N/2$ element of the ring $\mathcal{M}(\mathrm{Sp}_{2g}(\mathbb{Z}))$ of Siegel modular forms for $\mathrm{Sp}_{2g}(\mathbb{Z})$. This means that, assuming $g \geq 2$,

---

[5] This group is also called $\mathrm{Sp}_g(\mathbb{R})$ in the literature, and $\mathrm{Sp}_{2g}(\mathbb{Z})$ is often called Γ_g.

$$\Theta_\Lambda^{(g)}((AT+B)(CT+D)^{-1}) = (\det(CT+D))^{N/2}\,\Theta_\Lambda^{(g)}(T),$$

for $\begin{pmatrix} A & B \\ C & D \end{pmatrix} \in \mathrm{Sp}_{2g}(\mathbb{R})$.

More generally, the Siegel theta series of genus g for a unimodular lattice of dimension N is a weight $N/2$ element (with an appropriate character) in the ring $\mathcal{M}(\Gamma_{\vartheta,g})$ of Siegel modular forms for the so-called "theta-group" $\Gamma_{\vartheta,g}$.

This is the group $\Gamma_{\vartheta,g} \le \mathrm{Sp}_{2g}(\mathbb{Z})$, consisting of the elements of $\mathrm{Sp}_{2g}(\mathbb{Z})$ for which AB^{tr} and CD^{tr} are even, as in Definition 1.10.4 (cf. Freitag [176, p. 300], Mumford [375, p. 189]). Modulo 2, $\Gamma_{\vartheta,g}$ is an orthogonal group rather than a symplectic group.

There is a well-known connection between complete weight enumerators of self-dual codes over $\mathbb{F}_p$ for primes p and Siegel theta series of unimodular lattices. "Construction A" associates an N-dimensional unimodular lattice Λ_C with any code $C = C^\perp \le \mathbb{F}_p^N$. (This terminology was first introduced in [344]. See for example [133] for more about this and related constructions.) For a code $C \le \mathbb{F}_p^N$, let

$$\Lambda_C = \{(\lambda_1,\ldots,\lambda_N) \in \mathbb{Z}^N \mid (\lambda_1,\ldots,\lambda_N) \pmod{p} \in C\}$$

be the full preimage of C under the epimorphism $\mathbb{Z}^N \to (\mathbb{Z}/p\mathbb{Z})^N$. With respect to the scalar product $(,)_p : (x,y)_p := \frac{1}{p}\sum_{i=1}^N x_i y_i$ on $\mathbb{R}^N$, the dual lattice is

$$\Lambda_C^\# = \Lambda_{C^\perp},$$

which is the lattice associated to the dual code $C^\perp = \{x \in \mathbb{F}_p^N \mid \sum_{i=1}^N x_i c_i = 0 \text{ for all } c \in C\}$. In particular Λ_C is unimodular if and only if C is self-dual. If $p > 2$ then Λ_C is an odd lattice. If $p = 2$ then Λ_C is even if and only if the code C is doubly-even.

For any $x \in \mathbb{F}_p^{g\times 1}$ and $T \in \mathfrak{S}_g$ define

$$\vartheta_x(T) := \sum_{\lambda \in \mathbb{Z}^{g\times 1},\, \lambda \equiv x \bmod p} \exp(\frac{1}{p}\pi i \lambda^{tr} T\lambda).$$

Then we have:

Theorem 9.1.14. (Compare [133, Chap. 7,Theorem 3], Ebeling [164, Theorem 5.3], Herrmann [263], Runge [464].)

$$\Theta_{\Lambda_C}^{(g)}(T) = \mathrm{cwe}_g(C)(\vartheta_x(T), x \in \mathbb{F}_p^g),$$

Since the invariant rings of the associated Clifford-Weil groups are spanned by the complete weight enumerators of codes, we can define a ring homomorphism from the invariant ring of the Clifford-Weil group into the ring of Siegel-modular forms for the theta group $\Gamma_{\vartheta,g}$.

Definition 9.1.15. Let p be a prime, and let $\Theta_g : \mathrm{Inv}(\mathcal{C}_g(\rho(p^{\mathrm{E}}))) \to \mathcal{M}(\Gamma_{\vartheta,g})$ be defined by

$$\Theta_g(\mathrm{cwe}_g(C)) = \mathrm{cwe}_g(C)(\vartheta_x(T)), x \in \mathbb{F}_p^g ,$$

where C is a code of Type p^{E}.

Remark 9.1.16. The map Θ_g is a ring homomorphism. For Type 2_{II} codes we get $\Theta_g : \mathrm{Inv}(\mathcal{C}_g(\rho(2_{\mathrm{II}}))) \to \mathcal{M}(\mathrm{Sp}_{2g}(\mathbb{Z}))$. In fact, for Type 2_{II} codes and $g = 2$, Ozeki [394] showed that the image of this map spans $\mathcal{M}(\mathrm{Sp}_4(\mathbb{Z}))$. More precisely, we have the following analogue of Part (b) of Gleason's Theorem 6.4.2.

Theorem 9.1.17. (*i*) (Ozeki [394], Duke [161], based on the work of Igusa [289].) *The graded ring spanned by the genus-2 theta series of even unimodular lattices has Poincaré series* (6.4.5), *and as generators for this ring we may take the genus-2 theta series of the lattices obtained by applying Construction A to the codes mentioned in* (6.4.6).

(*ii*) (Runge [461, Theorem 6.2]) *Let $f(t)$ be the Molien series of $\mathcal{X}_3$ as given in Example 6.4.1(3). Then again Construction A yields a surjection onto genus-3 modular forms of weight divisible by 4, as described in Theorem 9.1.14. The theta-constants ϑ_x of genus 3 satisfy a homogeneous relation of degree 16. Therefore the graded ring spanned by the genus-3 theta series of even unimodular lattices has Poincaré series*

$$f(t) \cdot (1 - t^{16}) ,$$

and as generators for this ring we may take the genus-3 theta series of the lattices obtained by applying Construction A to the appropriate codes of Type 2_{II}.

Research Problem 9.1.18. *Find an analogue of Part (a) of Theorem 6.4.2.*

Note that the theta-group acts on the theta series of unimodular lattices with a character taking values in the set of eighth roots of unity (see for instance Freitag [176, Anhang II]). This makes it difficult to relate the ring spanned by the theta series to an invariant ring of a finite group. A plausible conjecture is that the graded ring spanned by the genus-2 theta series of odd unimodular lattices has Poincaré series

$$\frac{1 + t^{18}}{(1 - t)(1 - t^8)(1 - t^{12})(1 - t^{24})} \qquad [A097913]$$

(compare (6.3.8)). We also conjecture that as generators for this ring we may take the genus-2 theta series of the lattices obtained by applying Construction A to the codes mentioned in (6.3.13), except that the theta series for the dimension 2 lattice is to be replaced by the genus-2 theta series for the lattice $\mathbb{Z}$, which is

$$\Theta_{\mathbb{Z}}^{(2)}(T) \;=\; 1 \;+\; \sum_v e\!\left(\tfrac{1}{2}\,\mathrm{Tr}(Tvv^{tr})\right),$$

where v runs through all rank-1 matrices in $\mathrm{Mat}_2(\mathbb{Z})$ (an analogue of θ_3).

Concerning modular forms for the theta-group without a character, Runge showed that the ring of modular forms of even weight is the image of the invariant ring of a group of order 4608 strongly related to $\mathcal{C}_2$ under the mapping Θ_2 (see [464, §7]). His conjecture about the structure of the full ring of modular forms for $\Gamma_{\vartheta,2}$ was later proved by Kogiso and Tsushima [333].

Just as with multiweight enumerators, a Siegel theta series can be viewed as the theta series of a lattice over $\mathrm{Mat}_g(\mathbb{Z})$ (such a lattice necessarily being of the form Λ^g for some $\mathbb{Z}$-lattice Λ).

Jacobi-Siegel theta series and Riemann theta functions

The Siegel theta series combines well with Jacobi forms: in general, given a lattice Λ and a collection of vectors $v_0 \in \Lambda^k$, one can define the *Jacobi-Siegel theta series*

$$\Theta_{\Lambda,v_0}^{(g)}(T,Z) \;:=\; \sum_{v \in \Lambda^g} e\!\left(\tfrac{1}{2}\,\mathrm{Tr}(Tvv^{tr}) + \mathrm{Tr}(v_0 v^{tr} Z)\right),$$

where $T \in \mathfrak{S}_g$, $Z \in \mathrm{Mat}_{g \times k}(\mathbb{C})$.

As one might expect, the Jacobi-Siegel theta series of an even unimodular lattice is acted on by the group $(\mathbb{Z}^{2g})^k \rtimes \mathrm{Sp}_{2g}(\mathbb{Z})$.

A particularly important example is the case $\Lambda = \mathbb{Z}, k = 1, v_0 = 1$, in which case the Jacobi-Siegel theta series becomes what is called the *Riemann theta function*:

$$\Theta^{(g)}(T,Z) \;:=\; \sum_{v \in \mathbb{Z}^g} e\!\left(\tfrac{1}{2}\,v^{tr} T v + Z^{tr} v\right)$$

(Griffiths and Harris [205, p. 320], Mumford [375, Chapter II]). Note that in this sum v is a column vector. $\Theta^{(g)}(T,Z)$ is acted on by a subgroup of $\mathbb{R}^{2g} \rtimes \mathrm{Sp}_{2g}(\mathbb{R})$ isomorphic to a nonsplit extension $\mathbb{Z}^{2g}.\mathrm{Sp}_{2g}(\mathbb{Z})$.

Theorem 9.1.19. *For* $\begin{pmatrix} A & B \\ C & D \end{pmatrix} \in \mathrm{Sp}_{2g}(\mathbb{Z})$, $V, W \in \tfrac{1}{2}\mathbb{Z}^g$ *such that*

$$V_i \in \tfrac{1}{2}\,(C^{tr}A)_{ii} + \mathbb{Z}, \quad W_i \in \tfrac{1}{2}\,(D^{tr}B)_{ii} + \mathbb{Z},$$

the Riemann theta function satisfies the transformation law

$$\Theta^{(g)}\!\left((AT+B)(CT+D)^{-1}, (CT+D)^{-t}Z\right) =$$
$$\chi\,e\!\left(\tfrac{1}{2}\,V^{tr}TV + Z^{tr}V + \tfrac{1}{2}\,Z^{tr}(CT+D)^{-1}CZ\right)\Theta^{(g)}(T, Z+TV+W)$$
$$(9.1.15)$$

where $\chi^2/\det(CT + D)$ is a fourth root of unity. If $\det(C) \neq 0$, then $\chi =$

$$\det(\sqrt{-i(T + C^{-1}D)}) \sum_{v_0 \in C^{-1}\mathbb{Z}^g/\mathbb{Z}^g} e(\tfrac{1}{2}v_0^{tr}C^{tr}Av_0 - V^{tr}v_0 + \tfrac{1}{2}V^{tr}C^{-1}DV),$$

where for $T \in \mathfrak{S}_g$, $\sqrt{-iT}$ is the holomorphic choice of square root such that $\sqrt{I} = I$.

Proof. (Sketch.) For the first claim, we observe that it suffices to prove that such a transformation law holds for a set of generators of the group $\mathbb{Z}^{2g}.\mathrm{Sp}_{2g}(\mathbb{Z})$, and that it is compatible with the multiplication in that group. This is fairly straightforward; see, for instance Mumford [375] (which considers only the intersection of this group with $\mathrm{Sp}_{2g}(\mathbb{R})$, but the same arguments carry over). In particular, we note that the following transformations generate $\mathbb{Z}^{2g}.\mathrm{Sp}_{2g}(\mathbb{Z})$:

$$(T, Z) \mapsto (ATA^{tr}, A^{tr}Z) \text{ for } A \in \mathrm{GL}_g(\mathbb{Z}),$$

$$(T, Z) \mapsto (T + B, Z + V) \text{ for } B \in \mathrm{Sym}_g(\mathbb{Z}), V_i \in \tfrac{1}{2}B_{ii} + \mathbb{Z}, \text{ and}$$

$$(T, Z) \mapsto (-T^{-1}, -T^{-tr}Z).$$

For these generators we have

$$\Theta^{(g)}(ATA^{tr}, A^{tr}Z) = \Theta^{(g)}(T, Z),$$

$$\Theta^{(g)}(T + B, Z + V) = \Theta^{(g)}(T, Z),$$

and

$$\Theta^{(g)}(-T^{-1}, -T^{-tr}Z) = \det(\sqrt{-iT})\Theta^{(g)}(T, Z). \qquad (9.1.16)$$

The latter transformation is known as the *Jacobi identity*. The proof of (9.1.16) involves Fourier theory via the Poisson summation formula, and in particular requires the Fourier transform of a general Gaussian function. This explains the choice of the square root, since, for T in the Siegel upper half-plane,

$$\int e(vTv^{tr}/2)dv = \det(\sqrt{-iT})^{-1}. \qquad (9.1.17)$$

(To see this last equality, note that both sides are invariant under orthogonal changes of bases, and for positive semi-definite symmetric real matrices A we have

$$\int e(ivAv^{tr}/2)dv = \det(\sqrt{A})^{-1}.$$

Hence both sides of (9.1.17) agree for $T = iA$ imaginary and are holomorphic for T in the half-plane, and so must agree everywhere.)

For the second claim, we can write (following Shimura [483]—compare also the proof of Theorem 5.2.9):

$$\begin{pmatrix} A & B \\ C & D \end{pmatrix} = \begin{pmatrix} C^{-tr} & 0 \\ 0 & C \end{pmatrix} \begin{pmatrix} I & C^{tr}A \\ 0 & I \end{pmatrix} \begin{pmatrix} 0 & -I \\ I & 0 \end{pmatrix} \begin{pmatrix} I & C^{-1}D \\ 0 & I \end{pmatrix}. \qquad (9.1.18)$$

If $C \in \mathrm{GL}_g(\mathbb{Z})$, these are all in $\mathrm{Sp}_{2g}(\mathbb{Z})$, and the result is trivial; more generally, the first matrix has the effect of replacing $\Theta^{(g)}$ by a sum of theta functions:

$$\Theta^{(g)}(C^{-tr}TC^{-1}, C^{-tr}Z) = \sum_{v \in C^{-1}\mathbb{Z}^g} e(\tfrac{1}{2} v^{tr} T v + Z^{tr} v)$$
$$= \sum_{v_0 \in C^{-1}\mathbb{Z}^g / \mathbb{Z}^g} e(\tfrac{1}{2} v_0^{tr} T v_0 + Z^{tr} v_0) \, \Theta^{(g)}(T, Z + v_0).$$

The next two matrices then transform the theta function; after applying the remaining matrix, we can switch the order of summation and shift v_0 to separate the sums. The claim follows. $\qquad\square$

Remark. The sum

$$\sum_{v_0 \in C^{-1}\mathbb{Z}^g / \mathbb{Z}^g} e(\tfrac{1}{2} v_0^{tr} C^{tr} A v_0 - V^{tr} v_0 + \tfrac{1}{2} V^{tr} C^{-1} D V)$$

that appears above is essentially just a Gauss sum; in particular, it is of the form $|\det(C)|^{1/2}\zeta$ with $\zeta^8 = 1$, making the two claims of the theorem consistent.

In a certain sense, the Riemann theta function is the most general theta series, as any other Jacobi-Siegel theta series can be obtained as a special case:

Theorem 9.1.20. (Cf. Andrianov [9, Lemma 1.3.12] for the Siegel case.) *Let Λ be an N-dimensional lattice (not necessarily integral), and let $v_0 \in \Lambda^g$ be a collection of vectors. Let B be a generator matrix for Λ, with corresponding Gram matrix BB^{tr}. Then*

$$\Theta^{(g)}_{\Lambda, v_0}(T, Z) = \Theta^{(gN)}(T \otimes (BB^{tr}), \phi(Z v_0 B^{tr})),$$

where ϕ is the natural isometry, $\phi : \mathrm{Hom}(\mathbb{C}^N, \mathbb{C}^g) \to \mathbb{C}^g \otimes \mathbb{C}^N$.

Corollary 9.1.21. *Let Λ be an N-dimensional lattice with Gram matrix G. Then the ordinary theta series of Λ satisfies*

$$\Theta_\Lambda(\tau) = \Theta^{(N)}(\tau G, 0).$$

Thus for instance Theorem 9.1.4 is a special case of Theorem 9.1.19.

Proof. Fix Λ and $\begin{pmatrix} a & b \\ c & d \end{pmatrix}$ as in Theorem 9.1.4, let G and G' be Gram matrices for Λ and $\Lambda^{\#\Pi(d)}$, respectively, and assume $c \neq 0$. Then if we let C be a change of basis matrix from $\Lambda^{\#\Pi(d)}$ to its sublattice $c\Lambda^{\#}$, we find that

$$A = (a/c)GC, \quad B = (b/c)GCG', \quad D = (d/c)CG', \quad \text{and} \quad CG'C^{tr} = c^2 G^{-1},$$

and that

$$\begin{pmatrix} A & B \\ C & D \end{pmatrix}$$

is in the theta-group; the result follows readily.   □

If we relax the evenness constraints, the matrix need not be in the theta-group, so the resulting theta series can have $Z \neq 0$; this is how the shadow theta series arises.

In fact, it can be shown that if any transformation of the form

$$\Theta_{\Lambda_1}\left(\frac{az+b}{cz+d}\right) \propto \Theta_{\Lambda_2}(z),$$

where Λ_1 and Λ_2 are rational lattices, arises by specializing the transformation law for the Riemann theta function, then Λ_2 is similar to a partial dual of Λ_1. In other words, Theorem 9.1.4 is the most general result that can be so obtained.

Riemann theta functions with Harmonic coefficients

For completeness we add some remarks about theta functions with harmonic coefficients (see Mumford [376, §9]). Ordinary theta series with harmonic coefficients play an important role in the classification of even unimodular lattices, for instance in dimension 24 (see Venkov [533]). We discuss only the analogue for the Riemann theta function, but as we have already seen, it is straightforward to derive the ordinary analogues from this.

Lemma 9.1.22. *Let A be an $n \times n$ symmetric matrix, let B be a length n vector, and let P be an n-variable power series. Then*

$$P(\partial_z) \exp(\tfrac{1}{2}\, zAz^{tr} + Bz^{tr})$$
$$= \exp(\tfrac{1}{2}\, zAz^{tr} + Bz^{tr})[\exp(\partial_w A \partial_w^{tr})P(w)]_{w=zA+B},$$

assuming both sides converge. In particular, if

$$\partial_z A \partial_z^{tr} P(z) = 0,$$

then

$$P(\partial_z) \exp(\tfrac{1}{2}\, zAz^{tr} + Bz^{tr}) = \exp(\tfrac{1}{2}\, zAz^{tr} + Bz^{tr})P(zA+B).$$

Proof. Since the statement is linear in P, we may use Fourier theory to reduce to the case $P = \exp(zC^{tr})$ for some length n vector C. Since

$$\exp(C\partial_z^{tr})f(z) = f(z+C)$$

and

$$f(\partial_z) \exp(zC^{tr}) = f(C) \exp(zC^{tr}),$$

we find

$$\exp(C\partial_z^{tr}) \exp(1/2\, zAz^{tr} + zB^{tr})$$
$$= \exp(1/2\, zAz^{tr} + zB^{tr}) \exp(1/2\, CAC^{tr}) \exp((zA + B)C^{tr})$$
$$= \exp(1/2\, zAz^{tr} + Bz^{tr})[\exp(\partial_w A\partial_w^{tr}) \exp(wC^{tr})]_{w=zA+B}.$$

$\square$

Definition 9.1.23. Given a function $P : \mathbb{C}^g \to \mathbb{C}$, define

$$\Theta^{(g)}(T, Z; P) := \sum_{v \in \mathbb{Z}^g} P(v)e(1/2\, v^{tr}Tv + Z^{tr}v)$$

$$= P(\frac{\partial_z}{2\pi\sqrt{-1}})\Theta^{(g)}(T, Z). \tag{9.1.19}$$

Theorem 9.1.24. *We keep the notation of Theorem 9.1.19. If*

$$\partial_v^{tr}(CT + D)^{-1}C\partial_v P(v) = 0,$$

then

$$\Theta^{(g)}((AT + B)(CT + D)^{-1}, (CT + D)^{-t}Z; \tilde{P}) =$$
$$\chi e(1/2\, V^{tr}TV + Z^{tr}V + 1/2\, Z^{tr}(CT + D)^{-1}CZ)\, \Theta^{(g)}(T, Z + TV + W; P) \tag{9.1.20}$$

where

$$\tilde{P}(v) = P((CT + D)^{-1}(v - V - CZ)).$$

The Riemann theta function also satisfies the "heat equation"

$$\frac{\partial_T}{\pi\sqrt{-1}}\Theta^{(g)}(T, Z; P) = \frac{\partial_Z^{tr}\partial_Z}{(2\pi\sqrt{-1})^2}\Theta^{(g)}(T, Z; P). \tag{9.1.21}$$

9.1.6 Hilbert theta series

Let K be a totally real number field of degree d, and fix an order O of K. Then an *O-lattice* is simply a lattice Λ equipped with a homomorphism from O to $\mathrm{End}(\Lambda)$.

In particular, Λ is an O-module, and $\Lambda \otimes \mathbb{Q}$ may be regarded as a vector space of dimension N/d over K, where $N = \dim \Lambda$. The inner product on Λ can be expressed generically in the form $\mathrm{Tr}(vAv^{tr})$, where A is a totally positive definite symmetric matrix over K. The "*Hilbert upper half-plane*" $\mathfrak{H}$ is the subset of $K \otimes \mathbb{C}$ such that the imaginary part corresponding to every real embedding of K is positive (that is, $\mathfrak{H}$ is a product of d copies of the ordinary upper half-plane). Thus for any element $\tau \in \mathfrak{H}$ we can define an $N \times N$ symmetric matrix $T(\tau)$ by

$$\mathrm{Tr}_{(K \otimes \mathbb{C})/\mathbb{C}}(\tau v A v^{tr}) = v T(\tau) v^{tr}. \tag{9.1.22}$$

Then the *Hilbert theta series* of the O-lattice Λ is

$$\Theta_\Lambda^H(\tau) := \sum_{v \in \Lambda} e(1/2\, v T(\tau) v^{tr}). \tag{9.1.23}$$

The Hilbert theta series is obtained from the Riemann theta function by setting $Z = 0$, and satisfies a transformation law for a group related to the Hilbert modular group $\mathrm{SL}_2(O)$. In some cases, as we will now see, the Hilbert theta series is a Hilbert modular form for $\mathrm{SL}_2(O)$.

There is also a well-known analogue of "Construction A" for obtaining O-lattices from codes. It has been studied by Bayer-Fluckiger, Craig, Hirzebruch, Thompson, among others (cf. [133, Chap. 8], Ebeling [164, Chap. 5]). The following description is based on a letter from Hirzebruch to N.J.A.S. ([265], [164, Chap. 5]).

Let p be an odd prime and let $\zeta := e^{2\pi i/p}$. The cyclotomic field $\mathfrak{K} := \mathbb{Q}(\zeta)$ contains the totally real field $K := \mathbb{Q}(\zeta + \zeta^{-1})$ of degree $d = (p-1)/2$. The corresponding rings of integers are

$$O_{\mathfrak{K}} = \mathbb{Z}\zeta + \cdots + \mathbb{Z}\zeta^{p-1},$$
$$O_K = \mathbb{Z}(\zeta + \zeta^{-1}) + \cdots + \mathbb{Z}(\zeta^{(p-1)/2} + \zeta^{-(p-1)/2}). \tag{9.1.24}$$

The ideal $\mathfrak{p} = (1 - \zeta)$ in $O_{\mathfrak{K}}$ has norm p. For a self-dual code C of length n over $\mathbb{F}_p$, let $\Lambda_C \subset O_{\mathfrak{K}}^n$ be the full preimage of C under the epimorphism $O_{\mathfrak{K}}^n \to (O_{\mathfrak{K}}/\mathfrak{p})^n \cong (\mathbb{Z}/p\mathbb{Z})^n$. With respect to the inner product

$$\langle x, y \rangle := \frac{1}{p} \sum_{j=1}^n \mathrm{Tr}_{\mathfrak{K}/\mathbb{Q}}(x_j \bar{y}_j),$$

Λ_C is a self-dual even unimodular lattice of real dimension $N = n(p-1) = 2nd$. The Hilbert theta series of Λ_C is a Hilbert modular form of weight n for $\mathrm{SL}_2(O_K)$, which is symmetric in the sense that it is unchanged if the components of $\tau \in \mathfrak{H}$ are permuted by the Galois group.

The case $p = 5, K = \mathbb{Q}(\sqrt{5}), O_K = \mathbb{Z}[(1 + \sqrt{5})/2]$ is especially interesting, for now we have another analogue of Gleason's theorem, due to Hirzebruch ([265], [164]):

Theorem 9.1.25. *For $p = 5$, the graded ring of symmetric Hilbert modular forms for $\mathrm{SL}_2(O_K)$ has Poincaré series*

$$\frac{1}{(1 - t^2)(1 - t^6)(1 - t^{10})} \quad [A008672],$$

and as generators for this ring we may take the Hilbert theta series for the lattices obtained by applying Construction A to the self-dual codes c_2, c_6 and d_5^{2+} (or e_{10}^+) over $\mathbb{F}_5$ mentioned in (7.4.21), (7.4.22).

Although we will not go into details here, one can introduce higher genus Hilbert modular forms and Jacobi analogues thereof.

Also, one can replace the totally real field K by a totally complex field, or even by a quaternion algebra ramified at ∞. In genus 1 the resulting half-plane is just the Hilbert upper half-plane associated with the real subfield, but for higher genera the half-plane consists of matrices $X + iY$, where X and Y are now Hermitian and Y is positive definite.

For instance, a lattice with an automorphism of order 3 with no fixed points is a lattice over the order $\mathbb{Z}[\zeta_3]$. The corresponding genus-1 theta series however is just the ordinary theta series of this lattice. It is only in higher genera that the coordinatization shows up in the form of the theta series.

In the totally complex case we proceed differently (cf. [133, Chap. 2, §2.6]). Consider for example the ring O_K of integers in the field $K = \mathbb{Q}(\sqrt{-m})$ for an integer $m \geq 1$. An O_K-lattice Λ is now simply a discrete and co-compact subgroup of $\mathbb{C}^n$ which is also an O_K-module. The dual lattice is defined with respect to the Hermitian inner product:

$$\Lambda^{\#} := \{ u \in \mathbb{C}^n \mid u \cdot \bar{v} \in O_K \text{ for } v \in \Lambda \}.$$

There is an associated real lattice Λ^{real} of dimension $N = 2n$ obtained by mapping each $u \in \Lambda$ to $(\Re u, \Im u)$. The theta series is

$$\Theta_\Lambda(\tau) := \sum_{v \in \Lambda} e(\,^{1}/_{2} v \cdot \bar{v}) = \Theta_{\Lambda^{\text{real}}}(\tau).$$

For $m = 1$, O_K is the ring of Gaussian integers $\mathcal{G} := \mathbb{Z}[i]$, $i = \sqrt{-1}$, and for $m = 3$, O_K is the ring of Eisenstein integers $\mathcal{E} := \mathbb{Z}[\omega]$, $\omega := (1 + \sqrt{-3})/2$. The simplest self-dual examples are $\mathcal{G}$, for which the real version is $\mathbb{Z}^2$, and $\mathcal{E}$, for which the real version is, apart from a scale factor, the root lattice A_2. Then we have two further analogues of Gleason's theorem ([502], [133, Chap. 7]):

Theorem 9.1.26. *The graded ring spanned by the theta series of self-dual lattices over $\mathcal{G}$ has Poincaré series*

$$\frac{1}{(1 - t^2)(1 - t^8)} \qquad [A008621],$$

and as generators for this ring we may take $\theta_3(\tau)^2$ *and* $\Delta_8(\tau)$ *(see (9.1.8)).*

Theorem 9.1.27. *The graded ring spanned by the theta series of self-dual lattices over $\mathcal{E}$ has Poincaré series*

$$\frac{1}{(1 - t^2)(1 - t^{12})} \qquad [A097992],$$

and as generators for this ring we may take the theta series of $\mathcal{E}$ itself and of the Coxeter-Todd lattice K_{12}.

In Theorem 9.1.27, the theta series of $\mathcal{E}$ is

$$\phi_0(\tau) := \theta_2(\tau)\theta_2(3\tau) + \theta_3(\tau)\theta_3(3\tau)$$
$$= 1 + 6q^{1/2} + 6q^{3/2} + 6q^2 + 12q^{7/2} + \cdots ; \tag{9.1.25}$$

and instead of the theta series of K_{12} we may use the cusp form

$$\Delta_6(\tau) := q^{1/2} \prod_{j=1}^{\infty} (1 - q^{j/2})^6 (1 - q^{3j/2})^6$$
$$= q^{1/2} - 6q + 9q^{3/2} + 4q^2 + \cdots . \tag{9.1.26}$$

9.2 Positive definite form $\mathbb{R}$-algebras

So far we have seen examples of theta series for lattices over $\mathbb{Z}$, over $\mathrm{Mat}_n(\mathbb{Z})$ (i.e. Siegel theta series), over number fields (i.e. Hilbert theta series), as well as series relative to fixed vectors (i.e. Jacobi-Siegel theta series). The common thread is that in each case the ring acting on the lattice is an order in a semi-simple $\mathbb{R}$-algebra with a positive definite involution. In our setting this leads naturally to the notion of a positive definite form algebra.

Definition 9.2.1. Let $\rho := (V, \rho_M, \beta)$ be a representation of a twisted $\mathbb{R}$-algebra $(\mathcal{A}, M, \psi)$, where $\mathbb{R}$ denotes the real numbers (cf. Definition 3.3.1). Then ρ is said to be *positive definite* if $\beta \in \mathrm{Bil}(V, \mathbb{R})$ is positive definite. A twisted $\mathbb{R}$-algebra is *positive definite* if it admits a faithful positive definite representation. A form $\mathbb{R}$-algebra is *positive definite* if the underlying twisted $\mathbb{R}$-algebra is positive definite.

The following is surely well known:

Proposition 9.2.2. *If the twisted $\mathbb{R}$-algebra $(\mathcal{A}, M, \psi)$ has a positive definite faithful representation, then $(\mathcal{A}, M, \psi)$ is a product of twisted algebras of the form $\mathrm{Mat}_n(\mathbb{R})$, $\mathrm{Mat}_n(\mathbb{C})$, or $\mathrm{Mat}_n(\mathbb{H})$, where in each case the involution is given (up to rescaling) by conjugate transposition.*

Proof. We first observe that a positive definite representation $\rho := (V, \rho_M, \beta)$ is anisotropic, implying that $\mathcal{A}$ is semisimple. In fact, let $I := \mathrm{rad}(\mathcal{A})$ and assume that $I \neq 0$. Then (since $\mathcal{A}$ is Artinian), there is an $m > 1$ such that $I^m = 0$ and $I' := I^{m-1} \neq 0$. Since

$$\beta(I'V, I'V) = \beta(V, (I')^J I'V) = \beta(V, (I')^2 V) = 0,$$

$I'V$ is an isotropic subspace of V, and hence $I'V = 0$. Since ρ is faithful, $I' = 0$, a contradiction. Hence $\mathcal{A}$ is semisimple. The same calculation shows that the involution J preserves the simple components of $\mathcal{A}$, and so $\mathcal{A}$ is a direct sum of simple algebras with involution. The only positive definite simple algebras with involution are the ones stated in the proposition (see e.g. Plesken [410]). □

Since 2 is invertible in $\mathbb{R}$, a form $\mathbb{R}$-algebra satisfies $\Phi = \{\!\{M\}\!\} \oplus \ker(\lambda)$. Thus to define a representation of a form $\mathbb{R}$-algebra, it suffices to give a representation of the underlying twisted $\mathbb{R}$-algebra and to specify $\rho_\Phi(\ker(\lambda))$. The latter consists of linear functionals on V. Thus in particular, for $\phi \in \Phi$, $\rho_\Phi(\phi)$ will be a degree 2 polynomial on V with constant term 0. The hyperbolic co-unitary group $\mathfrak{U}(\mathcal{A}, \Phi)$ associated with the form $\mathbb{R}$-algebra $(\mathcal{A}, M, \psi, \Phi)$ has the structure

$$\mathfrak{U}(\mathcal{A}, \Phi) \cong (\ker(\lambda) \oplus \ker(\lambda)).U(\mathcal{A}, \{\!\{M\}\!\}),$$

where $U(\mathcal{A}, \{\!\{M\}\!\})$ is a product of the co-unitary groups corresponding to the simple summands of $\mathcal{A}$. If $\mathcal{A}$ is simple, say $\mathcal{A} \cong \mathrm{Mat}_n(K)$, where K is one of $\mathbb{R}$, $\mathbb{C}$ or $\mathbb{H}$, then

$$U(\mathcal{A}, \{\!\{M\}\!\}) = \left\{ g \in \mathrm{GL}_2(\mathcal{A}) \mid \overline{g}^{tr} \begin{pmatrix} 0 & -I_n \\ I_n & 0 \end{pmatrix} g = \begin{pmatrix} 0 & -I_n \\ I_n & 0 \end{pmatrix} \right\},$$

where $^-$ denotes the canonical involution of K, and so $U(\mathcal{A}, \{\!\{M\}\!\})$ is one of the following classical groups:

$\mathcal{A}$	$U(\mathcal{A}, \{\!\{M\}\!\})$
$\mathrm{Mat}_n(\mathbb{R})$	$\mathrm{Sp}_{2n}(\mathbb{R})$
$\mathrm{Mat}_n(\mathbb{C})$	$U_{n,n}(\mathbb{C})$
$\mathrm{Mat}_n(\mathbb{H})$	$SO^*_{2n}(\mathbb{C})$

(cf. Helgason [261, Table V, Chap. X, §6, p. 518], entries CI, $AIII$ and $DIII$; note that in each case the corresponding symmetric space is Hermitian). In the first two cases these are analogues of the co-unitary groups that we saw in §§7.3, 7.2 respectively.

As one might guess from our earlier discussion of the Riemann theta function, a particularly important example is the form $\mathbb{R}$-algebra $\mathrm{Mat}_n(\mathbb{R}, \mathbb{R})$, defined as follows:

$$\mathcal{A} = \mathrm{Mat}_n(\mathbb{R}), \ M = \mathcal{A}, \ \psi = \mathrm{id}, \ \Phi = \{\!\{M\}\!\} = \mathrm{Sym}_n(\mathbb{R}),$$
$$\{\!\{m\}\!\} = (m + m^{tr})/2, \ \lambda(\phi) = 2\phi \text{ for all } m \in M, \ \phi \in \Phi .$$

This is a matrix ring over the form ring $(\mathbb{R}, \mathbb{R})$ defined by $\mathcal{A} = M = \Phi = \mathbb{R}; \psi = \tau = \{\!\{ \ \}\!\} = \mathrm{id}; \lambda = 2\,\mathrm{id}$. Since λ is injective, the corresponding group is of course just

$$\mathfrak{U}(\mathcal{A}, M) = U(\mathcal{A}, M) = \mathrm{Sp}_{2n}(\mathbb{R}).$$

Also of interest is the form ring $(\mathbb{R}, \mathbb{R} \times \mathbb{R})$ defined by

$$\mathcal{A} = M = \mathbb{R}, \ \Phi = \mathbb{R} \times \mathbb{R}, \ \psi = \tau = \mathrm{id},$$
$$\{\!\{x\}\!\} = (x, 0), \ \lambda(x, y) = 2x, \ (x, y)[r] = (r^2x, ry),$$

for all $x, y, r \in \mathbb{R}$, with its matrix ring $\mathrm{Mat}_n(\mathbb{R}, \mathbb{R} \times \mathbb{R})$ given by

$$\mathcal{A} = \mathrm{Mat}_n(\mathbb{R}), \qquad M = \mathcal{A}, \psi = \mathrm{id}, \qquad\qquad \Phi = \mathrm{Sym}_n(\mathbb{R}) \oplus \mathbb{R}^n$$
$$\tau(m) = m^{tr}, \qquad\qquad \{\!|m|\!\} = ((m + m^{tr})/2, 0), \lambda(y, v) = 2y,$$
$$(y, v)[r] = (r^{tr} y r, vr)$$

for all $m \in M, (y, v) \in \Phi, r \in R$. Now $\ker(\lambda) \cong \mathbb{R}^n$ and the corresponding group is

$$\mathfrak{U}(\mathcal{A}, M) = \mathbb{R}^{2n} \rtimes \mathrm{Sp}_{2n}(\mathbb{R}).$$

For any form $\mathbb{R}$-algebra $(\mathcal{A}, \Phi)$, there is a canonical bijection between the set of n-dimensional representations of $(\mathcal{A}, \Phi)$ and the set of homomorphisms from $(\mathcal{A}, \Phi)$ to $\mathrm{Mat}_n(\mathbb{R}, \mathbb{R} \times \mathbb{R})$. (In particular, $\mathrm{Mat}_n(\mathbb{R}, \mathbb{R} \times \mathbb{R})$ has a canonical representation.)

9.3 Half-spaces

Using the notion of positive definite form $\mathbb{R}$-algebras we give a uniform definition of the half-spaces on which elliptic, Siegel and Jacobi modular forms are defined. We first have to introduce the appropriate notions of positive semi-definiteness and positive definiteness for elements of the algebra.

Definition 9.3.1. Let $(\mathcal{A}, M, \psi)$ be a positive definite twisted $\mathbb{R}$-algebra. Then $m \in M$ is called *positive semi-definite* if $\tau(m) = m$ and $\rho_M(m)$ is positive semi-definite for all positive definite representations ρ of $(\mathcal{A}, M, \psi)$.

Proposition 9.3.2. *Let $(\mathcal{A}, M, \psi)$ be a positive definite twisted $\mathbb{R}$-algebra and let $m \in M$ with $\tau(m) = m$. The following are equivalent:*
(a) m is positive semi-definite;
(b) $\rho_M(m)$ is positive semi-definite for some faithful positive definite representation (V, ρ_M, β) of $(\mathcal{A}, M, \psi)$;
(c) $m = \sum_{i=1}^{r} \psi(1)(a_i \otimes a_i)$ for some $r \geq 1$ and $a_1, \ldots, a_r \in \mathcal{A}$.
In (c), it suffices to take $r = 1$.

Proof. $(a) \Rightarrow (b)$ is clear.
$(c) \Rightarrow (a)$ A sum of positive semi-definite elements is semi-definite, so it is enough to show that $\psi(1)(a \otimes a) = \psi(a^J a)$ is positive semi-definite for all $a \in \mathcal{A}$. Let (V, ρ_M, β) be a positive definite representation. Then $\rho_M(\psi(a^J a))(x, y) = \beta(ax, ay)$, which is clearly positive semi-definite.
$(b) \Rightarrow (c)$ Since $m = \tau(m) \in M$, there are elements $a, b \in \mathcal{A}$ such that

$$m = \psi(a^J a) - \psi(b^J b),$$

and $ab = ba = 0$. (This can be seen by looking at the simple factors of $\mathcal{A}$.) If $b \neq 0$, there is an $x \in V$ with $ax = 0$ and $bx \neq 0$. Then $\rho_M(m)(x, x) = \beta(ax, ax) - \beta(bx, bx) < 0$, contradicting (b). So $b = 0$, and thus

$$m = \psi(a^J a) = \psi(1)(a \otimes a).$$

$\square$

Definition 9.3.3. Let $(\mathcal{A}, M, \psi)$ be a positive definite twisted $\mathbb{R}$-algebra. Then $m \in M$ is called *positive definite* if $\tau(m) = m$ and there is a faithful positive definite representation ρ such that $\rho_M(m)$ is positive definite.

Remark. Identifying M with $\mathcal{A}$ via ψ, the positive semi-definite elements of M form a closed cone in the finite-dimensional real vector space $\mathcal{A}$. The positive definite elements in M form the interior of this cone.

Definition 9.3.4. Let $(\mathcal{A}, M, \psi, \Phi)$ be a positive definite simple form $\mathbb{R}$-algebra. Then we define the *half-space*

$$\mathcal{H}(\Phi) := \{\phi \in \Phi \otimes \mathbb{C} \mid \Im(\lambda(\phi)) \in M \text{ is positive definite}\} .$$

Note that an element of $\mathcal{H}(\Phi)$ can be written in the form $T + Z$, where $T = X + iY$, $Z = u + iv$, $X, Y \in \{\!\{M\}\!\}$, $u, v \in \ker(\lambda)$ and $\lambda(Y)$ is positive definite. In the case of $\mathrm{Mat}_n(\mathbb{R}, \mathbb{R})$, this is simply the parameter space for the Riemann theta function; we take the convention that (T, Z) corresponds to the element $\{\!\{T\}\!\} + Z^{tr}$.

The hyperbolic co-unitary group has a natural action on the half-space $\mathcal{H}(\Phi)$ (cf. Eichler and Zagier [166]). For convenience we will restrict our attention to the case when $\mathcal{A}$ is a simple $\mathbb{R}$-algebra.

Let

$$g := \left(\begin{pmatrix} A & B \\ C & D \end{pmatrix}, \begin{pmatrix} \phi_1 & m \\ & \phi_2 \end{pmatrix} \right) \in \mathfrak{U}(\mathcal{A}, \Phi),$$

and let $T + Z \in \mathcal{H}(\Phi)$ as above. Then we define $g \circ (T + Z)$ on generators g as follows. If $\phi_1, \phi_2 \in \{\!\{M\}\!\}$, then

$$g \circ (T + Z) = (AT + D)(CT + D)^{-1} + Z(CT + D)^{-1},$$

where

$$(AT + B)(CT + D)^{-1} = \{\!\{\psi((A \cdot \psi^{-1}(\lambda(T))/2 + B)(C \cdot \psi^{-1}(\lambda(T))/2 + D)^{-1})\}\!\}$$

and

$$Z(CT + D)^{-1} = Z[(C \cdot \psi^{-1}(\lambda(T))/2 + D)^{-1}] .$$

If $\begin{pmatrix} A & B \\ C & D \end{pmatrix} = I_{2n}$, so $\phi_1, \phi_2 \in \ker(\lambda)$ and $m = 0$, then

$$g \circ (T + Z) = T + Z + \phi_1[\psi^{-1}(\lambda(T))/2] + \phi_2 .$$

Proposition 9.3.5. *This gives an action of $\mathfrak{U}(\mathcal{A}, \Phi)$ on $\mathcal{H}(\Phi)$.*

Proof. Let ρ be a faithful positive definite representation of $\mathcal{A}$. This can be viewed as an injection from $\mathcal{A}$ to the form $\mathbb{R}$-algebra $\mathrm{Mat}_n(\mathbb{R}, \mathbb{R})$ for some n. It therefore suffices to prove the proposition in that case. But this is simply the action of $\mathbb{R}^{2g} \rtimes \mathrm{Sp}_{2g}(\mathbb{R})$ on the arguments of the Riemann theta function. $\quad\square$

9.4 Form orders and lattices

Definition 9.4.1. An *order* in a form $\mathbb{R}$-algebra $(\mathcal{A}, M, \psi, \Phi)$ is a sub-form ring (O, M_O, ψ, Φ_O) such that O, M_O, Φ_O are lattices in $\mathcal{A}$, M, Φ respectively.

Remark. In particular, O is closed under the involution J and $M_O = \psi(O)$ is closed under τ. Moreover, $\{M_O\} \subset \Phi_O$ and $\lambda(\Phi_O) \subset M_O$.

Example 9.4.2. If $\mathcal{A} = \mathbb{R}$, $M = \mathbb{R}$, $\psi = \mathrm{id}$ and $\Phi = \{M\}$, then we may take $O = \mathbb{Z}$, $M_O = \mathbb{Z}$ and Φ_O to be either $\{\mathbb{Z}\}$ (in which case $\{\}$ is surjective) or $\Phi_O = \frac{1}{2}\{\mathbb{Z}\}$ (in which case λ is surjective). These two form orders will be used to define the Types of odd and even unimodular lattices in §9.5.

Example 9.4.3. More generally, if K is a totally real number field and

$$\mathcal{A} = K \otimes_{\mathbb{Q}} \mathbb{R}, \quad M = \mathcal{A}, \quad \psi = \mathrm{id}, \quad \Phi = \{M\},$$

then we can take O to be the ring of integers in K, $M_O = O$ and Φ_O to be any O-qmodule satisfying

$$\{M_O\} \subset \Phi_O \subset \frac{1}{2}\{M_O\}.$$

Just as in the case of codes over extension fields, when we obtain the form β as the trace of a form over the extension field, so here the typical representation of interest arises from a quadratic form over K and $\beta(v, w) = \widehat{\beta}(v, w)$ for $v, w \in K^N$ and some totally positive definite bilinear form $\widehat{\beta}$.

Example 9.4.4. Note that in the definition we did not require that the ambient $\mathbb{R}$-algebra be simple. In particular, we may consider an order in a product of number fields. An instance when this occurs is when we have a lattice Λ with an endomorphism σ satisfying $\sigma^a = n^{a/2} \, \mathrm{Id}$ for some a and n. For example, σ might be an automorphism ($n = 1$), or an n-modularity of Λ. This corresponds to the following form order:

$$\mathcal{A} = \mathbb{R}[z]/(z^a - n^{a/2}), \quad M = \mathcal{A}, \quad \tau(z) = \frac{n}{z}, \quad \psi = \mathrm{id},$$

$$\Phi = \{M\}, \quad O = \mathbb{Z}[z], \quad \Phi_0 = \{O\} \text{ or } \frac{1}{2}\{O\},$$

where the latter correspond to the two different evenness requirements. The corresponding isotropic self-dual lattices will be precisely the unimodular lattices with such an endomorphism σ.

Note here that $O \otimes \mathbb{Q}$ is in general a product of totally real and totally complex number fields, although O need not split as a product of orders in the respective fields. (For example, take $a = 2, n = 1$, for which $O = \mathbb{Z}[z]/(z^2 - 1)$.) Quebbemann and Rains [441] apply this idea to the study of modular lattices.

Example 9.4.5. If $\mathcal{A} = \mathbb{R}$, $M = \mathbb{R}$, $\psi = \mathrm{id}$ and $\Phi = \{\!\!\{M\}\!\!\} \oplus \mathbb{R}$, we may take $O = \mathbb{Z}$, $M_O = \mathbb{Z}$, as before, and

$$\Phi_O = \{\!\!\{\{\!\!\{\frac{k}{2}\}\!\!\} \oplus l \mid k, l \in \mathbb{Z}, k \equiv l \ (\mathrm{mod}\ 2)\}.$$

This is a canonical example of a form order, in the sense that for any N-dimensional lattice, there exists a representation of $\mathrm{Mat}_N(O, M_O, \Phi_O)$ with respect to which the lattice is isotropic self-dual (cf. Definition 9.4.8 below). This corresponds to the fact that any N-dimensional lattice is isomorphic to $\mathbb{Z}^N$.

Definition 9.4.6. Let $\rho := (V, \rho_M, \rho_\Phi, \beta)$ be a positive definite representation of a form $\mathbb{R}$-algebra. For a lattice Λ in V, the *dual lattice* of Λ with respect to β is defined to be

$$\Lambda^{\#,\beta} := \{v \in V \mid \beta(v, l) \in \mathbb{Z} \text{ for } l \in \Lambda\}.$$

The *β-determinant* of Λ is defined to be the determinant

$$\det{}_\beta(\Lambda) := \det{}_{1 \le i \le j \le n} \ (\beta(v_i, v_j)),$$

where $v_1, \ldots, v_n$ are a basis for Λ.

Proposition 9.4.7. *The β-determinants of Λ and $\Lambda^{\#,\beta}$ are related by*

$$\det{}_\beta(\Lambda) \det{}_\beta(\Lambda^{\#,\beta}) = 1.$$

If $\Lambda \subset \Lambda'$, then

$$\det{}_\beta(\Lambda) = \det{}_\beta(\Lambda')[\Lambda' : \Lambda]^2.$$

We can now define the Type of a lattice, analogous to the definition for codes given in §1.8.

Definition 9.4.8. Let $O := (O, M_O, \psi, \Phi_O)$ be an order in the form $\mathbb{R}$-algebra $\mathcal{A} := (\mathcal{A}, M, \psi, \Phi)$ and let $\rho := (V, \rho_M, \rho_\Phi, \beta)$ be a positive definite representation of $\mathcal{A}$. An *isotropic lattice of Type $\rho_{|O}$* is a lattice Λ in V^N which is an O-submodule such that
(a) $\rho_M(m)(v_1, v_2) \in \mathbb{Z}$ for $m \in M_O$, $v_1, v_2 \in \Lambda$, and
(b) $\rho_\Phi(\phi)(v) \in \mathbb{Z}$ for $\phi \in \Phi_O$, $v \in \Lambda$.
If Λ additionally satisfies
(c) $\Lambda^{\#,\beta} = \Lambda$,
then Λ is called an *(isotropic self-dual) lattice of Type $\rho_{|O}$*.

Remark. Since Λ is an O-submodule and $O = O^J$, (c) implies (a).

Proposition 9.4.9. *If Λ is an isotropic lattice of type $\rho_{|O}$, then $\det_\beta(\Lambda) \in \mathbb{Z}$. The β-determinant is 1 if and only if Λ is self-dual, that is, if $\Lambda^{\#,\beta} = \Lambda$.*

Proof. For the first claim, we have

$$\det_\beta(\Lambda) = \det_\beta(\Lambda^{\#,\beta})[\Lambda^{\#,\beta} : \Lambda]^2,$$

and thus

$$\det_\beta(\Lambda) = [\Lambda^{\#,\beta} : \Lambda].$$ □

For this reason, self-dual lattices are often called *unimodular*.

Given any lattice Λ which is also an O-submodule, there is a natural notion of theta series. Recall that a positive definite representation ρ of $\mathcal{A}$ can be viewed as a homomorphism from $(\mathcal{A}, \Phi)$ to the appropriate $\mathrm{Mat}_n(\mathbb{R}, \mathbb{R})$, and thus induces a corresponding map on the half-spaces. We rigidify this map by using a basis of Λ as our basis for $\mathbb{R}^n$. This immediately suggests the following definition: the *theta series* of Λ is

$$\Theta_\Lambda(H) = \Theta^{(n)}(\rho(H)) \text{ for } H \in \mathcal{H}(\Phi), \tag{9.4.1}$$

where $\Theta^{(n)}$ is the Riemann theta function defined on $\mathcal{H}(\mathrm{Mat}_n(\mathbb{R}, \mathbb{R}))$ by

$$\Theta^{(n)}(\{\!\{T\}\!\} + Z^t) := \Theta^{(n)}(T, Z). \tag{9.4.2}$$

Every kind of theta series considered in §9.1 can be obtained from this definition: ordinary theta series correspond to the order $\mathbb{Z}$ of $\mathbb{R}$, Jacobi theta series correspond to the order $(\mathbb{Z}, \mathbb{Z})$ in $(\mathbb{R}, \mathbb{R})$, Jacobi-Siegel theta series correspond to $\mathrm{Mat}_n(\mathbb{Z}, \mathbb{Z})$, and Hilbert theta series correspond to an order in a number field.

The transformation law for this theta series is the following:

Theorem 9.4.10. *Let Λ be a lattice of Type $\rho_{|O}$. Then for any element $g \in \mathfrak{U}(O, \Phi)$,*

$$\Theta_\Lambda(g \otimes H) = \zeta\sqrt{\det(CT + D)}\,\Theta_\Lambda(H),$$

where ζ is an 8-th root of unity and $CT + D$ is the element of $\mathrm{Sym}_N(\mathbb{C})$ corresponding to $\rho(g)$ and $\rho(H)$.

Proof. In order to deduce this from the transformation law for the Riemann theta function, it is necessary and sufficient that $\rho(g)$ be in the subgroup $\mathbb{Z}^{2N}.\mathrm{Sp}_{2N}(\mathbb{Z}) \subset \mathbb{R}^{2N} \rtimes \mathrm{Sp}_{2N}(\mathbb{Z})$. But this follows immediately from the fact that $\Lambda = \Lambda^\#$ is isotropic.   □

9.5 Even and odd unimodular lattices

Let $R(\mathbb{Z}_{\mathrm{II}})$ and $R(\mathbb{Z}_\mathrm{I})$ be the form orders in the form $\mathbb{R}$-algebra $(\mathbb{R}, \mathbb{R}, \mathrm{id}, \mathbb{R})$ defined as follows:

$$R(\mathbb{Z}_{\mathrm{II}}) := (\mathbb{Z}, M, \psi, \Phi),$$

where $M = \mathbb{Z}$, $\psi = \mathrm{id}$, $\Phi = \mathbb{Z}$, $\{\ \}$ is multiplication by 2 and λ is surjective, and

$$R(\mathbb{Z}_{\mathrm{I}}) := (\mathbb{Z}, M, \psi, \varPhi),$$

where $M = \mathbb{Z}$, $\psi = \mathrm{id}$, $\varPhi = \mathbb{Z}$, $\{\ \}$ is surjective and λ is multiplication by 2. Define the infinite representations $\rho(\mathbb{Z}_{\mathrm{II}}) = (V_{even}, (\rho_{even})_M, (\rho_{even})_\varPhi, \beta)$ and $\rho(\mathbb{Z}_{\mathrm{I}}) = (V_{odd}, (\rho_{odd})_M, (\rho_{odd})_\varPhi, \beta)$ by

$$
\begin{aligned}
V_{even} = V_{odd} &= \mathbb{R}, \\
(\rho_{even})_M(m)(x,y) &= mxy + \mathbb{Z} \text{ for } m \in M, \\
(\rho_{even})_\varPhi(\phi)(x) &= {}^{1}\!/\!_{2}\,\phi x^2 + \mathbb{Z} \text{ for } \phi \in \varPhi, \\
(\rho_{odd})_M &= (\rho_{even})_M, \\
(\rho_{odd})_\varPhi(\phi)(x) &= \phi x^2 + \mathbb{Z} \text{ for } \phi \in \varPhi.
\end{aligned}
$$

Then lattices of Type $\mathbb{Z}_{\mathrm{II}}$ are simply even unimodular (or Type II) lattices, and lattices of Type $\mathbb{Z}_{\mathrm{I}}$ are unimodular (or Type I) lattices (cf. [133]). Isotropic lattices of the two Types correspond to integral or even integral lattices respectively.

Note that λ is injective in both cases, so any element of the hyperbolic co-unitary group is already determined by its projection onto the first component. Using (5.2.12), (5.2.13) we find

$$\mathfrak{U}(\mathbb{Z}_{\mathrm{II}}) = \left\{ \begin{pmatrix} a & b \\ c & d \end{pmatrix} \in \mathbb{Z}^{2\times 2} \mid ad - bc = 1 \right\} = \mathrm{SL}_2(\mathbb{Z}),$$

and the *theta-group*

$$
\begin{aligned}
\Gamma_\vartheta := \mathfrak{U}(\mathbb{Z}_{\mathrm{I}}) &= \left\{ \begin{pmatrix} a & b \\ c & d \end{pmatrix} \in \mathbb{Z}^{2\times 2} \mid ad - bc = 1, ca \in 2\mathbb{Z}, db \in 2\mathbb{Z} \right\} \\
&= \left\langle \begin{pmatrix} 0 & 1 \\ -1 & 0 \end{pmatrix}, \begin{pmatrix} 1 & 2 \\ 0 & 1 \end{pmatrix} \right\rangle.
\end{aligned}
\tag{9.5.1}
$$

The latter is the group denoted by Θ in Schoeneberg [477, p. 85].

Similarly, for the corresponding Siegel modular groups, we have

$$
\begin{aligned}
\mathfrak{U}_g(\mathbb{Z}_{\mathrm{II}}) = \{ \begin{pmatrix} A & B \\ C & D \end{pmatrix} &\in \mathbb{Z}^{2g\times 2g} \mid \\
D^{tr}A - C^{tr}B = I_g, C^{tr}A &= A^{tr}C, D^{tr}B = B^{tr}D \} = \mathrm{Sp}_{2g}(\mathbb{Z})
\end{aligned}
$$

(cf. Freitag [176, Bemerkung 1.2]), and $\mathfrak{U}_g(\mathbb{Z}_{\mathrm{I}})$ is the theta-group of genus g

$$
\begin{aligned}
\mathfrak{U}_g(\mathbb{Z}_{\mathrm{I}}) = \{ \begin{pmatrix} A & B \\ C & D \end{pmatrix} &\in \mathrm{Sp}_{2g}(\mathbb{Z}) \mid \\
C^{tr}A = A^{tr}C \text{ and } D^{tr}B &= B^{tr}D \text{ even } \} = \Gamma_{\vartheta,g}
\end{aligned}
$$

(cf. Freitag [176, Bemerkung A.2.3]). The latter is an analogue of the theta-group introduced in (8.1.1) for codes over $\mathbb{Z}/m\mathbb{Z}$.

9.6 Gluing theory for codes

Gluing is a technique for building up codes or lattices using smaller codes or lattices as building blocks. It was briefly mentioned in §3.5.1. We will use this technique in the following three chapters, so it is appropriate to discuss it here.

For simplicity, we begin by discussing gluing theory in the context of classical codes, that is, linear codes over finite fields. There are analogous statements for lattices, which can usually be obtained simply by changing "code" to "lattice", "length" to "dimension", etc. At the end of this section we generalize the main theorem of gluing theory (Theorem 9.6.1) to apply to codes of any given Type. We will say more about gluing theory for lattices in the next section (§9.7).

Gluing theory is especially useful when one is attempting to classify all self-dual codes (or lattices) of a given length (or dimension). Typically one finds that there are many codes with low minimal distance and only a few with high minimal distance. Gluing theory is good at finding all the codes of low distance.

The first formal description of gluing theory appeared in Conway and Pless [121]. It has also been used in [123], [131], [133], [346], [347], etc.

Let $C_1, \ldots, C_t$ be self-orthogonal codes of lengths $n_1, \ldots, n_t$, over a finite field $\mathbb{F}$, with generator matrices $G_1, \ldots, G_t$. If C is a self-dual code with the generator matrix shown in Fig. 9.1 then we say that C is formed by *gluing* the *components* $C_1, \ldots, C_t$ together, and we write

$$C = (C_1 C_2 \ldots C_t)^+ \qquad (9.6.1)$$

to indicate this process. (Whenever possible the subcodes are chosen so that every minimal weight codeword of C belongs to one of the C_i.) The codewords in C which contain a nonzero linear combination of the rows of the matrix X are called *glue words*, since these hold the components together. A glue word has the form

$$u = u_1 u_2 \ldots u_t, \qquad (9.6.2)$$

where each glue element u_i has length n_i. Since C is self-dual, u_i is in $C_i^\perp$.

Let us choose coset representatives $a_0 = 0, a_1, \ldots, a_{s-1}$ for C_i in $C_i^\perp$, where $s = |C_i^\perp|/|C_i|$, so that

$$C_i^\perp = \bigcup_{j=0}^{s-1} (a_j + C_i).$$

Then we can assume that each u_i in (9.6.2) is one of $a_0, \ldots, a_{s-1}$.

As illustrations we give the two indecomposable Type 2_{I} self-dual codes of length 18 (see Tables 12.2 and 12.7), using the components and glue vectors from the list in §12.2. The first code, d_6^{3+}, is formed by gluing three copies of the component d_6 together:

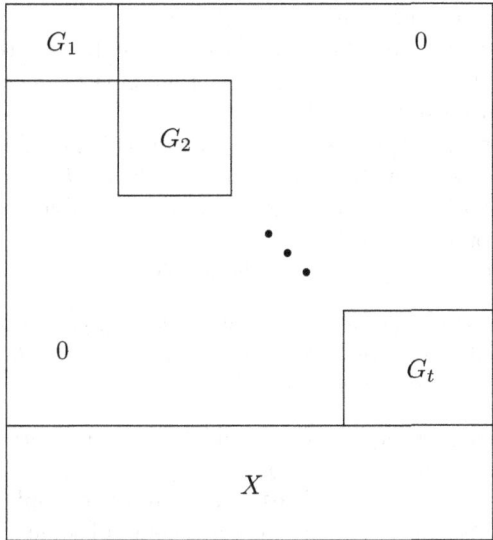

Fig. 9.1. Generator matrix G for a code formed by gluing components $C_1, \ldots, C_t$ together. G_i is a generator matrix for C_i, and X denotes the rest of the generator matrix for C.

1111 1111		
	1111 1111	
		1111 1111
010101 010110 000011	000011 010101 000011	010110 000011 000011

The three glue vectors shown are abc, cab and bbb.

The second example is a code that appears several times in this book, $(d_{10}e_7f_1)^+$. This is formed by gluing together d_{10}, e_7 and a "free" (or empty) component f_1:

1111
 1111
  1111
   1111		

	1110100 0111010 0011101	
0101010101 0101010110	0000000 1111111	1 0

The two glue vectors shown are $a0A$ and $cd0$.

Of course a self-*dual* code has no (nonzero) glue. If a self-orthogonal code C has a component B, say, which is self-dual, then C is a direct sum $C = B \oplus C'$, where C' is again self-orthogonal.

It may happen that there is a glue word in which only one u_i is nonzero, in which case we say that the component C_i has *self-glue*, and that u is a *self-glue vector*. So if C has a single component C_1 (say) with self-glue, we write $C = C_1^+$ (compare (9.6.1)).

A basic result of gluing theory is the following.

Theorem 9.6.1. *If a self-dual code C is formed by gluing together two codes C_1 and C_2 in such a way that there is no self-glue, then the quotient groups $C_1^\perp/C_1$ and $C_2^\perp/C_2$ are isomorphic.*

We omit the easy proof. The isomorphism is given by $u_1 + C_1 \mapsto -u_2 + C_2$ whenever there is a glue vector $u_1 u_2 \in C$.

The above discussion generalizes immediately to apply to codes of any given Type. Suppose $C_1, C_2, \ldots$ are isotropic codes of some fixed Type ρ. Then the isotropic codes containing $C_1 \oplus C_2 \oplus \cdots$ are precisely the isotropic codes in the quotient representation $N\rho/(C_1 \oplus C_2 \oplus \cdots)$. Theorem 9.6.1 generalizes as follows.

Theorem 9.6.2. *Fix a Type ρ. If a self-dual isotropic code C is obtained by gluing together two self-orthogonal isotropic codes C_1 in $N_1\rho$ and C_2 in $N_2\rho$ in such a way that there is no self-glue, then the quotient representations $N_1\rho/C_1$ and $N_2\bar{\rho}/C_2$ are isomorphic.*

9.7 Gluing theory for lattices

In order to present the gluing theory for lattices in the language of this chapter, we first need the appropriate notion of a quotient representation. This is in fact a finite representation of the relevant form order, rather than a positive definite representation. In other words, the quotient representation associated to a lattice defines a Type of codes.

Let $O := (O, M_O, \psi, \Phi_O)$ be an order in some form $\mathbb{R}$-algebra $(\mathcal{A}, M, \psi, \Phi)$. Then, in particular, O is a form ring (more precisely, a form $\mathbb{Z}$-algebra) and we can use the notion of finite representation of form rings. A *finite representation* $(V, \rho_M, \rho_\Phi, \beta)$ of O is a finite O-module V, together with a non-singular J-Hermitian $\mathbb{Q}/\mathbb{Z}$-valued bilinear form $\beta = \rho_M(\psi(1))$ on V (which defines the representation $\rho_M : M \to \mathrm{Bil}(V, \mathbb{Q}/\mathbb{Z})$), and a representation $\rho_\Phi : \Phi \to \mathrm{Quad}_0(V, \mathbb{Q}/\mathbb{Z})$.

Example 9.7.1. Let Λ be an even lattice, i.e. an isotropic lattice of Type $\rho(\mathbb{Z}_{\mathrm{II}})$ as defined in §9.5. Then Λ defines a *finite* representation $\rho^{(\Lambda)}$ of $\mathbb{Z}_{\mathrm{II}}$ as follows. The module $V_{\rho^{(\Lambda)}} = \Lambda^{\#,\beta}/\Lambda$, with $\rho_M^{(\Lambda)}(1)(v + \Lambda, w + \Lambda) = (v, w) + \mathbb{Z}$

and $\rho_\phi^{(\Lambda)}(1)(v+\Lambda) = 1/2\,(v,v) + \mathbb{Z}$ for all $v, w \in \Lambda^{\#,\beta}$. These maps are well-defined since Λ is even. The representation $\rho^{(\Lambda)}$ is called the *Type of* Λ. It only depends on the genus of Λ (as an integral lattice) and together with $\dim \Lambda$ determines the genus of Λ uniquely (again, since Λ is even). Since for any rational quadratic space (V, β) the isometry class of the quotient $\Gamma^{\#,\beta}/\Gamma$ does not depend on the choice of the maximal integral lattice Γ in V (see W. Scharlau [475, Theorem 5.3.4]), rational equivalence of lattices corresponds precisely to Witt equivalence of the associated finite representations.

To put this another way, the genus of an integral lattice can essentially be thought of as a Type, since the genus of Λ is determined by the quadratic form on $\Lambda^\#/\Lambda$, together with the ambient dimension (the real part of the genus), and knowledge of whether the lattice is even.

More generally, if Λ is an isotropic lattice of Type ρ_O, then there is a corresponding finite "quotient" representation $\rho^{(\Lambda)}$ on $\Lambda^{\#,\beta}/\Lambda$. Gluing theory for lattices now follows in the same way as gluing theory for codes, since any isotropic code in $\rho^{(\Lambda)}$ lifts to an isotropic lattice containing Λ; in particular, $\rho^{(\Lambda)}$ is Witt-null if and only if there exists an isotropic self-dual lattice containing Λ.

One of the best-known applications of gluing theory is to the classification of the 24-dimensional even unimodular lattices. Niemeier [389] originally carried out this classification using Kneser's method of neighbors.[6] Venkov [533] used theta series with harmonic coefficients to show that (with the exception of the Leech lattice), every such lattice has a finite index sublattice generated by "roots" (vectors of norm 2). Furthermore, for each such lattice, the component root lattices must all have the same Coxeter number h. It thus suffices to classify the root lattices that can occur, and determine how to glue them up to give even unimodular lattices. The root lattices are products of the following lattices, with the corresponding quotient representations shown in parentheses:

$$
\begin{aligned}
A_n :\ & (\mathbb{Z}/n\mathbb{Z}, \beta(x,y) = xy/n, \phi(1)(x) = x^2/2n)\,, \\
D_{2n} :\ & (\mathbb{Z}/2\mathbb{Z} \times \mathbb{Z}/2\mathbb{Z}, \\
& \beta((x_1,x_2),(y_1,y_2)) = n(x_1 - x_2)(y_1 - y_2)/2 + (x_1 y_2 + x_2 y_1)/2, \\
& \phi(1)((x_1,x_2)) = n(x_1 - x_2)^2/4 + x_1 x_2/2)\,, \\
D_{2n+1} :\ & (\mathbb{Z}/4\mathbb{Z}, \beta(x,y) = (2n+1)xy/4, \phi(1)(x) = (2n+1)x^2/8)\,, \\
E_6 :\ & (\mathbb{Z}/3\mathbb{Z}, \beta(x,y) = 2xy/3, \phi(1)(x) = x^2/3)\,, \\
E_7 :\ & (\mathbb{Z}/2\mathbb{Z}, \beta(x,y) = xy/2, \phi(1)(x) = -x^2/4)\,, \\
E_8 :\ & (\text{unimodular})\,.
\end{aligned}
\tag{9.7.1}
$$

---

[6] The correctness of Niemeier's enumeration was verified by Conway and Sloane [125] by using the mass formula (12.1.15).

Table 9.1. The 24-dimensional even unimodular lattices: the 23 Niemeier lattices and the Leech lattice.

Name	Components	h	Generators for glue code
α	D_{24}	46	[1]
β	$D_{16}E_8$	30	[10]
γ	E_8^3	30	[000]
δ	A_{24}	25	[5]
ϵ	D_{12}^2	22	[12],[21]
ζ	$A_{17}E_7$	18	[31]
η	$D_{10}E_7^2$	18	[110],[301]
θ	$A_{15}D_9$	16	[21]
ι	D_8^3	14	[(122)]
κ	A_{12}^2	13	[15]
λ	$A_{11}D_7E_6$	12	[111]
μ	E_6^4	12	[1(012)]
ν	$A_9^2D_6$	10	[240],[501],[053]
ξ	D_6^4	10	[even perms. of {0123}]
o	A_8^3	9	[(114)]
π	$A_7^2D_5^2$	8	[1112],[1721]
ρ	A_6^4	7	[1(216)]
σ	$A_5^4D_4$	6	[2(024)0],[33001],[30302],[30033]
τ	D_4^6	6	[111111],[0(02332)]
υ	A_4^6	5	[1(01441)]
ϕ	A_3^8	4	[3(2001011)]
χ	A_2^{12}	3	[2(11211122212)]
ψ	A_1^{24}	2	[1(00000101001100110101111)]
ω	Leech	0	—

Thus to specify one of the Niemeier lattices, it suffices to give the appropriate isotropic code in the corresponding direct sum of quotient representations. The resulting classification is shown in Table 9.1 (based on [133, Table 16.1]). The second column shows the component root lattices, the third column the Coxeter number h, and the last column gives the isotropic (or glue) code.

10

Maximal Isotropic Codes and Lattices

If a binary code has odd length N then it cannot be self-dual, and if N is even but not a multiple of 8 then the code can be (singly-even) self-dual but not doubly-even. However, for any length N, one may consider the *largest* isotropic (or self-orthogonal) codes (cf. §1.2) that can exist, and ask what can be said about their weight enumerators. The earliest results of this kind were given by Mallows and Sloane [366], for singly-even codes of odd length and doubly-even codes of lengths congruent to ± 1 mod 8 (see Theorem 10.3.1 and cases 1 and 7 of Theorem 10.2.1 below). The case of maximal self-orthogonal ternary codes containing the all-ones vector (see Theorem 10.4.2) was discussed by Mallows, Pless and Sloane [364, Theorem 9].[1] See also [499] and Mallows and Sloane [367]. In this chapter we show how our theory can be applied in this setting, to both codes and lattices. Replacing the *ad hoc* methods used in the earlier papers by a systematic approach enables us to correct some omissions in that work and obtain new results.

In §10.1 we develop the machinery needed to handle these codes and lattices. Then in the subsequent sections we describe the space of weight enumerators of maximal isotropic codes from the following families:

- doubly-even binary codes (Theorem 10.2.1)
- singly-even binary codes (Theorem 10.3.1)
- ternary codes (Theorem 10.4.1)
- ternary codes with $\mathbf{1}$ in the dual (Theorems 10.4.2 and 10.4.3)
- even additive trace-Hermitian self-orthogonal codes over $\mathbb{F}_4$ (Theorem 10.5.1)
- doubly-even codes over $\mathbb{Z}/4\mathbb{Z}$ (Theorem 10.6.1)

---

[1] The assertions made in the four lines following Eq. (5) of [364] are incorrect: Theorem 10.4.2 below gives the correct statement. Also, in Table 1 of that paper, the order of the monomial group for the ternary Golay code $\mathcal{G}_{12}$ should be $2^5 3^2 5\, 11$, and on the next line $4\mathcal{C}_3(12)$ should carry an asterisk.

In the second half of the chapter we use the results of the first half to describe the space of modular forms spanned by the theta series of

- maximal even lattices of determinant 3^k (Corollary 10.7.7)
- maximal even lattices of determinant 2^k (Theorem 10.7.14)

(Note that lattices of small determinant and modest length have been classified by several authors—see Conway and Sloane [127], [133] and the references therein.)

10.1 Maximal isotropic codes

If the representation ρ does not admit isotropic self-dual codes, then direct application of the theory developed in Chapter 5 tells us only that the Clifford-Weil group has no invariants. It turns out, however, that the main theorems can be used to obtain information about maximal self-orthogonal or isotropic codes (a maximal isotropic code is one with an anisotropic quotient representation, see Definition 3.5.3).

Theorem 10.1.1. *Let ρ be an anisotropic representation of the form ring (R, M, ψ, Φ) such that $\rho \oplus \overline{\rho}$ satisfies the Weight Enumerator Conjecture 5.5.2. Denote by $\mathbb{C}[V_\rho]^{\mathrm{Aut}(\rho)}$ the space of $\mathrm{Aut}(\rho)$-fixed points in $\mathbb{C}[V_\rho]$. Then*

$$\mathrm{End}_{\mathcal{C}(\rho)}(\mathbb{C}[V_\rho]) \cong \mathbb{C}[\mathrm{Aut}(\rho)].$$

In particular, $\mathcal{C}(\rho)$ acts irreducibly on $\mathbb{C}[V_\rho]^{\mathrm{Aut}(\rho)}$.

Proof. We use Schur's lemma that a $\mathbb{C}$-representation of a finite group is irreducible if and only if its endomorphism ring is $\mathbb{C}$. An endomorphism of the $\mathcal{C}(\rho)$-module $\mathbb{C}[V_\rho]$ is equivalent to an invariant of $\mathcal{C}(\rho)$ in $\mathbb{C}[V_\rho] \otimes \mathbb{C}[V_\rho]^* = \mathbb{C}[V_\rho \oplus V_{\overline{\rho}}]$ because the contragredient $\mathcal{C}(\rho)$-module $\mathbb{C}[V_\rho]^*$ is isomorphic to $\mathbb{C}[V_{\overline{\rho}}]$. By our assumption the space of $\mathcal{C}(\rho)$-invariants in $\mathbb{C}[V_\rho \oplus V_{\overline{\rho}}]$ is spanned by the self-dual isotropic codes in $\rho + \overline{\rho}$. Since ρ is anisotropic, any such code is a full subdirect product

$$C = C_\alpha := \{(v, \alpha(v)) \mid v \in V_\rho\},$$

for some automorphism $\alpha \in \mathrm{Aut}(\rho)$. Hence the self-dual isotropic codes in $\rho + \overline{\rho}$ are in bijection with $\mathrm{Aut}(\rho)$, proving that

$$\mathrm{End}_{\mathcal{C}(\rho)}(\mathbb{C}[V_\rho]) \cong \mathbb{C}[\mathrm{Aut}(\rho)]$$

as $\mathrm{Aut}(\rho)$-modules. The endomorphism ring of the subspace of $\mathrm{Aut}(\rho)$-fixed points in $\mathbb{C}[V_\rho]$ is then the image of $\mathbb{C}[\mathrm{Aut}(\rho)]$ under the trivial representation of $\mathrm{Aut}(\rho)$ and hence is isomorphic to $\mathbb{C}$.   □

Let us fix a finite representation ρ of the form ring (R, M, ψ, Φ). Then ρ induces an element of the Witt group (see §4.6), which in turn has up to equivalence a unique anisotropic representative ρ'. In particular, any maximal isotropic code C in the representation ρ gives rise to the quotient representation $\rho/C \cong \rho'$.

Consider the representation $\rho + \overline{\rho'}$ of (R, M, ψ, Φ). A self-dual isotropic code C^+ in this representation is obtained by gluing an isotropic code C in ρ to an isotropic code C' in ρ'. But since ρ' is anisotropic, $C' = 0$ and the code C must be maximal isotropic. Therefore C^+ is determined by C together with an isomorphism $\rho/C \to \rho'$, and the set of codes C^+ over a given C is bijective with the automorphism group $\mathrm{Aut}(\rho')$. Moreover, any two such codes over the same C are thus equivalent under some automorphism of ρ'.

In particular, if we symmetrize the full weight enumerator of C^+ under $\mathrm{Aut}(\rho')$, the result depends only on C. In fact, define a linear map

$$\Pi : \mathbb{C}[V_\rho + V_{\rho'}] \to \mathbb{C}[V_\rho], \quad e_{vw} \mapsto \begin{cases} 0 & \text{if } w \neq 0, \\ e_v & \text{if } w = 0, \end{cases} \tag{10.1.1}$$

for $v \in V_\rho, w \in V_{\rho'}$. Then

$$\Pi \left(\frac{1}{|\mathrm{Aut}(\rho')|} \sum_{g \in \mathrm{Aut}(\rho')} \mathrm{fwe}(C^+)^g \right) = \Pi \left(\mathrm{fwe}(C^+) \right) = \mathrm{fwe}(C). \tag{10.1.2}$$

Finally, since by Theorem 10.1.1 the Clifford-Weil group $\mathcal{C}(\overline{\rho'})$ acts irreducibly on the $\mathrm{Aut}(\rho')$-invariant subspace, the full weight enumerator of C in turn determines the $\mathrm{Aut}(\rho')$-symmetrized weight enumerator of C^+.

Theorem 10.1.2. *Let (R, Φ) be a form ring such that, for all finite representations of (R, Φ), the invariant space of the Clifford-Weil group is spanned by full weight enumerators of self-dual codes. For any finite representation ρ of (R, Φ) with anisotropic quotient ρ', the map Π of Equation (10.1.1) gives an isomorphism between the space of invariants of $\mathcal{C}(\rho + \overline{\rho'})$ fixed by $\mathrm{Aut}(\rho')$ and the subspace of $\mathbb{C}[V_\rho]$ spanned by the full weight enumerators of maximal isotropic codes in ρ.*

Proof. By assumption the space of vector invariants of $\mathcal{C}(\rho + \overline{\rho'})$ is spanned by the full weight enumerators of isotropic self-dual codes C^+ in $\rho + \overline{\rho'}$. We will establish the bijection on the level of codes.
(a) We first show that Π is surjective. Let C be a maximal isotropic code in ρ and choose an isomorphism $\varphi : \rho/C \to \rho'$. Then the code

$$C_\varphi^+ := \{(v, w) \in C^\perp \times V_{\overline{\rho'}} \mid \varphi(v + C) = w\} \tag{10.1.3}$$

is a self-dual isotropic code in $\rho + \overline{\rho'}$. The other self-dual isotropic codes in $\rho + \overline{\rho'}$ that contain $C + 0$ are the codes $C_{\alpha \circ \varphi}^+ = (\mathrm{id}, \alpha)C_\varphi^+$ for $\alpha \in \mathrm{Aut}(\rho')$. Since α fixes 0, the image under Π is independent of α:

$$\Pi(\text{fwe}(C_\varphi^+)) = \Pi(\text{fwe}(C_{\alpha\circ\varphi}^+)) = \Pi\Big(\frac{1}{|\text{Aut}(\rho')|} \sum_{\alpha\in\text{Aut}(\rho')} \text{fwe}(C^+)^\alpha\Big) = \text{fwe}(C).$$

This shows the surjectivity of Π.

(b) To construct an inverse of the map Π, we note that each isotropic self-dual code C^+ in $\rho + \overline{\rho'}$ is obtained from a maximal isotropic code C in V_ρ. Let C^+ be such a code and define $C := C^+ \cap V_\rho$. Then $\rho/C \cong \rho'$ and therefore C is maximal isotropic. $\qquad\square$

Of course, we can also symmetrize by appropriate subgroups of $\text{Aut}(\rho)$.

We will now relate these vector invariants to the more familiar notion of relative (polynomial) invariants (cf. §5.6.2). Let $\rho = N\rho_0$ be a finite representation of a form ring and let C be a maximal isotropic code in ρ. Symmetrizing by the permutation group S_N of ρ maps the full weight enumerator of C to the complete weight enumerator; the latter will turn out to be a relative polynomial invariant of $\mathcal{C}(\rho_0)$.

We therefore compose the map $\Pi : \mathbb{C}[V_{N\rho_0} + V_{\overline{\rho'}}] \to \mathbb{C}[V_{N\rho_0}] \cong \otimes^N \mathbb{C}[V_{\rho_0}]$ defined in (10.1.1) (where ρ' is the anisotropic representative of ρ) with symmetrization under S_N to obtain a map $\tilde{\Pi} : \mathbb{C}[V_{N\rho_0} + V_{\overline{\rho'}}] \to \text{Sym}_N(\mathbb{C}[V_{\rho_0}])$.

Lemma 10.1.3. *Let $i \in \mathbb{C}[V_{N\rho_0} + V_{\overline{\rho'}}]$ be invariant under $\mathcal{C}(\rho_0)$. Then $\tilde{\Pi}(i) \in \text{Inv}(\mathcal{C}(\rho_0), S)$ is a relative invariant of $\mathcal{C}(\rho_0)$, where S is the simple $\mathcal{C}(\rho_0)$-module $\mathbb{C}[V_{\overline{\rho'}}]^{\text{Aut}(\rho')}$.*

Proof. For $w_0 \in V_{\rho'}$, let $\Pi_{w_0} : \mathbb{C}[V_{N\rho_0} + V_{\overline{\rho'}}] \to \mathbb{C}[V_{N\rho_0}] \cong \otimes^N \mathbb{C}[V_{\rho_0}]$ be defined by

$$e_{vw} \mapsto \begin{cases} e_v & \text{if } w = w_0, \\ 0 & \text{otherwise}, \end{cases}$$

and let $\tilde{\Pi}_{w_0}$ be the composition of Π_{w_0} with symmetrization by S_N.

Let $i := \sum a_{v_1\ldots v_N w} e_{v_1\ldots v_N w} \in \mathbb{C}[V_{N\rho_0} + V_{\overline{\rho'}}]^{\mathcal{C}(\rho_0)}$. Then for $w_0 \in V_{\rho'}$ we set $p_{w_0}(x) := \tilde{\Pi}_{w_0}(i) = \sum a_{v_1\ldots v_N w_0} x_{v_1} \cdots x_{v_N}$. For $g \in \mathcal{C}(\rho_0)$ we have $i = i^g$ and hence

$$p_{w_0}(gx) = \tilde{\Pi}_{w_0}\Big(\sum a_{v_1\ldots v_N w} e_{gv_1\ldots gv_N w}\Big)$$
$$= \tilde{\Pi}_{w_0}\Big(\sum a_{v_1\ldots v_N w} e_{v_1\ldots v_N \rho'(g^{-1})w}\Big)$$
$$= \tilde{\Pi}_{\rho'(g)w_0}(i) = p_{\rho'(g)w_0}(x). \qquad (10.1.4)$$

Symmetrization by $\text{Aut}(\rho')$ finishes the proof. $\qquad\square$

Therefore, if $\mathcal{C}(\rho_0)$ satisfies the hypotheses of Theorem 10.1.2, we get a bijection between the space $\text{Inv}(\mathcal{C}(\rho_0), \mathbb{C}[V_{\rho'}]^{\text{Aut}(\rho')})_{e_0}$ and the space spanned by the complete weight enumerators of maximal isotropic codes in $N\rho_0$ (such that $N\rho_0$ is Witt equivalent to ρ'). In particular, we have the following remark:

Remark 10.1.4. Let ρ be a representation of a form ring (R, Φ), as in Theorem 10.1.2 and let n be the order of ρ in the Witt group $W(R, \Phi)$. For all $i \in \mathbb{Z}/n\mathbb{Z}$, let $I_i :=$

$\langle \mathrm{cwe}(C) \mid C$ is a maximal isotropic code in $j\rho$ for some $j \equiv i \pmod{n}\rangle_{\mathbb{C}}$.

Then $I_0 = \mathrm{Inv}(\mathcal{C}(\rho))$ is the ring of polynomial invariants of $\mathcal{C}(\rho)$, and, for each i, the space I_i is a module for I_0. Let ρ_i' be the anisotropic representative of $i\rho$ in the Witt group and let χ_i be the character of $\mathcal{C}(\rho)$ acting on $\mathbb{C}[V_{\overline{\rho_i'}}]^{\mathrm{Aut}(\rho_i')}$. The Molien series for I_i is

$$\mathrm{MS}_{I_i}(t) := \sum_{N=0}^{\infty} \dim((I_i)_N)t^N = \frac{1}{|\mathcal{C}(\rho)|} \sum_{g \in \mathcal{C}(\rho)} \frac{\overline{\chi_i}(g)}{\det(I - tg)}. \tag{10.1.5}$$

Proof. The space I_i is a module over I_0, since I_0 is spanned by complete weight enumerators of codes C_0 of Type ρ and for such a code C_0 in $dn\rho$ and a maximal isotropic code C_i in $(i + d'n)\rho$, the code $C_0 \perp C_i$ is clearly maximal isotropic in $(i + (d+d')n)\rho$. The dimension formula follows from the discussion above together with Remark 5.6.6.   □

Remark 10.1.5. If I_i is a free module over I_0, then its rank equals $\chi_i(1)$, the dimension of the fixed space of $\mathrm{Aut}(\rho_i')$.

Proof. In this case the rank of I_i is just $\lim_{t \to 1} \frac{\mathrm{MS}_{I_i}}{\mathrm{MS}_{I_0}}$, which is easily seen to be $\chi_i(1)$.   □

This remark applies in particular to the case when $\mathcal{C}(\rho)$ is a reflection group, since then the modules of relative invariants are free (Smith [506, §7.6], Stanley [511]).

To find codes that generate the I_0-module I_i, we can either guess (plenty of information is available, since codes of modest length have been well-studied, cf. Chap. 12), or we can apply the following *subtraction* procedure (see also [454, §11.3]):

Definition 10.1.6. Let $\rho = (V, \rho_M, \rho_\Phi, \beta)$ be a Type and let $C \leq V^N$ be an isotropic self-dual code in $N\rho$. For $m \leq N$ let C' be a maximal isotropic code in $m\rho$. We define a code $S_{C'}(C)$ as follows. Let $\tilde{C} := C' + (C \cap C'^{\perp})$, where $C'^{\perp}$ is the dual of $C' \oplus \{0^{N-m}\}$ in $N\rho$. Then $\tilde{C}$ is an isotropic self-dual code in $N\rho$, as shown in the proof of Lemma 4.6.7. We say that the code

$$S_{C'}(C) := S_{\{1,\dots,m\},C'}(C) := \{(c_{m+1}, \dots, c_N) \mid c \in \tilde{C}, (c_1, \dots, c_m) = \mathbf{0}\}$$

of length $N - m$ is obtained by *subtracting* C' from C.

In general there will be many different ways to do this depending on the choice of C' and the m coordinates. This technique will be used again in §12.2.

Remark 10.1.7. Since C' is maximal isotropic, an equivalent description of the subtracted code is $S_{C'}(C) = \{(c_{m+1}, \ldots, c_N) \mid c \in C, (c_1, \ldots, c_m) \in C'\}$.

Closely related to the subtraction construction is the construction of relative invariants with differential operators given in Corollary 5.6.8. By Remark 10.1.7, we may obtain the full weight enumerator of $S_{C'}(C)$ by applying the differential operator $\mathrm{fwe}_{C'}(\partial)$ with respect to the first m coordinates and then dividing by the number of $c \in C$ such that the first m coordinates of c are a given codeword in C', which is just the order of the intersection $|C \cap C'|$. Symmetrizing by S_N, we obtain the following theorem for the complete weight enumerators:

Theorem 10.1.8. Let $C \leq V^N$ be a self-dual isotropic code of Type ρ and let $C' \leq V^m$ be a maximal isotropic code in $m\rho$ for some length $m < N$. Let $D := \mathrm{cwe}_{C'}(\partial)$ be the differential operator associated with the complete weight enumerator of C'. Then

$$D(\mathrm{cwe}_C) = \sum_{1 \leq i_1 < \ldots < i_m \leq N} \sum_{C''} |\mathrm{Perm}(C'')| \cdot |C''_{i_1, \ldots, i_m} \cap C| \cdot \mathrm{cwe}(S_{i_1, \ldots, i_m, C''}(C)),$$

where the second sum is over all codes C'' in V^m that are permutation equivalent to C'.

Note that, for any anisotropic representation ρ' of the form ring (R, M, ψ, Φ), the vector e_0 is the unique (up to scalar multiples) vector in the simple $\mathcal{C}(\rho)$-module $S := \mathbb{C}[V_{\rho'}]^{\mathrm{Aut}(\rho')}$ that is fixed by the parabolic subgroup $P(\rho') = \langle d_\phi, m_u \mid u \in R^*, \phi \in \Phi \rangle$. Since $S^* = \mathbb{C}[V_{\overline{\rho}'}]^{\mathrm{Aut}(\overline{\rho}')}$ and $\mathcal{C}(\rho) = \overline{\mathcal{C}(\rho)}^{tr}$, any relative invariant $p \in \mathrm{Inv}(\mathcal{C}(\rho), \mathbb{C}[V_{\overline{\rho}'}]^{\mathrm{Aut}(\overline{\rho}')})_{e_0}$ yields a differential operator $D := \overline{p}(\partial)$ with

$$D(\mathrm{Inv}(\mathcal{C}(\rho))) \subset \mathrm{Inv}(\mathcal{C}(\rho), \mathbb{C}[V_{\rho'}]^{\mathrm{Aut}(\rho')})_{e_0}$$

(see Corollary 5.6.8). This is a useful technique for constructing relative invariants in both modules simultaneously, as will be illustrated in the examples below.

10.2 Maximal isotropic doubly-even binary codes

We now give several examples of interesting maximal isotropic codes. We start with doubly-even binary codes. For $N \leq 8$, there is a unique maximal isotropic code in the representation $N\rho(2_{\mathrm{II}})$. This is the zero code z_N for $N \leq 3$, d_N for $4 \leq N \leq 6$, e_7 and e_8 for $N = 7$ and 8. Since the automorphism group of the extended binary Golay code g_{24} acts 5-transitively on $\{1, \ldots, 24\}$, there is a unique maximal isotropic code of length $N = 19, \ldots, 24$ obtained from the Golay code by shortening (then adjoining appropriate vectors for

$N = 19, 20$), or equivalently by subtracting the unique maximal isotropic code of length $24 - N$; call this maximal isotropic code g_N. In particular, the weight enumerators of the zero codes of length N are in I_N for $N = 1, 2, 3$. They provide differential operators $D_N = \frac{\partial^N}{\partial x^N}$ such that $I_{8-N} = D_N(I_0)$ (see Corollary 5.6.8). The modules I_{8-N} then provide differential operators D_{8-N} to produce generators for I_N. For I_4, we may begin with the weight enumerator $x^4 + y^4$ of e_4, which yields the differential operator $D_4 = \frac{\partial^4}{\partial x^4} + \frac{\partial^4}{\partial y^4}$ such that $I_4 = D_4(I_0)$. To verify the completeness of the result, we compare the Molien series of the resulting I_0-modules with the series calculated using Remark 10.1.4. This yields the following theorem:

Theorem 10.2.1. *Consider the Type $\rho(2_{II})$ of doubly-even self-dual binary codes and let $\rho = N\rho(2_{II})$ be a representation of $R(2_{II})$. Now $\rho(2_{II})$ has order 8 in the Witt group, by Example 4.6.12, so there are eight possibilities for ρ'.*

$N \equiv 0 \pmod 8$. Then ρ is Witt-null, and a maximal isotropic code is self-dual. This case is covered by Gleason's Theorem 6.4.2. The space spanned by the complete weight enumerators of the maximal isotropic (hence self-dual) codes is the invariant ring of the Clifford-Weil group:

$$I_0 := \mathbb{C}[x^8 + 14x^4y^4 + y^8, \; x^4y^4(x^4 - y^4)^4], \qquad (10.2.1)$$

generated by the weight enumerators of e_8 and g_{24}, as described in (6.4.2), (6.4.3), (6.4.4).

$N \equiv 1 \pmod 8$. Then $\rho' = \rho(2_{II})$. The $\mathrm{Aut}(\rho')$-invariant subspace of $\mathbb{C}[V_{\overline{\rho'}}]$ is two-dimensional (since $\mathrm{Aut}(\rho') = 1$); we denote the generators by $u = e_0$ and $v = e_1$. The space of $S_N \times \mathrm{Aut}(\rho')$-symmetrized weight enumerators of the maximal isotropic codes C^+ is the linear subspace consisting of the elements of the appropriate degree in the graded I_0-module generated by $xu + yv$ and

$$(x^{17} + 17x^{13}y^4 + 187x^9y^8 + 51x^5y^{12})u$$
$$+ (y^{17} + 17y^{13}x^4 + 187y^9x^8 + 51y^5x^{12})v; \qquad (10.2.2)$$

taking the coefficient of u gives the space of weight enumerators of maximal isotropic codes of length $N \equiv 1 \bmod 8$. (In the remaining examples we will just give this space.) The Molien series for this space is $(t + t^{17})/((1 - t^8)(1 - t^{24}))$. An I_0-basis for I_1 is given by the weight enumerator, x, of the zero code of length 1 and the weight enumerator $x^{17} + 17x^{13}y^4 + 187x^9y^8 + 51x^5y^{12}$ of the $[17, 8, 4]$ code $(d_{10}e_7)^+$. The latter code was first found by Pless [414].

$N \equiv 2 \pmod 8$. Then $\rho' = 2\rho(2_{II})$, and $\mathrm{Aut}(\rho') = Z_2$. The $\mathrm{Aut}(\rho')$-invariant subspace of $\mathbb{C}[V_{\overline{\rho'}}]$ is thus three-dimensional, spanned (say) by u^2, uv, v^2. The space of S_N-symmetrized weight enumerators of maximal isotropic codes is the appropriate slice of the graded I_0-module generated by

$$(x^2,\; x^2y^4(x^4 - y^4),\; x^6y^4(x^4 + 3y^4)(x^4 - y^4)),\qquad (10.2.3)$$

and having Molien series $(t^2 + t^{10} + t^{18})/((1 - t^8)(1 - t^{24}))$. An I_0-basis for I_2 is given by the weight enumerators of d_2, d_{10} and d_{18} (cf. (2.4.17)).

$N \equiv 3 \pmod 8$. Then $\rho' = 3\rho(2_{\mathrm{II}})$ and $\mathrm{Aut}(\rho') = S_3$. The $\mathrm{Aut}(\rho')$-invariant subspace is four-dimensional, spanned (say) by u^3, u^2v, uv^2, v^3. The corresponding graded I_0-module is generated by

$$(x^3,\; x^3y^4(x^4 - y^4),\; x^7y^4(x^4 + 3y^4)(x^4 - y^4),$$
$$x^{11}y^8(5x^4 + 7y^4)(x^4 - y^4)),\qquad (10.2.4)$$

with Molien series $(t^3 + t^{11} + t^{19} + t^{27})/((1 - t^8)(1 - t^{24}))$. An I_0-basis for I_3 is given by the weight enumerators of d_3, d_{11}, g_{19} and d_4g_{23}.

$N \equiv 4 \pmod 8$. Then $\rho' = \rho(2_{\mathrm{II}})' = 4\rho(2_{\mathrm{II}})/\langle 1111 \rangle$ (see Example 4.6.12) and $\mathrm{Aut}(\rho') = S_3 \cong S_4/(Z_2 \times Z_2)$. The $\mathrm{Aut}(\rho')$-invariant subspace is two-dimensional, spanned (say) by u and v, transforming under the MacWilliams transform as $(u, v) \mapsto ((u + 3v)/2, (u - v)/2)$. The corresponding graded I_0-module is generated by

$$(x^4 + y^4,\; x^4y^4(x^4 - y^4)(x^8 - y^8)),\qquad (10.2.5)$$

with Molien series $(t^4 + t^{20})/((1 - t^8)(1 - t^{24}))$. An I_0-basis for I_4 is given by the weight enumerators of d_4 and g_{20}.

$N \equiv 5 \pmod 8$. Similar to the case $N \equiv 3 \pmod 8$.

Now $\rho' = 3\rho(2_{\mathrm{II}})$. The corresponding graded I_0-module is generated by

$$(x(x^4+y^4), xy^4(5x^4-y^4)(x^4-y^4), x^5y^4(3x^4+y^4)(x^4-y^4)^2, x^5y^4(x^4-y^4)^3),$$
$$(10.2.6)$$

with Molien series $(t^5 + t^{13} + 2t^{21})/((1 - t^8)(1 - t^{24}))$. An I_0-basis for I_5 is given by the weight enumerators of d_5, e_7d_6, e_7^3 and g_{21}.

$N \equiv 6 \pmod 8$. Similar to the case $N \equiv 2 \pmod 8$.

Now $\rho' = 2\rho(2_{\mathrm{II}})$ and the corresponding graded I_0-module is generated by

$$(x^2(x^4 + 3y^4),\; x^2y^4(x^4 - y^4)^2, x^6y^4(x^4 - y^4)^3),\qquad (10.2.7)$$

with Molien series $(t^6 + t^{14} + t^{22})/((1 - t^8)(1 - t^{24}))$. An I_0-basis for I_6 is given by the weight enumerators of d_6, d_{14} and g_{22}.

$N \equiv 7 \pmod 8$. Similar to the case $N \equiv 1 \pmod 8$.

Here $\rho' = \rho(2_{\mathrm{II}})$ and the corresponding graded I_0-module is generated by

$$(x^7 + 7x^3y^4,\; x^{23} + 506x^{15}y^8 + 1288x^{11}y^{12} + 253x^7y^{16}),\qquad (10.2.8)$$

with Molien series $(t^7 + t^{23})/((1 - t^8)(1 - t^{24}))$. An I_0-basis for I_7 is given by the weight enumerators of the $[7, 4, 3]$ Hamming code e_7 and the $[23, 12, 7]$ Golay code g_{23}.

Remark. When $N \equiv \pm 2 \pmod 8$, the set of vectors $v \in V_\rho$ such that $v00$ or $v11$ is in C^+ is a singly-even self-dual binary code uniquely determined by C. The set of vectors $v \in V_\rho$ such that $v01$ or $v10$ is in C^+ is then the shadow of C. The corresponding weight enumerators can be read off from the weight enumerator of C^+.

10.3 Maximal isotropic even binary codes

Theorem 10.3.1. *Type $\rho(2_\mathrm{I})$, singly-even self-dual binary codes. Let $\rho = N\rho(2_\mathrm{I})$ be a representation of the form ring $R(2_\mathrm{I})$. Since $\rho(2_\mathrm{I})$ has order 2 in the Witt group, there are two cases.*

$N \equiv 0 \pmod 2$. This is the other part of Gleason's Theorem 6.4.2. The invariant ring of $\mathcal{C}(\rho(2_\mathrm{I}))$ is

$$I_0 := \mathbb{C}[x^2 + y^2, \; x^2 y^2 (x^2 - y^2)^2], \tag{10.3.1}$$

spanned by the weight enumerators of i_2 and e_8, as described in (6.3.5), (6.3.6), (6.3.7).
$N \equiv 1 \pmod 2$. Then $\rho' = \rho(2_\mathrm{I})$, and $\mathrm{Aut}(\rho') = 1$; the $\mathrm{Aut}(\rho')$-invariants are two-dimensional. The corresponding graded I_0-module is generated by

$$(x, \; x^7 + 7x^3 y^4), \tag{10.3.2}$$

with Molien series $(t + t^7)/((1 - t^2)(1 - t^8))$. As an I_0-module this is spanned by the weight enumerators of d_1 and e_7.

10.4 Maximal isotropic ternary codes

Theorem 10.4.1. *Hamming weight enumerators of ternary maximal self-orthogonal codes. Let $\rho = N\rho(3)$ be a representation of $R(3)$. Since $\rho(3)$ has order 4 in the Witt group, there are four possibilities for N.*

$N \equiv 0 \pmod 4$. The invariant ring for $\mathcal{C}(\rho(3))$ is

$$I_0 := \mathbb{C}[x^4 + 8xy^3, \; y^3(x^3 - y^3)^3], \tag{10.4.1}$$

and is generated by the Hamming weight enumerators of the codes t_4 and g_{12} of §2.4.5, as described in (7.4.3), (7.4.4). This result is also due to Gleason (cf. Theorem 6.4.2, §7.4.1).
$N \equiv 1 \pmod 4$. Then $\rho' = \rho(3)$, $\mathrm{Aut}(\rho') = Z_2$, and the space of $\mathrm{Aut}(\rho')$-invariants is two-dimensional, generated by u and v transforming as $(u, v) \mapsto (u + 2v, u - v)/\sqrt{3}$. The corresponding graded I_0-module is generated by

$$(x, \; y^3(5x^3 + 4y^3)(x^3 - y^3)), \tag{10.4.2}$$

with Molien series $(t + t^9)/((1 - t^4)(1 - t^{12}))$. An I_0-basis for I_1 is given by the Hamming weight enumerators of the zero code of length 1 and the $[9, 4, 3]$ code e_3^{3+} (see below), where e_3 is the $[3, 1, 3]$ repetition code.
$N \equiv 2 \pmod 4$. Then $\rho' = 2\rho(3)$, $\mathrm{Aut}(\rho') = O_2^-(\mathbb{F}_3) = Z_2 \wr Z_2$. The space of $\mathrm{Aut}(\rho')$-invariants is thus generated by u^2, uv, v^2. The corresponding graded I_0-module is generated by

$$(x^2, \; y^3(x^3 - y^3), xy^3(x^3 - y^3)^2)\,, \qquad\qquad (10.4.3)$$

with Molien series $(t^2 + t^6 + t^{10})/((1 - t^4)(1 - t^{12}))$. An I_0-basis for I_1 is given by the Hamming weight enumerators of the zero code of length 2, e_3^2 and the $[10, 4, 6]$ shortened Golay code g_{10}.

$N \equiv 3 \pmod 4$. Similar to the case $N \equiv 1 \pmod 4$, but $\rho' = \overline{\rho(3)}$. The corresponding graded I_0-module is generated by

$$(x^3 + 2y^3, \; x^2y^3(x^3 - y^3)^2)\,, \qquad\qquad (10.4.4)$$

with Molien series $(t^3 + t^{11})/((1 - t^4)(1 - t^{12}))$. An I_0-basis for I_2 is given by the Hamming weight enumerators of e_3 and the $[11, 5, 6]$ shortened Golay code g_{11}.

The $[9, 4, 3]$ ternary code e_3^{3+} mentioned in the theorem has generator matrix

$$\begin{bmatrix} 1 & 1 & 1 & 0 & 0 & 0 & 0 & 0 & 0 \\ 0 & 0 & 0 & 1 & 1 & 1 & 0 & 0 & 0 \\ 0 & 0 & 0 & 0 & 0 & 0 & 1 & 1 & 1 \\ 1 & 2 & 0 & 1 & 2 & 0 & 1 & 2 & 0 \end{bmatrix}\,,$$

and $\mathrm{hwe}_{e_3^{3+}} = x^9 + 6\,x^6y^3 + 66\,x^3y^6 + 8\,y^9$ (Mallows, Pless and Sloane [364]).

Theorem 10.4.2. *Complete weight enumerators of ternary maximal self-orthogonal codes whose duals contain* **1**. *Let* $\rho = N\rho(3_1)$ *be a representation of* $R(3_1)$. *Now* $\rho(3_1)$ *has order 12 in the Witt group and so there are 12 possibilities for* N:

$N \equiv 0 \pmod{12}$. I_0 *is the invariant ring of the associated Clifford-Weil group* $\mathcal{C}(\rho(3_1))$, *with Molien series* $(1 + t^{24})/\sigma(t)$, *where* $\sigma(t) = (1 - t^{12})^2(1 - t^{36})$ *(cf. (7.4.5)). The invariant ring* I_0 *is spanned by the complete weight enumerators* $p_1 := \mathrm{cwe}(e_3^{4+})$, $p_2 := \mathrm{cwe}(g_{12})$, $p_3 := \mathrm{cwe}(S(36))$ *and* $p_4 := \mathrm{cwe}(XQ(\mathbb{F}_3, 23))$ *(cf. (7.4.6)). Let* $I_0' := \mathbb{C}[p_1, p_2, p_3]$ *be the polynomial ring spanned by the specified primary invariants. The modules* I_i *are free over* I_0'. *To describe the modules* I_i $(i = 1, \ldots, 11)$ *we will use the complete weight enumerators of the unique maximal isotropic codes of lengths* $1, \ldots, 6$ *which have* **1** *in their duals. These codes and their weight enumerators* q_N *are as follows (as usual* 0_N *denotes the zero code of length* N):

$$
\begin{aligned}
0_1 : \quad & q_1 := x_0\,, \\
0_2 : \quad & q_2 := x_0^2\,, \\
e_3 : \quad & q_3 := x_0^3 + x_1^3 + x_2^3\,, \\
e_3 0_1 : \quad & q_4 := x_0^4 + x_0 x_1^3 + x_0 x_2^3\,, \\
e_3 0_2 : \quad & q_5 := x_0^5 + x_0^2 x_1^3 + x_0^2 x_2^3\,, \\
e_3^2 : \quad & q_6 := x_0^6 + 2x_0^3 x_1^3 + 2x_0^3 x_2^3 + 2x_1^3 x_2^3 + x_1^6 + x_2^6\,.
\end{aligned}
\qquad (10.4.5)
$$

$N \equiv 1 \pmod{12}$. Here $\rho' = \rho(3_1)$, $\mathrm{Aut}(\rho) = 1$ and the Molien series of I_1 is $(t + 2t^{13} + 2t^{25} + t^{37})/\sigma(t)$. An I_0'-basis for I_1 is given by

$$a_1 := q_1, \qquad a_2 := A_2(\partial)(p_4), \; a_3 := A_2(\partial)(p_1 p_2),$$
$$a_4 := A_2(\partial)(p_3), \; a_5 := p_4 a_1, \qquad a_6 := p_4 a_2,$$

where the polynomial A_2 is defined below in (10.4.10).

$N \equiv 2 \pmod{12}$. Here $\rho' = 2\rho(3_1)$, $\mathrm{Aut}(\rho') = Z_2$ and the Molien series of I_2 is $(t^2 + 4t^{14} + 4t^{26} + 3t^{38})/\sigma(t)$. An I_0'-basis for I_2 is given by

$$b_1 := q_2, \qquad b_2 := B_1(\partial)(p_4), \; b_3 := B_1(\partial)(p_1 p_2), \; b_4 := B_1(\partial)(p_1^2),$$
$$b_5 := B_1(\partial)(p_2^2), \; b_6 := p_4 q_2, \qquad b_7 := B_1(\partial)(p_3), \quad b_8 := B_2(\partial)(p_3),$$
$$b_9 := B_3(\partial)(p_3), \; b_{10} := p_4 b_2, \qquad b_{11} := p_4 b_3, \qquad b_{12} := B_1(\partial)(p_4^2),$$

where the polynomials B_i are defined in (10.4.9).

$N \equiv 3 \pmod{12}$. Here $\rho' = 3\rho(3_1)/\langle(1,1,1)\rangle$. The underlying module

$$V_{\rho'} = \langle(1,1,1)\rangle^{\perp}/\langle(1,1,1)\rangle = \{(0,0,0), (1,2,0), (2,1,0)\}$$

has three elements, and $\mathrm{Aut}(\rho') = S_3/Z_3 = Z_2$. The fixed space $M_3 := \mathbb{C}[V_{\rho'}]^{\mathrm{Aut}(\rho')}$ has basis $(e_{(0,0,0)}, e_{(1,2,0)} + e_{(2,1,0)})$, and with respect to these generators the representation ρ_3 of $\mathcal{C}(\rho)$ on M_3 is given by

$$h \mapsto \frac{1}{\sqrt{3}} \begin{pmatrix} 1 & 1 \\ 2 & -1 \end{pmatrix}, d_{\phi_1} \mapsto 1, m_2 \mapsto 1, d_{[1]} \mapsto \mathrm{diag}(1, \zeta_3^2).$$

From Remark 10.1.4 we find that the Molien series of I_3 is $(t^3 + t^{15} + 2t^{27})/\sigma(t)$. An I_0'-basis for I_3 is given by

$$c_1 := q_3, \; c_2 := C_1(\partial)(p_4), \; c_3 := C_1(\partial)(p_3), \; c_4 := p_4 c_1,$$

where the polynomial C_1 is defined in (10.4.8).

$N \equiv 4 \pmod{12}$. Here $\rho' = 4\rho(3_1)/\langle(1,1,1,0)\rangle$ and $\mathrm{Aut}(\rho') = S_3/Z_3 = Z_2$. The representation of $\mathcal{C}(\rho)$ on the fixed space $\mathbb{C}[V_{\rho'}]^{\mathrm{Aut}(\rho')}$ is the tensor product of ρ_3 with the natural representation ρ of $\mathcal{C}(\rho)$. The Molien series of I_4 is $(t^4 + 4t^{16} + 4t^{28} + 3t^{40})/\sigma(t)$. An I_0'-basis for I_4 is given by

$$d_1 := q_4, \qquad d_2 := D_1(\partial)(p_4), \quad d_3 := D_1(\partial)(p_1 p_2),$$
$$d_4 := D_1(\partial)(p_1^2), \; d_5 := D_2(\partial)(p_3), \quad d_6 := p_4 d_1,$$
$$d_7 := D_1(\partial)(p_3), \; d_8 := D_1(\partial)(p_1 p_4), \; d_9 := D_1(\partial)(p_2 p_4),$$
$$d_{10} := p_4 d_2, \qquad d_{11} := p_4 d_3, \qquad d_{12} := D_1(\partial)(p_4^2),$$

where the polynomials D_i are defined in (10.4.7).

$N \equiv 5 \pmod{12}$. Here $\rho' = 5\rho(3_1)/\langle(1,1,1,0,0)\rangle$ and $\mathrm{Aut}(\rho') \cong D_8$. The fixed space $\mathbb{C}[V_{\rho'}]^{\mathrm{Aut}(\rho')}$ has dimension 9 and the Molien series of I_5 is $(t^5 + 6t^{17} + 6t^{29} + 5t^{41})/\sigma(t)$. An I_0'-basis for I_5 is given by

$$e_1 := q_5, \qquad e_2 := E_1(\partial)(p_4), \qquad e_3 := E_1(\partial)(p_1^2),$$
$$e_4 := E_1(\partial)(p_2^2), \qquad e_5 := E_2(\partial)(p_1p_2), \ e_6 := E_2(\partial)(p_4),$$
$$e_7 := E_3(\partial)(p_3), \qquad e_8 := p_4e_1, \qquad e_9 := E_1(\partial)(p_3),$$
$$e_{10} := E_2(\partial)(p_3), \quad e_{11} := E_1(\partial)(p_2^3), \quad e_{12} := E_1(\partial)(p_2p_4),$$
$$e_{13} := E_1(\partial)(p_1p_4), \ e_{14} := p_4e_2, \qquad e_{15} := p_4e_3,$$
$$e_{16} := p_4e_4, \qquad e_{17} := E_1(\partial)(p_4^2), \quad e_{18} := E_1(\partial)(p_1p_3),$$

where the polynomials E_i are defined in (10.4.6).

$N \equiv 6 \pmod{12}$. *Here* $\rho' = 6\rho(3_1)/\langle(1,1,1,0,0,0),(0,0,0,1,1,1)\rangle$ *and* $\mathrm{Aut}(\rho') = S_3/Z_3 \wr Z_2 = Z_2 \wr Z_2$. *The representation of* $C(\rho)$ *on the fixed space* $\mathbb{C}[V_{\rho'}]^{\mathrm{Aut}(\rho')}$ *is the symmetric square* $\mathrm{Sym}^2(\rho_3)$. *The Molien series of* I_6 *is* $(t^6 + 2t^{18} + 2t^{30} + t^{42})/\sigma(t)$. *An* I_0'-*basis for* I_6 *is given by*

$$f_1 := q_6, \quad f_2 := q_6(\partial)(p_1^2), \ f_3 := q_6(\partial)(p_4),$$
$$f_4 := p_4f_1, \ f_5 := q_6(\partial)(p_3), \ f_6 = p_4f_2.$$

$N \equiv -5 \pmod{12}$. *Similar to the case* $N \equiv 5 \pmod{12}$ *but with the conjugate representation. The Molien series of* I_7 *is* $(2t^7 + 6t^{19} + 6t^{31} + 4t^{43})/\sigma(t)$. *An* I_0'-*basis for* I_7 *is given by*

$$
\begin{aligned}
&E_1 := q_5(\partial)(p_1), &&E_2 := e_2(\partial)(p_4), &&E_3 := q_5(\partial)(p_1^2),\\
&E_4 := q_5(\partial)(p_2^2), &&E_5 := q_5(\partial)(p_1p_2), &&E_6 := e_2(\partial)(p_3),\\
&E_7 := e_2(\partial)(p_1p_4), &&E_8 := e_2(\partial)(p_2p_4), &&E_9 := q_5(\partial)(p_2^3),\\
&E_{10} := q_5(\partial)(p_3), &&E_{11} := p_4E_1, &&E_{12} := p_4E_2,\\
&E_{13} := q_5(\partial)(p_1^3), &&E_{14} := q_5(\partial)(p_1p_4), &&E_{15} := p_4E_3,\\
&E_{16} := p_4E_4, &&E_{17} := p_4E_5, &&E_{18} := q_5(\partial)(p_4^2).
\end{aligned}
\tag{10.4.6}
$$

$N \equiv -4 \pmod{12}$. *Using the conjugate representation to the one for* $N \equiv 4 \pmod{12}$, *we find that the Molien series of* I_8 *is* $(t^8 + 4t^{20} + 4t^{32} + 3t^{44})/\sigma(t)$. *An* I_0'-*basis for* I_8 *is given by*

$$
\begin{aligned}
&D_1 := q_4(\partial)(p_1), &&D_2 := q_4(\partial)(p_4), \ D_3 := q_4(\partial)(p_1^2), \ D_4 := q_4(\partial)(p_2^2),\\
&D_5 := d_2(\partial)(p_3), &&D_6 := D_1p_4, \qquad D_7 := q_4(\partial)(p_3), \ D_8 := q_4(\partial)(p_2p_4),\\
&D_9 := q_4(\partial)(p_1p_4), &&D_{10} := D_2p_4, \qquad D_{11} := D_3p_4, \qquad D_{12} := q_4(\partial)(p_4^2).
\end{aligned}
\tag{10.4.7}
$$

$N \equiv -3 \pmod{12}$. *Similar to the case* $N \equiv 3 \pmod{12}$ *but with* $\rho' = 3\rho(3_1)/\langle(1,1,1)\rangle$. *The Molien series of* I_9 *is* $(t^9 + t^{21} + 2t^{33})/\sigma(t)$. *An* I_0'-*basis for* I_9 *is given by*

$$C_1 := q_3(\partial)(p_1), \ C_2 := q_3(\partial)(p_4), \ C_3 := p_4C_1, \ C_4 := q_3(\partial)(p_3). \tag{10.4.8}$$

$N \equiv -2 \pmod{12}$. *Similar to the case* $N \equiv 2 \pmod{12}$ *but with* $\rho' = 2\rho(3_1)$. *The Molien series of* I_{10} *is* $(3t^{10} + 4t^{22} + 4t^{34} + t^{46})/\sigma(t)$. *An* I_0'-*basis for* I_{10} *is given by*

$$
\begin{aligned}
&B_1 := q_2(\partial)(p_1), &&B_2 := q_2(\partial)(p_2), \ B_3 := b_2(\partial)(p_2^2), \ B_4 := q_2(\partial)(p_4),\\
&B_5 := q_2(\partial)(p_1p_2), &&B_6 := q_2(\partial)(p_1^2), \ B_7 := b_2(\partial)(p_3), \ B_8 := b_2(\partial)(p_1p_3),\\
&B_9 := q_2(\partial)(p_1p_4), &&B_{10} := p_4B_2, \qquad B_{11} := p_4B_3, \qquad B_{12} := p_4B_4.
\end{aligned}
\tag{10.4.9}
$$

$N \equiv -1 \pmod{12}$. *Similar to the case* $N \equiv 1 \pmod{12}$ *but with* $\rho' = \overline{\rho(3_1)}$. *The Molien series of* I_{11} *is* $(2t^{11} + 2t^{33} + 2t^{35})/\sigma(t)$. *An* I_0'*-basis for* I_{11} *is given by*

$$A_1 := q_1(\partial)(p_1), \ A_2 := q_1(\partial)(p_2), \ A_3 := q_1(\partial)(p_4),$$
$$A_4 := a_2(\partial)(p_3), \ A_5 := q_1(\partial)(p_3), \ A_6 := p_4 A_2. \tag{10.4.10}$$

To find the space spanned by the Hamming weight enumerators of the codes in the last theorem, we set $x_0 = x, y_0 = z_0 = y$ in the complete weight enumerators. This gives the following:

Theorem 10.4.3. *Hamming weight enumerators of ternary maximal self-orthogonal codes whose duals contain* **1**. *For* $i \in \{0, \ldots, 11\}$, *let* I_i *and* ρ *be as in Theorem 10.4.2 and set*

$$\overline{I}_i := \{p(x, y, y) \mid p \in I_i\}.$$

Then $\overline{I}_i$ *is the vector space spanned by the Hamming weight enumerators of maximal isotropic codes in* $N\rho$ *with* $N \equiv i \pmod{12}$. *In the notation of Theorem 10.4.2 we have*

$$\overline{I}_0 = \mathbb{C}[p_1(x, y, y), p_2(x, y, y)]$$
$$= \mathbb{C}[x^{12} + 264x^6 y^6 + 440x^3 y^9 + 24y^{12}, (y(x - y)(x^2 + xy + y^2))^3]. \tag{10.4.11}$$

Then $\overline{I}_0$*-bases for the modules* I_i *are as follows:*

($\overline{I}_1$) $x, y^6 x(x - y)(x^2 + xy + y^2)(2x^3 + y^2)$,

($\overline{I}_2$) $x^2, x^2 y^9(x^3 - y^3), x^2 y^6(x^6 - y^6)$,

($\overline{I}_3$) $x^3 + 2y^3, y^6(x - y)^2(x^2 + xy + y^2)^2(7x^3 + 2y^3)$,

($\overline{I}_4$) $x^4 + 2xy^3, x^4 y^6(x^6 - y^6), xy^{12}(x^3 - y^3), x^4 y^9(x^3 - y^3)$,

($\overline{I}_5$) $x^5 + 2x^2 y^3, x^5 y^9(x^3 - y^3), x^8 y^6(x^3 - y^3), x^2 y^{12}(x^3 - y^3)$,

($\overline{I}_6$) $x^6 + 4x^3 y^3 + 4y^6, y^9(x - y)^2(x^2 + xy + y^2)^2(10x^3 - y^3)$,
$\qquad y^6(x - y)^2(x^2 + xy + y^2)^2(5x^6 + 10x^3 y^6 + 3y^6)$,

($\overline{I}_7$) $x^4(x^3 + 8y^3), xy^3(x^3 - y^3), xy^{15}(x^3 - y^3), x^4 y^{12}(x^3 - y^3)$,

($\overline{I}_8$) $x^8 + 4x^5 y^3 + 22x^2 y^6, y^{12} x^2(x - y)^2(x^2 + xy + y^2)^2$,
$\qquad y^9 x^2(x - y)^2(x^2 + xy + y^2)^2(x^3 + 2y^3)$,
$\qquad y^6 x^2(x - y)^2(x^2 + xy + y^2)^2(x^6 + 2x^3 y^3 + 3y^6)$,

($\overline{I}_9$) $x^9 + 6x^6 y^3 + 66x^3 y^6 + 8y^9, y^6(x - y)^4(x^2 + xy + y^2)^4(5x^3 + 4y^3)$,

($\overline{I}_{10}$) $x^4(x^6 + 80y^6), xy^6(x^3 - y^3), x^4 y^3(x^3 - y^3)$,

($\overline{I}_{11}$) $x^{11} + 132x^5 y^6 + 110x^2 y^9, y^3(x(x - y)(x^2 + xy + y^2))^2$.

Remark 10.4.4. (1) The case $N \equiv -1 \pmod{12}$ of the above two theorems was already given by Mallows, Pless and Sloane [364, Th. 9 and Cor. 10].

(2) We have not yet identified a list of specific codes that generate the modules I_i ($i = 1, \ldots, 11$) for the above two theorems, but this should not be difficult to find from the given bases (keeping in mind Theorem 10.1.8).

10.5 Maximal isotropic additive codes over $\mathbb{F}_4$

Theorem 10.5.1. *Even additive trace-Hermitian self-orthogonal codes over $\mathbb{F}_4$. Let $\rho = N\rho(2_{\mathrm{II}})'$ be the representation of the form ring $R(2_{\mathrm{II}})$ defined in Examples 1.12.5 and 4.6.12. Since $\rho(2_{\mathrm{II}})'$ has order 2 in the Witt group, there are two cases:*

$N \equiv 0 \pmod 2$. *Then ρ is Witt-null; the invariant ring of the Clifford-Weil group $\mathcal{C}(\rho(2_{\mathrm{II}})')$ is given by*

$$\mathbb{C}[x^2 + 3y^2, \ y^2(x^2 - y^2)^2], \tag{10.5.1}$$

with Molien series and invariants as in (7.6.22), (7.6.23).

$N \equiv 1 \pmod 2$. *Then $\rho' = \rho(2_{\mathrm{II}})'$ and $\mathrm{Aut}(\rho') = S_3$; the space of $\mathrm{Aut}(\rho')$-invariants is two-dimensional, generated by u and v transforming as $(u, v) \mapsto ((u + 3v)/2, (u - v)/2)$. The corresponding graded I_0-module is generated by*

$$(x, \ xy^2(x^2 - y^2)), \tag{10.5.2}$$

with Molien series $(t + t^5)/((1 - t^2)(1 - t^6))$. As an I_0-module this is generated by the Hamming weight enumerators of the codes i_1 and h_5, where h_5 is the shortened hexacode defined in Subsection 2.4.8.

10.6 Maximal isotropic codes over $\mathbb{Z}/4\mathbb{Z}$

Our last example deals with codes over the Galois ring $\mathbb{Z}/4\mathbb{Z}$. We will need these results later for the determination of the spaces of theta series of maximal even lattices of determinant 2^k.

Theorem 10.6.1. (*Type $4_{\mathrm{II}}^{\mathbb{Z}}$, symmetrized weight enumerators.*) *Consider the Type $\rho(4_{\mathrm{II}}^{\mathbb{Z}})$ of doubly-even self-dual codes over $\mathbb{Z}/4\mathbb{Z}$ and let $\rho = N\rho(4_{\mathrm{II}}^{\mathbb{Z}})$ be a representation of $R(4_{\mathrm{II}}^{\mathbb{Z}})$. Now $\rho(4_{\mathrm{II}}^{\mathbb{Z}})$ has order 8 in the Witt group (Example 4.6.13), so there are eight possibilities for ρ'. We will only calculate the symmetrized weight enumerators. The symmetrized Clifford-Weil group G is generated by*

$$\begin{pmatrix} 1/2 & 1 & 1/2 \\ 1/2 & 0 & -1/2 \\ 1/2 & -1 & 1/2 \end{pmatrix}, \ \mathrm{diag}(1, \zeta_8, -1),$$

and has order 768. The group is no longer unitary, but fixes the Hermitian form $\mathrm{diag}(1, 1/2, 1)$. For a polynomial $p \in \mathbb{C}[x, y, z]$, we define $\tilde{p}(x, y, z) := p(x, \frac{1}{2}y, z)$. From Lemma 5.6.7 we find that

$$\tilde{q}(\partial)(p) \in \mathrm{Inv}(G, S^*), \quad \text{for } q \in \mathrm{Inv}(G, S) \text{ and } p \in \mathrm{Inv}(G).$$

Let

$$a_1 := x, \ b_1 := x^2 + z^2, \ c_1 := x^3 + 3xz^2, \ d_1 := x^4 + z^4 + 6x^2z^2$$

be the Lee-symmetrized weight enumerators of the maximal isotropic codes in
$N\rho(4_{\mathrm{II}}^{\mathbb{Z}})$ *for $N = 1, 2, 3, 4$, respectively. (These are the codes consisting of all*
vectors in $\{0, 2\}^N$ with an even number of 2's.) We define the polynomials

$$e_1 := \tilde{c}_1(\partial)(p_1), \ f_1 := \tilde{b}_1(\partial)(p_1), \ g_1 := \tilde{a}_1(\partial)(p_1),$$

where $p_1 = \mathrm{swe}(\mathcal{O}_8)$ is the symmetrized weight enumerator (2.4.34) of the
octacode. The eight cases are as follows:

$N \equiv 0 \pmod 8$. *Then ρ is Witt-null, and a maximal isotropic code is self-*
dual. The space spanned by the complete weight enumerators of the max-
imal isotropic (hence self-dual) codes is the invariant ring of the Clifford-
Weil group already described in §8.2.4. The Molien series is

$$f := \mathrm{MS}_{\mathrm{swe}, \, 4_{\mathrm{II}}^{\mathbb{Z}}} = \frac{1 + t^{16}}{(1 - t^8)^2(1 - t^{24})},$$

with invariant ring

$$R_1 := \mathrm{Inv}_{\mathrm{swe}, \, 4_{\mathrm{II}}^{\mathbb{Z}}} = \frac{1, \, q_1}{p_1 \, p_2, \, p_3} = R_0 \oplus q_1 R_0,$$

where

$$R_0 := \mathbb{C}[p_1 \, p_2, \, p_3]$$

is the polynomial ring in the symmetrized weight enumerators of the three
codes $\mathcal{O}_8$, $\mathcal{K}_8$ and $\mathcal{K}_{24}$, and $q_1 = \mathrm{swe}(\mathcal{C}_{16})$. For convenience we denote the
Poincaré series of R_0 by

$$f_0 := \mathrm{MS}(R_0) = 1/((1 - t^8)^2(1 - t^{24})).$$

$N \equiv 1 \pmod 8$. *Then $\rho' = \rho(4_{\mathrm{II}}^{\mathbb{Z}})$, $\mathrm{Aut}(\rho') = 1$ and the space of relative*
invariants has Molien series

$$\frac{t + t^9}{(1 - t^8)^3} = (t + t^9)f + (t^9 + t^{17})f_0.$$

The R_1-module is generated by

$$a_1, \ a_2 := \tilde{g}_1(\partial)(p_1^2), \ a_3 := \tilde{g}_1(\partial)(q_1), \ a_4 := \tilde{g}_1(\partial)(p_3).$$

$N \equiv 2 \pmod 8$. *Here $\rho' = 2\rho(4_{\mathrm{II}}^{\mathbb{Z}})/\langle(2, 2)\rangle$ and the basis of $\mathbb{C}[V_{\rho'}]$ is in-*
dexed by $((0, 0), (1, 1), (2, 0), (3, 1))$. The automorphism group $\mathrm{Aut}(\rho')$ has
order 2 and has a 3-dimensional fixed space. Calculating the representation
of $\mathcal{C}(\rho(4_{\mathrm{II}}^{\mathbb{Z}}))$ on $\mathbb{C}[V_{\rho'}]^{\mathrm{Aut}(\rho')}$, and using (5.6.11), we find that the Molien
series is

$$\frac{t^2 + t^{10}}{(1 - t^8)^3} = (t^2 + t^{10})f + (t^{10} + t^{18})f_0.$$

The R_1-module is generated by

$$b_1, b_2 := \tilde{f}_1(\partial)(p_1^2), b_3 := \tilde{f}_1(\partial)(q_1), b_4 := \tilde{f}_1(\partial)(p_3).$$

$N \equiv 3 \pmod 8$. *Now* $\rho' = 3\rho(4_{\mathrm{II}}^{\mathbb{Z}})/\langle(2,2,0),(0,2,2)\rangle$ *and the basis of* $\mathbb{C}[V_{\rho'}]$ *is indexed by* $((0,0,0),(1,1,1),(2,2,2),(3,3,3))$. *The automorphism group* $\mathrm{Aut}(\rho')$ *has order 2 and has a 3-dimensional fixed space. The representation on* $\mathbb{C}[\rho']^{\mathrm{Aut}(\rho')}$ *is Galois conjugate to that on* $\mathbb{C}[\rho(4_{\mathrm{II}}^{\mathbb{Z}})]^{\mathrm{Aut}(\rho(4_{\mathrm{II}}^{\mathbb{Z}}))}$ *under replacing* ζ_8 *by* ζ_8^3. *One easily calculates the Molien series*

$$\frac{t^3 + t^{11}}{(1 - t^8)^3} = (t^3 + t^{11})f + (t^{11} + t^{19})f_0 \,.$$

The R_1-*module is generated by*

$$c_1, c_2 := \tilde{e}_1(\partial)(p_1^2), c_3 := \tilde{e}_1(\partial)(q_1), c_4 := \tilde{e}_1(\partial)(p_3) \,.$$

$N \equiv 4 \pmod 8$. *Here* $\rho' = 4\rho(4_{\mathrm{II}}^{\mathbb{Z}})/\langle(2,2,0,0),(0,2,2,0),(0,0,2,2)\rangle$ *and* $\mathrm{Aut}(\rho') \cong S_3$. *We find a 2-dimensional* $\mathrm{Aut}(\rho')$-*fixed space and Molien series*

$$\frac{t^4 + t^{12} + 2t^{20}}{(1 - t^8)^2(1 - t^{24})} = t^4 f + (t^{12} + t^{20})f_0 \,.$$

The R_1-*module is generated by*

$$d_1, d_2 := \tilde{d}_1(\partial)(q_1), d_3 := \tilde{d}_1(\partial)(p_3) \,.$$

$N \equiv 5 \pmod 8$. *Similar to the case* $N \equiv 3 \pmod 8$. *The Molien series is*

$$\frac{2t^5}{(1 - t^8)^3} = t^5 f + (t^5 + 2t^{13} + t^{21})f_0 \,,$$

and the R_1-*module is generated by*

$$e_1, e_2 := \tilde{c}_1(\partial)(p_2), e_3 := \tilde{c}_1(\partial)(q_1), e_4 := \tilde{c}_1(\partial)(p_1^2), e_5 := \tilde{c}_1(\partial)(p_3) \,.$$

$N \equiv 6 \pmod 8$. *Similar to the case* $N \equiv 2 \pmod 8$. *The Molien series is*

$$\frac{2t^6}{(1 - t^8)^3} = t^6 f + (t^6 + 2t^{14} + t^{22})f_0 \,,$$

and the R_1-*module is generated by*

$$f_1, f_2 := \tilde{b}_1(\partial)(p_2), f_3 := \tilde{b}_1(\partial)(q_1), f_4 := \tilde{b}_1(\partial)(p_1^2), f_5 := \tilde{b}_1(\partial)(p_3) \,.$$

$N \equiv 7 \pmod 8$. *Here* $\rho' = \overline{\rho(4_{\mathrm{II}}^{\mathbb{Z}})}$, $\mathrm{Aut}(\rho') = 1$ *and the space of relative invariants has Molien series*

$$\frac{2t^7}{(1 - t^8)^3} = t^7 f + (t^7 + 2t^{15} + t^{23})f_0 \,.$$

The R_1-*module is generated by*

$$g_1, g_2 := \tilde{a}_1(\partial)(p_2), g_3 := \tilde{a}_1(\partial)(q_1), g_4 := \tilde{a}_2(\partial)(p_1 q_1), g_5 := \tilde{a}_1(\partial)(p_3) \,.$$

We leave it to the reader to find the relevant codes for this example. Useful examples of maximal isotropic codes are the codes $d_N(\mathbb{Z}/4\mathbb{Z}) \leq 2\mathbb{Z}/4\mathbb{Z}^N$ consisting of the vectors in $\{0, 2\}^N$ that have an even number of 2's. Construction $A_{2\mathbb{Z}}$ (see Remark 10.7.3) applied to these codes yields the root lattices D_N.

10.7 Maximal even lattices

This section uses the above theory to calculate the spaces of modular forms spanned by theta series of maximal even lattices. The main tool here is the generalization of Construction A (cf. §9.1.5) given in Definition 10.7.3. A recent preprint by Böcherer [55] gives an analytic characterization of the spaces of modular forms containing the theta-series of certain classes of maximal even lattices using the theory of modular forms.

The (infinite) form ring for even unimodular lattices is the ring

$$\mathbb{Z}_{\mathrm{II}} := (\mathbb{Z}, M, \psi, \Phi) , \qquad (10.7.1)$$

where $M = \mathbb{Z}$, $\psi = \mathrm{id}$, $\Phi = \mathbb{Z}$, $\{\ \}$ is multiplication by 2 and λ is surjective (see §9.5). For any even lattice Λ we obtain a *finite* representation $\rho^{(\Lambda)}$ of $\mathbb{Z}_{\mathrm{II}}$, defined in Example 9.7.1, with underlying module $V_{\rho^{(\Lambda)}} = \Lambda^{\#}/\Lambda$.

Definition 10.7.1. Let Λ be an even lattice and write $\Lambda^{\#}/\Lambda = \{\Lambda, l_2 + \Lambda, \ldots, l_d + \Lambda\}$. The *vector-valued theta series* of Λ is

$$\boldsymbol{\Theta}_{\Lambda}(z) := (\Theta_{\Lambda}(z), \Theta_{l_2+\Lambda}(z), \ldots, \Theta_{l_d+\Lambda}(z)) = \sum_{v \in V_{\rho^{(\Lambda)}}} \Theta_v(z) e_v , \qquad (10.7.2)$$

which is a holomorphic map from the upper half plane to $\mathbb{C}[V_{\rho^{(\Lambda)}}]$.

Then $\mathrm{SL}_2(\mathbb{Z})$ acts projectively on such functions, as follows (see Borcherds [67], Bruinier [87], Weil [546]): $S := \begin{pmatrix} 0 & 1 \\ -1 & 0 \end{pmatrix}$ acts as

$$\boldsymbol{\Theta}_{\Lambda}(z) \mapsto \left(\sqrt{\frac{i}{z}}\right)^{\dim(\Lambda)} \boldsymbol{\Theta}_{\Lambda}\left(-\frac{1}{z}\right) ,$$

and $T := \begin{pmatrix} 1 & 1 \\ 0 & 1 \end{pmatrix}$ acts as the substitution $z \mapsto z + 1$.

We claim that this action is isomorphic to that of a certain Clifford-Weil group $\mathcal{C}(\rho^{(\Lambda)})$ on $\mathbb{C}[V_{\rho^{(\Lambda)}}]$. To define this group note that there is an ideal $I \trianglelefteq \mathbb{Z}_{\mathrm{II}}$ such that $\mathbb{Z}_{\mathrm{II}}/I$ is a finite form ring of which $\rho^{(\Lambda)}$ is a finite representation. Then define $\mathcal{C}(\rho^{(\Lambda)})$ to be the corresponding Clifford-Weil group. (This is independent of the choice of I since there is a largest such I.)

We are mainly interested in maximal even lattices Λ for which $|\Lambda^{\#}/\Lambda|$ is a prime power. If so, the scalar matrices in the associated Clifford-Weil group $\mathcal{C}(\rho^{(\Lambda)})$ have order dividing 8, and this projective representation yields a group epimorphism from a certain covering group

$$\widetilde{\mathrm{SL}_2(\mathbb{Z})} := Z_4 . \mathrm{SL}_2(\mathbb{Z})$$

with center Z_8. Here $\widetilde{\mathrm{SL}_2(\mathbb{Z})}$ is generated by $\tilde{S}$ and $\tilde{T}$, where $\tilde{S}$ and $\tilde{T}$ map onto S and T respectively under the natural epimorphism $\widetilde{\mathrm{SL}_2(\mathbb{Z})} \to \mathrm{SL}_2(\mathbb{Z})$, and where $(ST)^3$ generates the center of $\widetilde{\mathrm{SL}_2(\mathbb{Z})}$. Note that the preimages $\tilde{S}$ and $\tilde{T}$ are not uniquely determined by this condition; this is reflected in the fact that (10.7.6) below does not contain Bruinier's [87] multiplier $\sqrt{i}^{\dim(\Lambda)}$.

Theorem 10.7.2. *Let Λ be a maximal even lattice such that $|\Lambda^\#/\Lambda|$ is a power of some prime p. Then there is an epimorphism*

$$\Delta : \widetilde{\mathrm{SL}_2(\mathbb{Z})} \to \mathcal{C}(\rho_\Lambda) \tag{10.7.3}$$

such that

$$\Theta_\Gamma g = \Theta_\Gamma \rho^{(\Lambda)}(\Delta(g)) \tag{10.7.4}$$

for all $g \in \widetilde{\mathrm{SL}_2(\mathbb{Z})}$ and all lattices Γ with $\rho^{(\Gamma)} = \rho^{(\Lambda)}$.

In fact $\Delta : \widetilde{\mathrm{SL}_2(\mathbb{Z})} \to \mathcal{C}(\rho^{(\Lambda)})$ *is defined by* $\Delta(\tilde{T}) := d_\phi$ *with*

$$\rho^{(\Lambda)}(d_\phi)(e_v) = \exp(2\pi i \rho_\Phi^{(\Lambda)}(1)(v)) \, e_v \,, \tag{10.7.5}$$

and $\Delta(\tilde{S}) := h_{1,1,1}$ *is the MacWilliams transformation with*

$$\rho^{(\Lambda)}(h_{1,1,1})(e_v) = \frac{1}{|V_{\rho^{(\Lambda)}}|^{1/2}} \sum_{w \in V_{\rho^{(\Lambda)}}} \exp(2\pi i \rho_M^{(\Lambda)}(1)(w,v)) \, e_w \,. \tag{10.7.6}$$

Note that Δ is essentially the Weil representation attached to the quadratic module $(\Lambda^*/\Lambda, \rho_\Phi^{(\Lambda)}(1))$ (cf. Bruinier [87], Nobs [391]).

Proof. The group $\mathcal{C}(\rho^{(\Lambda)})$ is a projective representation of $\mathfrak{U}(\mathbb{Z}_{\mathrm{II}}/I)$, for some ideal I of finite index in the form ring $\mathbb{Z}_{\mathrm{II}}$. The maximal such ideal I contains an ideal of the form $I_q := (q\mathbb{Z}, q\mathbb{Z}, \mathrm{id}, q\mathbb{Z})$, where $q = p$ if p is odd and $q = 4$ if $p = 2$, so we may regard $\rho^{(\Lambda)}$ as a projective representation of

$$\mathfrak{U}(\mathbb{Z}_{\mathrm{II}}/I_q) = \mathfrak{U}(\mathbb{Z}/q\mathbb{Z}, \mathbb{Z}/q\mathbb{Z}) = \mathrm{SL}_2(\mathbb{Z}/q\mathbb{Z}) \,.$$

Reduction modulo q is a surjection from $\mathrm{SL}_2(\mathbb{Z})$ onto $\mathrm{SL}_2(\mathbb{Z}/q\mathbb{Z})$ (this is true for any q) and thus Δ maps $\widetilde{\mathrm{SL}_2(\mathbb{Z})}$ onto a subgroup of $\mathcal{C}(\rho^{(\Lambda)})$ that coincides with $\mathcal{C}(\rho^{(\Lambda)})$ modulo scalars. On the other hand, since q is a prime power, 1 is the only symmetric idempotent in the form ring $(\mathbb{Z}/q\mathbb{Z}, \mathbb{Z}/q\mathbb{Z})$ and therefore $h_{1,1,1}$ is the only possible MacWilliams transformation. By Theorem 5.4.6, $(\Delta(\tilde{S})\Delta(\tilde{T}))^3$ is a scalar element γ in $\mathcal{C}(\rho^{(\Lambda)})$. By Lemma 5.4.14 the scalar elements in $\mathcal{C}(\rho^{(\Lambda)})$ are precisely the o-th roots of unity, where o is the order of $\rho^{(\Lambda)}$ in the Witt group $W(\mathbb{Z}/q\mathbb{Z}, \mathbb{Z}/q\mathbb{Z})$. Direct calculations show that the character group of this Witt group is generated by γ. This establishes the surjectivity of Δ. That the actions of $\mathrm{SL}_2(\mathbb{Z})$ and the Clifford-Weil group are compatible can also be checked by direct calculations. $\qquad\square$

In particular, it follows that $\text{Aut}(\rho^{(\Lambda)})$ commutes with the $\text{SL}_2(\mathbb{Z})$-action (this is true even if we do not have surjectivity). We may thus define $\text{Aut}(\rho^{(\Lambda)})$-symmetrized vector-valued theta series and, as above, the corresponding action of $\text{SL}_2(\mathbb{Z})$ on the $\text{Aut}(\rho^{(\Lambda)})$-fixed space is then irreducible, since Λ is a maximal even lattice and hence $\rho^{(\Lambda)}$ is anisotropic.

In order to apply this theorem we define a generalized version of Construction A, namely Construction A_Λ where Λ is an even lattice. The original Construction A as in §9.1.5 is then $A_{\sqrt{2}\mathbb{Z}}$.

Definition 10.7.3. Let Λ be an even $\mathbb{Z}$-lattice of dimension d and let $C \leq \rho_\Lambda^N$ be an isotropic self-orthogonal code in $N\rho_\Lambda$. Then the lattice

$$A_\Lambda(C) := \{(x_1,\ldots,x_N) \in (\Lambda^\#)^N \mid (x_1+\Lambda,\ldots,x_N+\Lambda) \in C\} \quad (10.7.7)$$

is an even lattice of dimension Nd (as one can readily see from the definition of ρ_Λ in Example 9.7.1).

The theta series of $A_\Lambda(C)$ is given by the following generalization of Theorem 9.1.14. We only state it for genus 1 since that is all we shall need. For $X \in \Lambda^\#/\Lambda$ and $\tau \in \mathbb{H}$ let

$$\vartheta_X(\tau) := \sum_{z \in X} \exp(2\pi i \tau \rho_\Phi^{(\Lambda)}(1)(z)). \quad (10.7.8)$$

Then we have:

Theorem 10.7.4.

$$\Theta_{A_\Lambda(C)}(\tau) = \text{cwe}_C(\vartheta_X \mid X \in \Lambda^\#/\Lambda).$$

Proof. Let $\Gamma := A_\Lambda(C)$. Then $\Lambda^N \subset \Gamma \subset (\Lambda^\#)^N$ and

$$\Gamma = \bigcup_{c \in C} (c_1,\ldots,c_N) = \bigcup_{c \in C} (\overline{c_1},\ldots,\overline{c_N}) + \Lambda^N,$$

where $\overline{c_j} \in c_j \subset \Lambda^\#$ is a representative of the residue class $c_j = \overline{c_j} + \Lambda$. Hence

$$\begin{aligned}
\Theta_\Gamma(\tau) &= \sum_{\gamma \in \Gamma} \exp(2\pi i \tau (\sum_{j=1}^N (\gamma_j,\gamma_j))) \\
&= \sum_{c \in C} \sum_{\gamma \in c} \prod_{j=1}^N \exp(\pi i \tau \rho_\Phi^{(\Lambda)}(1)(\gamma_j)) \\
&= \text{cwe}_C(\vartheta_X \mid X \in \Lambda^\#/\Lambda).
\end{aligned}$$

$\square$

It may happen (for instance when $\Lambda = A_2$ or $2\mathbb{Z}$) that $\vartheta_X = \vartheta_Y$ for two different cosets in $\Lambda^\#/\Lambda$. In this case it is clearly enough to work with the appropriate symmetrized weight enumerator instead with the complete weight enumerator.

We easily see that for primes p and even lattices Λ of even level p, Construction A_Λ commutes with the representation Δ defined in Theorem 10.7.2.

As an application, we calculate the modules spanned by the theta series of maximal even lattices of determinant p^k, where $p = 3$ or $p = 2$.

10.7.1 Maximal even lattices of determinant 3^k

The first remark is well-known (see for instance Kneser [318, (2.10)]) but quite useful:

Remark 10.7.5. Let Λ be an even lattice of odd determinant. Then any Gram matrix for Λ is congruent modulo 2 to a skew-symmetric matrix of odd determinant (by multiplying the entries below the diagonal by -1 and replacing the entries on the diagonal by 0). In particular, the dimension of Λ is necessarily even.

Let Λ be a maximal even lattice of determinant 3^k. By the maximality, $\rho^{(\Lambda)}$ is anisotropic. In particular $k \leq 2$. Moreover, if $\dim(\Lambda) = 2n$, then $\rho^{(\Lambda)}$ is the anisotropic representation corresponding to maximal isotropic ternary codes of length n. Indeed, Construction A_{A_2} induces a map from the appropriate space of Hamming weight enumerators of maximal isotropic ternary codes to the space of symmetrized vector-valued theta series of maximal even lattices in the given genus.

Theorem 10.7.6. Let ρ be an anisotropic representation of the form ring $R(3)$. Then $\rho \sim m\rho(3)$ for $m \in \{0, 1, 2, 3\}$. The map from the relative invariants of $C(\rho(3))$ generated by the maximal isotropic codes in $(4N + m)\rho(3)$ to the space of modular forms for $\Gamma_0(3)$ generated by the theta series of maximal even lattices Λ of dimension $2(4N + m)$ with $\rho_\Lambda = m\rho(3)$ defined by

$$\mathrm{hwe}(C) \mapsto \Theta_{A_{A_2}(C)}$$

is an isomorphism.

Proof. By Theorem 10.7.2 the map is well defined. Injectivity follows because the theta series ϑ_x for $x \in A_2^\# / A_2$ are algebraically independent. To show the surjectivity, consider the action of $\mathrm{SL}_2(\mathbb{Z})$ on the space of modular forms M for $\pm\Gamma(3) := \langle \Gamma(3), -I_2 \rangle$, where the character is trivial on $\Gamma(3)$ and is $(-1)^k$ for $-I_2$ on the space of modular forms of weight k. By the dimension formula of Eholzer and Skoruppa ([165, p. 129], [497]), the ring M is a polynomial ring in two generators of weight 1. It is easy to see that

$$M = \mathbb{C}[\Theta_{A_2}(\tau), \frac{1}{2}(\Theta_{A_2^\#}(\tau) - \Theta_{A_2}(\tau)) = \frac{1}{2}(\Theta_{A_2}(\frac{\tau}{3}) - \Theta_{A_2}(\tau))]$$

(for the last identity see [133, p. 103, Eqs. (18) and (19)], where Θ_{A_2} is denoted by ϕ_0). With respect to this basis the generators of $\widetilde{\mathrm{SL}_2(\mathbb{Z})}$ act as

$$\Delta(\tilde{S}) = \frac{1}{\sqrt{3}} \begin{pmatrix} 1 & 2 \\ 1 & -1 \end{pmatrix}, \quad \Delta(\tilde{T}) = \mathrm{diag}(1, \exp(2\pi i/3)).$$

Hence $\Delta(\widetilde{\mathrm{SL}_2(\mathbb{Z})}) = \langle \Delta(\tilde{S}), \Delta(\tilde{T}) \rangle = C(\rho(3))$ in its action on the variables x, y of the Hamming weight enumerators of ternary codes. At the level of

Hamming weight enumerators, Construction A_{A_2} maps x to Θ_{A_2} and y to $\Theta_{A_2^{\#}} - \Theta_{A_2}$, and therefore the corresponding relative invariants of $\mathcal{C}(\rho(3))$ (which are spanned by the Hamming weight enumerators of the appropriate maximal isotropic codes) to the relative invariants of $\mathrm{SL}_2(\mathbb{Z})$ (which are linear combinations of theta series of maximal even lattices in the appropriate genus). □

For any root lattice R that is an orthogonal direct summand of the root sublattice of one of the 24 Niemeier lattices L (see Table 9.1), we define

$$\mathrm{Comp}(R) := \{l \in L \cap R^{\perp}\} \qquad (10.7.9)$$

to be the orthogonal complement of R in L. Since L is even and unimodular, we have $\rho^{(\mathrm{Comp}(R))} = \overline{\rho^{(R)}}$. Note that, although the isomorphism type of the lattice $\mathrm{Comp}(R)$ depends on the choice of L, its theta series does not, as one easily verifies by direct computation. The Poincaré series for a space M of modular forms will be denoted by

$$\mathrm{PS}(M) := \sum_{m=-\infty}^{\infty} a_m t^m,$$

where $a_m := \dim\langle f \in M \mid$ weight of $f = m\rangle$. Construction A relates the Molien series of the space of relative invariants of a suitable Clifford-Weil group to the Poincaré series of the corresponding space of modular forms generated by the theta series of suitable maximal even lattices.

By applying Construction A_{A_2} to the maximal isotropic codes in Theorem 10.4.1 and using Theorem 10.7.6, we obtain the following corollary, which is the main result of this subsection.

Corollary 10.7.7. *For $m \in \{0, 1, 2, 3\}$, let M_m be the $\mathbb{C}$-vector space spanned by the theta series of maximal even lattices of dimension $N \in 2m + 8\mathbb{Z}$ and determinant 3^k. The spaces M_m and their Poincaré series are as follows:*

$m = 0$. The lattices are even unimodular lattices and, from Theorem 9.1.2,

$$M_0 := M(1) := \mathbb{C}[\Theta_{E_8}, \Theta_{\Lambda_{24}}], \qquad (10.7.10)$$

with Poincaré series

$$\mathrm{PS}(M(1)) := \frac{1}{(1 - t^4)(1 - t^{12})}.$$

$m = 1$. M_1 is generated as an M_0-module by the theta series of the lattices A_2 and $\mathrm{Comp}(E_6)$ (in the notation described above), with Poincaré series

$$\mathrm{PS}(M_1) = \frac{t + t^9}{(1 - t^4)(1 - t^{12})} = (t + t^9)\,\mathrm{PS}(M(1)).$$

$m = 2$. M_2 is generated by the theta series of the lattices $A_2 \perp A_2$, $E_6 \perp E_6$ and $\mathrm{Comp}(A_2 \perp A_2)$, with Poincaré series

$$\mathrm{PS}(M_2) = \frac{t^2 + t^6 + t^{10}}{(1 - t^4)(1 - t^{12})} = (t^2 + t^6 + t^{10})\,\mathrm{PS}(M(1))\,.$$

$m = 3$. M_3 is generated by the theta series of the lattices E_6 and $\mathrm{Comp}(A_2)$, with Poincaré series

$$\mathrm{PS}(M_3) = \frac{t^3 + t^{11}}{(1 - t^4)(1 - t^{12})} = (t^3 + t^{11})\,\mathrm{PS}(M(1))\,.$$

10.7.2 Maximal even and integral lattices of determinant 2^k

In this section we will determine the spaces of modular forms generated by the theta series of maximal integral and maximal even lattices of determinant 2^k. These spaces are $M(1)$-modules, where $M(1)$ is defined in (10.7.10).

We will not use the same method as in the case of determinant 3^k, since this cannot easily be applied here. Also, some of the genera cannot be obtained directly from Construction A. Instead we will use the following idea:

Proposition 10.7.8. *If Λ is a maximal even lattice, then either Λ is also maximal integral or Λ is the even sublattice of an odd maximal integral lattice.*

Proof. If Λ is not maximal integral, it is contained in a maximal integral lattice, and is contained in the even sublattice thereof. Since Λ is maximal even, the proposition follows.   □

We will therefore consider maximal integral lattices of determinant 2^k. It is easy to see that such a lattice is either unimodular or of determinant 2. In particular, it has level 1 or 2.

The next lemma follows from [453, Corollary 1].

Lemma 10.7.9. *Let Λ be an integral lattice of level 2. Then Θ_Λ is a modular form for*

$$\frac{1}{2}\Gamma_0(8) := \left\{ \begin{pmatrix} a & 2b \\ 4c & d \end{pmatrix} \in \mathrm{SL}_2(\mathbb{Z}) \mid a, b, c, d \in \mathbb{Z} \right\},$$

for a certain character depending on the dimension and on the square-class of the determinant of Λ.

Passing to an appropriate index 2 subgroup

$$G := \left\{ \begin{pmatrix} a & 2b \\ 4c & d \end{pmatrix} \in \frac{1}{2}\Gamma_0(8) \mid a \equiv d \equiv 1 \pmod 4 \right\},$$

to eliminate the dependence on the square class, we obtain a group G with ring of modular forms $\mathcal{M}(G) \cong \mathbb{C}[\Theta_\mathbb{Z}, \Theta_{A_1}]$ (see Eholzer and Skoruppa [165]).

The theta series of a level-2 integral lattice must be in this ring, and moreover must be a linear combination of terms in which the degrees with respect to $\Theta_{\mathbb{Z}}$ and with respect to Θ_{A_1} have the same parity. In other words, the theta series is in one of the four modules $\mathbb{C}[\Theta_{\mathbb{Z}}^2, \Theta_{A_1}^2]$, $\mathbb{C}[\Theta_{\mathbb{Z}}^2, \Theta_{A_1}^2]\Theta_{\mathbb{Z}}$, $\mathbb{C}[\Theta_{\mathbb{Z}}^2, \Theta_{A_1}^2]\Theta_{A_1}$, $\mathbb{C}[\Theta_{\mathbb{Z}}^2, \Theta_{A_1}^2]\Theta_{\mathbb{Z}}\Theta_{A_1}$, depending in an obvious way on the square class of the determinant and the parity of the dimension.

Lemma 10.7.10. *For $N \in \mathbb{N}$, let X_N be the $\mathbb{C}$-vector space spanned by the theta series of integral lattices of dimension N and determinant 2, and similarly let Y_N be spanned by the theta series of unimodular lattices of dimension N. If N is odd (resp. even) then Construction A_{A_1} yields a bijection between X_N (resp. Y_N) and the space of relative invariants of $\mathcal{C}(\rho(2_\mathrm{I}))$ spanned by the weight enumerators of maximal self-orthogonal binary codes of length N.*

Proof. Let $\widetilde{\Gamma_\vartheta} \leq \widetilde{\mathrm{SL}_2(\mathbb{Z})}$ be the full preimage of the theta-group $\Gamma_\vartheta = \langle S, T^2 \rangle \leq \mathrm{SL}_2(\mathbb{Z})$ defined in (9.5.1). Let $\Delta : \widetilde{\Gamma_\vartheta} \to \mathcal{C}(\rho(2_\mathrm{I}))$ be the representation obtained from Construction A_{A_1}, as in Theorem 10.7.2. With respect to the basis

$$(\Theta_{A_1}(\tau) = \Theta_{\mathbb{Z}}(2\tau), \Theta_{A_1^\#}(\tau) - \Theta_{A_1}(\tau) = \Theta_{\mathbb{Z}}(\tau/2) - \Theta_{\mathbb{Z}}(2\tau))$$

we obtain

$$\Delta(\tilde{S}) = \frac{1}{\sqrt{2}}\begin{pmatrix} 1 & 1 \\ 1 & -1 \end{pmatrix}, \quad \Delta(\tilde{T}^2) = \mathrm{diag}(1, -1).$$

The images generate the Clifford-Weil group $\mathcal{C}(\rho(2_\mathrm{I}))$. Let $K := \ker(\Delta)$. Then $|\widetilde{\Gamma_\vartheta}/K| = |\mathcal{C}(\rho(2_\mathrm{I}))| = 16$.

We claim that the ring of modular forms $\mathcal{M}(K)$ is the polynomial ring $\mathbb{C}[\Theta_{\mathbb{Z}}(2\tau), \Theta_{\mathbb{Z}}(\tau/2)]$. Indeed, both generators are clearly modular forms for K, and they generate a (Galois) extension of $\mathbb{C}[\Theta_{\mathbb{Z}}^2, \Theta_{E_8}]$ of degree 16. The precise equations are most easily given in terms of

$$f(\tau) := \theta_3(2\tau) = \Theta_{A_1}(\tau) \text{ and } g(\tau) := \theta_2(2\tau) = \Theta_{l+A_1}(\tau),$$

where $l + A_1 \in A_1^\#/A_1$ is the non-trivial coset. Since the lattices $\mathbb{Z}^2$ and E_8 are obtained by applying Construction A_{A_1} to the repetition code i_2 and the Hamming code e_8, we have

$$\Theta_{\mathbb{Z}}^2 = f^2 + g^2, \quad \Theta_{E_8} = f^8 + 14f^4 g^4 + g^8.$$

Hence $\mathbb{C}(f, g)$ is a Galois extension over the fixed field $\mathbb{C}(\Theta_{\mathbb{Z}}^2, \Theta_{E_8})$ of $\mathcal{C}(\rho(2_\mathrm{I}))$, with Galois group $\mathcal{C}(\rho(2_\mathrm{I}))$ of order 16. Since $\mathbb{C}[f, g]$ is a polynomial ring, it is integrally closed, and so is equal to the full ring of modular forms of the group K. The surjectivity of Construction A_{A_1} now follows as in Theorem 10.7.6. $\square$

Theorem 10.7.11. *Let Λ be a maximal integral lattice of dimension N and determinant 2^k. Then $k \leq 1$. If $k = 0$, then Λ is unimodular and*

$$\Theta_\Lambda \in \mathcal{M}(\Gamma_\vartheta) = \mathbb{C}[\Theta_\mathbb{Z}, \Theta_{E_8}].$$

If $k = 1$, then

$$\Theta_\Lambda \in \mathcal{M}(\Gamma_\vartheta)\Theta_{A_1} \oplus \mathcal{M}(\Gamma_\vartheta)\Theta_{E_7}.$$

Proof. It is easy to see that $k \leq 1$, and the case of unimodular lattices ($k = 0$) is already covered by Theorem 9.1.2. Therefore it is enough to treat the case $k = 1$. As in Lemma 10.7.10, let X_N be the $\mathbb{C}$-vector space spanned by the theta series of integral lattices of dimension N and determinant 2. For odd N, the result now follows from Lemma 10.7.10.

For even dimensions N, we use the fact that $N - 1$ and $N + 1$ are both odd and

$$X_{N-1}\Theta_\mathbb{Z} \subset X_N, \; X_N\Theta_\mathbb{Z} \subset X_{N+1}.$$

On the other hand, an element of X_N can be expressed as a polynomial in $\Theta_\mathbb{Z}$ and Θ_{A_1}, and must be a multiple of $\Theta_\mathbb{Z}\Theta_{A_1}$. We claim that $X_N = \Theta_\mathbb{Z}X_{N-1}$. We consider the four congruence classes modulo 8 in turn.

$N \equiv 0 \pmod 8$. Then $\dim X_{N-1} = \frac{N}{4}$ and $\dim X_{N+1} = \frac{N}{4} + 1$, but X_{N+1} contains $\Theta_{E_8}^{N/8}\Theta_{A_1}$, which cannot be in the image of multiplication by $\Theta_\mathbb{Z}$, since it is not a multiple of $\Theta_\mathbb{Z}$. Therefore $\dim X_N = \frac{N}{4}$ and $X_N = X_{N-1}\Theta_\mathbb{Z}$.

$N \equiv 2, 4 \pmod 8$. Here $\dim X_{N-1} = \dim X_{N+1} = \lfloor \frac{N}{4} \rfloor + 1$, so multiplication by $\Theta_\mathbb{Z}$ gives an isomorphism in both cases.

$N \equiv 6 \pmod 8$. Here $\dim X_{N-1} = (N - 2)/4$ and $\dim X_{N+1} = (N + 2)/4$, but X_{N+1} contains $\Theta_{E_8}^{(N-6)/8}\Theta_{E_7}$, which is not divisible by $\Theta_\mathbb{Z}$. $\square$

We now use Proposition 10.7.8 to obtain generators for the $M(1)$-modules spanned by the theta series of maximal even lattices in a given genus. If Λ is an odd lattice, then the theta series of the even sublattice of Λ is $\frac{1}{2}(\Theta_\Lambda(\tau) + \Theta_\Lambda(\tau + 1))$. To obtain generating sets from the $\mathcal{M}(\Gamma_\vartheta)$-bases in Theorem 10.7.11 we need an $M(1)$-basis for the $M(1)$-module $\mathcal{M}(\Gamma_\vartheta)$.

Theorem 10.7.12. $B := (1, \Theta_\mathbb{Z}, \ldots, \Theta_\mathbb{Z}^{23})$ *is an $M(1)$-basis for $\mathcal{M}(\Gamma_\vartheta)$.*

Proof. We first show that the elements generate $\mathcal{M}(\Gamma_\vartheta)$. It is clearly enough to show the existence of $a_1, a_2, a_3 \in M(1)$ such that $\Theta_\mathbb{Z}^{n+24} = a_3\Theta_\mathbb{Z}^{n+16} + a_2\Theta_\mathbb{Z}^{n+8} + a_1\Theta_\mathbb{Z}^n$ for all $n \geq 0$, and for this it is sufficient to establish the result for $n = 0$. From the identities in [133, page 103 ff.] we find that

$$\Theta_{E_8} = \frac{1}{2}(\theta_2^8 + \theta_3^8 + \theta_4^8), \; \Delta_{24} = (\frac{1}{2}\theta_2\theta_3\theta_4)^8, \; \Theta_\mathbb{Z}^4 = \theta_2^4 + \theta_4^4.$$

In particular (recall $\theta_3 = \Theta_\mathbb{Z}$), $\Delta = \frac{1}{2^8}\Theta_\mathbb{Z}^8(\Theta_\mathbb{Z}^8 - \Theta_{E_8})^2$, which yields the desired $M(1)$-linear dependence of $1, \Theta_\mathbb{Z}^8, \Theta_\mathbb{Z}^{16}, \Theta_\mathbb{Z}^{24}$, as well as the polynomial

$$P(t) := t^{24} - 2\Theta_{E_8}t^{16} + \Theta_{E_8}^2 t^8 - 256\Delta \in \mathbb{C}[\Theta_{E_8}, \Delta][t]$$

satisfying $P(\Theta_{\mathbb{Z}}) = 0$. To show the linear independence of B over the field of fractions $\mathbb{C}(\Theta_{E_8}, \Delta)$, we reduce the polynomial ring $\mathbb{C}[\Theta_{E_8}, \Delta]$ modulo the prime ideal (Θ_{E_8}). In the quotient ring the polynomial

$$\overline{P}(t) = t^{24} - 256\Delta \in \mathbb{C}[\Delta][t] = (\mathbb{C}[\Theta_{E_8}, \Delta]/(\Theta_{E_8}))[t]$$

is an Eisenstein polynomial for the prime element $\Delta \in \mathbb{C}[\Delta]$ and therefore irreducible. Hence $P(t)$ is also irreducible and therefore is the minimal polynomial of $\Theta_{\mathbb{Z}}$ over the ring $M(1)$.   □

Corollary 10.7.13. *Let*

$$\Omega_D := \langle \Theta_{D_n} \mid n = 0, 1, \ldots \rangle_{M(1)}.$$

Then $\{\Theta_{D_n} \mid n = 0, \ldots, 23, n \neq 12\}$ *is an $M(1)$-basis for Ω_D.*

Proof. Let $P(t)$ be the minimal polynomial of θ_3 over $M(1)$, as in the proof of Theorem 10.7.12. Then $\theta_4(\tau)$ is also a zero of P. Since D_n is the even sublattice of $\mathbb{Z}^n$, we have

$$\Theta_{D_n}(\tau) = \frac{1}{2}(\Theta_{\mathbb{Z}}^n(\tau) + \Theta_{\mathbb{Z}}^n(\tau + 1)) = \frac{1}{2}(\theta_3^n + \theta_4^n)(\tau).$$

Noting that $\Theta_{D_{12}} = \Theta_{D_4}\Theta_{E_8}$, Theorem 10.7.12 implies that the given set spans the $M(1)$-module Ω_D. To see the linear independence of this set we consider the ring $F := \mathbb{C}[\Theta_{E_8}, \Theta_{\mathbb{Z}}][\theta_4]$. The minimal polynomial of θ_4 over $\mathbb{C}[\Theta_{E_8}, \Theta_{\mathbb{Z}}]$ is

$$t^8 - \Theta_{\mathbb{Z}}^4 t^4 - \Theta_{E_8} + \Theta_{\mathbb{Z}}^8.$$

Therefore the set

$$B_2 := \{\theta_3^i \theta_4^j \mid 0 \leq i \leq 23, 0 \leq j \leq 7\}$$

is a basis for F over $M(1)$.

We now express the Θ_{D_n} as linear combinations of these basis vectors in order to establish their linear independence. Clearly any $M(1)$-linear relation among the Θ_{D_n} is of the form

$$a_1\Theta_{D_n} + a_2\Theta_{D_{n+8}} + a_3\Theta_{D_{n+16}} = 0, \quad a_1, a_2, a_3 \in M(1),$$

for some n between 0 and 7. If n is not divisible by 4, then the B_2-expansion of Θ_{D_n} starts with $\frac{1}{2}\Theta_{\mathbb{Z}}^n$. Therefore Θ_{D_n}, $\Theta_{D_{n+8}}$, $\Theta_{D_{n+16}}$ are $M(1)$-linearly independent in this case. For $n = 0$, direct calculations show that Θ_{D_8} starts with $\frac{1}{2}\Theta_{\mathbb{Z}}^8$ and $\Theta_{D_{16}}$ with $\frac{1}{2}\Theta_{\mathbb{Z}}^{16}$. Therefore again these are linearly independent. For $n = 4$ one finds that $\Theta_{D_4}\Theta_{E_8} = \Theta_{D_{12}}$ and Θ_{D_4} and $\Theta_{D_{20}}$ are $M(1)$-linearly independent (which also follows from Theorem 10.2.1).   □

Theorem 10.7.14. *Let Λ be an N-dimensional maximal even lattice of determinant 2^k. Then Λ belongs to one of the genera shown[2] in Table 10.1. The space of modular forms spanned by the theta series of any of these genera has an $M(1)$-basis given by the theta series of the lattices shown in the final column. The third column gives the dimensions of the generators.*

Table 10.1. Genera of N-dimensional maximal even lattices of determinant 2^k, together with lattices whose theta series are an $M(1)$-basis for the corresponding spaces of modular forms.

$N \bmod 8$	Genus	Dimensions	Lattices
0	$\mathrm{II}_{N,0}(1)$	0	$\{0\}$
	$\mathrm{II}_{N,0}(2 \times 4)$	$8, 8, 16,$	$D_7 A_1, 2\mathbb{Z} \oplus E_7, D_{15} A_1,$
		$16, 24, 24$	$D_9 E_7, D_{23} A_1, D_{17} E_7$
1	$\mathrm{II}_{N,0}(2_{\mathrm{I}}^1)$	$1, 17$	$A_1, (D_{10} E_7)^+$
	$\mathrm{II}_{N,0}(4^1)$	$1, 9, 17$	$2\mathbb{Z}, D_9, D_{17}$
2	$\mathrm{II}_{N,0}(2_{\mathrm{I}}^2)$	$2, 10, 18$	D_2, D_{10}, D_{18}
	$\mathrm{II}_{N,0}(2 \times 4)$	$2, 10, 10,$	$2\mathbb{Z} \oplus A_1, D_9 A_1, A_3 E_7,$
		$18, 18, 26$	$D_{17} A_1, D_{11} E_7, D_{19} E_7$
3	$\mathrm{II}_{N,0}(2^3)$	$3, 11, 19,$	$A_1^3, D_{10} A_1, D_{18} A_1,$
		27	$\mathrm{Comp}(A_1) D_4$
	$\mathrm{II}_{N,0}(4^{-1})$	$3, 11, 19$	$A_3 \cong D_3, D_{11}, D_{19}$
4	$\mathrm{II}_{N,0}(2_{\mathrm{II}}^{-2})$	$4, 20$	$D_4, \mathrm{Comp}(D_4)$
	$\mathrm{II}_{N,0}(2 \times 4)$	$4, 12, 12,$	$A_3 A_1, D_{11} A_1, D_5 E_7,$
		$20, 20, 28$	$D_{19} A_1, D_{13} E_7, D_{21} E_7$
5	$\mathrm{II}_{N,0}(2^3)$	$5, 13, 21, 21$	$A_1 D_4, E_7 D_6, E_7^3, \mathrm{Comp}(A_1^3)$
	$\mathrm{II}_{N,0}(4^{-1})$	$5, 13, 21$	D_5, D_{13}, D_{21}
6	$\mathrm{II}_{N,0}(2_{\mathrm{I}}^2)$	$6, 14, 22$	$D_6, D_{14}, \mathrm{Comp}(A_1^2)$
	$\mathrm{II}_{N,0}(2 \times 4)$	$6, 14, 14,$	$D_5 A_1, D_{13} A_1, D_7 E_7,$
		$22, 22, 30$	$D_{21} A_1, D_{15} E_7, D_{23} E_7$
7	$\mathrm{II}_{N,0}(2_{\mathrm{I}}^1)$	$7, 23$	$E_7, \mathrm{Comp}(A_1)$
	$\mathrm{II}_{N,0}(4^1)$	$7, 15, 23$	D_7, D_{15}, D_{23}

Proof. Let us first treat the genera $\mathrm{II}_{N,0}(2_{\mathrm{I}}^1)$ for $N \equiv \pm 1 \pmod 8$. In these cases the maximal even lattices are also maximal integral. Since Construction A_{A_1} maps doubly-even codes to even lattices, the theorem follows from Lemma 10.7.10 and Theorem 10.2.1. For the unimodular genera $\mathrm{II}_{N,0}(1)$ ($N \equiv 0 \pmod 8$) the result is also clear. In all other cases, the lattices are not maximal

---

[2] Using the notation for genera given in [133, Chapter 15].

integral and hence are obtained as even sublattices of (odd) maximal integral lattices of the appropriate determinant (1 or 2). We obtain generators for the spaces by applying Proposition 10.7.8 and Theorem 10.7.12 to Theorem 10.7.11. For even N, the lattices D_N are obtained by applying Construction A_{A_1} to the codes d_N. In particular, we see that Construction A_{A_1} yields a surjection from the appropriate space of relative invariants of $\mathcal{C}(\rho(2_{\mathrm{II}}))$ to the space spanned by the theta series of maximal even lattices. This surjection is also injective since the two theta series Θ_{A_1} and $\Theta_{A_1^\#} - \Theta_{A_1}$ are algebraically independent. Applying it to the bases given in Theorem 10.2.1 yields the $M(1)$-bases in the cases $\mathrm{II}_{N,0}(2_{\mathrm{I}}^2)$ ($N \equiv 2,6 \pmod 8$), $\mathrm{II}_{N,0}(2^3)$ ($N \equiv 3,5 \pmod 8$), and $\mathrm{II}_{N,0}(2_{\mathrm{II}}^{-2})$ ($N \equiv 4 \pmod 8$). For the cases $\mathrm{II}_{N,0}(4^1)$ ($N \equiv \pm 1 \pmod 8$)) and $\mathrm{II}_{N,0}(4^{-1})$ ($N \equiv \pm 3 \pmod 8$)) we just take the appropriate even sublattices D_N of the lattices $\mathbb{Z}^N$ to get a generating set. By Corollary 10.7.13 these are also linearly independent over $M(1)$.

For the genera $\mathrm{II}_{N,0}(2 \times 4)$, N even, the generators are given by the theta series of $D_{N-1} \perp A_1$ and $D_{N-7} \perp E_7$, which are the even sublattices of the maximal integral lattices of determinant 2 given in Theorem 10.7.11. The $M(1)$-linear independence of

$$TD := \{\Theta_{D_N}\Theta_{A_1}, \Theta_{D_N}\Theta_{E_7} \mid N \in \{1, \ldots, 23\}, N \text{ odd }\}$$

may be seen as follows. We express the theta-series of the lattices $\mathbb{Z}^N A_1$ and $\mathbb{Z}^N E_7$ ($N \in \{1, \ldots, 23\}$, N odd) in terms of $\theta_3(\tau)$ and $\theta_3(2\tau)$ using the relations given in [133, Chapter 4], and then divide by the common factor $\theta_3(2\tau) = \Theta_{A_1}$. The resulting series can be expressed as polynomials $p(\theta_3, \theta_4)$ in θ_3 and θ_4, and the theta-series of the corresponding even sublattices are just

$$\Theta_{A_1}(p(\theta_3, \theta_4) + p(\theta_4, \theta_3))/2.$$

It follows from an easy Gröbner basis computation that as an $M(1)$-module, $\mathbb{C}[\theta_3, \theta_4]$ is free on

$$B_3 := \{\theta_3^i \theta_4^j \mid 0 \leq i \leq 23, 0 \leq j \leq 7\}.$$

We now can express the series s/Θ_{A_1} with $s \in TD$ as $M(1)$ linear combinations of the series in B_3. The resulting matrix has full rank, thus establishing the linear independence of the theta-series in TD.   □

Corollary 10.7.15. *Construction A_{A_1} yields a bijection between the $\mathbb{C}$-vector space spanned by the theta series of maximal even lattices of dimension N, level 2 and determinant 2^{N-2k} for some $k \in \mathbb{Z}$ and the space of relative invariants of $\mathcal{C}(\rho(2_{\mathrm{II}}))$ spanned by the weight enumerators of maximal isotropic codes in $N\rho(2_{\mathrm{II}})$ given in Theorem 10.2.1.*

11

Extremal and Optimal Codes

A basic problem in coding theory is to find codes with large minimal distance (Hamming, Lee, or Euclidean distance, as appropriate). In order to decide if a particular code is good, it is necessary to know how good comparable codes could be; that is, for a given length and dimension, what is the optimal minimal distance? For general codes, this question is discussed in many references—see for example [361, Chap. 17], and Chapters 4 (by Brouwer [84]), 5 (by Levenshtein [352]), and 6 (by Litsyn [348]) of *The Handbook of Coding Theory* [426]. In the present book, of course, we are interested in self-dual codes. As one might imagine, the constraint of self-duality usually leads to stronger bounds.

Essentially all of the upper bounds we will be discussing are obtained by the linear programming approach, which is discussed in §11.1.1. This is based on the fact that the weight enumerators of the code and the dual code must both have nonnegative coefficients. For a self-dual code these weight enumerators are of course equal.

Section 11.1 deals with upper bounds and §11.2 with lower bounds. The final section, §11.3, contains tables giving the highest minimal distance of self-dual codes of the main Types, as far as this is known at the present time. The problem of enumerating self-dual codes of modest length is the subject of Chapter 12.

We follow [454] in using *extremal* to indicate a code (or weight enumerator) which has the highest minimal distance permitted by the appropriate linear programming bound, and *optimal* to indicate a code which has the actual highest minimal distance of any code of the given Type and length (for in general no extremal code may exist). The precise definition of extremal is given in §§11.1.1, 11.1.4 (see Table 11.1). An extremal code is automatically optimal.

Shortage of space prevents us from giving analogous tables for self-dual lattices. Fortunately there is an excellent survey by R. Scharlau and Schulze-Pillot [473]. This should be supplemented with the new upper bounds obtained in [451].

11.1 Upper bounds

11.1.1 Extremal weight enumerators and the linear programming bound

It follows from the main theorems in Chapter 5 that the complete weight enumerator of a self-dual isotropic code of a given Type ρ lies in the invariant ring of the associated Clifford-Weil group $\mathcal{C}(\rho)$. For small alphabets, this invariant ring can often be calculated explicitly. Investigation of the homogeneous elements in this invariant ring usually gives good upper bounds on the minimal distance of codes of the given Type. The strategy is based on the next theorem. The point is that if C is a self-dual isotropic code of length N of the given Type, we can read off the minimal Hamming distance from its complete weight enumerator cwe(C): we simply look for the term $\neq x_0^N$ of highest degree in the variable x_0. If this term involves x_0^j (say), the code has minimal Hamming distance $d = N - j$.

Equivalently, if I_1 is the ideal generated by the variables $\{x_v \mid v \in V, v \neq 0\}$, then the minimal Hamming distance is the highest power of I_1 which contains cwe$(C) - x_0^N$. So in particular the following theorem holds, where W is the complete weight enumerator of the code C. This is a direct generalization of [454, Theorem 24].

Theorem 11.1.1. *Let $\rho := (V, \rho_M, \rho_\Phi, \beta)$ be a finite representation of a form ring and let $\mathcal{C}(\rho)$ be the associated Clifford-Weil group. If there exists a self-dual isotropic code $C \leq V^N$ of Type ρ and length N with minimal Hamming distance d, then there is a homogeneous polynomial $W \in \mathrm{Inv}(\mathcal{C}(\rho))$ of degree N (with nonnegative integer coefficients) such that*

$$W \equiv x_0^N \pmod{I_d}, \tag{11.1.1}$$

where I_d is the ideal in $\mathbb{C}[x_v \mid v \in V]$ generated by all monomials in the variables $\{x_v \mid v \in V \setminus \{0\}\}$ of total degree d.

Definition 11.1.2. A homogeneous polynomial $W \in \mathrm{Inv}(\mathcal{C}(\rho))$ of degree N with the highest minimal Hamming distance d permitted by the theorem (without imposing the condition that the coefficients be nonnegative) is called an *extremal complete weight enumerator*. More formally, using the notation of the theorem, let $d = d(N) = d(N)(\rho)$ be as large as possible subject to

$$E(N) := (x_0^N + I_d) \cap \mathrm{Inv}(\mathcal{C}(\rho))_N \neq \emptyset. \tag{11.1.2}$$

Then the elements of $E(N)$ are the extremal complete weight enumerators of length N. A code C of Type ρ and length N is said to be *extremal* (with respect to $\mathrm{Inv}(\mathcal{C}(\rho))$) if it has minimal Hamming distance $d(N)$. Note in particular that this definition places no constraints on the coefficients of the extremal complete weight enumerators. Certainly if there is a negative coefficient then no extremal code can exist. We will say exactly what extremal means for the main Types in §11.1.4.

If we add in the nonnegativity constraints we obtain what we call the complete weight enumerator version of the linear programming bound for codes of Type ρ and length N. The reason for this name is best explained by an example. For an arbitrary binary code of length N and minimal distance d, not necessarily self-dual, we know that the Hamming weight enumerator

$$\text{hwe}(C)(x,y) = \sum_{i=0}^{N} A_i x^{N-i} y^i$$

must satisfy the following constraints: $A_i \geq 0$, for $i = 0, \ldots, N$,

$$A_k^{\perp} = \frac{1}{|C|} \sum_{i=0}^{N} A_i P_k(i) \geq 0, \text{ for } k = 0, \ldots, N$$

(see (2.3.8)), $A_0 = 1, A_1 = \ldots = A_{d-1} = 0$; and furthermore the size of the code is equal to $\text{hwe}(C)(1,1) = \sum_{i=0}^{N} A_i$. (In fact, thanks to Delsarte [142], [361] we know that these constraints also apply to nonlinear binary codes.) We may therefore use linear programming—the simplex method, say—to choose real numbers $\{A_i\}$ so as to maximize $\sum_{i=0}^{N} A_i$ subject to these constraints. If the maximum is M_d, we can infer that a code of minimal distance d can have at most M_d codewords: this is the linear programming bound. A code of specified size M can have minimal distance at most $d_0 = \max\{d \mid M_d \geq M\}$. For self-dual codes we know M in advance, but the above constraints are still valid.

Remark 11.1.3. In principle, one could also consider the *integer* programming bound, in which integrality is also taken into account. However, in contrast to linear programming, integer programming is NP-hard, and thus extremely difficult to apply systematically. In any event, the gains from integrality seem to be relatively small.

Note that the definition of extremality depends on the Type. For instance an extremal Type 2_{II} code might not be extremal when considered as a Type 2_{I} code. In particular, for Type 2_{I} codes, the bounds on the minimal distance given by Theorem 11.1.1 are quite weak and the corresponding notion of extremality is too strong (see §11.1.4, where we give a better definition of "extremal" for this Type).

Clearly the complete weight enumerator of an extremal code is an extremal complete weight enumerator. On the other hand, an extremal complete weight enumerator need not be the complete weight enumerator of a code, for there may not exist a code with that weight enumerator.

Definition 11.1.2 can be generalized to define extremal polynomials with respect to any subring of a polynomial ring (not necessarily the invariant ring of a finite group):

Definition 11.1.4. Let

$$R = \bigoplus_{N=0}^{\infty} R_N \leq \mathbb{C}[x, y_1, \ldots, y_n]$$

be a graded subring of a complex polynomial ring, where R_N denotes the set of homogeneous polynomials in R of degree N. For $d \in \mathbb{N}$, let $I_d \trianglelefteq \mathbb{C}[x, y_1, \ldots, y_n]$ be the ideal generated by the monomials of total degree d in the variables $y_1, \ldots, y_n$. For each degree N with $R_N \neq \emptyset$ there is a maximal $d = d(N)$ such that

$$E(N) := (x^N + I_d) \cap R_N \neq \emptyset.$$

The elements of $E(N)$ are called *R-extremal polynomials* with respect to the variable x.

Hence extremal complete weight enumerators are the extremal elements of $\mathrm{Inv}(\mathcal{C}(\rho))$ with respect to the variable x_0.

Usually it is easier to work with Hamming weight enumerators than complete weight enumerators, so we replace $\mathrm{Inv}(\mathcal{C}(\rho))$ by $\mathrm{Inv}^{\mathrm{Ham}}(\mathcal{C}(\rho))$ (as defined in Definition 5.7.13) or—if possible—directly calculate the invariant ring of the symmetrized Clifford-Weil group (cf. §5.7).

The relevance of Definition 11.1.4 arises in the following context. Sometimes $\mathrm{Inv}(\mathcal{C}(\rho))$ is hard to compute, yet there is no suitable symmetrization that commutes with the action of $\mathcal{C}(\rho)$. When this happens we may try to replace $\mathrm{Inv}^{\mathrm{Ham}}(\mathcal{C}(\rho))$ by a larger ring $\mathrm{Inv}(G)$, where G is the symmetrization of a subgroup of $\mathcal{C}(\rho)$.

However, one must remember that information is lost whenever we work with a symmetrized weight enumerator, in particular because the nonnegativity and integrality conditions are now imposed only on the appropriate sums of the coefficients. For instance [382, Theorem 21] shows the advantage of working with complete weight enumerators and that using additional constraints— besides those from linear programming—may yield stronger bounds.

A result analogous to Theorem 11.1.1 holds for Hamming weight enumerators, and extremality is defined similarly. This is the Hamming weight enumerator version of the linear programming bound.

Definition 11.1.5. Let $d = d^{\mathrm{Ham}}(N) = d^{\mathrm{Ham}}(N)(\rho)$ be as large as possible subject to

$$E^{\mathrm{Ham}}(N) := (x^N + \mathbb{C}[x, y]y^d) \cap \mathrm{Inv}^{\mathrm{Ham}}(\mathcal{C}(\rho))_N \neq \emptyset. \tag{11.1.3}$$

Then the elements of $E^{\mathrm{Ham}}(N)$ are called *extremal Hamming weight enumerators*. A code C of Type ρ and length N is said to be *Hamming-extremal* with respect to $\mathrm{Inv}(\mathcal{C}(\rho))$ if it has minimal Hamming distance $d^{\mathrm{Ham}}(N)$.

Clearly $d^{\mathrm{Ham}}(N)(\rho) \geq d(N)(\rho)$, and a Hamming-extremal code can only exist if equality holds.

There are analogous definitions of *Lee-extremal* or *Euclidean-extremal* codes, obtained by replacing the invariant ring of the Clifford-Weil group by the appropriate symmetrization (which need not be the invariant ring of a group).

Though they are not used to define extremality, one might also try to use higher genus weight enumerators to establish the nonexistence of extremal codes. Usually the invariant rings of the higher-genus Clifford-Weil groups are much more complicated and are difficult to calculate, but this may sometimes yield stronger upper bounds. For lattices, the strategy of using Siegel modular forms of genus 2 was successfully applied by Nebe and Venkov [386] to show that for $s = 11$ and $s = 13$ the minimal norm of an even s-modular lattice in dimension 12 is at most 6.

For 2_I codes, we can add further constraints from shadow theory (§1.12), since the weight enumerator of the shadow code must also have nonnegative integral coefficients. The following is the direct generalization of [454, Theorem 25] to our situation:

Theorem 11.1.6. *Let $\rho := (V, \rho_M, \rho_\Phi, \beta)$ be a finite representation of a form ring $R = (R, M, \psi, \Phi)$, let $R' = (R, M, \psi, \Phi')$ be a sub-form ring of R and let $\rho' := \rho_{|R'}$ be the restriction of ρ to R'. Thus (ρ, ρ') is a shadow pair in the sense of Remark 1.12. Let $\mathcal{C}(\rho')$ be the associated Clifford-Weil group. For all $\phi \in \Phi \setminus \Phi'$ and all $W \in \mathrm{Inv}(\mathcal{C}(\rho'))$, let $S_\phi(W)$ be the $\rho_\Phi(\phi)$-shadow of W, obtained by replacing the variable x_v by*

$$\sum_{w \in V} \exp(2\pi i(\beta(w, v) - \rho_\Phi(\phi)(v))) \, x_w \,,$$

for all $v \in V$, and then dividing by $|V|^{N/2}$ (see Corollary 2.2.9). If there exists a self-dual isotropic code $C \leq V^N$ of Type ρ' and length N with Hamming minimal distance d, then there is a homogeneous polynomial $W \in \mathrm{Inv}(\mathcal{C}(\rho'))$ of degree N with nonnegative integer coefficients such that

$$W \equiv x_0^N \pmod{I_d},$$

where I_d is the ideal in $\mathbb{C}[x_v \mid v \in V]$ generated by all monomials in the variables $\{x_v \mid x \in V \setminus \{0\}\}$ of total degree d. Furthermore, for all $\phi \in \Phi \setminus \Phi'$, the $\rho_\Phi(\phi)$-shadow $S_\phi(W)$ has nonnegative integer coefficients.

A similar theorem holds when $\mathrm{Inv}(\mathcal{C}(\rho'))$ is replaced by $\mathrm{Inv}^{\mathrm{Ham}}(\mathcal{C}(\rho'))$.

11.1.2 Self-dual binary codes, 2_{II} and 2_I

In this and the following section we give some applications of the above theorems. The binary case is the most fruitful—on one hand these are the most important codes, and the alphabet is small enough that the Clifford-Weil groups can be easily calculated. Also here the Hamming and complete weight enumerators coincide. And—perhaps the most important point—here the rings of invariants are polynomial rings (cf. §5.6.1), which simplifies the calculations.

Direct application of Theorem 11.1.1 gives a surprisingly good bound for Type 2_{II} codes – although not the best asymptotic bound known, see Theorem 11.1.25. In fact, most of the bounds given here and in §11.1.3 can be improved when N is large. We discuss asymptotic bounds in §11.1.5.

As our starting point we take Theorem 6.4.2, which states that the Hamming weight enumerator of a Type 2_{II} code C lies in the polynomial ring

$$R := \mathbb{C}[x^8 + 14x^4y^4 + y^8, x^4y^4(x^4 - y^4)^4] = \mathrm{Inv}(\mathcal{C}(\rho(2_{II}))).$$

If C has length N, $\mathrm{hwe}_C(x, y)$ has degree N. Since the two generators of R are algebraically independent, the subspace of R of homogeneous polynomials of degree N has dimension $D = \lfloor \frac{N}{24} \rfloor + 1$. This allows us to choose the first D coefficients of $\mathrm{hwe}_C(x, y)$ arbitrarily, so we choose them to be 1 followed by $\lfloor \frac{N}{24} \rfloor$ 0's. It follows that there is a unique extremal Hamming weight enumerator

$$W_N^*(x, y) \in E(N) = (x^N + y^{4D}\mathbb{C}[x, y]) \cap R.$$

Since the weights of Type 2_{II} codes are divisible by 4, $W_N^*(x, y)$ has minimal distance $d(N)(2_{II})$ at least $4D$. The minimal distance of any Type 2_{II} code of length N is therefore bounded above by the minimal distance of W_N^*. It is conceivable that the coefficient of y^{4D} in the extremal weight enumerator might be zero, in which case we would have $d(N)(2_{II}) > 4D$. But this does not happen:

Theorem 11.1.7. (Mallows and Sloane [365].) *The first nonzero coefficient of $W_N^*(1, y) - 1$ occurs precisely at degree $4D$; in particular, the minimal distance of a Type 2_{II} self-dual code of length N is at most $4\lfloor N/24 \rfloor + 4$.*

In fact it is possible to use the Bürmann-Lagrange theorem to derive an explicit formula for the number of words of weight $4D$ in the extremal enumerator.

Theorem 11.1.8. (Bürmann-Lagrange.) *Let $f(x)$ and $g(x)$ be formal power series, with $g(0) = 0$ and $g'(0) \neq 0$. If coefficients κ_{ij} are defined by*

$$x^j f(x) = \sum_{0 \leq i} \kappa_{ij} g(x)^i,$$

then

$$\kappa_{ij} = \frac{1}{i}[\text{coeff. of } x^{i-1} \text{ in } [jx^{j-1}f(x) + x^j f'(x)]\left(\tfrac{x}{g(x)}\right)^i].$$

For proof and generalizations, see Whittaker and Watson [547, p. 133], Good [195], Sack [468], [469], [470].

Applying this to $W_N^*(1, y)$ leads to the following result:

Theorem 11.1.9. (Mallows and Sloane [365].) *Let* $\mu = \lfloor N/24 \rfloor$ *and let*

$$W_N^*(x,y) =: x^N + \sum_{i=\mu+1}^{N/4} A_{4i}^* x^{N-4i} y^{4i} \in \mathrm{Inv}(\mathcal{C}(\rho(2_{\mathrm{II}})))_N \qquad (11.1.4)$$

be the extremal weight enumerator. Then $A_{4\mu+4}^*$, *the first nonzero coefficient of* $W_N^*(1,y) - 1$, *is given by:*

$$\binom{N}{5}\binom{5\mu-2}{\mu-1} / \binom{4\mu+4}{5}, \quad \text{if } N = 24\mu, \qquad (11.1.5)$$

$$\frac{1}{4}N(N-1)(N-2)(N-4)\frac{(5\mu)!}{\mu!(4\mu+4)!}, \quad \text{if } N = 24\mu+8, \qquad (11.1.6)$$

$$\frac{3}{2}N(N-2)\frac{(5\mu+2)!}{\mu!(4\mu+4)!}, \quad \text{if } N = 24\mu+16, \qquad (11.1.7)$$

and is never zero.

For the details of the proof, see [365] or [361, Chap. 19]. For Type 2_{I} codes, applying the same method yields:

Theorem 11.1.10. (Mallows and Sloane [365].) *The minimal distance of a Type* 2_{I} *self-dual code is at most* $2\lfloor N/8 \rfloor + 2$.

Again there are explicit formulae for the number of words of minimal weight: see [361, Chap. 19, Problem 12].

It can also be shown (see Mallows, Odlyzko and Sloane [363]) that the bounds of Theorems 11.1.7 and 11.1.10 can be met for at most finitely many N: in fact, that paper shows that the next coefficient ($A_{4\mu+8}^*$) *after* the leading nonzero coefficient in the extremal enumerator becomes negative for sufficiently large N. Furthermore, for any constant α, the minimal distance can be within α of the bound only finitely often. For Type 2_{II} codes, for instance, it was shown in [363] that $A_{4\mu+8}^*$ first goes negative when N is around 3720. Ma and Zhu [354] and Zhang [566] have determined precisely when $A_{4\mu+8}^*$ first goes negative, and have obtained similar results for several other families. The following result incorporates the work of several authors.

Theorem 11.1.11. ([566]) *For Type* 2_{I} *let* $c = 2, \mu = \lfloor N/8 \rfloor$, *and for Type* 2_{II} *let* $c = 4, \mu = \lfloor N/24 \rfloor$. *The coefficient* $A_{c(\mu+2)}^*$ *in the extremal Hamming weight enumerator* $W_N^* \in E(N)$ *is negative if and only if*

Type 2_{I}: $N = 8i \, (i \geq 4), \ 8i+2 \, (i \geq 5), \ 8i+4 \, (i \geq 6), \ 8i+6 \, (i \geq 7)$;

Type 2_{II}: $N = 24i \, (i \geq 154), \ 24i+8 \, (i \geq 159), \ 24i+16 \, (i \geq 164)$.

In particular, the first time $A_{4\mu+8}^*$ *goes negative for Type* 2_{II} *codes is at length* $N = 24 \times 154 = 3696$.

It is possible that other coefficients in the extremal weight enumerator may go negative before this.

For Type 2_I codes, the bound of Theorem 11.1.10 is especially weak. Ward [540] has shown that the minimal distance can be $2\lfloor N/8 \rfloor + 2$ precisely when N is one of 2, 4, 6, 8, 12, 14, 22 or 24. This suggests that the bound can be greatly strengthened, which is indeed the case. Conway and Sloane [130] showed that $d \leq 2\lfloor (N+6)/10 \rfloor$ for $N > 72$, and Ward ([543], [544]) established $d \leq N/6 + O(\log N)$. In fact, a bound analogous to that of Theorem 11.1.7 holds for Type 2_I codes. The key to proving this fact is the observation that we have not yet used the shadow enumerator. The result is the following:

Theorem 11.1.12. ([442]) *Suppose C is an $[N, N/2, d]$ Type 2_I code. Then $d \leq 4\lfloor N/24 \rfloor + 4$, except when $N \equiv 22$ (mod 24), when $d \leq 4\lfloor N/24 \rfloor + 6$. If N is a multiple of 24, any code meeting the bound is of Type 2_{II}. If $N \equiv 22$ (mod 24), any code meeting the bound can be obtained by shortening a Type 2_{II} code of length $N + 2$ that also meets the bound.*

Proof. (Sketch) Let $W := \mathrm{hwe}(C)$ be the weight enumerator of C. Then, by Theorem 6.4.2, $W(x, y)$ lies in the ring $\mathrm{Inv}(\mathcal{C}(\rho(2_I))) = \mathbb{C}[x^2 + y^2, x^2 y^2 (x^2 - y^2)^2]$; consequently we can write

$$W(1, y) = \sum_j a_j y^{2j}$$

$$= \sum_i c_i (1 + y^2)^{N/2 - 4i} (y^2 (1 - y^2)^2)^i .$$

Applying the shadow transform $S(x, y) = 2^{-N/2}(W(x + y, i(x - y))$ (see (2.3.9)), we get

$$S(1, y) = \sum_j b_j y^{2j+t}$$

$$= \sum_i c_i (2y)^{N/2 - 4i} (-(1 - y^4)/2)^i ,$$

where $t = (N/2 \bmod 4)$. Suppose C had minimal distance $4\lfloor N/24 \rfloor + 6$. This fact determines c_i for $0 \leq i \leq 2\lfloor N/24 \rfloor + 2$, and in particular $c_{2\lfloor N/24 \rfloor + 2}$. On the other hand, we can also express $c_{2\lfloor N/24 \rfloor + 2}$ as a linear combination of the b_j for small j. It turns out that these two expressions for $c_{2\lfloor N/24 \rfloor + 2}$ are incompatible; in particular, we find that a certain nonnegative linear combination of the b_j is negative.

Rather than give the (somewhat messy) details of the proof, we will simply show how one can use the Bürmann-Lagrange theorem to compute the coefficients in these linear combinations. For instance, to compute $c_{2\lfloor N/24 \rfloor + 2}$, we note that

$$\sum_i c_i (1 + y^2)^{N/2 - 4i} (y^2 (1 - y^2)^2)^i = 1 + O(y^{4\lfloor N/24 \rfloor + 6}) .$$

Dividing both sides by $(1 + y^2)^{N/2}$ and setting $y = \sqrt{Y}$, we get

$$\sum_i c_i \left(\frac{Y(1-Y)^2}{(1+Y)^4}\right)^i = (1+Y)^{-N/2} + O(Y^{2\lfloor N/24\rfloor+3}).$$

We can then apply Theorem 11.1.8, with

$$f(Y) = (1+Y)^{N/2}, \quad g(Y) = Y(1-Y)^2(1+Y)^{-4},$$

to obtain

$$c_i = \frac{1}{i}[\text{coeff. of } Y^{i-1} \text{ in } [\tfrac{d}{dY}(1+Y)^{-N/2}]\left((1+Y)^4(1-Y)^{-2}\right)^i]$$

$$= \frac{-N}{2i}[\text{coeff. of } Y^{i-1} \text{ in } (1+Y)^{-N/2-1+4i}(1-Y)^{-2i}]$$

$$= \frac{-N}{2i}[\text{coeff. of } Y^{i-1} \text{ in } (1+Y)^{-N/2-1+6i}(1-Y^2)^{-2i}].$$

In particular, for $i = 2\lfloor N/24\rfloor + 2$, $c_{2\lfloor N/24\rfloor+2}$ is equal to $\frac{-N}{4\lfloor N/24\rfloor+4}$ times the coefficient of $Y^{2\lfloor N/24\rfloor+1}$ in

$$(1+Y)^{-N/2+12\lfloor N/24\rfloor+11}(1-Y^2)^{-4\lfloor N/24\rfloor-4}.$$

It follows that $c_{2\lfloor N/24\rfloor+2} \leq 0$, with equality only when $N \equiv 22 \pmod{24}$, since all coefficients of any power series of the form $(1 + Y)^a(1 - Y^2)^{-b}$ are positive whenever $a, b > 0$.

Similarly, we find that the coefficients of the expansion of $c_{2\lfloor N/24\rfloor+2}$ in terms of the b_j are positive. This proves the bound, except when $N \equiv 22 \pmod{24}$; for the proof that the bound holds in that case, and that a code meeting the bound is of Type 2_{II} if $N \equiv 0 \pmod{24}$, the reader is referred to [442]. $\qquad\square$

For N up to a certain point (at least 200, but almost certainly much higher), this bound is the best that can be achieved by solving the linear programming problem by computer. However, for larger N we can certainly do better by linear programming—see Theorem 11.1.25.

11.1.3 Some other types

Results similar to Theorem 11.1.7 hold for other families. The proof of the following result again applies Theorem 11.1.1 to the Hamming weight enumerators.

Theorem 11.1.13. *The minimal Hamming distance of a code of length N and Type 3 is at most $3\lfloor N/12\rfloor + 3$; of Type 4^H at most $2\lfloor N/6\rfloor + 2$; and of Type $4_{\text{II}}^{\text{H}+}$ at most $2\lfloor N/6\rfloor + 2$. The minimal Hamming distance of a code of any of the Types 4^E, 4^{H+}, q^H or q^E is at most $\lfloor N/2\rfloor + 1$.*

Note that the last bound mentioned in the theorem is simply the Singleton bound obtained from the ring $\mathbb{C}[x^2 + (q-1)y^2, y(x-y)]$. As we have already remarked in §5.7, this is not the correct ring (that is, the smallest ring containing all Hamming enumerators of self-dual codes) and should be replaced by $\mathrm{Inv}^{\mathrm{Ham}}(\mathcal{C})$ where $\mathcal{C}$ is the associated Clifford-Weil group. However, the latter ring is very difficult to determine in general. Even when we know this smaller ring (e.g. for $q = 4$ or $q = 5$), because the ring is no longer free, it is much more difficult to use. In particular, it is no longer the case that we may choose the leading coefficients in the weight enumerator arbitrarily. This leads to the extremal enumerator not being unique, making it difficult to determine its first nonzero coefficient. Of course, for small lengths N, explicit calculations with these rings lead to better upper bounds on the minimal Hamming distance of a self-dual code (see for example Gaborit, Pless, Solé and Atkin [186]).

Similarly, any attempt to apply analogous arguments to Types $4^{\mathbb{Z}}$ or $m^{\mathbb{Z}}$ will run into the problem that, in these cases, we are primarily interested in the minimal Lee distance or Euclidean norm, forcing us to work with symmetrized weight enumerators. These are more difficult to deal with than the Hamming weight enumerator. A partial solution to this problem is provided by Theorem 11.1.17 below.

As in the binary case, the bounds of Theorem 11.1.13 can be met for at most finitely many N. The following result incorporates the work of several authors.

Theorem 11.1.14. ([566]) *For Type 3 let $c = 3, \mu = \lfloor N/12 \rfloor$, and for Type 4^{H} let $c = 2, \mu = \lfloor N/6 \rfloor$. The coefficient $A^*_{c(\mu+2)}$ in the extremal Hamming weight enumerator for codes of length N is negative if and only if*

$$\text{Type 3:} \quad N = 12i \, (i \geq 70), \ 12i + 4 \, (i \geq 75), \ 12i + 8 \, (i \geq 78) \, ;$$

$$\text{Type } 4^{\mathrm{H}}\text{:} \quad N = 6i \, (i \geq 17), \ 6i + 2 \, (i \geq 20), \ 6i + 4 \, (i \geq 22) \, .$$

Again it is possible that other coefficients in the extremal weight enumerator may go negative before this. In the case of ternary (Type 3) codes, for example, the extremal Hamming weight enumerator contains a negative coefficient for lengths 72, 96, 120 and all $N \geq 144$.

For codes of Type 4^{E}, Gaborit et al. [186] have given a stronger bound than that in Theorem 11.1.13.

Theorem 11.1.15. (Gaborit, Pless, Solé and Atkin [186]). *Let C be a code of Type 4^{E} and length N. The minimal Lee distance of C is at most $4\lfloor \frac{N}{12} \rfloor + 4$. This expression is also an upper bound on the minimal Hamming distance.*

Proof. The map (2.4.20) induces an isometry from $(\mathbb{F}_4^N, \text{Lee distance})$ to $(\mathbb{F}_2^{2N}, \text{Hamming distance})$. Since N is even, the first claim follows from Theorem 11.1.12. The second claim follows because the Hamming distance cannot exceed the Lee distance.   □

There is also an analogue of Theorems 11.1.12 and 11.1.13 for Type I codes from family 4^{H+}. Again this is stronger than the result in Theorem 11.1.13.

Theorem 11.1.16. ([443]) *If C is an additive self-dual code of Type 4^{H+}, length N and minimal Hamming distance d, then $d \leq 2\lfloor N/6 \rfloor + 2$, except when $N \equiv 5$ (mod 6), when $d \leq 2\lfloor N/6 \rfloor + 3$. If N is a multiple of 6, then any code meeting the bound is of Type 4_{II}^{H+}.*

We conjecture that if $N \equiv 1$ (mod 6) then the bound can be strengthened to $d \leq 2\lfloor N/6 \rfloor + 1$. This is certainly true for $N \leq 25$, and may be provable for all N using integer programming.

We next discuss codes over $\mathbb{Z}/4\mathbb{Z}$.

Theorem 11.1.17. (Bonnecaze, Solé, Bachoc and Mourrain [66].) *A code C of Type $4_{II}^{\mathbb{Z}}$ and length N has minimal Euclidean norm at most $8\lfloor \frac{N}{24} \rfloor + 8$.*

The proof uses C to define an even unimodular N-dimensional lattice $\Lambda(C) := \{\frac{1}{2}u \in \mathbb{R}^n : u \ (\text{mod } 4) \in C\}$, and examines its theta series.

As usual, one can derive an analogue for Type I codes:

Theorem 11.1.18. ([453]) *Suppose C is a code of Type $4^{\mathbb{Z}}$ and length N. The minimal Euclidean norm of C is at most $8\lfloor \frac{N}{24} \rfloor + 8$, except when $N \equiv 23$ (mod 24), in which case the bound is $8\lfloor \frac{N}{24} \rfloor + 12$. If equality holds in the latter bound, then C is a shortened version of a code of Type $4_{II}^{\mathbb{Z}}$ and length $N+1$.*

There are analogous results for Hamming and Lee distances.

Theorem 11.1.19. ([450].) *Suppose C is a Type $4^{\mathbb{Z}}$ code of length $N = 24m + \ell$, where $1 \leq \ell \leq 24$. Then the minimal Hamming distance of C is at most*

$$4m + f(\ell), \tag{11.1.8}$$

where f is given in Theorem 11.1.21, and the minimal Lee distance of C is at most

$$8m + g(\ell), \tag{11.1.9}$$

where g is given by the following table:

ℓ :	1	2	3	4	5	6	7	8	9	10	11	12	13	14	15	16	17	18	19	20	21	22	23	24
g :	2	2	2	4	2	4	6	8	2	4	4	4	4	6	6	8	8	6	8	8	8	10	12	

Remark 11.1.20. The key ingredient in the proof is to observe that the Hamming distance of a self-dual code over $\mathbb{Z}/4\mathbb{Z}$ is equal to the dual distance of an associated doubly-even binary code, and to use Theorem 11.1.21. The Lee distance can then be bounded by the Euclidean norm and twice the Hamming weight.

For lengths $N \leq 24$ the bound of (11.1.8) agrees with the results obtained by solving the linear programming bound by computer (and also with the minimal distance of the optimal codes). The bound of (11.1.9) agrees for $N \leq 48$ (and probably for much larger N) with the results obtained by solving the linear programming bound (Dougherty, Harada and Oura [160], as extended in [446]) by computer, except for the cases $\ell = 7$ and 8 (and $N \leq 104$), when the linear programming bound is 2 less than (11.1.9). It is not clear how to prove the stronger bound, however. For $\ell = 24$, it appears that a code meeting the bound must be of Type $4_{\mathrm{II}}^{\mathbb{Z}}$ (all weights divisible by 8, [160]), but again a general proof appears difficult.

The following result was also established in [450].

Theorem 11.1.21. ([450].) *If C is an isotropic code in $N\rho(2_{\mathrm{II}})$ with length $N = 24m+\ell$, then the minimal distance of $C^{\perp}$ is bounded above by $4m+f(\ell)$, where f is given by the following table:*

$$\ell : 1\ 2\ 3\ 4\ 5\ 6\ 7\ 8\ 9\ 10\ 11\ 12\ 13\ 14\ 15\ 16\ 17\ 18\ 19\ 20\ 21\ 22\ 23\ 24$$
$$f : 1\ 1\ 1\ 2\ 1\ 2\ 3\ 4\ 1\ \ 2\ \ \ 2\ \ \ 2\ \ \ 2\ \ \ 3\ \ \ 3\ \ \ 4\ \ \ 4\ \ \ 4\ \ \ 3\ \ \ 4\ \ \ 5\ \ \ 6\ \ \ 7\ \ \ 8$$

A code of length

$$24m+\ 1\ 2\ 3\ 4\ 6\ 7\ 8\ 14\ 17\ 18\ 21\ 22\ 23\ 24$$

meeting the bound must have dimension

$$12m+\ 0\ 0\ 0\ 1\ 2\ 3\ 4\ 6\ 8\ 8\ 9\ 10\ 11\ 12$$

respectively.

For the lengths not listed, the dimension of a code meeting the bound is in general not determined.

We briefly mention the analogous bound for unimodular lattices.

Theorem 11.1.22. ([453]) *Suppose Λ is a unimodular lattice of dimension N and minimal norm μ. Then $\mu \leq 2\lfloor N/24 \rfloor + 2$, except when $N = 23$, when $\mu = 3$ can be attained.*

Remark 11.1.23. For even unimodular lattices this was obtained by Mallows, Odlyzko and Sloane [363], and implicitly by Siegel [495]. Comparison with Theorem 11.1.12 above suggests that when $N \equiv 0 \pmod{24}$, a lattice meeting the bound must be even. This was shown by Gaulter [187]. The lattice in dimension 23 with minimal norm $\mu = 3$ is the shorter Leech lattice [133]; note however that in contrast to the code case, this is the only exception to the bound.

11.1.4 A new definition of extremality

We will define a code to be *extremal* (or, for Type $4^{\mathbb{Z}}$ codes, *norm-extremal*) if it meets the appropriate bound given in Table 11.1.

These are the best explicit bounds that have been obtained to date using linear programming (although better *asymptotic* bounds are known, see the next section). It is worth mentioning that partial results on the limits of the linear programming bound for not necessarily self-dual codes have been obtained by Samorodnitsky [472].

Table 11.1. Upper bounds used to define extremality. Bounds are for Hamming distance unless indicated otherwise. The length N must be a multiple of ν and all weights (Hamming weights unless indicated otherwise) must be divisible by c.

Type	ν, c	Upper bound	Theorem
2_{I}	$2, 2$	$4\lfloor \frac{N}{24}\rfloor + 4 + \epsilon$, where $\epsilon = -2$ if $N = 2, 4$ or 6, $\epsilon = 2$ if $N \equiv 22 \pmod{24}$, and $\epsilon = 0$ otherwise	11.1.12
2_{II}	$8, 4$	$4\lfloor \frac{N}{24}\rfloor + 4$	11.1.7
3	$4, 3$	$3\lfloor \frac{N}{12}\rfloor + 3$	11.1.13
4^{H}	$2, 2$	$2\lfloor \frac{N}{6}\rfloor + 2$	11.1.13
4^{E}	$2, 1$	$4\lfloor \frac{N}{12}\rfloor + 4$	11.1.15
$4^{\mathrm{E}}_{\mathrm{II}}$	$4, 4^{(a)}$	Lee distance $4\lfloor \frac{N}{12}\rfloor + 4$	11.1.15
$4^{\mathrm{H}+}$	$1, 1$	$2\lfloor \frac{N}{6}\rfloor + 2 + \epsilon'$, where $\epsilon' = -1$ if $N = 1$, $\epsilon' = 1$ if $N \equiv 5 \pmod{6}$, and $\epsilon' = 0$ otherwise	11.1.16
$4^{\mathrm{H}+}_{\mathrm{II}}$	$2, 2$	$2\lfloor \frac{N}{6}\rfloor + 2$	11.1.13
q^{H}	$2, 1$	$\lfloor \frac{N}{2}\rfloor + 1$	11.1.13
q^{E}	$2^{(b)}, 1$	$\lfloor \frac{N}{2}\rfloor + 1$	11.1.13
$4^{\mathbb{Z}}$	$1, 1$	norm $8\lfloor \frac{N}{24}\rfloor + 8 + \epsilon''$, where $\epsilon'' = 4$ if $N \equiv 23 \pmod{24}, \epsilon'' = 0$ otherwise	11.1.18
$4^{\mathbb{Z}}_{\mathrm{II}}$	$8, 8^{(c)}$	norm $8\lfloor \frac{N}{24}\rfloor + 8$	11.1.17
	$^{(a)}$Lee weight. $^{(b)}$Or 4 if $q \equiv 3 \pmod 4$. $^{(c)}$Euclidean norm.		

For Types $2_{\mathrm{II}}, 3, 4^{\mathrm{H}}, 4^{\mathrm{E}}, q^{\mathrm{H}}, q^{\mathrm{E}}$ this definition of extremality agrees with the historical usage [361], [454]. For codes of Type $4^{\mathbb{Z}}_{\mathrm{II}}$ it agrees with the definition given by Bonnecaze et al. [66]. For Type 2_{I} codes, however, "extremal" has generally been used to mean a code meeting the much weaker bound of Theorem 11.1.10; in the light of Theorem 11.1.12, it seems appropriate to change the definition. There should be an analogous concept of *Lee-extremal* for Type $4^{\mathbb{Z}}$ codes, but at present we do not know what this is. Of course, the bounds of Theorem 11.1.18 also apply to Lee distance. But this is not a satisfactory bound, since it is not even tight at length 24, where the highest attainable Lee distance is 12 rather than 16 (see Table 11.11).

In contrast, we call a code *optimal* if it has the highest minimal distance of any self-dual code of that length. An extremal code is automatically optimal.

The fact that, from Theorem 11.1.12, an extremal binary code of length a multiple of 24 must be Type 2_{II} suggests that these codes are likely to be particularly nice. Indeed, we have the following result, which is a consequence of the Assmus-Mattson theorem (cf. [361, Chap. 6]; Pless, Huffman and Brualdi [427, Theorem 11.14]; Tonchev [524, §5]). See also Harada, Kitazume and Munemasa [245] and Harada, Munemasa and Tonchev [251].

Theorem 11.1.24. *Let C be an extremal Type 2_I code of length $24m$. Then the codewords of C of any given weight form a 5-design.*

Similarly, the supports of the minimal codewords of an extremal ternary code of length $12m$ form a 5-design. For codewords of higher weight, the natural incidence structure is *almost* a 5-design, except that it may have repeated blocks. Again, for an extremal Type 4^{H+} code of length $6m$, the supports with multiplicities of the codewords of any fixed weight form a 5-design with repeated blocks. Gulliver and Harada [217], Harada [233] and Shin, Kumar and Helleseth [484] have shown that the $\mathbb{Z}/4\mathbb{Z}$-lift of the Golay code g_{24} also yields 5-designs. More generally, one can show that the words of any fixed symmetrized type, in any of the 13 Lee-optimal Type $4^{\mathbb{Z}}$ codes of length 24 (see §11.3.6) form a colored 5-design, possibly with repeated blocks (Bonnecaze, Rains and Solé [65]). Other papers dealing with designs in extremal codes are Bachoc and Gaborit [24], Kim and Pless [307] and Tanabe [517], [518]. A notion of shadow-extremal code has been defined in the literature, although we will not discuss it here—see Bachoc and Gaborit [23], Elkies [167] and Gaborit [181].

11.1.5 Asymptotic upper bounds

Krasikov and Litsyn have shown that the bound of Theorem 11.1.7 can be strengthened slightly for large N. This is the best asymptotic bound presently known for Type 2_{II} codes.

Theorem 11.1.25. (Krasikov and Litsyn [335], [336].) *Let C_i be a sequence of Type 2_{II} codes with minimal distance d_i and length N_i tending to infinity. Then*

$$\limsup_{i \to \infty} \frac{d_i}{N_i} \leq (1 - 5^{-1/4})/2 = 0.1656298476\ldots. \tag{11.1.10}$$

The proof uses a variant of the linear programming bound.

Reference [451] gives the following generalization of this result (and a simpler proof). For the hypotheses, compare the Gleason-Pierce Theorem 2.5.

Theorem 11.1.26. ([451]) *Let q and c be such that either $q > 1$ and $c = 1$, or $(q, c) \in \{(2, 2), (2, 4), (3, 3), (4, 2)\}$. Let C_i be a sequence of self-dual codes of any Type over an alphabet of size q with Hamming distance d_i, all Hamming weights divisible by c and length N_i tending to infinity. Then*

$$\limsup_{i \to \infty} \frac{d_i}{N_i} \leq \frac{q-1}{q}(1 - (c+1)^{-1/c}).$$

Corollary 11.1.27. ([451]) *This implies the following asymptotic upper bounds on* d/N :

> *Type* 2_{II} : $(1 - 5^{-1/4})/2 = 0.1656298476\ldots$
> *Type* 3 : $(2 - 2^{1/3})/3 = 0.2466929834\ldots$
> *Types* 4^{H} *or* $4_{\text{II}}^{\text{H+}}$: $(3 - 3^{1/2})/4 = 0.3169872982\ldots$
> *Type* q^{H} *or* q^{E} : $1/2 - 1/(2q)$.

Remark 11.1.28. For Type 2_{II} this is the bound of Theorem 11.1.25, while for the last four Types this is worse than the bounds of McEliece, Rodemich, Rumsey and Welch ([370], [361, Chap. 17]) and Aaltonen ([1], [2]). For Type 3 this appears to be the strongest asymptotic bound known.

This strongly suggests that these bounds are not optimal, even for Types 2_{II} and 3. Presumably the correct bound needs to combine this approach with the approach used to prove the McEliece et al. bound.

Again, this bound is relatively weak for Types 2_{I} and $4^{\text{H+}}$, because it has not taken the shadow into account. We have:

Theorem 11.1.29. ([451]) *Let* C_i *be a sequence of codes of Type* 2_{I} *with Hamming distance* d_i *and length* N_i *tending to infinity. Then*

$$\limsup_{i \to \infty} d_i/N_i \leq (1 - 5^{-1/4})/2.$$

For Type $4^{\text{H+}}$, *we instead have*

$$\limsup_{i \to \infty} d_i/N_i \leq (3 - 3^{1/2})/4.$$

Again we mention the analogous bound for unimodular lattices.

Theorem 11.1.30. ([451]) *Let* Λ_i *be a sequence of unimodular lattices of minimal norm* μ_i *and dimension* N_i *tending to infinity. Then*

$$\limsup_{i \to \infty} \mu_i/N_i \leq 0.0833210664\ldots.$$

Remark 11.1.31. The number $0.0833210664\ldots$ can be expressed as $\tau_0/(2\pi i)$, where τ_0 is the unique imaginary zero of the Eisenstein series

$$E_2(\tau) = 1 - 24 \sum_{m=1}^{\infty} \sigma_1(m)e^{2\pi i \tau m}.$$

Note that, even for even unimodular lattices, this improves on the old bound [363] of $1/12$. For odd unimodular lattices the old bound from [363] was $1/8$.

Analogues of both theorems hold for strongly k-modular lattices for $k \in \{2, 3, 5, 6, 7, 11, 14, 15, 23\}$—see [451], [453]. As with codes, this bound is eventually worse (for $k > 3$) than the corresponding generic bound of Kabatiansky and Levenshtein ([294]; [133, Chap. 9]).

This again suggests that the bound of Theorem 11.1.30 is not tight.

There are also asymptotic improvements to Theorems 11.1.17, 11.1.18 and 11.1.19:

Theorem 11.1.32. ([451]) *Let C_i be a sequence of codes of Type $4^{\mathbb{Z}}$ or $4^{\mathbb{Z}}_{\mathrm{II}}$ with Hamming distance d_i, Lee distance L_i, Euclidean norm n_i and length N_i tending to infinity. Then*

$$\limsup_{i \to \infty} d_i/N_i \leq (1 - 5^{-1/4})/2 = 0.1656298476\ldots,$$
$$\limsup_{i \to \infty} L_i/N_i \leq 1 - 5^{-1/4} = 0.3312596950\ldots,$$
$$\limsup_{i \to \infty} n_i/N_i \leq 2 - 2\lambda = 0.3332625492\ldots,$$

where $\lambda = 0.8333687\ldots$ is the positive real root of the polynomial

$$11\lambda^{16} + 2112\lambda^{14} - 8525\lambda^{12} + 15048\lambda^{10} - 15218\lambda^8 + 9552\lambda^6 - 3718\lambda^4 + 828\lambda^2 - 81.$$

Research Problem 11.1.33. *What is the true asymptotic behavior of*

$$\overline{\lim}_{i \to \infty} \frac{d_i}{N_i} ?$$

A weaker version is: what is the best bound that can be proved by linear programming?

It is also shown in [451] that the error in the above bounds grows faster than $\Omega(\sqrt{N})$; that is, there are only finitely many values of N for which the minimal distance (or norm) is within any fixed constant times $\sqrt{N}$ of the bound. It may be possible to use that argument to improve the bound on $\limsup d/N$ if one could obtain good explicit constants for the relevant asymptotic estimates.

11.2 Lower bounds

There are two ways to obtain lower bounds on the optimum minimal distance of a code of length N. The first way, naturally, is simply to construct a good code. Just as for general linear codes, there is also a nonconstructive lower bound, analogous to the Gilbert-Varshamov bound (cf. [361], Pless, Huffman and Brualdi [427, Theorems 3.1, 3.4, 3.5]).

We first consider the case of self-dual binary codes.

Theorem 11.2.1. (Thompson [521], MacWilliams, Sloane and Thompson [362].) *Let N be any positive even integer. Let d_{GV} be the largest integer d such that*

$$\sum_{\substack{0 < i < d \\ 2|i}} \binom{N}{i} < 2^{N/2-1} + 1. \tag{11.2.1}$$

Then there exists a Type 2_I code of length N and minimal distance at least d_{GV}.

Proof. If we can show that the expected number of nonzero vectors of weight less than d_{GV} in a *random* self-dual code of length N is less than 1, it will immediately follow that there exists *some* self-dual code of length N with no such vectors.

   Let us therefore compute the average weight enumerator of the set of self-dual codes. Consider the group G of binary matrices that preserve the quadratic form I. On the vector space of even weight vectors, modulo the all-ones vector, the quadratic form becomes symplectic, and the group acts as the full symplectic group. In particular, it is therefore transitive on nonzero vectors of even weight, modulo $\mathbf{1}$. It follows that the expected number of vectors of weight $2i$ in a random code must be proportional to $\binom{N}{2i}$, except for $i = 0$ or $i = N/2$. Thus the average weight enumerator has the form

$$\overline{W}(x, y) = ax^N + b \sum_{1 \le i \le N/2-1} \binom{N}{2i} x^{N-2i} y^{2i} + cy^N$$

$$= ax^N + cy^N + b\left(\frac{1}{2}(x+y)^N + \frac{1}{2}(x-y)^N - x^N - y^N\right).$$

Since every self-dual binary code contains the 0 vector and the all 1's vector, $\overline{W}(1, 0) = \overline{W}(0, 1) = 1$; since every self-dual code contains a total of $2^{N/2}$ vectors, $\overline{W}(1, 1) = 2^{N/2}$. Solving for a, b, and c, we find:

$$\overline{W}(x, y) = x^N + y^N + \frac{1}{2^{N/2-1} + 1} \sum_{1 \le i \le N/2-1} \binom{N}{2i} x^{N-2i} y^{2i}.$$

Thus the average number of nonzero vectors of weight less than d is

$$\frac{1}{2^{N/2-1} + 1} \sum_{\substack{0 < i < d \\ 2|i}} \binom{N}{i}. \qquad \square$$

Corollary 11.2.2. (Thompson [521], MacWilliams, Sloane and Thompson [362].) *There exists an infinite sequence of $[N_i, N_i/2, d_i]$ Type 2_I codes such that N_i tends to infinity and*

$$\liminf_{i \to \infty} \frac{d_i}{N_i} \ge H_2^{-1}(\frac{1}{2}) = 0.11002786\ldots, \tag{11.2.2}$$

where $H_q(x)$ is the q-ary entropy function

$$H_q(x) = x \log_q(q-1) - x \log_q(x) - (1-x) \log_q(1-x). \qquad (11.2.3)$$

Proof. Take the logarithm of both sides of (11.2.1), divide by N, and let N tend to infinity. The resulting inequality is $H_2(\delta) \leq \frac{1}{2}$, as desired.   □

There are similar results for Type 2_{II} codes.

Theorem 11.2.3. (Thompson [521], MacWilliams, Sloane and Thompson [362].) *Let N be any positive multiple of 8. Let d_{GV} be the largest integer d such that*

$$\sum_{\substack{0 < i < d \\ 4 \mid i}} \binom{N}{i} < 2^{N/2-2} + 1. \qquad (11.2.4)$$

Then there exists a Type 2_{II} code of length N and minimal distance at least d_{GV}.

Proof. Again we compute the average weight enumerator. The key observation is that the function $\frac{1}{2} \operatorname{wt}(v)$ induces a quadratic form on the space of even weight vectors modulo the all-ones vector. The group of matrices that preserve this quadratic form is transitive on the kernel of this quadratic form; that is, on vectors of weight divisible by 4, modulo **1**. This allows us to write down the average weight enumerator:

$$\overline{W}_{\text{II}}(x, y) = x^N + y^N + \frac{1}{2^{N/2-2} + 1} \sum_{0 < i < N/4} \binom{N}{4i} x^{N-4i} y^{4i}. \qquad (11.2.5)$$

     □

Asymptotically, this agrees with Corollary 11.2.2 (as well as the Gilbert-Varshamov bound). For finite N, it is actually (slightly) stronger. That is, the constraint that the code be Type 2_{II} makes it easier to find good codes.

Similar arguments prove:

Theorem 11.2.4. ([454]) *For each of the Types 2_{I}, 2_{II}, 3, 4^{H}, 4^{E}, $4_{\text{I}}^{\text{H}+}$, $4_{\text{II}}^{\text{H}+}$, q^{H} and q^{E} there exists a sequence of self-dual codes with minimal distance d_i and length N_i tending to infinity, satisfying*

$$\liminf_{i \to \infty} \frac{d_i}{N_i} \geq H_q^{-1}(\tfrac{1}{2}) \qquad (11.2.6)$$

(see (11.2.3)), where q is the alphabet size, respectively $2, 2, 3, 4, 4, 4, 4, q, q$ in the nine cases.

The result for families q^{H} and q^{E} was first given by Pless and Pierce [429]. Similar results hold for Type $4^{\mathbb{Z}}$ codes:

Theorem 11.2.5. ([454]) *There exists a family of Type $4_{II}^{\mathbb{Z}}$ codes with length N_i tending to infinity, such that*

$$\liminf_{i\to\infty} \frac{l_i}{2N_i} \geq H_q^{-1}(\frac{1}{2}), \tag{11.2.7}$$

where l_i is the minimal Lee distance of the ith code.

Theorem 11.2.6. ([454]) *There exists a family of Type $4_{II}^{\mathbb{Z}}$ codes with length N_i tending to infinity, such that*

$$\liminf_{i\to\infty} \frac{E_i}{N_i} \geq 0.34737283\ldots, \tag{11.2.8}$$

where E_i is the minimal Euclidean norm of the ith code.

It is possible, although far less elementary, to compute the average theta series of a genus of lattices, in particular of unimodular and even unimodular lattices. This gives analogous lower bounds—we omit the details, and just give one example. The average theta series of an even unimodular lattice of dimension $N = 4k \equiv 0 \pmod 8$ is

$$\sum_\Lambda \frac{1}{\mathrm{Aut}(\Lambda)} \Theta_\Lambda(q) = 1 + \frac{(-1)^k 4k}{B_{2k}} \sum_{m=1}^\infty \sigma_{2k-1}(m)q^m, \tag{11.2.9}$$

the Eisenstein series of weight $2k$.

11.3 Tables of extremal self-dual codes

In this section we will summarize what is presently known about extremal and optimal codes of the most important Types.

In the following tables we have tried, for each length N in the range of the table, to list all known codes with the specified minimal distance, or otherwise to indicate how many extremal codes are known. In all of these tables, a period after a list of codes indicates that the list is complete. Whenever possible we have attempted to name at least one extremal code. These tables are updated and extended versions of those in [454]. In some cases the table captions include the identification numbers of the corresponding sequences in [504].

Other tables of extremal codes can be found in Gaborit [180], Gaborit, Huffman, Kim and Pless [183], Gaborit and Otmani [185], Huffman [279], and [133, Chapter 7].

11.3.1 Binary codes

Type 2_I codes meeting the $d \leq 2\lfloor N/8 \rfloor + 2$ bound of Theorem 11.1.10 (the old definition of extremal) were completely classified by Ward [540] (completing the work begun in Mallows and Sloane [365], Pless [414], Pless and Sloane [431]): the final result is that such codes exist if and only if n is 2 (i_2), 4 (i_2^2), 6 (i_2^3), 8 (e_8), 12 (d_{12}^+), 14 (e_7^{2+}), 22 (g_{22}^+) or 24 (g_{24}) – compare Tables 12.2 and 12.7. In each case the code is unique.

However, there are many more Type 2_I codes that are extremal in the new sense (see Table 11.1), and they have certainly not yet been fully classified. It is known (Theorem 11.1.11) that extremal Type 2_{II} codes do not exist for lengths ≥ 3952 and presumably a similar bound could be obtained for extremal Type 2_I codes in the new sense.

Table 11.2 shows the highest minimal distance of binary self-dual codes of lengths $N \leq 72$.

Remark 11.3.1. Table 11.2 is based on earlier tables in Fig. 19.2 of [361], Conway and Sloane [130], Dougherty, Gulliver and Harada [153], Gaborit [180] and Gaborit and Otmani [185], [454]. It also includes results from Bilous [49], Bouyuklieva and Harada [79], Conway and Pless [121], Dalan [136], Dontcheva and Harada [147], [148], Gaborit [181], Harada [236], [229], Harada and Kim [242], Harada, Kitazume, Munemasa and Venkov [246], Harada and Munemasa [249], [250], Harada, Munemasa and Tonchev [251], Houghten, Lam, Thiel and Parker [269], Huffman and Tonchev [285], King [309], Nishimura [390], Scharlau and Schomaker [474] and Tsai and Jiang [530], In Gaborit [181] the author points out that all the $[N + 2, (N + 2)/2, d]$ Type 2_I codes may be obtained from the $[N, N/2, d]$ codes, and uses this to classify the $[36, 18, 8]$ codes.

In the table d_I (resp. d_{II}) denotes the highest minimal distance of any strictly Type 2_I (resp. Type 2_{II}) self-dual code.

The fourth column of Table 11.2 lists the known codes having the indicated minimal distance. The enumeration for lengths $N \leq 32$ will be discussed in §12.2. When N is a multiple of 8, a semicolon separates the Type 2_I and 2_{II} codes.

In the years since the manuscript of Conway and Sloane [130] was first circulated, a large number of sequels have been written, supplying additional examples of self-dual codes in the range of Table 11.2. These papers sometimes constructed extremal codes for lengths where none were known, but more often supply additional examples of extremal codes with different weight enumerators. The present bibliography includes all the references known to us. It was not possible to mention all these in the table, so instead we list them here. This list also includes a number of older papers. Readers interested in extremal self-dual codes, especially those of Type 2_I, in the range of the table (and beyond) should therefore consult the following: [86], [88], [89], [72], [73], [77], [80], [151], [153], [155], [156], [208], [209], [210], [211], [214], [221], [228], [229], [230], [241], [242] [255], [258] [271], [284], [285], [286], [295], [308], [399], [404], [419], [424], [436], [465], [466], [522], [523], [526], [527], [528], [555], [556], [557], [560], [561], [562], [563].

Table 11.2. Highest minimal distance of binary self-dual codes [A105674, A105675, A105685].

N	d_{I}	d_{II}	$Codes$
2	2		i_2.
4	2		i_2^2.
6	2		i_2^3.
8	2	4	i_2^4; e_8.
10	2		i_2^5, $e_8 i_2$.
12	4		d_{12}^+.
14	4		e_7^{2+}.
16	4	4	d_8^{2+}; d_{16}^+, e_8^2.
18	4		d_6^{3+}, $(d_{10}e_7 f_1)^+$.
20	4		7 codes. (Table 12.2)
22	6		g_{22}.
24	6	8	h_{24}^+; g_{24}.
26	6		f_{13}^2. [121]
28	6		3 codes. [121]
30	6		13 codes. [121], [124]
32	8	8	3 codes. [130]; 5 codes. (Table 12.3)
34	6		938 codes. [49]
36	8		41 codes. [180]
38	8		≥ 900 [130], [242], [180], [181]
40	8	8	≥ 3199 [136]; ≥ 12579 [309]
42	8		≥ 6137 [136], [153]
44	8		≥ 14016 [136], [153]
46	10		1 ([130], [251])
48	10	12	≥ 74 [246]; 1 ($XQ(\mathbb{F}_2, 47)$) [269]
50	10		≥ 2910 [79], [249]
52	10		≥ 499 [285]
54	10		≥ 54
56	10 or 12	12	?; ≥ 1151 [236]
58	10		≥ 80 [153]
60	12		≥ 18 [147], [148], [180], [242], [530]
62	12		≥ 20 [234], [148]
64	12	12	≥ 22 ([180], [250], [390]); ≥ 3270 [153]
66	12		≥ 3
68	12		≥ 65
70	12 or 14		? [229], [474]
72	12 or 14	12 or 16	?; ?

Note that if we do not distinguish between Type 2_{I} and Type 2_{II} codes, but just ask what is the highest minimal distance of a binary self-dual code, then the answer is known for all $N \leq 68$.

The symbol $XQ(\mathbb{F}_q, p)$ in any of these tables indicates an extended quadratic residue code of length $p + 1$ over the field $\mathbb{F}_q$ as defined in Section 2.4 (see also [361], van Lint and MacWilliams [351]). Both quadratic residue codes and double circulant codes provide many examples of good self-dual codes (cf. [361, Chapter 16], Pless, Huffman and Brualdi [427, §12], Ward [544]). There are two basic types of binary double circulant codes, having generator matrices either of the form

$$
\left[
\begin{array}{ccccc|cccccc}
1 & & & & & 0 & 1 & 1 & 1 & 1 \\
& 1 & & & & 1 & & & & \\
& & 1 & & & 1 & & R & & \\
& & & 1 & & 1 & & & & \\
& & & & 1 & 1 & & & &
\end{array}
\right]
\tag{11.3.1}
$$

or

$$
\left[
\begin{array}{ccccc|c}
1 & & & & & \\
& 1 & & & & \\
& & 1 & & & R \\
& & & 1 & & \\
& & & & 1 &
\end{array}
\right],
\tag{11.3.2}
$$

where R is a circulant matrix with first row r (say). (11.3.1) is used only when the length is a multiple of 4. Such codes and their generalizations to other fields have been studied by many authors, including Beenker [31], Bhargava and Nguyen [48], Gaborit [179], Gulliver and Harada [209], [210], [211], [213], [220], etc., Huffman [279], Huffman and Yorgov [286], Karlin [296], MacWilliams [357], [358], [361, Chap. 16], Poli and Rigoni [437], Ruseva [465], Tonchev and Raev [525], Ventou and Rigoni [535], Yorgov [555] [556], [557].

Table 11.3, based on Conway and Sloane [130] and Moore [374], gives a selection of double circulant binary codes. The first column gives the name of the codes, following [130], and the last column gives r, the initial row of R, in hexadecimal. The codes marked (∗) are not necessarily optimal. The minimal distance of the last two codes in the table was determined by Moore [374]. For these two codes r has 1's at the squares modulo 43 and 67, respectively. For an extension of this table see Gulliver and Harada [220].

We see from Table 11.2 that there are extremal Type 2_I codes (in the new sense) that are not also Type 2_{II} codes at lengths

$$2, 4, 6, 12, 14, 16, 18, 20, 22, 32, 36, 38, 40, 42, 44, 46, 60, 62, 64, 66, 68;$$

that such codes do not exist at length

$$8, 10, 24, 26, 28, 30, 34, 48, 50, 52, 54, 58, 72; \tag{11.3.3}$$

and that their existence at lengths 56 and 70 is at present an open question. The nonexistence of the Type 2_I codes of the lengths listed in (11.3.3) (other than 24 and 48) is established by imposing the extra condition that the shadow enumerator must have integral coefficients.

Concerning extremal Type 2_{II} codes, with $d = 4\lfloor N/24 \rfloor + 4$, these exist for the following values of N:

Table 11.3. Double circulant binary self-dual codes.

Name	N	k	d	Type	Form	r (hexadecimal)
g_{22}	22	11	6	I	(11.3.2)	97
g_{24}	24	12	8	II	(11.3.1)	B7
$A_{26} = f_{13}^{2+}$	26	13	6	I	(11.3.2)	5F7
A_{28} =D1	28	14	6	I	(11.3.1)	8D
D2	34	17	6	I	(11.3.2)	1ECE
D3	36	18	8	I	(11.3.1)	2C6B
D4	38	19	8	I	(11.3.2)	5793
D5	40	20	8	II	(11.3.2)	57EB
D6	40	20	8	I	(11.3.2)	11E35
D7	40	20	8	I	(11.3.2)	B393
D8	44	22	8	I	(11.3.1)	5E6B5
D9	50	25	10	I	(11.3.2)	31C4D
D10	52	26	10	I	(11.3.1)	57F69D
D11	56	28	12	II	(11.3.1)	ADF1FF
D12	58	29	10	I	(11.3.2)	D5A89B
D12a	58	29	10	I	(11.3.2)	2DD1D3
D13	60	30	12	I	(11.3.1)	3EF6B77
D14	64	32	12	II	(11.3.1)	427BD0B
D15	64	32	12	I	(11.3.2)	2EF3DD75
D16	66	33	12	I	(11.3.2)	B2D97D9
D17	68	34	12	I	(11.3.2)	1F5C885F
D18*	72	36	12	I	(11.3.2)	2B8795E5
D19*	74	37	12	I	(11.3.2)	1439372C7
D20*	82	41	12	I	(11.3.2)	A464B919B
H86	86	43	16	I	(11.3.2)	7F7101712E2
M88	88	44	16	II	(11.3.1)	
M136	136	68	24	II	(11.3.1)	

$$8, 16, 24, 32, 40, 48, 56, 64, 80, 88, 104, 136,$$

but their existence at lengths 72, 96, 112, ... is open. For lengths 8, 24, 32, 48, 80 and 104 we can use extended quadratic residue codes, and for lengths 40, 56, 64, 88, 136 we can use double circulant codes (see Table 11.3).

The extremal Type 2_{II} codes of lengths 8 and 24 have been known to be unique for a long time (Pless [412]). Houghten, Lam, Thiel and Parker [269] showed that the code of length 48 is also unique (see also Huffman [271]). This is the extended quadratic residue code $XQ(\mathbb{F}_2, 47)$, which is generated by **1**

and
$$1(01111011110010101110010011011000101011000010000),$$

with 1's at the nonzero squares modulo 47. It is worth remarking that there are at least three extremal even unimodular lattices in dimension 48 ([380]). As Table 11.2 shows, if $N \geq 40$ is congruent to 8 or 16 (mod 24) there are often large numbers of extremal codes. It is easy to find $[72, 36, 12]$ Type 2_{II} codes, for example $XQ(\mathbb{F}_2, 71)$; Dougherty, Gulliver and Harada [153] show that there are at least 33 inequivalent codes with these parameters. Other $[72, 36, 12]$ Type 2_{II} codes have been constructed by Dontcheva [146]. Dontcheva and Harada [149] give further examples of extremal $[80,40,16]$ Type 2_{II} codes.

Concerning the existence of self-dual codes with a specified minimal distance, from the work of Baartmans and Yorgov [18], Conway and Sloane [130], Dougherty, Gulliver and Harada [153], Gaborit, Pless, Solé and Atkin [186], Gulliver, Harada and Kim [221], Harada [229], [234] and W. Scharlau and Schomaker [474] we know that:

Self-dual codes with minimal distance
$d \geq 6$ exist precisely for $N \geq 22$;
$d \geq 8$ exist precisely for $N = 24, 32$ and $N \geq 36$;
$d \geq 10$ exist precisely for $N \geq 46$;
$d \geq 12$ exist for $N = 48$, 56 and $N \geq 60$; and do not exist for all other values of N. (As pointed out in [130], the $[58, 29, 12]$ self-dual code claimed by Bhargava and Nguyen [48] is an error.)
$d \geq 14$ exist for $N \geq 76$; perhaps for $N = 70, 72, 74$; and do not exist for all other values of N;
$d \geq 16$ exist for $N = 80, 86, 88, 92, 96, 100\text{-}112$ and $N \geq 120$ (and possibly for other values of N).

Of course, every question mark in the tables in this chapter, and every line where the number of optimal codes is not known (indicated by the absence of a period) is a potential research problem. But to emphasize one of the most important open problems, we state (cf. Table 11.2, [498]):

Research Problem 11.3.2. *Does there exist a $[72, 36, 16]$ Type 2_{II} code?*

11.3.2 Type 3: Ternary codes

Table 11.4 shows the highest minimal Hamming distance of ternary self-dual codes of lengths $N \leq 72$.

Remarks on Table 11.4

For the entries at lengths $N \leq 24$, see the discussion in §12.3. Extremal codes exist at lengths 4, 8, 12, 16, 20, 24, 28, 32, 36, 40, 44, 48, 52, 56, 60 and

Table 11.4. Type 3: Highest minimal Hamming distance d_H of ternary self-dual codes [A105676].

N	d_H	$Codes$
4	3	t_4.
8	3	t_4^2.
12	6	g_{12}.
16	6	f_8^{2+}.
20	6	6 codes. [432]
24	9	$XQ(\mathbb{F}_3, 23), S(24)$. [346]
28	9	≥ 32 [106], [231], [337], [278]
32	9	≥ 239 [278]
36	12	$\geq 1\ (S(36))$
40	12	≥ 20 [542], [141], [231], [278]
44	12	≥ 8 [231]
48	15	$\geq 2\ (XQ(\mathbb{F}_3, 47), S(48))$
52	15	≥ 1 [185]
56	15	≥ 1
60	18	$\geq 2\ (XQ(\mathbb{F}_3, 59), S(60))$
64	18	≥ 1 [31], [141]
68	15 or 18	?
72	18	$\geq 1\ (XQ(\mathbb{F}_3, 71)\ [134])$

64. Extremal codes do not exist at lengths 72, 96, 120 and all $N \geq 144$, because then the extremal Hamming weight enumerator contains a negative coefficient. The existence of extremal codes in the remaining cases ($N = 68$, 76, ..., 140) is undecided.

In Table 11.4 $S(m)$ denotes a Pless double circulant (or "symmetry") code of length m (see Pless [413], [415], [425]; Blake [51], [52]; Ito [290]; Mallows, Pless and Sloane [364]; [133, Chap. 3, §2.10]; [361, Chap. 16]). A $[28, 14, 9]_3$ code was discovered by Cheng and R. Scharlau [106]. Another such code was given by Kschischang and Pasupathy [337], namely the negacyclic code generated by the polynomial $(x^2 + x - 1)(x^6 - x^4 + x^3 + x^2 - 1)(x^6 - x^5 - x - 1)$, i.e. by the vectors

$$(2002021222020010000000000000)_-\,,$$

where the subscript $-$ indicates that the code is negacyclic (that is, closed under the operation "cycle the coordinates and then negate the first coordinate"). Huffman [278] shows that there are at least 14 inequivalent $[28, 14, 9]_3$ codes with nontrivial automorphisms of odd order.

Ward [542] and Dawson [141] independently discovered that $[40, 20, 12]_3$ codes can be constructed using generator matrices of the form $[I_{20}\, H_{20}]$, where H_{20} is a Hadamard matrix of order 20. There are three distinct Hadamard matrices of this order, and Dawson shows that all three produce $[40, 20, 12]_3$ codes. Harada [231] shows that these three codes are inequivalent. Dawson also shows that the same construction using the Paley-Hadamard matrix of order 32 leads to a $[64, 32, 18]_3$ self-dual code. A $[64, 32, 18]_3$ code B_{24} (equivalent to Dawson's) had been constructed earlier by Beenker [31].

The codes of length 32, 44, 56 and 68 can be obtained by "subtracting" (see §12.2) a copy of t_4 from a code of length 4 greater.

Table 11.4 also includes results from Coppersmith and Seroussi [134], Gaborit and Otmani [185], Harada [231], Leon, Pless and Sloane [346], Pless, Sloane and Ward [432].

Other constructions for ternary self-dual codes can be found in Ozeki [396].

Table 11.5. Type 4^E: Highest minimal Hamming distance d_H of Euclidean linear self-dual codes over $\mathbb{F}_4$ [A105677].

N	d_H	Codes
2	2	i_2.
4	3	RS_4.
6	3	$RS_4 i_2$.
8	4	3 codes. E.g. $\mathbb{F}_4 \otimes e_8$ [301], [360]
10	4	4 codes. [301]
12	6	≥ 1 code, e.g. $XQ(\mathbb{F}_4, 11)$
14	6	≥ 5 codes [186], [42], [185]
16	6	≥ 1 code [186], [42], [185]
18	6 or 7	≥ 1 code [185]
20	8	≥ 1 code, e.g. $XQ(\mathbb{F}_4, 19)$ [42]
22	8	≥ 1 code [185]
24	$8 - 10$	≥ 1 code, e.g. $\mathbb{F}_4 \otimes g_{24}$ [185], [360]
26	$8 - 10$	≥ 1 code [42], [185]
28	$9 - 11$	≥ 1 code, e.g. $XQ(\mathbb{F}_4, 27)$ [351]
30	$10 - 12$	≥ 1 code [185]
32	$10 - 12$	≥ 1 code [42], [185]

11.3.3 Types 4^E and 4^E_{II}: Euclidean self-dual codes over $\mathbb{F}_4$

Table 11.5 shows the highest minimal Hamming distance d_H of Euclidean self-dual codes over $\mathbb{F}_4$ of lengths $N \leq 32$, and Table 11.6 shows the highest

Table 11.6. Highest minimal Hamming (d_H) and Lee (d_L) distances of Type 4_{II}^E Euclidean linear self-dual codes over $\mathbb{F}_4$.

N	d_H	Codes	d_L	Codes
4	3	RS_4.	4	RS_4.
8	4	$\mathbb{F}_4 \otimes e_8$.	4	$\mathbb{F}_4 \otimes e_8$. [42], [360], [382]
12	6	$XQ(\mathbb{F}_4, 11)$. [382], [42]	8	$XQ(\mathbb{F}_4, 11)$. [42]
16	6	?	8	≥ 1 code [42]
20	8	$XQ(\mathbb{F}_4, 19)$	8	≥ 1 code, e.g. $XQ(\mathbb{F}_4, 19)$ [42]
24	8	$\mathbb{F}_4 \otimes g_{24}$	8 or 12	$\mathbb{F}_4 \otimes g_{24}$ has $d_L = 8$ [360]
28	?	?	12	≥ 1 code [42]
32	?	?	12	≥ 1 code [42]

minimal Hamming (d_H) and Lee (d_L) distances of Type 4_{II}^E Euclidean self-dual codes over $\mathbb{F}_4$ of lengths $N \leq 32$.

Remark 11.3.3. Tables 11.5 and 11.6 are based on earlier tables of Betsumiya, Gulliver, Harada and Munemasa [42], Gaborit, Pless, Solé and Atkin [186], Gaborit and Otmani [185], Kim, Kim and Kim [301], and Nebe, Quebbemann, Rains and Sloane [382]. The extended quadratic-residue code of length 28 is from van Lint and MacWilliams [351].

Double circulant codes can be used to achieve the values of d_H shown in Table 11.5 for lengths $N = 14, 16, 26, 32$ and the values of d_L shown in Table 11.6 for $N = 16, 28, 32$ ([186], [42], [185]).

We see that extremal codes of Type 4^E exist at lengths $N = 8, 10, 20, 22$, do not exist at lengths $N = 2, 4, 6, 12, 14, 16, 18, 24, 26, 28$, and their existence at lengths $N = 30$ and 32 is presently undecided. Hamming-extremal codes of Type 4_{II}^E exist at all lengths $N = 4, 8, 12, \ldots, 24$. Lee-extremal codes of Type 4_{II}^E exist at all lengths $N = 4, 8, 12, \ldots, 32$, except possibly for length 24. Lee-extremal codes also exist at lengths 44 and 71 ([42], [186], [282]).

11.3.4 Type 4^H: Hermitian linear self-dual codes over $\mathbb{F}_4$

Table 11.7 shows the highest minimal Hamming distance of Hermitian self-dual codes over $\mathbb{F}_4$ of lengths $N \leq 32$.

Remark 11.3.4. Table 11.7 is an updated version of tables in [185] and [454]. For the entries at lengths $N \leq 16$, see the discussion in §12.4. Table 11.7 also includes results from Conway, Pless and Sloane [123], Danielsen and Parker [139], Gulliver [207], Huffman [275], [277], Kim, Kim and Kim [301] and Östergård, [392].

Extremal codes exist at lengths 2, 4, 6, 8, 10, 14, 16, 18, 20, 22, 28 and 30. They do not exist at lengths 12, 24, 102, 108, 114, 120, 122 and all $N \geq 126$

Table 11.7. Type 4^H: Highest minimal Hamming distance d_H of Hermitian linear self-dual codes over $\mathbb{F}_4$ [A105678, A105686].

N	d_H	Codes
2	2	i_2.
4	2	i_2^2.
6	4	h_6.
8	4	e_8.
10	4	d_{10}^+, e_5^{2+}.
12	4	5 codes. (Table 12.10)
14	6	q_{14}.
16	6	4 codes. [123]
18	8	S_{18}. [280], [360]
20	8	2 codes. [280].
22	8	≥ 46 codes [275], [277], [301]
24	8	≥ 17 codes [207], [301]
26	8	H_{26} (≥ 49 codes) [207], [301], [392]
28	10	H_{28} (≥ 3 codes) [275], [277]
30	12	$XQ(\mathbb{F}_4, 29)$ [360]
32	10 or 12	H_{32} and G_{32} have $d = 10$

(the larger N being eliminated by the presence of negative coefficients in the extremal Hamming weight enumerator). The remaining lengths $(26, 32, 34, \ldots)$ are undecided.

The $[18, 9, 8]_4$ code S_{18} generated by

$$1(1\omega\overline{\omega}\omega\omega\overline{\omega\omega\omega\omega\omega}\omega\omega\omega\overline{\omega}\omega)$$

has a number of interesting properties (see Conway and Sloane [128]; Cheng and Sloane [107]; MacWilliams, Odlyzko, Sloane and Ward [360]; Pless [420]). S_{18} has weak automorphism group $Z_3 \times (PSL_2(16).4)$, of order 48960 [107] and is the unique $[18, 9, 8]_4$ code (Huffman [280]).

The long-standing question of the existence of a $[24, 12, 10]_4$ code was settled in the negative by Lam and Pless [339] (see also Huffman [273]). The code $\mathbb{F}_4 \otimes g_{24}$ is an example of a $[24, 12, 8]_4$ code.

11.3.5 Types 4^{H+} and 4_{II}^{H+}: Trace-Hermitian additive self-dual codes over $\mathbb{F}_4$

Table 11.8 gives the highest minimal Hamming distance of additive codes over $\mathbb{F}_4$ (that is, of Type 4^H or 4_{II}^{H+}) of lengths $N \leq 30$ that are self-dual with respect to the trace inner product.

Table 11.8. Type 4^{H+}: Highest minimal Hamming distance d_H of trace-Hermitian additive self-dual codes over $\mathbb{F}_4$ [A016729, A105687].

N	d_H	Codes	N	d_H	Codes
1	1	i_1.	16	6	≥ 7 codes [123], [183]
2	2	i_2.	17	7	
3	2	d_3^+.	18	8	S_{18}
4	2	3 codes.	19	7	
5	3	h_5.	20	8	≥ 2 codes [280]
6	4	h_6.	21	8	c_{21}
7	3	4 codes. [139], [183]	22	8	≥ 38 codes [275]
8	4	5 codes. E.g. e_8 ([183])	23	$8-9$	c_{23} has $d = 8$
9	4	8 codes. E.g. c_9 ([183])	24	$8-10$	$\mathbb{F}_4 \otimes g_{24}$ has $d = 8$
10	4	120 codes. [23], [139], [183]	25	$8-9$	c_{25} has $d = 8$
11	5	1 code. [96], [183]	26	$8-10$	
12	6	1 code. (z_{12})	27	$9-10$	
13	5	≥ 5 codes [183]	28	10	
14	6	q_{14}	29	11	
15	6	c_{15}	30	12	$XQ(\mathbb{F}_4, 29)$

Table 11.9. Generators for cyclic trace-Hermitian additive codes over $\mathbb{F}_4$.

Code	Generator
c_9	$(\omega 10100101)$
c_{15}	$(\omega 11010100101011)$
c_{21}	$(\overline{\omega}\overline{\omega}1\omega 00111101011011000), (10111001011100101100)$
c_{23}	$(\omega 0101111000000001111010)$
c_{25}	$(111010\omega 010111000000000000)$

Remark 11.3.5. Table 11.8 is an updated version of a table in Calderbank, Rains, Shor and Sloane [96]. It incorporates information from the following papers: Bachoc and Gaborit [23], Conway, Pless and Sloane [123], Gaborit, Huffman, Kim and Pless [183], Höhn [266] and Huffman [275], [280]. See also §13.6.

We see that extremal Type 4^{H+} codes exist at lengths 1–6, 8–12, 14–18, 20–22 and 28–30, and do not exist at lengths 7, 13 and 19. Lengths 23–27 are undecided. For lengths $23, 24, 25$, the codes mentioned have $d_H = 8$.

Many of the entries are the same as in the table of Hermitian self-dual codes, Table 11.7. The codes d_N^+ are defined in §12.5, h_6 is the hexacode, and h_5 is the $[5, 2.5, 3]_{4+}$ shortened hexacode defined in Subsection 2.4.8. Also, c_9, c_{15}, c_{21}, c_{23}, c_{25} are cyclic codes with generators shown in Table 11.9. If no

Table 11.10. Type 4_{II}^{H+}: Highest minimal Hamming distance d_H of even trace-Hermitian additive self-dual codes over $\mathbb{F}_4$ [A106169].

N	d_H	Number	References
2	2	1.	[183]
4	2	2.	[183]
6	4	1.	[183]
8	4	3.	[183]
10	4	19.	[23]
12	6	1.	[183]
14	6	≥ 490	[23]
16	6	≥ 4	[183]
18	8	≥ 1	[23]
20	8	≥ 2	[275], [280]
22	8	≥ 38	[275], [280]
24	8 or 10	?	

name is given, the code can be obtained by shortening a code of length one greater.

Table 11.10, based on the same sources as Table 11.8, gives the highest minimal Hamming distance of Type 4_{II}^{H+} additive codes over $\mathbb{F}_4$ of lengths $N \leq 24$; that is, codes that are self-dual with respect to the trace inner product and have even Hamming weights. Extremal codes exist for all $N \leq 22$; their existence at length 24 is undecided (but unlikely in view of the result of Lam and Pless [339] mentioned above).

11.3.6 Type $4^{\mathbb{Z}}$: Self-dual codes over $\mathbb{Z}/4\mathbb{Z}$

Table 11.11 gives the exact highest minimal Hamming distance, Lee distance and Euclidean norm of self-dual codes of Type $4^{\mathbb{Z}}$ of lengths $N \leq 24$. This is based on Conway and Sloane [131]; Dougherty, Harada and Solé [160]; Fields, Gaborit, Leon and Pless [174]; Huffman [281]; Pless, Leon and Fields [428]; and [446]. The columns headed # give the number of extremal codes.

Table 11.11. Type $4^{\mathbb{Z}}$: Highest minimal Hamming distance (d_H), Lee distance (d_L) and Euclidean norm (Norm) of self-dual codes over $\mathbb{Z}/4\mathbb{Z}$, and the numbers of codes with these parameters [A066016, A066017, A105681, A105688, A105682, A105689].

Length	Hamming			Lee			Norm		
N	d_H	code	#	d_L	code	#	Norm	code	#
1	1	i_1	1	2	i_1	1	4	i_1	1
2	1	i_1^2	1	2	i_1^2	1	4	i_1^2	1
3	1	i_1^3	1	2	i_1^3	1	4	i_1^3	1
4	2	$D_4^{\oplus}$	1	4	$D_4^{\oplus}$	1	4	i_1^4	2
5	1	$D_4^{\oplus}i_1$	2	2	$D_4^{\oplus}i_1$	2	4	i_1^5	2
6	2	$D_6^{\oplus}$	1	4	$D_6^{\oplus}$	1	4	i_1^6	3
7	3	E_7^+	1	4	E_7^+	1	4	i_1^7	4
8	4	$\mathcal{O}_8$	2	6	$\mathcal{O}_8$	1	8	$\mathcal{O}_8$	1
9	1	$\mathcal{O}_8 i_1$	11	2	$\mathcal{O}_8 i_1$	11	4	i_1^9	11
10	2	$D_4^{\oplus}D_6^{\oplus}$	5	4	$D_4^{\oplus}D_6^{\oplus}$	5	4	i_1^{10}	16
11	2	$D_4^{\oplus}E_7^+$	3	4	$D_4^{\oplus}E_7^+$	3	4	i_1^{11}	19
12	2	$D_4^{\oplus}\mathcal{O}_8$	39	4	$D_4^{\oplus}\mathcal{O}_8$	39	8	[174]	19
13	2	$D_6^{\oplus}E_7^+$	8	4	$D_6^{\oplus}E_7^+$	8	4	i_1^{13}	66
14	3	$(E_7^+)^2$	4	6	[174]	1	8	[174]	35
15	3	$E_7^+\mathcal{O}_8$	47	6	[174]	15	8	[174]	28
16	4	$\mathcal{O}_8^2$	≥ 1	8	C_{16}	≥ 5	8	$\mathcal{O}_8^2$	≥ 5
17	4	C_{17}	62	6	C_{17}	≥ 17	8	C_{17}	≥ 17
18	4	C_{18}	66	8	C_{18}	7	8	C_{18}	≥ 39
19	3	G_{19}	≥ 1	6	G_{19}	≥ 1	8	G_{19}	≥ 1
20	4	G_{20}	≥ 1	8	G_{20}	≥ 1	8	G_{20}	≥ 1
21	5	G_{21}	384	8	G_{21}	384	8	G_{21}	≥ 384
22	6	G_{22}	≥ 19367	8	G_{22}	≥ 19367	8	G_{22}	≥ 19367
23	7	G_{23}	$\geq 1.72 \times 10^6$	10	G_{23}	30	12	G_{23}	≥ 30
24	8	G_{24}	$\geq 1.47 \times 10^8$	12	G_{24}	13	16	G_{24}	≥ 50

Remark 11.3.6. The length 16 code C_{16} in Table 11.11 is given in [428], where it is called 5_f5. It has $|\,\mathrm{WAut}(C_{16})| = 2^{5+10}3^2 5.7$ and generator matrix

$$\begin{bmatrix} 1\,1\,1\,1\,1\,1\,1\,1\,1\,1\,1 & 1\,1\,1\,1\,1 \\ 1\,0\,1\,1\,1\,1\,1\,1\,0\,0\,0 & 0\,1\,0\,0\,0 \\ 1\,1\,0\,1\,0\,0\,1\,1\,1\,1\,0 & 0\,0\,1\,0\,0 \\ 1\,1\,1\,0\,1\,0\,1\,0\,1\,0\,1 & 0\,0\,0\,1\,0 \\ 0\,0\,0\,0\,1\,1\,1\,1\,1\,1\,1 & 0\,0\,0\,0\,1 \\ \hline 0\,0\,0\,0\,0\,2\,0\,0\,0\,0\,0 & 2\,2\,0\,0\,2 \\ 0\,0\,0\,0\,0\,0\,2\,0\,0\,0\,0 & 2\,2\,2\,2\,2 \\ 0\,0\,0\,0\,0\,0\,0\,2\,0\,0\,0 & 0\,2\,2\,0\,2 \\ 0\,0\,0\,0\,0\,0\,0\,0\,2\,0\,0 & 0\,0\,2\,2\,2 \\ 0\,0\,0\,0\,0\,0\,0\,0\,0\,2\,0 & 2\,0\,2\,0\,2 \\ 0\,0\,0\,0\,0\,0\,0\,0\,0\,0\,2 & 2\,0\,0\,2\,2 \end{bmatrix}.$$

The codes C_{17} and C_{18} mentioned in Table 11.11 have generator matrices

$$\begin{bmatrix} 1\,0\,0\,0\,0\,0\,0\,0\,1\,2\,1\,1\,1\,1\,1\,2 \\ 0\,1\,0\,0\,0\,0\,0\,0\,1\,1\,0\,0\,0\,1\,2\,2\,0 \\ 0\,0\,1\,0\,0\,0\,0\,0\,1\,3\,2\,0\,0\,2\,3\,2\,2 \\ 0\,0\,0\,1\,0\,0\,0\,0\,1\,3\,0\,0\,0\,2\,0\,3\,0 \\ 0\,0\,0\,0\,1\,0\,0\,0\,0\,1\,3\,3\,3\,3\,1\,1\,2 \\ 0\,0\,0\,0\,0\,1\,0\,0\,0\,0\,0\,3\,1\,0\,2\,0\,3 \\ 0\,0\,0\,0\,0\,0\,1\,0\,0\,0\,3\,2\,1\,0\,0\,0\,3 \\ 0\,0\,0\,0\,0\,0\,0\,1\,0\,0\,1\,3\,2\,0\,0\,0\,3 \\ 0\,0\,0\,0\,0\,0\,0\,0\,2\,2\,2\,2\,2\,0\,0\,0\,0 \end{bmatrix}$$

and

$$\begin{bmatrix} 1\,0\,0\,3\,0\,0\,0\,0\,0\,3\,2\,0\,3\,0\,2\,0\,0\,0 \\ 0\,1\,0\,0\,3\,0\,0\,0\,0\,0\,3\,2\,2\,3\,0\,0\,0\,0 \\ 0\,0\,1\,0\,0\,3\,0\,0\,0\,2\,0\,3\,0\,2\,3\,0\,0\,0 \\ 0\,0\,0\,1\,0\,0\,3\,0\,0\,0\,0\,0\,3\,2\,0\,3\,0\,2 \\ 0\,0\,0\,0\,1\,0\,0\,3\,0\,0\,0\,0\,0\,3\,2\,2\,3\,0 \\ 0\,0\,0\,0\,0\,1\,0\,0\,3\,0\,0\,0\,2\,0\,3\,0\,2\,3 \\ 0\,0\,0\,0\,0\,0\,1\,3\,0\,0\,3\,3\,0\,3\,3\,3\,2\,1 \\ 0\,0\,0\,0\,0\,0\,0\,1\,3\,3\,0\,3\,3\,0\,3\,1\,3\,2 \\ 0\,0\,0\,0\,0\,0\,0\,0\,2\,0\,2\,2\,0\,2\,2\,0\,2\,0 \\ 0\,0\,0\,0\,0\,0\,0\,0\,2\,2\,2\,2\,2\,2\,2\,2 \end{bmatrix},$$

and weak automorphism groups of orders 576 and 144, respectively.

G_{24} was defined in (2.4.38), and G_{19} through G_{23} are shortened versions of it.

Note from the table that, including G_{24}, there are 13 codes of length 24 with optimal Lee distance 12. These all have the property that their reduction modulo 2 is the binary Golay code g_{24}, and their image under Construction $A_{2\mathbb{Z}}$ is the Leech lattice. In particular, these codes are actually of Type $4_{II}^{\mathbb{Z}}$. These codes were classified in [446].

Besides the norm-extremal codes of length 8, 12, 14–24 shown in the table, there are also norm-extremal codes of lengths 32 and 48 obtained by lifting binary extended quadratic residue codes to $\mathbb{Z}/4\mathbb{Z}$. The code of length 32 has minimal Lee distance 14 and minimal norm 16. Pless and Qian [430] have shown that the code of length 48 has minimal Lee distance 18 and minimal norm 24.

Further examples of good self-dual codes over $\mathbb{Z}/4\mathbb{Z}$ may be found in Bonnecaze, Calderbank and P. Solé [60]; Bonnecaze, Gaborit, Harada, Kitazume and Solé [63]; Bonnecaze, Solé, Bachoc and Mourrain [66]; Calderbank and Sloane [99]; Dougherty, Harada and Solé [160]; Gaborit and Harada [182]; Gulliver and Harada [212], [217]; Harada [232]; Harada and Kitazume [243], Huffman [281]; Pless and Qian [430]; Pless, Solé and Qian [433]; [446].

11.3.7 Other types

There has been some work done on finding extremal codes of Types q^E and q^H for values of $q \geq 5$. See Arasu and Gulliver [11], Araya and Harada [12], Betsumiya, Georgiou, Gulliver, Harada and Koukouvinos [38], Gaborit [179], Georgiou and Koukouvinos [189], [188], Gulliver and Harada [216], [219], Harada [237], Harada and Kharaghani [239], [240], Kim and Lee [304], Leon, Pless and Sloane [347], Pless and Tonchev [435].

Betsumiya, Gulliver and Harada [41] and Betsumiya and Harada [45] have determined the optimal Hamming distance of self-dual codes of Type 2^{lin} (cf. §7.2.1) for lengths $N \leq 30$ ([A106160, A106161]). That is, we look for a binary linear (but not necessarily self-orthogonal) code C for which $\min\{d_C, d_{C^\perp}\}$ is as large as possible. See also the references in §12.7.

12

Enumeration of Self-Dual Codes

In this chapter we discuss what is currently known about the enumeration of self-dual codes of the most important Types. The main tool for these enumerations are the mass formulae given in §12.1.

12.1 The mass formulae

Let ρ be a finite representation $\rho := (V, \rho_M, \rho_\Phi, \beta)$ of the form ring (R, M, ψ, Φ), and let G denote the corresponding group of weak equivalences (cf. §1.11). If C is a self-dual code of Type ρ and length N, the number of codes that are weakly equivalent to C is $|G|/|\operatorname{WAut}(C)|$. For many of our Types it is possible to determine *ab initio* the total number T_N of distinct codes of length N. Then

$$T_N = \sum_{\text{inequivalent } C} \frac{|G|}{|\operatorname{WAut}(C)|},$$

where the sum is over all weakly inequivalent codes. In other words

$$\sum_{\text{inequivalent } C} \frac{1}{|\operatorname{WAut}(C)|} = \frac{T_N}{|G|}. \tag{12.1.1}$$

This is called a *mass formula*. In contrast, it is completely intractable to give the number of inequivalent codes, except for very small lengths. This illustrates one of the general principles of enumeration: when counting equivalence classes of objects with automorphisms, it is almost always a good idea to weight them by the reciprocal of the order of their automorphism group. Mass formulae are useful when trying to find all weakly inequivalent codes of length N: one builds up a list of known inequivalent codes, and when the left-hand side of (12.1.1) reaches the value given on the right-hand side, one knows the list is complete.

The following are the values of $T_N/|G|$ for the main Types (T_N is given before the solidus, $|G|$ after it, and no cancellation has been performed).

$$2_{\mathrm{I}} : \prod_{i=1}^{\frac{1}{2}N-1} (2^i + 1) \, / \, N! \quad (N \equiv 0 \bmod 2) \tag{12.1.2}$$

$$2_{\mathrm{II}} : 2 \prod_{i=1}^{\frac{1}{2}N-2} (2^i + 1) \, / \, N! \quad (N \equiv 0 \bmod 8) \tag{12.1.3}$$

$$3 : 2 \prod_{i=1}^{\frac{1}{2}N-1} (3^i + 1) \, / \, \{2^N N!\} \quad (N \equiv 0 \bmod 4) \tag{12.1.4}$$

$$4^{\mathrm{E}} : \prod_{i=1}^{\frac{1}{2}N-1} (4^i + 1) \, / \, \{2\,N!\} \quad (N \equiv 0 \bmod 2) \tag{12.1.5}$$

$$4^{\mathrm{E}}_{\mathrm{II}} : \prod_{i=0}^{\frac{1}{2}N-2} (4^i + 1) \, / \, \{2\,N!\} \quad (N \equiv 0 \bmod 4) \tag{12.1.6}$$

$$4^{\mathrm{H}} : \prod_{i=0}^{\frac{1}{2}N-1} (2^{2i+1} + 1) \, / \, \{2 \cdot 3^N N!\} \quad (N \equiv 0 \bmod 2) \tag{12.1.7}$$

$$4^{\mathrm{H+}} : \prod_{i=1}^{N} (2^i + 1) \, / \, \{6^N N!\} \tag{12.1.8}$$

$$4^{\mathrm{H+}}_{\mathrm{II}} : 2 \prod_{i=1}^{N-1} (2^i + 1) \, / \, \{6^N N!\} \quad (N \equiv 0 \bmod 2) \tag{12.1.9}$$

$$q^{\mathrm{H}} : \prod_{i=0}^{\frac{1}{2}N-1} (q^{i+\frac{1}{2}} + 1) \, / \, \{(\log_p q)(\sqrt{q} - 1)(\sqrt{q} + 1)^N N!\} \quad (N \equiv 0 \bmod 2) \tag{12.1.10}$$

$$q^{\mathrm{H+}} : \prod_{i=1}^{N} (\sqrt{q}^{\,i} + 1) \, / \, \{(\log_p \sqrt{q})(\sqrt{q} - 1)(\sqrt{q}(q-1))^N N!\} \tag{12.1.11}$$

$$q^{\mathrm{H+}}_1 : \prod_{i=1}^{N-1} (\sqrt{q}^{\,i} + 1) \, / \, \{(\log_p \sqrt{q})(\sqrt{q} - 1)^2 \sqrt{q}^{\,N} N!\} \tag{12.1.12}$$

$$q^{\mathrm{E}} : b \prod_{i=1}^{\frac{1}{2}N-1} (q^i + 1) \, / \, \{(\log_p q) \frac{q-1}{2} 2^N N!\} \quad (N \equiv 0 \bmod 2) \tag{12.1.13}$$

where $b = 1$ if q is even, 2 if q is odd

$$4^{\mathbb{Z}}: \quad \sum_{k=0}^{N/2} \tau_{N,k}\, 2^{k(k+1)/2} \; / \; \{2^N N!\}, \qquad (12.1.14)$$

where $\tau_{N,k}$, the number of binary self-orthogonal $[N, k]$ codes with all weights divisible by 4, is equal to 1 if $k = 0$, and otherwise is given by

$$\prod_{i=0}^{k-1} \frac{2^{N-2i-2} + 2^{\lfloor \frac{N}{2} \rfloor - i - 1} - 1}{2^{i+1} - 1} \quad \text{if } N \equiv \pm 1 \bmod 8,$$

$$\prod_{i=0}^{k-1} \frac{2^{N-2i-2} - 1}{2^{i+1} - 1} \quad \text{if } N \equiv \pm 2 \bmod 8,$$

$$\prod_{i=0}^{k-1} \frac{2^{N-2i-2} - 2^{\lfloor \frac{N}{2} \rfloor - i - 1} - 1}{2^{i+1} - 1} \quad \text{if } N \equiv \pm 3 \bmod 8,$$

$$\left[\prod_{i=0}^{k-2} \frac{2^{N-2i-2} + 2^{\frac{N}{2} - i - 1} - 2}{2^{i+1} - 1} \right] \cdot \left[\frac{1}{2^{k-1}} + \frac{2^{N-2k} + 2^{\frac{N}{2} - k} - 2}{2^k - 1} \right] \quad \text{if } N \equiv 0 \bmod 8,$$

$$\left[\prod_{i=0}^{k-2} \frac{2^{N-2i-2} - 2^{\frac{N}{2} - i - 1} - 2}{2^{i+1} - 1} \right] \cdot \left[\frac{1}{2^{k-1}} + \frac{2^{N-2k} - 2^{\frac{N}{2} - k} - 2}{2^k - 1} \right] \quad \text{if } N \equiv 4 \bmod 8.$$

There is a similar but even more complicated formula for T_N for self-dual codes over $\mathbb{Z}/4\mathbb{Z}$ with Euclidean norms divisible by 8, see Gaborit [178].

Remark 12.1.1. The individual groups are described in §2.3. Note that in (12.1.10) the order of the group has a factor $\log_p(q)$ from the global Galois automorphisms, $(\sqrt{q} + 1)^N$ from multiplication of each coordinate by an element x in $\mathbb{F}_q^*$ with $x^{\sqrt{q}+1} = x\bar{x} = 1$, and global multiplication by an arbitrary $x \in \mathbb{F}_q^*$ contributes another factor of $(\sqrt{q} - 1)$.

Formulae (12.1.2)–(12.1.13) are based on various sources including [361, Chap. 9, §6], Höhn [266], MacWilliams, Sloane and Thompson [362], Pless [411], and [361, Chap. 19]. Equation (12.1.6) is from Munemasa [377] and Gaborit, Pless, Solé and Atkin [186], and (12.1.14) is from Gaborit [178]. Other mass formulae have been given by Betsumiya, Ling and Nemenzo [47]. As has been pointed out by several authors, some of these formulae were already known from classical geometry—see Feng and Dai [172], Munemasa [377], Pless [411], Ray-Chaudhuri [455], Segre [478], Wan [537], [538].

Proof. We first give two proofs of (12.1.2). (i) Let $\sigma_{N,k}$ denote the number of $[N, k]$ self-orthogonal codes C containing $\mathbf{1}$. Any such C can be extended to an $[N, k + 1]$ self-orthogonal code D by adjoining any vector of $C^\perp \setminus C$, and any D will arise $2^k - 1$ times from different C's. So we have $\sigma_{N,1} = 1$,

$$\frac{\sigma_{N,k+1}}{\sigma_{N,k}} = \frac{2^{N-2k} - 1}{2^k - 1},$$

and $\sigma_{N,N/2}$ gives (12.1.2). (ii) A more sophisticated proof can be obtained by observing that the Euclidean inner product induces a symplectic geometry structure on the space of even weight vectors modulo $\mathbf{1}$. A self-dual code is then a maximally isotropic subspace. The number of maximally isotropic subspaces of a symplectic geometry of dimension $2k$ is $\Pi_{i=1}^{k}(2^i + 1)$ (Brouwer, Cohen and Neumaier [85, §9.4]), and we obtain (12.1.2) by noting that our symplectic geometry has dimension $N - 2$.

Similarly, a Type $\rho(2_{\mathrm{II}})$-code is a maximally totally singular subspace of the orthogonal geometry $N\rho(2_{\mathrm{II}})/\langle \mathbf{1}\rangle$ of dimension $N - 2$, which leads to (12.1.3). Equations (12.1.4), (12.1.5), (12.1.9), (12.1.13) are also obtained via orthogonal geometry, (12.1.8) via symplectic geometry, and (12.1.7) and (12.1.10) via unitary geometry. For (12.1.14) we refer the reader to Gaborit [178].   □

As an illustration of the use of these mass formulae, suppose we are trying to find all binary self-dual codes of length 8. We immediately find two codes, $i_2 \oplus i_2 \oplus i_2 \oplus i_2$, where $i_2 = [11]$, and the Hamming code e_8, and then it appears that there are no others. To prove this, we compute the (weak) automorphism groups of these two codes: they have orders $2^4 4! = 384$ and $8 \cdot 7 \cdot 6 \cdot 4 = 1344$, respectively. We also calculate $T_8/|G| = 3 \cdot 5 \cdot 9/8! = 3/896$ from (12.1.2), and see that indeed

$$\frac{1}{384} + \frac{1}{1344} = \frac{3}{896},$$

verifying that this enumeration is complete. We will return to this in the following section.

There are also formulae that give the total number of self-dual codes containing a fixed self-orthogonal vector or code—see [361, Chapter 19].

The formulae given in §11.2 for average weight enumerators can also be viewed as mass-type formulae. We note that there are formulae (albeit much more complicated) for the masses $\sum \frac{1}{\mathrm{Aut}(\Lambda)}$ of arbitrary genera of lattices. For instance, if $N = 2k \equiv 0 \pmod 8$, then the total mass of even unimodular lattices in dimension N is

$$\frac{|B_k|}{2k} \prod_{j=i}^{k-1} \frac{|B_{2j}|}{4j}. \tag{12.1.15}$$

12.2 Enumeration of binary self-dual codes

We begin with some remarks about the automorphism groups of glued codes. One advantage of the gluing method is that it makes it much easier to find the weak automorphism group of a self-dual code C. We will denote this group by $G(C)$ rather than $\mathrm{WAut}(C)$ in this section. It is essential that every weak automorphism of C takes the set of component codes $C_1, \ldots, C_t$ to itself. We will always choose the components so that this is true.

This being the case, any weak automorphism in $G(C)$ will effect some permutation of the C_i, so that $G(C)$ will have a normal subgroup G_{01} consisting of just those elements for which this permutation is trivial. The group of permutations of the components that are realized in this way we call $G_2(C)$—it is isomorphic to the quotient group $G(C)/G_{01}$.

Let $G_0(C)$ be the normal subgroup of G_{01} consisting of those automorphisms which, for every i, send each glue element u_i into a vector in the same coset $u_i + C_i$, i.e. which fix the glue elements modulo the components. Then $G_{01}/G_0(C)$ is isomorphic to a group acting on the glue elements of each component: we call this group $G_1(C)$. Thus the full group $G(C)$ is compounded of the groups $G_0(C)$, $G_1(C)$ and $G_2(C)$, and has order

$$| \operatorname{WAut}(C)| = |G(C)| = |G_0(C)||G_1(C)||G_2(C)|. \qquad (12.2.1)$$

Also $G_0(C)$ is the direct product of the groups $G_0(C_i)$ (the subgroup of the weak automorphism group of C_i that fixes all cosets of $C_i^\perp/C_i$). But in general $G_1(C)$ is only a subgroup of the direct product of the $G_1(C_i)$, and therefore must be computed directly for each C.

The enumeration of binary self-dual codes of length $N \leq 32$ has been carried out in a series of papers: Pless [414] for $N \leq 20$; Conway (unpublished) for Type 2_{II} of length 24; Pless and Sloane [431] for $N = 22, 24$; Conway and Pless [121] and Pless [416] for $N = 26$ to 30 and Type 2_{II} of length 32 (some errors in the last two references are corrected in Conway, Pless and Sloane [124]), Bilous and van Rees [50] for Type 2_{I} and length 32, and Bilous [49] for length 34. Conway and Sloane [130] had earlier shown that there are precisely three inequivalent $[32, 16, 8]$ extremal Type 2_{I} codes. The enumeration of Bilous and van Rees at length 32 has been confirmed by Munemasa (unpublished). These results are summarized in Table 12.1. (The entry in column (a) at length $N = 32$ was given incorrectly as 76 in Conway and Pless [121] and as 74 in [454]. See the entries marked with daggers in Table 12.4.)

It is an immediate consequence of the mass formula (12.1.3) that there are at least 17493 inequivalent Type 2_{II} codes of length 40. The exact number is presently unknown. King [309] has studied the mass of Type 2_{II} codes of length 40 and minimal distance 4 and 8.

We will now describe these codes, drawing heavily from the tables in [124]. The following self-orthogonal codes from §2.4.1 will be used as components.

d_4 : [1111], glue: $a := 0011$, $b := 0101$, $c := 0110$, $|G_0| = 4$, $G_1 = S_3$ on $\{a, b, c\}$, $|G_1| = 6$.

d_{2m} $(m \geq 3)$:

$$[111100\ldots, 00111100\ldots, \ldots, \ldots 001111], \qquad (12.2.2)$$

glue: $a := 0101\ldots01$, $b := 0000\ldots11$, $c := 0101\ldots10$, $|G_0| = 2^{N-1}N!$, $|G_1| = 2$ (swap a and c)

Table 12.1. Number of binary self-dual codes of length N:
(a) indecomposable, Type 2_{II} [A106162],
(b) indecomposable or decomposable, Type 2_{II} [A106163],
(c) indecomposable, Type 2_I but not 2_{II} [A106164],
(d) indecomposable or decomposable, Type 2_I but not 2_{II} [A106165],
(e) indecomposable, Types 2_I or 2_{II}, = (a) + (c) [A003178],
(f) indecomposable or decomposable, Types 2_I or 2_{II}, = (b) + (d) [A003179].
[Sequences A106166, A106167, A110193 are closely related.]

N	(a)	(b)	(c)	(d)	(e)	(f)
0	1	1	0	0	1	1
2	–	–	1	1	1	1
4	–	–	0	1	0	1
6	–	–	0	1	0	1
8	1	1	0	1	1	2
10	–	–	0	2	0	2
12	–	–	1	3	1	3
14	–	–	1	4	1	4
16	1	2	1	5	2	7
18	–	–	2	9	2	9
20	–	–	6	16	6	16
22	–	–	8	25	8	25
24	7	9	19	46	26	55
26	–	–	45	103	45	103
28	–	–	148	261	148	261
30	–	–	457	731	457	731
32	75	85	2448	3210	2523	3295
34	–	–	20786	24147	20786	24147

e_7 : [(1110100)], glue: $a := 1111111$, $G_0 = L_3(2)$, $|G_0| = 168$, $|G_1| = 1$.
e_8: no glue.

f_N : If some coordinate positions contain very few codewords, it is often best to regard these places as containing the *free* (or *empty*) component $f_N = \{0^N\}$. In this case we label the coordinate positions by A, B, C, ..., and use ABD for example to denote the glue word 110100 .... Also $|G_0| = 1$.

The above components are important in view of the following decomposition theorem for binary codes with low minimal distance.

Theorem 12.2.1. (*a*) *If a self-orthogonal binary code C has minimal distance 2 then $C = i_2^k \oplus C'$, where C' has minimal distance at least 4. (b) If a self-orthogonal binary code C is generated by words of weight 4 then C is a direct sum of copies of the codes d_{2m} $(m \geq 2)$, e_7 and e_8.*

Proof. (a) Suppose C contains a word of weight 2, say $u := 1100\ldots$. Then any other word $v \in C$ must meet u evenly, so begins $00\ldots$ or $11\ldots$. Hence $C = B \oplus C'$ where $B = [11]$. (b) A set of mutually self-orthogonal words of weight 4 whose supports are linked is easily seen to be either a d_{2m} for some $m \geq 2$, or an e_7 or e_8. $\qquad\square$

Remark 12.2.2. (1) Suppose C is a self-dual binary code with minimal distance 4, and let C' be the subcode generated by words of weight 4. Then C' is as described in part (b) of the theorem, and C can be regarded as being obtained by gluing C' to some other subcode C'' (the latter may be the free component f_N).

(2) Generalizing part (a) of the theorem, it is easy to show that any self-orthogonal code over a field $\mathbb{F}_q$ with length $N > 2$ and minimal distance 2 is decomposable (Conway, Pless and Sloane [123, Theorem 3]).

Fig. 12.1. Generator matrix for the odd Golay code h_{24}^+.

1 1 1 1	1 1 1 1	0 0 0 0	0 0 0 0	0 0 0 0	0 0 0 0
1 1 1 1	0 0 0 0	1 1 1 1	0 0 0 0	0 0 0 0	0 0 0 0
1 1 1 1	0 0 0 0	0 0 0 0	1 1 1 1	0 0 0 0	0 0 0 0
1 1 1 1	0 0 0 0	0 0 0 0	0 0 0 0	1 1 1 1	0 0 0 0
1 1 1 1	0 0 0 0	0 0 0 0	0 0 0 0	0 0 0 0	1 1 1 1
1 1 0 0	0 0 0 0	0 0 0 0	1 1 0 0	1 0 1 0	1 0 1 0
0 0 0 0	1 1 0 0	0 0 0 0	1 0 1 0	1 1 0 0	1 0 1 0
0 0 0 0	0 0 0 0	1 1 0 0	1 0 1 0	1 0 1 0	1 1 0 0
1 0 1 0	0 0 0 0	0 0 0 0	1 0 1 0	1 0 0 1	1 0 0 1
0 0 0 0	1 0 1 0	0 0 0 0	1 0 0 1	1 0 1 0	1 0 0 1
0 0 0 0	0 0 0 0	1 0 1 0	1 0 0 1	1 0 0 1	1 0 1 0
1 0 0 0	1 0 0 0	1 0 0 0	1 0 0 0	1 0 0 0	1 0 0 0

The following are some additional components that will be used in Table 12.2.

The code g_{24-m} ($m = 0, 2, 3, 4, 6, 8$) is obtained by taking the words of the Golay code g_{24} that vanish on m digits (and then deleting those digits). For the $[16,5,8]$ first-order Reed-Muller code g_{16} the 8 digits must be a special octad, while for g_{18} they must be an umbral hexad (see [133] for terminology). For $0 \leq m \leq 6$, g_{24-m} is a $[24 - m, 12 - m, 8]$ code.

The $[24,11,8]$ *half-Golay code* h_{24} consists of the Golay codewords that intersect a given tetrad evenly.

The *odd Golay code* h_{24}^+ is the $[24, 12, 6]$ Type 2_I code generated by h_{24} and an appropriate vector of weight 6. Alternatively, the odd Golay code may be obtained as follows. Let $v \in \mathbb{F}_2^{24}$ be a fixed vector of weight 4, say $v := 1^4 0^{24}$. Then $h_{24}^+ := \{u \in g_{24} : \text{wt}(u \cap v) \text{ even}\} \bigcup \{u + v : u \in g_{24}, \text{wt}(u \cap v) \text{ odd}\}$, with generator matrix shown in Fig. 12.1. This code has weight enumerator $x^{24} + 64x^{18}y^6 + 375x^{16}y^8 + 960x^{14}y^{10} + 1296x^{12}y^{12} + 960x^{10}y^{14} + 375x^8y^{16} + 64x^8y^{18} + y^{24}$, and $\mathrm{Aut}(h_{24}^+)$ is the "sextet group" $2^6{:}3.S_6$, of order $2^{10}3^35 = 138240$ (Pless and Sloane [431], [133, p. 309]).

The first 11 rows of the above matrix generate h_{24}; if the last row is replaced by

$$\boxed{0\ 1\ 1\ 1 \mid 1\ 0\ 0\ 0 \mid 1\ 0\ 0\ 0 \mid 1\ 0\ 0\ 0 \mid 1\ 0\ 0\ 0 \mid 1\ 0\ 0\ 0}$$

we get the Golay code g_{24} itself; and if the last row is replaced by

$$\boxed{1\ 1\ 1\ 1 \mid 0\ 0\ 0\ 0 \mid 0\ 0\ 0\ 0 \mid 0\ 0\ 0\ 0 \mid 0\ 0\ 0\ 0 \mid 0\ 0\ 0\ 0}$$

we get $(d_4^6)^+$.

Under the action of $\mathrm{Aut}(g_{24})$ there are two distinct ways to select tetrads $t := \{c, d, e, f\}$, $u := \{a, b, e, f\}$, $v := \{a, b, c, d\}$ so that $t + u + v = 0$, depending on whether $\{a, b, \ldots, f\}$ is a special hexad or an umbral hexad (see Fig. 12.2). Correspondingly there are two [24,10,8] *quarter-Golay codes* q_{24}^+, q_{24}^-, consisting of the codewords of g_{24} that intersect all of t, u, v evenly. We refer to Conway and Pless [121] and Conway, Pless and Sloane [124] for a description of the glue vectors for these codes.

Fig. 12.2. Two choices for a hexad (special or umbral), used to define the two [24, 10, 8] quarter-Golay codes q_{24}^+ and q_{24}^-.

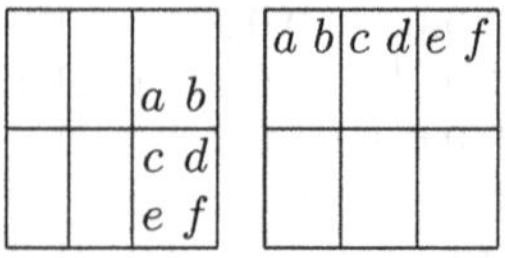

Our first table (Table 12.2) lists all indecomposable binary self-dual codes of length $N \leq 22$, together with the indecomposable Type 2_{II} codes of length 24, using the $+$ notation of (9.6.1). For these codes (and for most of those in the following tables) there is only one way to glue the specified components together without introducing additional minimal-weight words. We have therefore omitted the glue words from the table. (However, more information about these codes, including the glue words, will be given in Table 12.7.)

The next table (Tables 12.3 and 12.4) gives the full list of all 85 (decomposable (†) or indecomposable) Type 2_{II} codes of length 32. This table is taken from Conway, Pless and Sloane [124] and is a corrected version of the table in Conway and Pless [121]. The codes are labeled from C1 to C85 in the first column (using the same order as in [121] and [124]). The second column gives the components (omitting the superscripts "+" to save space).

The third and fourth columns give the orders of the groups $G_1(C)$ and $G_2(C)$, and the fifth column gives the order of the full group, using (12.2.1), where $|G_0(C)|$ is the product of the orders of the $G_0(C_i)$ for the components.

Table 12.2. Indecomposable binary self-dual codes of length $N \leq 24$ [A003178]. Here § indicates a Type 2_{II} code. For length 24 only Type 2_{II} codes are listed.

N	Components
2	i_2.
8	$e_8^§$.
12	d_{12}^+.
14	e_7^{2+}.
16	$d_{16}^§, d_8^{2+}$.
18	$(d_{10}e_7 f_1)^+, d_6^{3+}$.
20	$d_{20}^+, (d_{12}d_8)^+, (d_8^2 d_4)^+, (e_7^2 d_6)^+, (d_6^3 f_2)^+, d_4^{5+}$.
22	$g_{22}^+, (d_{14}e_7 f_1)^+, (d_{10}^2 f_2)^+, (d_{10}d_6^2)^+, (d_8 e_7 d_6 f_1)^+,$
	$(d_8 d_6^2 f_2)^+, (d_6^2 d_4^2 f_2)^+, (d_4^4 f_3^2)^+$.
24§	$g_{24}, d_{24}^+, d_{12}^{2+}, (d_{10}e_7^2)^+, d_8^{3+}, d_6^{4+}, d_4^{6+}$.

The latter are given in Table 12.6. The next column gives A_4, the number of codewords of weight 4. The weight enumerator of the code is then (from Theorem 6.4.2)

$$(x^8 + 14x^4 y^4 + y^8)^4 - (56 - A_4)x^4 y^4 (x^4 - y^4)^4 (x^8 + 14x^4 y^4 + y^8) .$$

The last four columns give the number of self-dual codes (the "children") of lengths 30, 28, 26, 24 that arise from the code.

To save space, we have omitted the glue vectors from Tables 12.3 and 12.4. In many cases they are uniquely determined by the components, and in any case they can be found in full in [121], with corrections in [124].

The enumeration in Tables 12.3 and 12.4 has been subjected to many checks, including the verification of the mass formula

$$\sum \frac{1}{|\operatorname{Aut}(C)|} = \frac{3912661228896364123}{532283035423762022400}$$

(in agreement with (12.1.3)).

Remark 12.2.3. There are just five Type 2_{II} codes of length 32 with minimal distance 8: the extended quadratic residue code $C81 = XQ(\mathbb{F}_2, 31)$, generated by

$$(10010001101101111100010101110000)1 ;$$

the second-order Reed-Muller code $C82 = r_{32}$, generated by

$$(11100100000100000011000000000000)1 ;$$

and the three codes $C83 = g_{16}^{2+}$, $C84 = f_4^{8+}$ and $C85 = f_2^{16+}$. Explicit generator matrices for the last three are shown in Fig. 12.3.

Table 12.3. Doubly-even self-dual (or Type 2_{II}) binary codes of length 32 (Part 1). A dagger (†) indicates a decomposable code.

Code	Components	$\|G_1\|$	$\|G_2\|$	$\|G\|$	A_4	n_{30}	n_{28}	n_{26}	n_{24}
C1	d_{32}	1	1	$2^{30}3^6 5^3 7^2 11{\cdot}13$	120	2	2	1	1
C2†	$d_{24}e_8$	1	1	$2^{27}3^6 5^2 7^2 11$	80	4	3	2	2
C3	$d_{20}d_{12}$	1	1	$2^{26}3^6 5^3 7$	60	5	4	2	2
C4	$d_{18}e_7^2$	1	2	$2^{22}3^6 5{\cdot}7^3$	50	5	3	2	1
C5†	d_{16}^2	1	2	$2^{29}3^4 5^2 7^2$	56	3	2	1	1
C6†	$d_{16}e_8^2$	1	2	$2^{27}3^4 5{\cdot}7^3$	56	5	3	2	2
C7	$d_{16}d_8^2$	1	2	$2^{27}3^4 5{\cdot}7$	40	6	4	2	2
C8	$d_{14}d_{10}e_7 f_1$	1	1	$2^{20}3^4 5^2 7^2$	38	11	5	3	2
C9	$d_{14}d_6^3$	1	6	$2^{20}3^6 5{\cdot}7$	30	6	4	2	1
C10†	$d_{12}^2 e_8$	1	2	$2^{25}3^5 5^2 7$	44	5	3	2	2
C11	$d_{12}^2 d_8$	1	2	$2^{25}3^5 5^2$	36	6	4	2	2
C12	$d_{12}d_8^2 d_4$	1	2	$2^{24}3^4 5$	28	11	7	2	2
C13	$d_{12}e_7^2 d_6$	1	2	$2^{19}3^5 5{\cdot}7^2$	32	9	5	3	1
C14	$d_{12}d_6^3 f_2$	1	6	$2^{19}3^6 5$	24	9	4	2	1
C15	$d_{12}d_4^5$	1	120	$2^{22}3^3 5^2$	20	5	3	1	1
C16	$d_{10}^3 f_2$	1	6	$2^{22}3^4 5^3$	30	5	2	1	1
C17	$d_{10}^2 d_6^2$	1	4	$2^{22}3^4 5^2$	26	7	4	2	1
C18†	$d_{10}e_8 e_7^2$	1	2	$2^{20}3^4 5{\cdot}7^3$	38	8	4	3	2
C19	$d_{10}d_8 e_7 d_6 f_1$	1	1	$2^{19}3^4 5{\cdot}7$	26	17	7	4	2
C20	$d_{10}d_8 d_6^2 f_2$	1	2	$2^{20}3^4 5$	22	15	6	3	2
C21	$d_{10}d_6^2 d_4^2 f_2$	1	4	$2^{19}3^3 5$	18	16	6	2	1
C22	$d_{10}d_4^4 f_6$	6	24	$2^{19}3^3 5$	14	9	3	1	1
C23	$d_{10}g_{22}$	2	1	$2^{15}3^3 5^2 7{\cdot}11$	10	4	2	1	1
C24†	e_8^4	1	24	$2^{27}3^5 7^4$	56	2	1	1	1
C25†	$e_8 d_8^3$	1	6	$2^{25}3^5 7$	32	5	3	2	2
C26†	$e_8 d_6^4$	1	24	$2^{21}3^6 7$	26	5	3	2	1
C27†	$e_8 d_4^6$	3	720	$2^{22}3^4 5{\cdot}7$	20	4	2	1	1
C28†	$e_8 g_{24}$	1	1	$2^{16}3^4 5{\cdot}7^2 11{\cdot}23$	14	3	1	1	1
C29	d_8^4	1	24	$2^{27}3^5$	24	3	2	1	1
C30	d_8^4	1	8	$2^{27}3^4$	24	4	2	1	1
C31	$d_8^3 d_4^2$	1	6	$2^{23}3^4$	20	6	3	1	1
C32	$d_8^2 e_7^2 f_2$	1	4	$2^{20}3^4 7^2$	26	10	3	2	1
C33	$d_8^2 d_6^2 f_4$	1	4	$2^{20}3^4$	18	14	4	2	1
C34	$d_8^2 d_4^4$	1	16	$2^{24}3^2$	16	8	4	1	1

Interrelations between types 2_I and 2_{II}

Suppose for concreteness that C is a Type 2_I code of length 26 with doubly-even subcode C_0. Then we obtain a Type 2_{II} code B (say) of length 32 by gluing C_0 to d_6, as follows. Write $C_0^\perp = C_0 \cup C_1 \cup C_2 \cup C_3$, as in §1.12, where

Table 12.4. Doubly-even self-dual (or Type 2_{II}) binary codes of length 32 (Part 2).

Code	Components	$\|G_1\|$	$\|G_2\|$	$\|G\|$	A_4	n_{30}	n_{28}	n_{26}	n_{24}
C35	$d_8 e_7 d_6^2 d_4 f_1$	1	2	$2^{18} 3^4 7$	20	18	7	3	1
C36	$d_8 d_6^4$	1	8	$2^{21} 3^5$	18	7	4	2	1
C37	$d_8 d_6^2 d_6 d_4 f_2$	1	2	$2^{18} 3^4$	16	22	9	3	1
C38	$d_8 d_6^2 d_4^2 f_4$	1	4	$2^{18} 3^3$	14	20	7	2	1
C39	$d_8 d_6 d_4^3 f_6$	2	6	$2^{17} 3^3$	12	17	6	2	1
C40	$d_8 d_4^6$	1	48	$2^{22} 3^2$	12	7	4	1	1
C41	$d_8 d_4^4 f_8$	2	24	$2^{18} 3^2$	10	12	4	1	1
C42	$d_8 d_4^2 g_{16}$	36	2	$2^{17} 3^3$	8	9	3	1	1
C43	$d_8 h_{24}$	1	1	$2^{16} 3^4 5$	6	5	2	1	1
C44	$e_7^4 d_4$	1	24	$2^{17} 3^5 7^4$	29	4	2	1	0
C45	$e_7^2 d_6^3$	1	6	$2^{16} 3^6 7^2$	23	6	3	2	0
C46	$e_7 d_6^3 d_4 f_3$	1	6	$2^{15} 3^5 7$	17	13	4	2	0
C47	$e_7 d_6 d_4^4 f_3$	1	24	$2^{17} 3^3 7$	14	12	4	2	0
C48	$e_7 d_4^4 f_9$	18	24	$2^{15} 3^4 7$	11	7	2	1	0
C49	$e_7 d_4 g_{21}$	6	1	$2^{12} 3^4 5 \cdot 7^2$	8	6	2	1	0
C50	$d_6^5 f_2$	1	10	$2^{16} 3^5 5$	15	6	2	1	0
C51	$d_6^4 d_4^2$	1	48	$2^{20} 3^5$	14	6	3	1	0
C52	$d_6^4 d_4 f_2^2$	1	8	$2^{17} 3^4$	13	13	4	1	0
C53	$d_6^4 f_8$	2	24	$2^{16} 3^5$	12	8	2	1	0
C54	$d_6^3 d_4^3 f_2$	1	6	$2^{16} 3^4$	12	12	5	1	0
C55	$d_6^3 d_4^2 f_6$	1	6	$2^{14} 3^4$	11	15	3	1	0
C56	$d_6^2 d_4^4 f_2^2$	1	8	$2^{17} 3^2$	10	16	4	1	0
C57	$d_6^2 d_4^4 f_4$	2	16	$2^{19} 3^2$	10	13	4	1	0
C58	$d_6^2 d_4^3 f_2 f_6$	1	12	$2^{14} 3^3$	9	18	4	1	0
C59	$d_6^2 d_4^2 f_{12}$	12	4	$2^{14} 3^3$	8	14	3	1	0
C60	$d_6^2 g_{20}$	4	2	$2^{15} 3^3 5$	6	6	2	1	0
C61	$d_6 d_4^2 d_4^3 f_3^2$	1	12	$2^{15} 3^2$	8	19	6	1	0
C62	$d_6 d_4^4 f_{10}$	2	8	$2^{15} 3$	7	21	4	1	0
C63	$d_6 d_4^3 f_{14}$	8	6	$2^{13} 3^2$	6	18	4	1	0
C64	$d_6 d_4^2 g_{18}$	36	2	$2^{10} 3^4$	5	12	3	1	0
C65	$d_6 d_4 g_{16} f_6$	72	1	$2^{12} 3^3$	4	14	4	1	0
C66	$d_6 f_{13}^2$	5616	2	$2^8 3^4 13$	3	6	2	1	0
C67	d_4^8		6 1344	$2^{23} 3^2 7$	8	2	1	0	0
C68	d_4^8		1 1152	$2^{23} 3^2$	8	3	1	0	0

$C = C_0 \cup C_2$, the shadow of C is $C_1 \cup C_3$, and $C_i = u_i + C_0$ for $i = 1, 2, 3$. Then B is generated by

Table 12.5. Doubly-even self-dual (or Type 2_{II}) binary codes of length 32 (Part 3).

Code	Components	$|G_1|$	$|G_2|$	$|G|$	A_4	n_{30}	n_{28}	n_{26}	n_{24}
C69	d_4^8	1	336	$2^{20}3\cdot7$	8	2	1	0	0
C70	$d_4^6 f_8$	4	48	$2^{18}3$	6	7	2	0	0
C71	$d_4^6 f_8$	1	48	$2^{16}3$	6	9	2	0	0
C72	$d_4^4 d_4 f_{12}$	6	24	$2^{14}3^2$	5	10	2	0	0
C73	$d_4^5 f_{12}$	1	60	$2^{12}3\cdot5$	5	6	1	0	0
C74	$d_4^4 g_{16}$	8	24	$2^{18}3$	4	7	2	0	0
C75	$d_4^4 f_{16}$	8	8	2^{14}	4	14	2	0	0
C76	$d_4^3 g_{18} f_2$	8	6	$2^{10}3^2$	3	13	2	0	0
C77	$d_4^2 q_{24}^+$	6	2	$2^{15}3^2$	2	6	1	0	0
C78	$d_4^2 q_{24}^-$	3	2	$2^{10}3^2$	2	8	1	0	0
C79	$d_4 f_4^6$	16	72	$2^{11}3^2$	2	8	1	0	0
C80	$d_4 f_7^4$	168	8	$2^8 3\cdot7$	1	8	2	0	0
C81	$XQ(\mathbb{F}_2,31)$	1	1	$2^5 3\cdot5\cdot31$	0	1	0	0	0
C82	r_{32}	1	1	$2^{15}3^2 5\cdot7\cdot31$	0	1	0	0	0
C83	g_{16}^2	20160	2	$2^{15}3^2 5\cdot7$	0	2	0	0	0
C84	f_4^8	256	336	$2^{12}3\cdot7$	0	2	0	0	0
C85	f_2^{16}	2	11520	$2^9 3^2 5$	0	3	0	0	0

$$
\begin{array}{|c|c|}
\hline
C_0 & \\
\hline
 & d_6 \\
\hline
u_1 & a \\
\hline
u_2 & b \\
\hline
u_3 & c \\
\hline
\end{array}
\qquad (12.2.3)
$$

This is a special case of the following construction. Let C, D be any strictly Type 2_{I} codes, of lengths n_1 and n_2, respectively, with $C_0^\perp = \cup_{i=0}^3 C_i$, $D_0^\perp = \cup_{i=0}^3 D_i$. Then $B = \cup_{i=0}^3 C_i \times D_i$ is self-dual if $n_1 + n_2 \equiv 0 \bmod 4$), and is Type 2_{II} if $n_1 + n_2 \equiv 0 \bmod 8$). The weight enumerator of B is then

$$
\sum_{i=0}^3 W_{C_i}(x,y)W_{D_i}(x,y) \ .
$$

Several constructions in the literature (Brualdi and Pless [86, Theorems 1, 2], Dougherty, Gulliver and Harada [153, Theorem 3.1], for example) are special cases of this construction. In (12.2.3) we have $D = i_2^3$.

In this way any Type 2_{I} code of length 26 leads to a unique (up to equivalence) Type 2_{II} code of length 32. Conversely, all Type 2_{I} codes of length 26 can be obtained by subtracting i_2^3 from a Type 2_{II} code of length 32.

More generally, suppose B is a Type 2_{II} code of length N. We choose a copy of $D = i_2^m$ so that $D_0 = d_{2m} \subset B$. Then we obtain a Type 2_{I} code of length $N - 2m$ by taking the vectors v such that $vw \in B$ for some $w \in D$.

Fig. 12.3. Generator matrices for the $[32, 16, 8]$ Type 2_{II} binary codes $C83 = g_{16}^{2+}$, $C84 = f_4^{8+}$ and $C85 = f_2^{16+}$.

```
11101000111010000000000000000000
10110100101101000000000000000000
10011010100110100000000000000000
10001101100011010000000000000000
00000000000000011011000110110000
00000000000000010101100101011100
00000000000000010010110100010110
00000000000000010001011100010111
11011000110110001101100000000000
10101100101011001010110000000000
10010110100101101001011000000000
10001011100010111000101100000000
00000000111010001110100011101000
00000000101101001011010010110100
00000000100110101001101010011010
00000000100011011000110110001101
```

```
11101000000000001110100011101000
10110100000000001011010010110100
10011010000000001001101010011010
10001101000000001000110110001101
00000000111010001110100010110100
00000000101101001011010010011010
00000000100110101001101010001101
00000000100011011000110111000110
11011000110110001101100000000000
10101100101011001010110000000000
10010110100101101001011000000000
10001011100010111000101100000000
11011000101100010000000011011000
10101100110110000000000010101100
10010110101011000000000010010110
10001011100101100000000010001011
```

```
10000000000000001111100010001000
01000000000000001111010001000100
00100000000000001110010001000010
00010000000000001111000100010001
00001000000000001000111110001000
00000100000000001001111010000100
00000010000000001011110010000010
00000001000000001111100010001
00000000100000001000100011111000
00000000010000001000100111110100
00000000001000001000101111110010
00000000000100000001000111110001
00000000000010001000100010001111
00000000000001000100010001001111
00000000000000100010001000101111
00000000000000010001000100011111
```

This is the subtraction process already discussed in §10.1. Every Type 2_I code of length $N - 2m$ can be obtained in this way by starting with a unique Type 2_{II} code and subtracting an appropriate d_{2m}. Of course any Type 2_{II} code is a direct summand of some Type 2_{II} code of any greater length.

Table 12.6. The groups G_0 for the components mentioned in Tables 12.2, 12.3 and 12.4.

Component	G_0	$|G_0|$
d_{2m}	$2^{m-1}.S_m$	$2^{m-1}m!$
e_7	$PSL_3(2)$	168
e_8	$AGL_3(2)$	1344
f_n	1	1
g_{16}	2^4	16
g_{18}	Z_3	3
g_{20}	M_{20}	$2^6 3{\cdot}5$
g_{21}	M_{21}	$2^6 3^2 5{\cdot}7$
g_{22}	M_{22}	$2^7 3^2 5{\cdot}7{\cdot}11$
g_{24}	M_{24}	$2^{10} 3^3 5{\cdot}7{\cdot}11{\cdot}23$
h_{24}	$2^6 \rtimes 3S_6$	$2^9 3^3 5$
q_{24}^+	$2^6.(S_3 \times 2^2)$	$2^9 3$
q_{24}^-	$2^2 \times S_4$	$2^5 3$

Table 12.7 shows all indecomposable or decomposable self-dual codes of Types 2_I or 2_{II} and lengths $N \leq 22$ with minimal distance $d \geq 4$, as obtained by subtracting suitable codes d_{2m} from one of the codes in Tables 12.3 and 12.4. The second column indicates the parent code in Tables 12.3 and 12.4 and the d_{2m} to be subtracted. The next two columns give the components, with a § to indicate a Type 2_{II} code, and the name (if any) given to this code in [414] or [431]. The remaining columns give the orders of the glue groups G_1 and G_2, the weight distribution, and generators for the glue.

Table 12.8 gives the self-dual codes (both Type 2_I and Type 2_{II}) of length 24 and minimal distance $d \geq 4$.

A complete list of all Type 2_I or Type 2_{II} self-dual codes of lengths $N \leq 24$ can be obtained by forming direct sums of the codes in Tables 12.7 and 12.8 in all possible ways with the codes i_2^m ($m = 0, 1, \ldots$).

There are over 1000 self-dual codes of lengths 26 through 30 (see Table 12.1, [121], [124]). The highest minimal distance is 6, and there are respectively 1, 3 and 13 codes with $d = 6$ of lengths 26, 28 and 30.

12.3 Type 3: Ternary self-dual codes

Ternary self-dual codes of lengths $N \leq 20$ (and the maximal self-orthogonal codes of lengths $N \leq 19$, $N \not\equiv 0 \pmod 4$) have been enumerated by Pless [412] and Mallows, Pless and Sloane [364] for $N \leq 12$, Conway, Pless and Sloane [123] for $N \leq 16$, and Pless, Sloane and Ward [432] for $N \leq 20$. Leon,

Table 12.7. Indecomposable or decomposable self-dual codes of Types 2_I or 2_{II} of lengths $N \leq 22$ and $d \geq 4$.

N	Code	Compts.	Name	$|G_1|$	$|G_2|$	A_4	A_6	A_8	A_{10}	A_{12}	Glue
0	C1(d_{32})	i_0	-	1	1						-
8	C2(d_{24})	e_8	A_8	1	1	14	0	1			-
12	C3(d_{20})	d_{12}	B_{12}	1	1	15	32	15	0	1	a
14	C4(d_{18})	e_7^2	D_{14}	1	2	14	49	49	14	0	dd
16	C5(d_{16})	d_{16}	E_{16}	1	1	28	0	198	0	28	a
	C6(d_{16})	e_8^2	A_8^2	1	2	28	0	198	0	28	-
	C7(d_{16})	d_8^2	F_{16}	1	2	12	64	102	64	12	(ab)
18	C8(d_{14})	$d_{10}e_7f_1$	I_{18}	1	1	17	51	187	187	51	aoA, cd-
	C9(d_{14})	d_6^3	H_{18}	6	9	75	171	171	75	$(abc), bbb$
20	C3(d_{12})	d_{20}	J_{20}	1	1	45	0	210	512	210	a
	C10(d_{12})	$d_{12}e_8$	A_8B_{12}	1	1	29	32	226	448	226	a-
	C11(d_{12})	$d_{12}d_8$	K_{20}	1	1	21	48	234	416	234	(ab)
	C12(d_{12})	$d_8^2d_4$	S_{20}	1	2	13	64	242	384	242	$(ab)x, bby$
	C13(d_{12})	$e_7^2d_6$	L_{20}	1	2	17	56	238	400	238	doa, ddd
	C14(d_{12})	$d_6^3f_2$	R_{20}	1	6	9	72	246	368	246	$aaaA, cccB, (abc)$-
	C15(d_{12})	d_4^5	M_{20}	1	120	5	80	250	352	250	$(ooxyx)$
22	C8(d_{10})	$d_{14}e_7f_1$	N_{22}	1	1	28	49	246	700	700	aoA, bdA
	C16(d_{10})	$d_{10}^2f_2$	P_{22}	1	2	20	57	270	676	676	$(ao)*, cc$-
	C17(d_{10})	$d_{10}d_6^2$	Q_{22}	1	2	16	61	282	664	664	aoc, oaa, bbb
	C18(d_{10})	$e_8e_7^2$	E_8D_{14}	1	2	28	49	246	700	700	$-dd$
	C19(d_{10})	$d_8e_7d_6f_1$	R_{22}	1	1	16	61	282	664	664	$odbA, boaA, aob$-
	C20(d_{10})	$d_8d_6^2f_2$	S_{22}	1	2	12	65	294	652	652	$baoA, aooAB, abb$-, occ-
	C21(d_{10})	$d_6^2d_4^2f_2$	T_{22}	1	4	8	69	306	640	640	$aoxoA, ooyyAB, aayo$-, $bozx$-, $obxz$-
	C22(d_{10})	$d_4^4f_6$	U_{22}	6	24	4	73	318	628	628	$oxyzBC, ozxyAC, ooxxAE, oyoyAD, ozzoAF, xxxx$-, $yyyy$-
	C23(d_{10})	g_{22}	G_{22}	2	1	0	77	330	616	616	1

Pless and Sloane [346] give a partial enumeration of the self-dual codes of length 24, making use of the complete list of Hadamard matrices of order 24, and show that there are precisely two codes with minimal distance 9 (cf. Table 11.4).

We will make use of the following components.

e_3: [111], glue: $\pm a$, $a = 120$. If the coordinates are labeled 1, 2, 3 then G_0 is generated by $(1, 2, 3)$ and $(1, 2)$ diag$\{-1, -1, -1\}$ and has order 6; $|G_1| = 2$.

t_4 is the $[4, 2, 3]_3$ tetracode, and g_{12} is the $[12, 6, 6]_3$ ternary Golay code, see §2.4.1.

Table 12.8. Indecomposable or decomposable self-dual codes of Types 2_I or 2_{II} of length $N = 24$ and $d \geq 4$.

Code	Components	Name	d	Code	Components	Name	d
$C2(e_8)$	$d_{24}\S$	E_{24}	4	$C32(d_8)$	$d_8 e_7^2 f_2$	J_{24}	4
$C6(e_8)$	$d_{16}e_8\S$	–	4	$C33(d_8)$	$d_8 d_6^2 f_4$	R_{24}	4
$C7(d_8)$	$d_{16}d_8$	H_{24}	4	$C34(d_8)$	$d_8 d_4^4$	T_{24}	4
$C10(e_8)$	$d_{12}^2\S$	A_{24}	4	$C35(d_8)$	$e_7 d_6^2 d_4 f_1$	P_{24}	4
$C11(d_8)$	d_{12}^2	–	4	$C26(e_8)$	$d_6^4\S$	D_{24}	4
$C12(d_8)$	$d_{12}d_8 d_4$	I_{24}	4	$C36(d_8)$	d_6^4	Q_{24}	4
$C18(e_8)$	$d_{10}e_7^2\S$	B_{24}	4	$C37(d_8)$	$d_6^2 d_6 d_4 f_2$	S_{24}	4
$C19(d_8)$	$d_{10}e_7 d_6 f_1$	K_{24}	4	$C38(d_8)$	$d_6^2 d_4^2 f_2$	U_{24}	4
$C20(d_8)$	$d_{10}d_6^2 f_2$	N_{24}	4	$C39(d_8)$	$d_6 d_4^3 f_6$	W_{24}	4
$C24(e_8)$	$e_8^3\S$	–	4	$C27(e_8)$	$d_4^6\S$	F_{24}	4
$C25(d_8)$	$e_8 d_8^2$	–	4	$C40(d_8)$	d_4^6	V_{24}	4
$C25(e_8)$	$d_8^3\S$	C_{24}	4	$C41(d_8)$	$d_4^4 f_8$	X_{24}	4
$C29(d_8)$	d_8^3	L_{24}	4	$C42(d_8)$	$d_4^2 g_{16}$	Y_{24}	4
$C30(d_8)$	d_8^3	M_{24}	4	$C43(d_8)$	h_{24}	Z_{24}	6
$C31(d_8)$	$d_8^2 d_4^2$	O_{24}	4	$C28(e_8)$	$g_{24}\S$	G_{24}	8

g_{10} is the $[10, 4, 6]_3$ code consisting of the vectors u such that $00u \in g_{12}$. If x and y are chosen so that $11x \in g_{12}$, $12y \in g_{12}$, then the glue words for g_{10} can be taken to be $\pm x$, $\pm y$, $\pm x \pm y$. $|G_0| = 360$, $|G_1| = 8$.

p_{13}: Let $Q_0, Q_1, \ldots, Q_{12}$ be the points of a projective plane of order 3, labeled so that the 13 lines are represented by the cyclic shifts $t_0, t_1, \ldots, t_{12}$ of the vector t_0 given by

$$Q_0 \ Q_1 \ Q_2 \ Q_3 \ Q_4 \ Q_5 \ Q_6 \ Q_7 \ Q_8 \ Q_9 \ Q_{10} \ Q_{11} \ Q_{12}$$
$$1 \ \ 1 \ \ 0 \ \ 1 \ \ 0 \ \ 0 \ \ 0 \ \ 0 \ \ 0 \ \ 1 \ \ 0 \ \ \ 0 \ \ \ 0$$

([361, p. 695], Conway and Sloane [128]). The vectors $t_0, \ldots, t_{12}$ generate a $[13, 7, 4]_3$ code $p_{13}^\perp$. The dual is p_{13}, a $[13, 6, 6]_3$ self-orthogonal code consisting of the vectors $\sum_{i=0}^{12} a_i t_i$ with $a_i \in \mathbb{F}_3$ and $\sum a_i = 0$, and having weight distribution $A_0 = 1$, $A_6 = 156$, $A_9 = 494$, $A_{12} = 78$. $G_0(p_{13}) \cong \mathrm{PGL}_3(3)$, of order 5616, $|G_1(p_{13})| = 2$. The glue words are $\pm t_0$.

The indecomposable self-dual codes of lengths $N \leq 16$ are shown in Table 12.9. H_8 denotes a suitably normalized version of the Hadamard matrix of order 8.

The analogue of Theorem 12.2.1 is: any self-orthogonal ternary code generated by words of weight 3 is a direct sum of copies of e_3 and t_4. A technique for classifying self-orthogonal codes generated by words of weight 6 (using "center sets") is given in Pless, Sloane and Ward [432].

Table 12.9. Indecomposable ternary self-dual codes of lengths $N \leq 20$ [A105510].

N	Components	$|G_0|$	$|G_1|$	$|G_2|$	d	Glue
4	t_4	48	1	1	3	–
8	–					
12	e_3^{4+}	6^4	2	24	3	$aaa0, 0\bar{a}aa$
	g_{12}	190080	1	1	6	–
16	$(e_3^4 f_4)^+$	$6^4.1$	8	24	3	$(a000)(2111)$
	$(e_3^2 g_{10})^+$	$6^2.360$	4	2	3	$a0x, 0ay$
	$(e_3 p_{13})^+$	$6 \cdot 5616$	2	1	3	at_0
	f_8^{2+}	1^2	$2^7.168$	2	6	$[I|H_8]$
20	17 codes					
	– see [432]					

12.3.1 Types 4^E and 4_{II}^E: Euclidean self-dual codes over $\mathbb{F}_4$

Codes of Type 4^E have been classified for lengths $N \leq 10$ and Type 4_{II}^E for lengths $N \leq 16$. This is the work of Betsumiya [35], Betsumiya, Harada and Munemasa [46], Gaborit, Pless, Solé and Atkin [186] and Kim, Kim and Kim [301]. Up to permutation-equivalence, the numbers of distinct codes are as follows:

$$N : \; 2 \; 4 \; 6 \; 8 \; 10 \; 12 \; 14 \quad 16$$
$$4^E : \; 1 \; 1 \; 3 \; 6 \; 17 \; 63 \; 404 \; 9858 \quad [A106158]$$
$$4_{II}^E : \; - \; 1 \; - \; 2 \; - \; 7 \; - \quad 48 \quad [A106159]$$

The second line refers to codes of Type 4^E that are not of Type 4_{II}^E. The Type 4_{II}^E codes of lengths 4 and 8 are RS_4, RS_4^2 and $\mathbb{F}_4 \otimes e_8$ (compare Tables 11.5 and 11.6).

12.4 Type 4^H: Hermitian self-dual codes over $\mathbb{F}_4$

These have been classified for lengths $N \leq 16$ (Conway, Pless and Sloane [123])—see Table 12.10.

We will make use of the following components.

d_{2m} $(m \geq 2)$: generated by (12.2.2). There are 16 cosets of d_{2m} in $d_{2m}^\perp$, and as glue words we choose 0, $\omega^\nu a$, $\omega^\nu b$, $\omega^\nu c$, $\omega^\nu d$, $\omega^\nu e$, $\nu \in \{0,1,2\}$, where

$$a := 1010\ldots1010$$
$$b := 0000\ldots0011$$
$$c := 1010\ldots1001$$
$$d := 1010\ldots10\omega\bar{\omega}$$
$$e := 1010\ldots10\bar{\omega}\omega$$

Also $|G_0| = 2^{m-1}m!$, $|G_1| = 36$ $(m = 2)$, or 12 $(m \geq 3)$.

$e_5 : [\omega\overline{\omega}\overline{\omega}\omega0, 0\omega\overline{\omega}\overline{\omega}\omega]$, glue: $\omega^\nu 1$, $\nu \in \{0, 1, 2\}$, $G_0 = A_5$, of order 60, $|G_1| = 6$.

h_6 is the hexacode, $\mathbb{F}_4 \otimes e_7$, $\mathbb{F}_4 \otimes e_8$ are $\mathbb{F}_4$-versions of the Hamming codes in §2.4.1, and 1_N is the $[N, 1, N]_4$ repetition code.

Table 12.10. Type 4^{H}: indecomposable Hermitian codes over $\mathbb{F}_4$ of lengths $N \leq 16$ [A001647].

N	Components	$|G_0|$	$|G_1|$	$|G_2|$	d	Glue
2	i_2	12	1	1	2	—
4	—	—	—	—	—	—
6	h_6	2160	1	1	4	—
8	e_8	8064	1	1	4	—
10	d_{10}^+	$2^4{\cdot}5!$	6	1	4	d
	e_5^{2+}	60^2	6	2	4	11
12	d_{12}^+	$2^5{\cdot}6!$	6	1	4	a
	$(e_7e_5)^+$	$60{\cdot}168$	6	1	4	11
	d_6^{2+}	24^2	6	2	4	(bd)
	d_4^{3+}	4^3	54	6	4	$(0de)$
14	d_{14}^+	$2^6{\cdot}7!$	6	1	4	d
	e_7^{2+}	168^2	6	2	4	11
	$(d_8e_5f_1)^+$	$2^3.4!60$	6	1	4	$d01, e10$
	$(e_5^2d_4)^+$	$4{\cdot}60^2$	18	2	4	$01d, 10e$
	$(d_8d_6)^+$	$2^34!2^23!$	6	1	4	ab, bd
	$(d_6^2f_2)^+$	$(2^2.3!)^2$	6	2	4	$(d0)11, bb\omega\overline{\omega}$
	$(d_6d_4^2)^+$	$4^22^23!$	18	2	4	$bbb, a0d, cd0$
	$(d_4^3f_2)^+$	4^3	6	6	4	$aa011, 0aaww, \ddot{b}\ddot{b}0w\overline{\omega}, 0\ddot{b}\ddot{b}\overline{\omega}1$
	$(d_4^21_6)^+$	4^2	108	2	4	$b00011\omega\omega, a00\overline{\omega}\overline{\omega}110,$
						$0b01\omega01\omega, 0a011\overline{\omega}\overline{\omega}0$
	$XQ(\mathbb{F}_4, 13)$	6552	1	1	6	—
16	31 codes					
	(see [123])					

Remark 12.4.1. (1) The group orders differ slightly form those in [123], since here we are using the weak automorphism group, which includes conjugation.

(2) The dots and double-dots in the glue column indicate multiplication by ω or ω^2, respectively.

(3) The unique distance 6 code at length 14 is the $[14, 7, 6]_4$ extended quadratic residue code $XQ(\mathbb{F}_4, 13)$, generated by

$$1(1\omega\overline{\omega}\omega\omega\overline{\omega}\,\overline{\omega}\,\overline{\omega}\omega\omega\overline{\omega}\omega)\ .$$

(4) The analogue of Theorem 12.2.1 is: (a) any self-orthogonal code with minimal distance 2 has i_2 as a direct summand; (b) any self-orthogonal code generated by words of weight 4 is a direct sum of copies of d_4, d_6, d_8, ..., e_5, h_6, e_7 and e_8.

12.5 Type 4^{H+}: Trace-Hermitian additive codes over $\mathbb{F}_4$

These have been classified up to length 12 by Danielsen and Parker [139], extending earlier work of Calderbank, Rains, Shor and Sloane [96] and Höhn [266]. The results are summarized in Table 12.11.

Table 12.11. Number of Type 4^{H+} codes of length N, from Danielsen and Parker [139]:
(a) indecomposable, Type 4_{II}^{H+} [A110302],
(b) indecomposable or decomposable, Type 4_{II}^{H+} [A110306],
(c) indecomposable, Types 4^{H+} or 4_{II}^{H+} [A090899],
(d) indecomposable or decomposable, Types 4^{H+} or 4_{II}^{H+} [A094927].

N	(a)	(b)	(c)	(d)
0	1	1	1	1
1	–	–	1	1
2	1	1	1	2
3	–	–	1	3
4	1	2	2	6
5	–	–	4	11
6	4	6	11	26
7	–	–	26	59
8	14	21	101	182
9	–	–	440	675
10	103	128	3132	3990
11	–	–	40457	45144
12	2926	3079	1274048	1323363

The analogue of Theorem 12.2.1 is the following. Let d_N be the code of length N generated by all even-weight binary vectors ($N \geq 2$), and let $i_2 = [11, \omega\omega]$. Then any trace self-orthogonal additive code over $\mathbb{F}_4$ generated by words of weight 2 is a direct sum of copies of i_2, d_2, d_3, d_4, ...

d_N^+ (mentioned in Table 11.8) is the code of length N, containing 2^N words, generated by d_N and $\omega\omega\ldots\omega$.

12.6 Type $4^{\mathbb{Z}}$: Self-dual codes over $\mathbb{Z}/4\mathbb{Z}$

These have been classified for lengths up to 16 in the following papers: Conway and Sloane [131] for $N \leq 9$, Fields, Gaborit, Leon and Pless [174] for $N \leq 15$, and Pless, Leon and Fields [428] and Duursma, Greferath and Schmidt [162] for Type $4_{\mathrm{II}}^{\mathbb{Z}}$ codes of length 16. The codes of length $N \leq 8$ are shown in Fig. 2.1 in Chapter 2. See also the tables of Fields [173].

In this section we will present enough component codes to state the analogue of Theorem 12.2.1.

The smallest self-dual code is $i_1 = \{0, 2\}$. If a self-orthogonal code C contains a vector of the form $2^1 0^{N-1}$ then $C = i_1 \oplus C'$ is decomposable. The next-simplest possible vectors are "tetrads", of type $\pm 1^4 0^{N-4}$. We list a number of self-orthogonal codes that are generated by tetrads; t denotes the total number of tetrads in the code.

The first four codes have the property that the associated binary code $C^{(1)}$ is the self-dual code d_{2m} of (12.2.2).

$\mathcal{D}_{2m}$ $(m \geq 2)$ is generated by the tetrads $11130 \ldots 0$, $0011130 \ldots 0$, $\ldots$, $0 \ldots 01113$; $|\mathcal{D}_{2m}| = 4^{m-1}$, $|\mathrm{WAut}(\mathcal{D}_{2m})| = 2 \cdot 4!$ $(m = 2)$ or $2^2 \cdot 2^m$ $(m > 2)$, $t = 2(m-1)$. $\mathcal{D}_{2m}^{\perp}/\mathcal{D}_{2m}$ is a group of type 4^2 with generators $v_1 = 0101 \ldots 01$, $v_2 = 00 \ldots 0011$.

$\mathcal{D}_{2m}^{O}$ $(m \geq 2)$ is generated by $\mathcal{D}_{2m}$ and the tetrad $1300 \ldots 0011$ (or equivalently the vector $2020 \ldots 20$); $|\mathcal{D}_{2m}^{O}| = 4^{m-1}2$, $|\mathrm{WAut}(\mathcal{D}_{2m}^{O})| = 2^2 \cdot 8$ $(m = 2)$ or $2 \cdot 2^{m-1} \cdot 2m$ $(m > 2)$, $t = 2m$. $(\mathcal{D}_{2m}^{O})^{\perp}/\mathcal{D}_{2m}^{O}$ is a cyclic group of order 4 generated by v_1 (if m is odd), or a 4-group generated by v_1 and $2v_2$ (if m is even).

$\mathcal{D}_{2m}^{+}$ $(m \geq 2$, but note that $\mathcal{D}_4^{+} \cong \mathcal{D}_4^{O})$ is generated by $\mathcal{D}_{2m}$ and $2v_2$; $|\mathcal{D}_{2m}^{+}| = 4^{m-1}2$, $|\mathrm{WAut}(\mathcal{D}_{2m}^{+})| = 2^m \cdot 2^{m+1}$, $t = 4(m-1)$. $(\mathcal{D}_{2m}^{+})^{\perp}/\mathcal{D}_{2m}^{+}$ is a 4-group generated by $2v_1$ and v_2.

$\mathcal{D}_{2m}^{\oplus}$ $(m \geq 2)$ is the self-dual code generated by $\mathcal{D}_{2m}^{O}$ and $\mathcal{D}_{2m}^{+}$; $|\mathcal{D}_{2m}^{\oplus}| = 4^{m-1}2^2$, $|\mathrm{WAut}(\mathcal{D}_{2m}^{\oplus})| = 2^3 \cdot 4!$ $(m = 2)$ or $2^m \cdot 2^m \cdot 2m$ $(m > 2)$, $t = 4m$. We have already noted in §2.4.9 that there are two permutation-inequivalent versions of $\mathcal{D}_4^{\oplus}$, with generator matrices shown in (2.4.30). $\mathcal{D}_4^{\oplus}$ (in either version) has swe $= x^4 + 6x^2z^2 + z^4 + 8y^4$.

$\mathcal{E}_7$ is generated by 1003110, 1010031, 1101003; $|\mathcal{E}_7| = 4^3$, $|\mathrm{WAut}(\mathcal{E}_7)| = 2 \cdot 4!$, $t = 8$. $\mathcal{E}_7^{\perp}/\mathcal{E}_7$ is a cyclic group of order 4 generated by 3111111.

$\mathcal{E}_7^{+}$ is the self-dual code generated by $\mathcal{E}_7$ and 2222222 (or equivalently by all cyclic shifts of 3110100); $|\mathcal{E}_7^{+}| = 4^3 2$, $|\mathrm{WAut}(\mathcal{E}_7^{+})| = 2 \cdot 168$, $t = 14$, swe $= x^7 + z^7 + 14y^4(x^3 + z^3) + 7x^3z^3(x + z) + 42xy^4z(x + z)$. For both $\mathcal{E}_7$ and $\mathcal{E}_7^{+}$ the associated binary code $C^{(1)}$ is the Hamming code e_7.

$\mathcal{E}_8$ is the self-dual code generated by $0u$, $u \in \mathcal{E}_7$ and 30001011. An equivalent generator matrix has already been given in (2.4.36). $|\mathcal{E}_8| = 4^4$, $g = 8 \cdot 2 \cdot 4! = 384$, $t = 16$, swe $= x^8 + 16y^8 + z^8 + 16y^4(x^4 + z^4) + 14x^4z^4 + 48xy^4z(x^2 + z^2) + 96x^2y^4z^2$.

Theorem 12.6.1. *Any self-orthogonal code over* $\mathbb{Z}/4\mathbb{Z}$ *generated by vectors of the form* $\pm 1^4 0^{N-4}$ *is weakly equivalent to a direct sum of copies of the codes*

$$\mathcal{D}_{2m}, \ \mathcal{D}_{2m}^{O}, \ \mathcal{D}_{2m}^{+}, \ \mathcal{D}_{2m}^{\oplus}(m = 1, 2, \ldots), \ \mathcal{E}_7, \ \mathcal{E}_7^{+}, \ \mathcal{E}_8 \ .$$

The (somewhat complicated) inclusions between the codes mentioned in the theorem can be seen in Fig. 2.1 of Chapter 2.

12.7 Other enumerations

Self-dual codes over $\mathbb{F}_5$ have been studied by Harada and Östergård [253] and Leon, Pless and Sloane [347], and self-dual codes over $\mathbb{F}_7$ by Harada and Östergård [252]. See also the references in §11.3.7.

In connection with Chapter 10, we should mention that some results on the classification of self-*orthogonal* codes over $\mathbb{F}_2$, $\mathbb{F}_3$ and $\mathbb{F}_4$ have been obtained by Mallows, Pless and Sloane [364], Pless [414], Bouyukliev, Bouyuklieva, Gulliver and Östergård [70] and Bouyukliev and Östergård [71].

13

Quantum Codes

Quantum codes, and especially the additive and symplectic constructions thereof, were one of the reasons for our initial interest in the Clifford-Weil group. General quantum codes behave in many ways like classical self-orthogonal codes, and there are strong connections with the theory of self-dual codes. In particular, Theorems 11.1.12, 11.1.16, 11.1.26 and 11.1.32 of Chapter 11 are all based on phenomena first noticed for quantum codes (more precisely, for codes of Type 4^{H+}, but as we will see, the two are extremely closely related). There is also a direct connection: the natural group of equivalences acting on a symplectic quantum code is exactly the complex Clifford group $\mathcal{X}_m = \mathcal{C}_m(\rho(2_{\mathrm{II}}))$ (cf. Theorem 6.2.1).

In this chapter we give a brief discussion of quantum codes and their constructions and bounds. In §13.1 we define quantum codes, state what it means mathematically for a quantum code to be able to correct certain classes of errors, and give several examples. In §13.2 we give our symplectic construction of quantum codes from Calderbank, Rains, Shor and Sloane [96]. These are also called additive or stabilizer codes. This construction is the most general one known for quantum codes (very few good nonadditive or nonsymplectic codes have ever been constructed).

In §13.3 we define various weight enumerators attached to quantum codes, and show how the problem of finding bounds on quantum codes can be formulated as a linear programming problem (see in particular Theorem 13.3.4). This leads to the "Purity Conjecture" 13.3.5, still an open problem. Section 13.4 presents bounds on quantum codes, obtained both by linear programming and by other methods. Nonbinary codes are discussed in §13.5, and the final section, §13.6, gives a table of lower and upper bounds on the optimal minimal distance d in any additive $[[N, k, d]]$ binary code.

There have been literally hundreds of papers written about quantum codes and quantum information theory, and we have only cited the papers that are explicitly needed for the chapter. To have given a more complete list of references would have swollen the bibliography out of proportion.

We will discuss only the mathematical aspects of quantum error-correcting codes. For further information, especially about the physical background, the reader is referred to any of the excellent books on quantum computers, such as Chen and Brylinski [105], Kitaev, Shen and Vyalyi [311], Nielsen and Chuang [388], Peres [406], Pittenger [409], or Stolze and Suter [515].

13.1 Definitions

We work in a (finite-dimensional) complex Hilbert space V which is a tensor product of N smaller spaces. Often $\dim V = 2^N$. Single elements of V will represent pure states of the quantum computer. A quantum code Q will then be a subspace of V.[1]

The partial trace operator plays a key role in the formal definition of a quantum code.

Definition 13.1.1. Let V and W be complex Hilbert spaces and let ρ be a Hermitian (or self-adjoint) operator on $V \otimes W$. The *partial trace* $\mathrm{Tr}_W(\rho)$ is the unique Hermitian operator on V such that

$$\mathrm{Tr}(\rho(\rho_V \otimes 1)) = \mathrm{Tr}(\mathrm{Tr}_W(\rho)\rho_V)$$

for all Hermitian operators ρ_V on V, where Tr is the usual trace. The existence and uniqueness of $\mathrm{Tr}_W(\rho)$ follows from the nondegeneracy of the trace inner product.

Remark. In quantum mechanics, a state is a positive semi-definite Hermitian operator of trace 1. If ρ is a state, then $\mathrm{Tr}_W(\rho)$ is the induced state on V if we ignore W.

Let $\{e_i\}$ be an orthonormal basis for V, i.e. a basis satisfying $e_i^\dagger e_j = \delta_{i,j}$, where $\dagger$ denotes adjoint or conjugate transpose. Similarly let $\{f_i\}$ be an orthonormal basis for W. We will think of e_i and f_i as column vectors.

The Hermitian operator ρ can be written as

$$\sum_{i,j,k,l} \rho_{ij,kl}(e_i \otimes f_j)(e_k \otimes f_l)^\dagger$$

for uniquely determined complex numbers $\rho_{ij,kl}$. In other words, ρ takes $e_k \otimes f_l$ to $\sum_{i,j} \rho_{ij,kl}\, e_i \otimes f_j$, and extends by linearity to the whole space $V \otimes W$. Then

$$\mathrm{Tr}_W(\rho) = \sum_{i,j,k} \rho_{ij,kj}(e_i e_k^\dagger).$$

If ρ is represented by a block matrix, we obtain $\mathrm{Tr}_W(\rho)$ by replacing each block by its trace.

---

[1] Unlike classical codes, quantum codes are always linear.

Definition 13.1.2. Let $V = V_1 \otimes V_2 \otimes \cdots \otimes V_N$ be a complex Hilbert space. A *quantum code* Q is any subspace of V. The dimension of the code is the dimension of the subspace. Let α be a subset of $\{1, \ldots, N\}$. We say that Q *can correct the erasure of* α if, for all unit vectors $u \in Q$, the partial trace $\text{Tr}_{\overline{\alpha}}(uu^{\dagger})$ is independent of u. Here $\text{Tr}_{\overline{\alpha}}$ is an abbreviation for

$$\text{Tr}_{V_{j_1} \otimes V_{j_2} \otimes \cdots},$$

and $\overline{\alpha} = \{j_1, j_2, \ldots\}$ is the complement of α in $\{1, \ldots, N\}$. Note that

$$\text{Tr}_{\emptyset} = \text{Id}, \text{Tr}_{\{1\ldots N\}} = \text{Tr} \text{ and } \text{Tr}_{\alpha} \text{Tr}_{\beta} = \text{Tr}_{\alpha \cup \beta} \text{ if } \alpha \cap \beta = \emptyset.$$

An equivalent form of the condition for Q to correct the erasure of α is that, for any Hermitian operator φ of trace 1 supported entirely on Q (i.e. such that the image $\text{Im}(\varphi) \subset Q$), the partial trace $\text{Tr}_{\overline{\alpha}}(\varphi)$ is independent of φ. This follows from the linearity of the partial trace. Note that the Hermitian operators supported on Q are exactly the linear combinations of rank 1 projections onto unit vectors in Q.

Remark. The physical interpretation of Definition 13.1.2 is that if all we know about a physical system is that it is a state in Q, then restricting the state to $V_\alpha := \bigotimes_{i \in \alpha} V_i$ gives no further information about the state.

If $\dim Q = 1$ the above definition is vacuous (since there is an essentially unique choice for u), but there is a stronger notion that applies.

Definition 13.1.3. Let V, Q, α be as above. Then Q is *pure with respect to* α if for all unit vectors $u \in Q$, $\text{Tr}_{\overline{\alpha}}(uu^{\dagger})$ is proportional to the identity. (In fact it will then be equal to $\frac{1}{\dim V_\alpha} \text{Id}$.)

The code Q has *minimal distance* $\geq d$ (resp. is *pure of minimal distance* $\geq d$) if Q corrects erasures of α (resp. is pure with respect to α) for all $|\alpha| < d$. A code has minimal distance exactly d if it has minimal distance $\geq d$ but not $\geq d + 1$.

Because the condition for correcting the erasure of α is vacuous when $\dim Q = 1$, we adopt the convention that saying that a 1-dimensional code Q can correct α implies that Q is pure with respect to α. Similarly, when we speak of the minimal distance of a 1-dimensional code, we will always mean the pure minimal distance.

A *self-dual* quantum code is a quantum code of dimension 1.

Although we will not discuss it here, there is also a notion of "*t-error correcting*" quantum code, which in particular satisfies $d \geq 2t + 1$.

Example 13.1.4. Let $V = V_1 \otimes V_2$, where $\dim V_1 = \dim V_2 = 2$, $V_1 = \langle e_0, e_1 \rangle$, $V_2 = \langle f_0, f_1 \rangle$. Let $Q = \langle e_{00} + e_{11} \rangle$ where we write e_{00} for $e_0 \otimes f_0$, e_{11} for $e_1 \otimes f_1$, etc. Since Q has dimension 1, its ordinary minimal distance is vacuous, but it has pure minimal distance 2. To show this, we must consider all α with $|\alpha| = 1$, but since multiplication by scalars of norm 1 has no effect, and

$\dim Q = 1$, the unit vector $u \in Q$ is essentially unique. We have $Q = \langle u \rangle$ where $u = (1/\sqrt{2})(1,0,0,1)^{tr}$,

$$uu^\dagger = \frac{1}{2}\begin{pmatrix} 1 & 0 & 0 & 1 \\ 0 & 0 & 0 & 0 \\ 0 & 0 & 0 & 0 \\ 1 & 0 & 0 & 1 \end{pmatrix},$$

and the partial traces $\mathrm{Tr}_{V_1}(uu^\dagger)$ and $\mathrm{Tr}_{V_2}(uu^\dagger)$ are both $\frac{1}{2}\begin{pmatrix} 1 & 0 \\ 0 & 1 \end{pmatrix}$ (by taking the traces of the 2×2 blocks), as required.

Example 13.1.5. Let e_7 be the (classical) $[7,3,4]$ Hamming code and set

$$v_0 = \sum_{b \in e_7} e_b, \quad v_1 = \sum_{b \in e_7} e_{b+1} \in \mathbb{C}[\mathbb{F}_2^7] = \otimes^7(\mathbb{C}^2).$$

Then $Q := \langle v_0, v_1 \rangle$ has minimal distance 3 and is pure.

Proof. We must show that for any two-element set α, i.e. for any pair of coordinates, and any $v \in Q$, $\mathrm{Tr}_{\overline{\alpha}}(vv^\dagger)$ is proportional to the identity. Since $\mathrm{Aut}\, e_7$ is doubly transitive, it suffices to check this for $\alpha = \{1,2\}$. We leave the details to the reader. □

This result also follows from the following more general construction.

Example 13.1.6. *The Calderbank-Shor-Steane construction* (Calderbank and Shor [97], Steane [513], Calderbank, Rains, Shor and Sloane [96, Theorem 9]). Let C be any binary self-orthogonal code of length N (with $C \subseteq C^\perp$), let d_P be the minimal distance of $C^\perp$ and let d be the minimal weight of $C^\perp \setminus C$. For each coset $\kappa = a + C$ of $C^\perp/C$, set

$$v_\kappa := \sum_{b \in \kappa} e_b.$$

Then $Q := \langle v_\kappa \mid \kappa \in C^\perp/C \rangle$ is a quantum code in $\otimes^N(\mathbb{C}^2)$ with minimal distance d and pure minimal distance d_P. (This in turn will follow from Theorem 13.2.4 below, using the observation from §7.2.1 that Type $4^{\mathrm{H}+}$ is a sub-Type of 2^{lin}.)

Example 13.1.7. The following code Q is a two-dimensional subspace of $\otimes^5(\mathbb{C}^2)$ with minimal distance 3. In the notation to be introduced in the next section, it is a $[[5,1,3]]$ code. This is the smallest N for which an $[[N,1,3]]$ code exists (see Table 13.3 below). As generators we may take (following Nielsen and Chang [388, pp. 468–469]) the two vectors

$e_{00000} - e_{11000} - e_{01100} - e_{00110} - e_{00011} - e_{10001} + e_{10100} + e_{01010}$
$\quad + e_{00101} + e_{10010} + e_{01001} - e_{01111} - e_{10111} - e_{11011} - e_{11101} - e_{11110},$
$e_{11111} - e_{00111} - e_{10011} - e_{11001} - e_{11100} - e_{01110} + e_{01011} + e_{10101}$
$\quad + e_{11010} + e_{01101} + e_{10110} - e_{10000} - e_{01000} - e_{00100} - e_{00010} - e_{00001}.$

(Note that these vectors are fixed under cyclic shifts of the subscripts, and the subscripts in the second vector are the binary complements of those in the first vector.)

13.2 Additive and symplectic quantum codes

This section describes the most important construction for quantum codes, the additive or symplectic (or stabilizer) construction.

We first consider the case of "binary" quantum codes, where the ambient Hilbert space is $V = \otimes^N(\mathbb{C}^2)$ and $\dim V = 2^N$. Here the tensor factors $\mathbb{C}^2$ are often called *quantum bits*, or *qubits* for short. If Q is a K-dimensional subspace of V with minimal distance $\geq d$, we will refer to it as an $((N, K, d))$ code. (If $K = 1$ then d refers to the pure minimal distance, as the ordinary minimal distance is irrelevant.)

In this situation we say that two quantum codes are *globally equivalent* if one can be obtained from the other by applying unitary operators separately in each tensor factor and then permuting the tensor components. They are *locally equivalent* if one can be obtained from the other just by applying unitary operators in each tensor factor. There are analogous notions of local and global *automorphism groups*.

The space of Hermitian operators on $\mathbb{C}^2$ is spanned by the four Pauli operators

$$\mathrm{id} := \begin{pmatrix} 1 & 0 \\ 0 & 1 \end{pmatrix}, \sigma_x := \begin{pmatrix} 0 & 1 \\ 1 & 0 \end{pmatrix}, \sigma_y := \begin{pmatrix} 0 & -i \\ i & 0 \end{pmatrix}, \sigma_z := \begin{pmatrix} 1 & 0 \\ 0 & -1 \end{pmatrix}. \quad (13.2.1)$$

Multiplicatively, these matrices generate a group of order 16 with center $\left\langle \begin{pmatrix} i & 0 \\ 0 & i \end{pmatrix} \right\rangle$. Abstractly, this is a central product of a quaternion group Q_8 of order 8 and a cyclic group Z_4. This group is a subgroup of the group $\mathcal{E}(\mathbb{F}_2)$ of (5.3.4) (differing only in the scalar subgroup).

Similarly, we obtain a group $E(V)$ of (non-Hermitian) operators on V which is generated by the Hermitian operators corresponding to arbitrary tensor products of N of the Pauli matrices (13.2.1). The properties of $E(V)$ are summarized in the next remark.

Remark 13.2.1. Let $V := \otimes^N(\mathbb{C}^2)$.

(a) The group $E(V) \leq \mathrm{GL}(V)$ is isomorphic to $2^{1+2N} \mathsf{Y} Z_4$.

(b) $E(V)/Z(E(V)) \cong \mathbb{F}_2^{2N}$, and the commutator $[x, y] := x^{-1}y^{-1}xy$ defines a symplectic form on $E(V)/Z(E(V))$ taking values in $E(V)' = \langle -\mathrm{id}\rangle$. This defines a symplectic form on $\mathbb{F}_2^{2N} \cong E(V)/Z(E(V))$ and thus a natural map from $\mathrm{Aut}(E(V))$ to $\mathrm{Sp}_{2N}(\mathbb{F}_2)$, which is in fact surjective with kernel $\mathrm{Inn}(E(V))$, the group of inner automorphisms of $E(V)$.

(c) The normalizer of $E(V)$ is

$$N_{\mathrm{GL}(V)}(E(V)) \cong \langle \mathbb{C}^* \,\mathrm{id}, \mathcal{X}_N \rangle \,,$$

isomorphic to the central product of the complex Clifford group $\mathcal{X}_N$ with the group of scalar matrices.

(d) $E(V)$ acts on $E(V)$ by conjugation. Since the commutator subgroup $E(V)'$ is contained in the center $Z(E(V))$, this action is trivial on $E(V)/Z(E(V))$. More precisely we have $x^y = \pm x$ for all $x, y \in E(V)$, where the sign is $+$ if and only if the symplectic form from (a) is 0 on (x, y).

(e) $E(V)$ is an irreducible matrix group, and its elements span $\mathrm{End}(V)$. If we choose one from each class modulo $Z(E(V)) = \langle i\,\mathrm{id}\rangle$, we obtain a basis for $\mathrm{End}(V)$. Partial traces act correctly on this basis, since the basis is consistent with tensor products, and the trace of a basis element $\notin Z(E(V))$ is zero. In particular, for $\alpha \subset \{1, \ldots, N\}$ the map

$$\mathrm{End}(V) \to \mathrm{End}(V), \ \psi \mapsto \mathrm{Tr}_\alpha(\psi) \otimes \mathrm{id}_{V_\alpha}$$

is a diagonal idempotent; the image is the span of the elements of the form $\sigma_1 \otimes \ldots \otimes \sigma_N$, where the σ_i are Pauli matrices, with $\sigma_i = I_2$ if $i \in \alpha$.

Definition 13.2.2. Given a commutative subgroup $C \subseteq E(V)$ that contains no scalar elements other than the identity, the *"additive" quantum code $Q(C)$* is the subspace of V fixed by C.

In order to determine what erasures $Q(C)$ can correct, we first need to understand the structure of such subgroups C.

Remark 13.2.3. (a) Let $C \leq E(V)$ be a commutative subgroup as in the definition above. Then $CZ(E(V))/Z(E(V))$ is a totally isotropic subspace of the symplectic space $E(V)/Z(E(V)) \cong \mathbb{F}_2^{2N}$. Hence C corresponds to a self-orthogonal code in the representation $N\rho(4^{\mathrm{H}+})$. In a slight abuse of previous notation, we will refer to this as a self-orthogonal code of Type $4^{\mathrm{H}+}$.

(b) Conversely, a totally isotropic subspace determines such a commutative group C uniquely, up to multiplication by a character.

(c) In particular, it follows that two groups C corresponding to a given totally isotropic subspace are conjugate in $E(V)$, and thus give rise to locally equivalent quantum codes.

(d) Because $E(V)$ is a normal subgroup of the complex Clifford group, the image of an additive code under an element of the complex Clifford group is again an additive code. Thus for these codes we have a third notion of equivalence: namely, two codes are equivalent if there is an element of the complex Clifford group that takes one to the other.

By abuse of notation, we will identify C with the corresponding totally isotropic subspace; this has the effect of identifying codes in the same $E(V)$ orbit.

Since C is a subgroup of $(\mathbb{F}_2^2)^N$, we obtain a notion of *weight* on elements of C: if $x \in (\mathbb{F}_2^2)^N, x = (x_1, x_2, \ldots, x_N), x_i \in \mathbb{F}_2^2$, we define $\mathrm{wt}(x)$ to be the number of i such that $x_i \neq 0$. This is the usual Hamming weight if we identify $\mathbb{F}_2^2$ with $\mathbb{F}_4$. The symplectic inner product can also be expressed in terms of $\mathbb{F}_4$, as

$$(x, y) = \sum_i \mathrm{Tr}\, x_i \overline{y}_i \,.$$

Thus the code C is a self-orthogonal code of Type $4^{\mathrm{H}+}$, that is, a trace-Hermitian code over $\mathbb{F}_4$.

Theorem 13.2.4. *([96]) Let C be a self-orthogonal code of length N, Type $4^{\mathrm{H}+}$, containing 2^k codewords. Suppose the minimal weight of a nonzero element of $C^\perp$ is d_P, and suppose the minimal weight of $C^\perp \setminus C$ is d. Then $Q(C)$ has dimension 2^{N-k}, minimal distance d and pure minimal distance d_P.*

Proof. We identify C and $C^\perp$ with corresponding subgroups of $E(V)$ and write the group law in C and $C^\perp$ multiplicatively. The group C has 2^k characters, which are permuted by the action of $E(V)$; so each character occurs with the same multiplicity in the natural representation of $C \leq E(V) \leq \mathrm{GL}(V)$. Thus $Q(C)$ has dimension 2^{N-k}.

Let $\varphi \in \mathrm{End}(V)$ be a Hermitian operator supported on $Q(C)$ with trace 1. Since the image of φ is contained in the fixed space $Q(C)$ of C, we have

$$g\varphi = \varphi = \varphi g \text{ for all } g \in C\,.$$

The elements that commute with C are precisely the linear combinations of elements in $C^\perp$. Therefore

$$\varphi = \sum_{g \in C^\perp} f(g) g \in \mathrm{End}(V)\,,$$

where $f : C^\perp \to \mathbb{C}$ is a function on $C^\perp \leq E(V)$. The condition that $h\varphi = \varphi$ for all $h \in C$ is equivalent to $f(hg) = f(g)$ for all $h \in C, g \in C^\perp$, and the condition that $\mathrm{Tr}(\varphi) = 1$ implies $f(1) = 1/\dim V$. If $\alpha \subseteq \{1, \ldots, N\}$ then $\mathrm{Tr}_{\overline{\alpha}}(\varphi) \otimes \mathrm{id}_{V_{\overline{\alpha}}}$ is the corresponding sum over $g \in C^\perp$ (now viewed as an additive code in $\mathbb{F}_4^N$) with support contained in α, since the trace of a non-identity Pauli matrix is 0. Thus $\mathrm{Tr}_{\overline{\alpha}}(\varphi)$ is independent of φ if and only if no element of $C^\perp \setminus C$ has support contained in α. Similarly, $\mathrm{Tr}_{\overline{\alpha}}(\varphi)$ is proportional to the identity for all such φ if and only if no element of $C^\perp \setminus \{0\}$ has support contained in α. $\qquad\square$

Quantum codes $Q(C)$ constructed in this way are called *additive* quantum codes. They were first constructed by Calderbank, Rains, Shor and Sloane [95], [96] and independently by Gottesman [197]. They are also called *stabilizer* codes, although we prefer the term "additive", as reflecting the nature of the code C.

An additive quantum code $Q(C)$ which is a 2^k-dimensional subspace of $\mathbb{C}^{2^N}$ with minimal distance $\geq d$ will be called an $[[N, k, d]]$ code. (As usual, if $k = 0$ then d refers to the pure minimal distance, as the ordinary minimal distance is irrelevant.) Such a code provides a mapping of k qubits into N qubits. The code $Q(C)$ constructed in Theorem 13.2.4 is therefore an $[[N, N - k, d]]$ code. Of course an $[[N, k, d]]$ code is automatically an $((N, 2^k, d))$ code.

The above construction is surprisingly powerful. In fact it is only for minimal distance 2 that quantum codes are known which are better than the best additive codes. The first such example was constructed by Rains, Hardin, Shor and Sloane [452] (see the next section), and others are given in [444].

More generally, given a finite abelian group A which decomposes as a direct sum $A = A_1 \oplus A_2 \oplus \cdots \oplus A_N$, we have a $\mathbb{Q}/\mathbb{Z}$-valued symplectic form on $A \times A^*$. The corresponding group $E(\mathbb{C}[A])$ is that generated by translations and multiplication by characters. Again, a commutative subgroup of $E(\mathbb{C}[A])$, i.e. an isotropic code in $A \times A^*$, induces a "symplectic" quantum code, and the above theory applies. In particular, codes of Type q^{H+} (§7.5.1, §7.6.1) give rise in this way to quantum codes (Knill [319], [320]; Knill and Laflamme [321], Knill, Laflamme and Zurek [322]; and [447]).

13.3 Hamming weight enumerators

If $Q = Q(C)$ is an additive code, then we can define the weight enumerator of Q to be essentially the weight enumerator of C. It turns out that this can be generalized to non-additive codes (Shor and Laflamme [487]; [447]).

We first consider the additive case in more detail. Let $P_{Q(C)}$ be the orthogonal projection onto the space $Q(C)$. This can be written as

$$P_{Q(C)} = \frac{1}{|C|} \sum_{x \in C} x \in \text{End}(V). \tag{13.3.1}$$

(Note that C is a commutative subgroup of $E(V)$.) Thus we may determine whether a given element $v \in \mathbb{F}_4^N$ is in C by calculating

$$\left\| \text{Tr}(E_v P_{Q(C)}) \right\|^2,$$

where E_v is any element of $E(V)$ lying above $v \in \mathbb{F}_4^N$. Indeed, since all elements $\neq 1$ in $E(V)$ have trace 0, this quantity is 0 if $v \notin C$, and is $\left(\frac{\dim(V)}{|C|}\right)^2 = (\dim Q(C))^2$ if $v \in C$. We can thus obtain the weight enumerator of an additive quantum code $Q = Q(C)$ from the formula

$$\frac{1}{(\dim Q)^2} \sum_{v \in \mathbb{F}_4^N} \left\| \text{Tr}(E_v P_Q) \right\|^2 x^{N - \text{wt}(v)} y^{\text{wt}(v)}$$

which is indeed the weight enumerator of C. We will use this same formula in the general case:

Definition 13.3.1. Let Q be a binary quantum code. Then the *weight enumerator* $A_Q(x, y)$ is defined to be

$$\sum_{v \in \mathbb{F}_4^N} \left\| \operatorname{Tr}(E_v P_Q) \right\|^2 x^{N - \operatorname{wt}(v)} y^{\operatorname{wt}(v)}. \tag{13.3.2}$$

We also define polynomials

$$B_Q(x, y) = A_Q\left(\frac{x + 3y}{2}, \frac{x - y}{2}\right), \quad S_Q(x, y) = A_Q\left(\frac{x + 3y}{2}, \frac{y - x}{2}\right), \tag{13.3.3}$$

called the *dual* and *shadow* weight enumerators.

Theorem 13.3.2. (Shor and Laflamme [487]; [443], [447].) *(a) For any binary quantum code Q, all three of the polynomials A_Q,*

$$B_Q - \frac{1}{\dim Q} A_Q \tag{13.3.4}$$

and S_Q have nonnegative coefficients.
(b) The polynomials A_Q, B_Q, S_Q are invariant under global equivalence of quantum codes.
(c) If $\dim Q = 1$ then $A_Q = B_Q$.
(d) The minimal distance of Q is the degree of the lowest degree term in

$$B_Q(1, y) - \frac{1}{\dim Q} A_Q(1, y). \tag{13.3.5}$$

The pure minimal distance of Q is the degree of the lowest degree term in

$$B_Q(1, y) - \dim Q. \tag{13.3.6}$$

Note that we always have

$$A_Q(x, y) = (\dim Q)^2 x^N + \cdots,$$
$$B_Q(x, y) = \dim Q \, x^N + \cdots. \tag{13.3.7}$$

Proof. (Sketch) The idea behind the proof is the following. The coefficients of A_Q, $B_Q - \frac{1}{\dim Q} A_Q$ and S_Q can be expressed as variances of random variables, and moreover for $B_Q - \frac{1}{\dim Q} A_Q$ the random variables are constant precisely when the corresponding erasures can be corrected. $\square$

As a consequence of Theorem 13.3.2, all the results that can be established for additive quantum codes using linear programming can also be proved for unrestricted quantum codes. With this in mind, there is some merit in directly studying the question of whether such polynomials A, B, S exist at all, giving rise to the following problem.

Linear programming problem for quantum codes. ([448].) Determine the triples (N, K, d) of positive integers such that there exist homogeneous polynomials $A(x, y)$, $B(x, y) = A\left(\frac{x+3y}{2}, \frac{x-y}{2}\right)$ and $S(x, y) = A\left(\frac{x+3y}{2}, \frac{y-x}{2}\right)$ of degree N satisfying

i) A, $B - \frac{1}{K}A$ and S have nonnegative coefficients,
ii) $A(1,y) = K^2 + O(y)$, and
iii) $B(1,y) - \frac{1}{K}A(1,y) = O(y^d)$.

A triple (N, K, d) is called *feasible* if such polynomials exist, and *purely feasible* if we can replace iii) by the condition

iii')$B(1,y) - K = O(y^d)$.

Theorem 13.3.3. ([448].) *If the triple (N, K, d) is feasible then so is (N, K', d) for all $1 \leq K' \leq K$.*

Proof. (Sketch) The reason we expect monotonicity to hold is that, given a code of dimension K, we can always take a subcode of dimension K'. It turns out that one can give an explicit formula just in terms of A_Q for the average weight enumerator of a uniform random subcode of dimension K': this is

$$\widehat{A}(x,y) = \frac{K'(K'K - 1)}{K^3 - K}A(x,y) + \frac{K'(K - K')}{K^3 - K}B(x,y).$$

For an arbitrary $A(x,y)$ satisfying the quantum linear programming problem for (N, K, d), $\widehat{A}$ is easily verified to satisfy the problem for (N, K', d).   □

It turns out that if one expresses $A(x,y)$ in terms of eigenfunctions of the MacWilliams transform, this random subcode operation becomes diagonal. In particular, the quantum linear programming problem can be rephrased in terms of the polynomials

$$C(x,y) = \frac{A(x,y) + B(x,y)}{K^2 + K}, \tag{13.3.8}$$

$$D(x,y) = \frac{A(x,y) - B(x,y)}{K^2 - K}. \tag{13.3.9}$$

Theorem 13.3.4. ([448].) *If there exists an $((N, K, d))$ quantum code $(K > 1)$, then there exist homogeneous polynomials $C(x,y)$ and $D(x,y)$ satisfying the equations*

$$C(x,y) = C\Big(\frac{x + 3y}{2}, \frac{x - y}{2}\Big),$$

$$D(x,y) = -D\Big(\frac{x + 3y}{2}, \frac{x - y}{2}\Big),$$

$$C(1,0) = 1,$$

$$C(1,y) - D(1,y) = O(y^d),$$

as well as the inequalities

$$C(x,y) - \frac{K-1}{2K}(C(x,y) - D(x,y)) \geq 0,$$

$$C(x,y) - D(x,y) \geq 0,$$

$$C\left(\frac{x+3y}{2}, \frac{y-x}{2}\right) \geq 0,$$

$$D\left(\frac{x+3y}{2}, \frac{y-x}{2}\right) \geq 0.$$

The last four inequalities just mean that all coefficients of the polynomials are nonnegative. The proof is straightforward. The point of this restatement is that now K appears in only one place in the new linear programming problem, and in a manifestly monotonic way.

In particular, we see that this monotonicity continues to hold even if we allow K to be an arbitrary real number ≥ 1. Thus for each pair N, d, there exists a maximal $K \geq 1$ such that (N, K, d) is feasible: call it $K(N, d)$. Experimentally, whenever $K(N, d) > 1$, the corresponding solutions to the quantum linear programming problem are especially nice, in that they actually satisfy the "pure" version of the problem. This gives rise to the so-called "Purity Conjecture", formally stated for the first time in Calderbank, Rains, Shor and Sloane [96]:

Conjecture 13.3.5. If $K(N, d) > 1$, and $A(x, y)$ satisfies the quantum linear programming constraints for $N, K(N, d), d$, then

$$B(1, y) - K(N, d) = O(y^d). \tag{13.3.10}$$

In particular, this conjecture (which we have verified for $N \leq 100$ using MAGMA) would imply that $K(N, d)$ is also the maximal K for which the triple N, K, d is purely feasible, or in other words that the two quantum linear programming problems give the same answers for $K > 1$. Then bounds such as Ashikhmin and Litsyn's [14] generalization of the Aaltonen bound [2], which in the purely additive case apply directly to the dual code $C^\perp$, would also apply to impure general quantum codes.

For emphasis, we state:

Research Problem 13.3.6. *Prove the Purity Conjecture!*

Remark. It is worth mentioning that $K(N, d)$ need not even be a rational number. This happens for the first time at $N = 89$, $d = 30$, when $K(N, d) = 7.55185\ldots = (a + \sqrt{b})/c$, where

$$a = 134912798981, \ b = 15216175495257713186569, \ c = 34199106681.$$

When is $K(N, d)$ an integer? The above discussion suggests that quantum codes of dimension $K(N, d)$ should be particularly interesting. Of course this is only possible if $K(N, d)$ is an integer, so we now discuss when this happens.

We begin with the case $d = 2$. In this case $K(N, 2)$ is always an integer. It is shown in [444] that $K(2m, 2) = 2^{2m-2}$ and a corresponding additive code exists (as well as non-additive codes for $m > 2$). If N is odd, the only time that $K(N, 2)$ can be achieved is for $N = 5$, when $K(5, 2) = 6$ (and the code must be nonadditive).

Such a code had already been constructed by Rains, Hardin, Shor and Sloane [452]. This is one of the few instances where a nonadditive code is known and known to be better than any additive code. This $((5, 6, 2))$ code Q, that is, a 6-dimensional subspace of $\mathbb{C}^{32}$, may be defined as follows. As basis for $\mathbb{C}^{32}$ we take vectors $e_{00000}, e_{00001}, \ldots, e_{11111}$, but abbreviate them to $00000, \ldots, 11111$; the cyclic group Z_5 acts on these subscripts. An explicit basis for Q consists of

$$00000 - \sum_{g \in Z_5} 00011^g + \sum_{g \in Z_5} 00101^g - \sum_{g \in Z_5} 01111^g,$$

together with the vectors v^g, $g \in Z_5$, where

$$v = 00001 - 00010 - 00100 - 01000 - 10000 + 00111$$
$$- 01110 - 11100 + 11001 + 10011 - 01011 + 10110$$
$$- 01101 + 11010 - 10101 - 11111.$$

This code Q is preserved by a group of order 3840. We refer the reader to [452] for a proof that Q has minimal distance 2, that is, that it can correct one erasure, and for a discussion of its other properties and the method by which it was found. Unfortunately, the technique used to find this code has so far proved infeasible in all other cases.

Next we consider the case $d > 2$. Up to $N = 100$ there are four families of examples when $K(N, d)$ is an integer.

1. There are 16 cases $N = 6m - 1$, $d = 2m + 1$, $K(N, d) = 2$, for $m = 1, 2, \ldots, 16$. One can show that such a $((6m - 1, 2, 2m + 1))$ code $Q_m^{(1)}$ exists if and only if a (pure) $((6m, 1, 2m + 2))$ code exists. (An extremal code of Type $4^{\text{H}+}$ induces such a code.) The first of these, $Q_1^{(1)}$, exists (based on the shortened hexacode) and is unique and additive. Additive versions of $Q_2^{(1)}$, $Q_3^{(1)}$ and $Q_5^{(1)}$ are known, but in the remaining twelve cases the existence of $Q_m^{(1)}$ is at present undecided. For $m > 16$ there are negative coefficients.

2. The infinite family of *quantum Hamming codes* with $N = (4^m - 1)/3$, $d = 3$ and $K(N, 3) = 2^{N-2m}$, constructed from $\mathbb{F}_4$-linear Hamming codes.

3. There are seven further cases when K is a power of 2: $K(17, 4) = 2^9$: an additive code exists; $K(18, 3) = 2^{12}$, $K(16, 4) = 2^8$ and $K(22, 4) = 2^{14}$: existence is undecided, but the codes cannot be additive; $K(27, 5) = 2^{15}$, $K(28, 6) = 2^{14}$ and $K(40, 13) = 2^6$: existence is undecided.

4. Finally, there are eight cases when $K(N, d)$ is not a power of 2. The existence is undecided in each case. The parameters of these putative codes are $((10, 24, 3))$, $((13, 40, 4))$, $((21, 7168, 4))$, $((24, 49152, 4))$, $((22, 384, 6))$,

$((22, 56, 7))$, $((24, 24, 8))$, $((39, 24, 13))$. Can these be constructed in the same way that the $((5, 6, 2))$ code above was constructed? We do not know.

Research Problem 13.3.7. *Settle the remaining cases in the above list— especially the existence of codes with parameters* $((24, 1, 10))$ *(from* $Q_4^{(1)}$*),* $((10, 24, 3))$, $((24, 24, 8))$, $((39, 24, 13))$.

13.4 Linear programming bounds

In this section we give explicit linear programming bounds for quantum codes, analogous to those in Chapter 11. We only consider binary (but not necessarily additive) quantum codes.

Knill and Laflamme [321] have shown that an $((N, K, d))$ quantum code must satisfy the following version of the Singleton bound (cf. [361]):

$$K \leq 2^{\frac{N}{2} - d + 1}. \tag{13.4.1}$$

Gottesman [197] showed that any "nondegenerate" $((N, K, 2t+1))$ code must satisfy the sphere-packing bound

$$\sum_{j=0}^{t} 3^j \binom{N}{j} \leq \frac{2^N}{K}. \tag{13.4.2}$$

Ashikhmin and Litsyn [14] later used linear programming to establish (13.4.2) for any quantum code.

The next result is a companion to Theorem 11.1.16.

Theorem 13.4.1. ([442].) *If Q is an $((N, 1, d))$ self-dual quantum code then*

$$d \leq 2 \left\lfloor \frac{N}{6} \right\rfloor + 2,$$

except when $N \equiv 5 \pmod{6}$, when $d \leq 2\lfloor N/6 \rfloor + 3$. If Q is an $((N, K, d))$ code with $K > 1$, then

$$d \leq 2 \left\lfloor \frac{N+1}{6} \right\rfloor + 1,$$

except when $N \equiv 4 \pmod{6}$, when $d \leq 2\lfloor (N+1)/6 \rfloor + 2$.

Concerning asymptotic bounds, so far these have mostly been obtained for pure distance. However, as already mentioned in the previous section, if the Purity Conjecture 13.3.5 holds then these bounds will also apply to the minimal distance. Ashikhmin and Litsyn [14] give several bounds on both kinds of distance. The best asymptotic bound on pure distance (and on minimal distance for codes of sufficiently high rate) that they obtain is an analogue of the McEliece et al. bound for classical codes ([361, Chap. 17]; McEliece, Rodemich, Rumsey and Welch [370], Aaltonen [2]). The bound is however extremely complicated, even to state, and we refer the reader to [14] for details. There is also an analogue of Theorem 11.1.29.

Theorem 13.4.2. ([451].) *Let d_N be the largest minimal distance of a quantum code of length N. Then*

$$\limsup_{N \to \infty} \frac{d_N - \frac{3-\sqrt{3}}{4} N}{\sqrt{N}} = -\infty.$$

The preceding bounds are all based on linear programming. A few other bounds are known.

In [444] it is shown that, if Q is a $((2m + 1, K, 2))$ code for some m and K, then

$$K < 4^{m-1}\left(2 - \frac{1}{m}\right), \quad m > 2. \tag{13.4.3}$$

This is a very slight improvement on the linear programming bound, which gives the corresponding weak inequality.

In the case of putative self-dual codes with parameters $((7, 1, 4))$ and $((13, 1, 6))$, one can use higher-order analogues of weight enumerators to show that no such codes exist. In fact the linear equations imposed by the minimal distance already fail to have solutions, obviating the need to use any inequalities.

Other bounds on the pure minimal distance have been given by several authors including Ashikhmin and Litsyn [14] and Cleve, Knill and Laflamme [321]. But much more remains to be done.

Research Problem 13.4.3. (*a*) *For classical codes there are a variety of techniques available for proving the nonexistence of codes with specified parameters. A key open problem is to find analogous techniques—not based on linear programming—for quantum codes. For example, there is not even a good "brute force" method known, analogous to exhaustive search-type methods for classical codes. Can the powerful methods of Jaffe [292], [293] be adapted to the quantum case?*
(*b*) *The bound of Theorem 13.4.2 is weaker than what would be obtained if the Ashikhmin and Litsyn [14] bound were applicable to ordinary minimal distance. What is the best bound on ordinary minimal distance that can be obtained from linear programming? (Compare Research Problem 11.1.33.)*

13.5 Other alphabets

Although binary quantum codes are certainly the most studied, some things are known about the nonbinary case, most notably the analogue of the linear programming problem.

We begin by defining the appropriate weight enumerators. For $i = 1, \ldots, N$ let τ_i be the operator on the Hilbert space of operators on $V = V_1 \otimes \cdots \otimes V_N$ defined by

$$\tau_i(\varphi) := \frac{1}{\dim V_i} \operatorname{Tr}_i(\varphi) \otimes 1_{V_i},$$

where Tr_i is the partial trace, and let $\overline{\tau}_i = 1 - \tau_i$. Then given a quantum code $Q \subseteq V$ and a subset $\alpha \subseteq \{1, \ldots, N\}$, we let

$$a_Q(\alpha) := \operatorname{Tr}\left(\left(\prod_{i \notin \alpha} \tau_i \prod_{i \in \alpha} \overline{\tau}_i P_Q\right)^2\right) \tag{13.5.1}$$

where P_Q is the orthogonal projection onto the subspace Q. The weight enumerator of Q, $A_Q(x_1, y_1, x_2, y_2, \ldots, x_N, y_N)$, is defined to be

$$\sum_{\alpha \subseteq \{1, \ldots, N\}} a_Q(\alpha) \prod_{i \notin \alpha} x_i \prod_{i \in \alpha} y_i. \tag{13.5.2}$$

Note that $a_Q(\alpha)$ is manifestly nonnegative, and thus the corresponding assertion in the following theorem is immediate.

Theorem 13.5.1. ([443], [449].) *Let $Q \subseteq \mathbb{C}^{a_1} \otimes \mathbb{C}^{a_2} \otimes \cdots \otimes \mathbb{C}^{a_N}$ be a quantum code. Then there exist multilinear polynomials $A_Q(x_1, y_1, \ldots, x_N, y_N)$,*

$$B_Q(x_1, y_1, \ldots, x_N, y_N) := A_Q\left(\frac{x_1 + (a_1^2 - 1)y_1}{a_1}, \frac{x_1 - y_1}{a_1},\right.$$
$$\left. \ldots, \frac{x_N + (a_N^2 - 1)y_N}{a_N}, \frac{x_N - y_N}{a_N}\right) \tag{13.5.3}$$

and

$$S_Q(x_1, y_1, \ldots, x_N, y_N) := A_Q\left(\frac{(a_1 - 1)x_1 + (a_1 + 1)y_1}{a_1}, \frac{y_1 - x_1}{a_1},\right.$$
$$\left. \ldots, \frac{(a_N - 1)x_N + (a_N + 1)y_N}{a_N}, \frac{y_N - x_N}{a_N}\right) \tag{13.5.4}$$

with the following properties:

(a) A_Q, $B_Q - \frac{1}{K} A_Q$ and S_Q have nonnegative coefficients.
(b) $A_Q(1, 0, 1, 0, \ldots, 1, 0) = K^2$ and $B_Q(1, 0, 1, 0, \ldots, 1, 0) = K$.
(c) Let α be a subset of $\{1, \ldots, N\}$. Then Q can correct the erasure of α if and only if, for all subsets $\beta \subseteq \alpha$, the coefficient of $\prod_{i \notin \beta} x_i \prod_{i \in \beta} y_i$ in $B_Q - \frac{1}{K} A_Q$ is 0. Similarly, Q is pure with respect to α if and only if the same condition holds for $B_Q - K \prod_{i=1}^{N} x_i$.
(d) If Q is self-dual, then $B_Q = A_Q$, and $S_Q(-x, y) = S_Q(x, y)$.

The proof is similar to that of Theorem 13.3.2.

Remark. Given a (classical) self-orthogonal code of symplectic type, we have seen in §13.2 that there exists a corresponding quantum code, to which the above theorem associates three polynomials A_Q, B_Q and S_Q. A_Q and B_Q are simply the full Hamming weight enumerators of C and $C^\perp$ respectively (up to scalar multiples $\dim Q$ and $(\dim Q)^2$ respectively); however S_Q (which we may regard as a generalized shadow enumerator) is new.

In particular, this has consequences for classical codes of any Type (e.g. q^{H+}) that has a symplectic Type as a sub-Type (cf. Remark 1.8.6), for instance Hermitian self-dual codes over a field $\mathbb{F}_q$, since q^{H+} is a sub-Type of q^H. This produces new inequalities for the corresponding linear program, which could in principle be used to obtain new bounds for classical codes. Note that the coefficients of S_Q, although nonnegative, are in general not integers when $q > 2$.

As an example, for a Hermitian self-dual code C of Type q^H where q is a square, we have

$$S(x,y) = \mathrm{hwe}(C)\left(\frac{(\sqrt{q}-1)x + (\sqrt{q}+1)y}{\sqrt{q}}, \frac{y-x}{\sqrt{q}}\right). \qquad (13.5.5)$$

This agrees with the shadow enumerator of C (viewed as a code of Type 4^{H+}) when $q = 4$ (cf. (2.3.32)). If C is the extended quadratic-residue code of length 6 over $\mathbb{F}_9$, for instance, with

$$\mathrm{hwe}(C) = x^6 + 120x^2y^4 + 240xy^5 + 368y^6,$$

we find that

$$S(x,y) = \frac{16}{27}(x^6 + 15x^4y^2 + 75x^2y^4 + 17y^6).$$

Research Problem 13.5.2. *Is there a combinatorial interpretation for the coefficients of the generalized shadow enumerator (13.5.4)?*

Remark. If Q is an arbitrary K-dimensional quantum code in $V_1 \otimes V_2 \otimes \cdots \otimes V_N$ we can obtain a self-dual quantum code $\widehat{Q}$ in the Hilbert space $V_1 \otimes V_2 \otimes \cdots \otimes V_N \otimes \mathbb{C}^K$ simply by choosing orthonormal bases $v_1, \ldots, v_K$ for Q and $e_1, \ldots, e_K$ for $\mathbb{C}^K$, and taking

$$v_1 \otimes e_1 + v_2 \otimes e_2 + \cdots + v_K \otimes e_K$$

as the generator. The weight enumerator of $\widehat{Q}$ is

$$A_{\widehat{Q}} := \frac{1}{K}A_Q x_{N+1} + \left(B_Q - \frac{1}{K}A_Q\right)y_{N+1}. \qquad (13.5.6)$$

The positivity of $B_Q - \frac{1}{K}A_Q$ thus follows from the corresponding statement for $\widehat{Q}$.

In terms of $\widehat{Q}$, the code Q can correct erasures of α if and only if $\mathrm{Tr}_{\overline{\alpha}}(P_{\widehat{Q}})$ is of the form $\varphi \otimes \mathrm{id}_{\mathbb{C}^\kappa}$ for some operator φ. In fact φ is precisely the partial trace of a unit vector in Q (which by assumption is independent of the choice of the unit vector). In particular, for any subset $\beta \subseteq \alpha$, the coefficient of $\prod_{i \notin \beta} x_i \prod_{i \in \beta} y_i \cdot y_{N+1}$ in $A_{\widehat{Q}}$ must be zero, and thus the appropriate coefficients of $B_Q - \frac{1}{K} A_Q$ vanish. So to obtain a higher-order analogue of weight enumerators for quantum codes and the associated inequalities it suffices to have such a generalization for self-dual quantum codes over mixed alphabets.

13.6 A table of quantum codes

Tables 13.3 and 13.4, based on the tables in Calderbank, Rains, Shor and Sloane [96], with some updates from Grassl [198], give lower and upper bounds on the optimal minimal distance d in any additive $[[N, k, d]]$ binary code. For $k = 0$ the codes are self-dual and are assumed to be pure. These table, with explicit codes, have been incorporated in the most recent version (2.12) of the computer algebra system MAGMA [68], [100], which also includes support for constructing and computing minimal distances of additive quantum codes.

All unmarked upper bounds in the tables come from the linear programming bound of Theorem 13.3.4. (A few of these bounds can also be obtained from Eq. 13.4.2 or from Theorem 13.4.1.)

Unmarked lower bounds are based on the following result:

Theorem 13.6.1. (Calderbank, Rains, Shor and Sloane [96, Th. 6].) *Suppose an $[[N, k, d]]$ code exists.*

(a) If $k > 0$ then an $[[N + 1, k, d]]$ code exists.
(b) If the code is pure and $N \geq 2$ then an $[[N - 1, k + 1, d - 1]]$ code exists.
(c) If $k > 1$ or if $k = 1$ and the code is pure, then an $[[N, k - 1, d]]$ code exists.
(d) If $N \geq 2$ then an $[[N - 1, k, d - 1]]$ code exists.
(e) If $N \geq 2$ and the associated code C contains a vector of weight 1 then an $[[N - 1, k, d]]$ code exists.

Note in particular that, except in the $k = 0$ column, once a particular value of d has been achieved, the same value holds for all lower entries in the same column using part (a) of the above theorem.

The labels in the tables have the following meaning:

A. A code meeting this upper bound must be impure (this follows from integer programming).
B. A special upper bound given in [96, §7]. These bounds do not apply to nonadditive codes, for which the upper bound must be increased by 1.

Most of the lower bounds in the tables are obtained by exhibiting a specific quantum code. The following symbols indicate the self-orthogonal $[N, N - k, d]_{4+}$ trace-Hermitian code C used in the construction (cf. Theorem 13.2.4). All except those marked 'y' or 'z' are taken from [96].

Table 13.1. Cyclic quantum codes

Quantum code	Generators for additive code
$[[15,0,6]]$	$(\omega 11010100101011)$
$[[21,0,8]]$	$(\overline{\omega}\overline{\omega}1\omega 00111101011011000), (101110010111001011100)$
$[[21,5,6]]$	$(\omega 0\overline{\omega}\overline{\omega}\omega\overline{\omega}\overline{\omega}0\omega 100100001001), (101110010111001011100)$
$[[23,0,8]]$	$(\omega 0101111000000001111010)$
$[[23,12,4]]$	$(\overline{\omega}\overline{\omega}\omega\overline{\omega}\omega 11\overline{\omega}11\omega 1\omega 1011000000)$
$[[25,0,8]]$	$(111010\omega 010111000000000000)$

Table 13.2. Linear quasi-cyclic quantum codes

Quantum code	Generators for linear code
$[[14,0,6]]$	$(1000000)\ (\overline{\omega}1\overline{\omega}\omega 00\omega)$
$[[14,2,5]]$	$(1000001)\ (1\omega 101\omega 1)$
$[[14,8,3]]$	$(1011100)\ (1\overline{\omega}\omega\omega 10\overline{\omega})$
$[[15,5,4]]$	$(10000)\ (11\overline{\omega}00)\ (11\omega\omega 0)$
$[[18,6,5]]$	$(110000)\ (101\overline{\omega}00)\ (11\omega 1\omega 0)$
$[[20,10,4]]$	$(10000)\ (1\overline{\omega}100)\ (1111\omega)\ (11\overline{\omega}\omega\overline{\omega})$
$[[25,15,4]]$	$(10000)\ (1\omega 1\omega 0)\ (0101\overline{\omega})\ (1\omega\overline{\omega}\omega 1)\ (10\omega\omega 0)$
$[[28,14,5]]$	$(\omega\omega\overline{\omega}1000)\ (\overline{\omega}0\overline{\omega}1000)\ (1\overline{\omega}\overline{\omega}1\omega\overline{\omega}0)\ (\overline{\omega}\omega\overline{\omega}\omega\omega 00)$
$[[30,20,4]]$	$(11100)\ (10\omega 00)\ (11\overline{\omega}\omega 0)\ (1\omega 1\omega\overline{\omega})\ (10\omega 10)\ (1\omega 100)$
$[[40,30,4]]$	$(001\omega\omega)\ (011\omega 1)\ (0010\overline{\omega})\ (001\omega 1)\ (00101)$
	$(1\omega 1\omega\overline{\omega})\ (111\overline{\omega}\omega)\ (01\omega 1\overline{\omega})$

a. The $[6,3,4]_4$ hexacode h_6 (§2.4.6).
b. A self-dual code of Type 4^{H}.
c. A cyclic additive code from the list in Table 13.1.
d. A $[[25,1,9]]$ code obtained by concatenating ([96, §4]) the $[[5,1,3]]$ code of Example 13.1.7 with itself.
e. The $[12,6,6]_{4+}$ dodecacode z_{12} (§2.4.8).
f. An $[[8,3,3]]$ code, discovered independently by Calderbank, Rains, Shor and Sloane [95], Gottesman [197] and Steane [514]. The $[8,2.5,4]_{4+}$ additive code may be generated by the vectors $(01\omega\omega\overline{\omega}1\overline{\omega})0$, 11111111, $\omega\omega\omega\omega\omega\omega\omega\omega$. Exhaustive search shows that this $\mathbb{F}_4$ code is essentially unique.
g. A quasi-cyclic linear code found by Gulliver [207] (see also Gulliver and Kim [222]), from the list in Table 13.2. For quasi-cyclic codes we use the notation introduced in §2.4, where the symbols inside the brackets have to be permuted simultaneously.

h. A quantum Hamming code, as mentioned above.

i. From the $[12, 4, 6]_4$ and $[14, 4, 8]_4$ linear codes with generator matrices

$$\begin{bmatrix} 0\,0\,0\,0\,0\,0\,1\,1\,1\,1\,1\,1 \\ 0\,0\,1\,1\,1\,1\,0\,0\,1\,1\,\omega\,\omega \\ 0\,1\,0\,1\,\omega\,\overline{\omega}\,0\,1\,0\,\omega\,1\,\omega \\ 1\,0\,0\,1\,\overline{\omega}\,\omega\,0\,1\,\omega\,0\,\omega\,1 \end{bmatrix}$$

and

$$\begin{bmatrix} 0\,0\,0\,0\,0\,0\,1\,1\,1\,1\,1\,1\,1\,1 \\ 0\,0\,1\,1\,1\,1\,0\,0\,0\,0\,1\,1\,1\,1 \\ 0\,1\,0\,1\,\omega\,\overline{\omega}\,0\,1\,\omega\,\overline{\omega}\,0\,1\,\omega\,\overline{\omega} \\ 1\,0\,0\,1\,\overline{\omega}\,\omega\,0\,1\,\overline{\omega}\,\omega\,1\,0\,\omega\,\overline{\omega} \end{bmatrix}$$

respectively. Their weak automorphism groups have orders 720 and 8064, respectively, and both groups act transitively on the coordinates. The first of these can be obtained from the $u|u + v$ construction (cf. [96, Theorem 12]) applied to the unique $[[6, 4, 2]]$ and $[[6, 0, 4]]$ codes.

j. A $[[17, 9, 4]]$ code, for which the corresponding $[17, 4, 12]_4$ code C is a two-weight linear code of class TF3 (Calderbank and Kantor [93]). The columns of the generator matrix of C represent the 17 points of an ovoid in $PG(3, 4)$. Both C and $C^\perp$ are cyclic, a generator for $C^\perp$ being $1\omega1\omega10^{12}$. The weight distribution of C is $A_0 = 1$, $A_{12} = 204$, $A_{16} = 51$, and its weak automorphism group has order 48960.

k. The $[[16, 4, 5]]$ extended cyclic code spanned by

$$(\overline{\omega}\,\overline{\omega}\,0\,\omega\,1\,\omega\,111100111)0,$$

together with the vectors of all 1's and all ω's.

s. *Shortening* an existing code using the following theorem or its analogue for additive codes:

Theorem 13.6.2. (Calderbank, Rains, Shor and Sloane [96, Th. 7].) *Suppose a linear $[[N, k, d]]$ code exists with associated $[N, (N - k)/2, d']_4$ code C. Then there exists a linear $[[N - m, k', d']]$ code with $k' \geq k - m$ and $d' \geq d$, for any m such that there exists a codeword of weight m in the dual of the binary code generated by the supports of the codewords of C.*

We apply this to the following codes: the $[[21, 15, 3]]$ or $[[85, 77, 3]]$ Hamming codes, the $[[32, 25, 3]]$ Gottesman code ([96, Th. 10]), or the $[[40, 30, 4]]$ code given in Table 13.2.

u. From the $u|u + v$ construction ([96, Theorem 12]).

v. The following $[17, 3, 12]_{4+}$ code with trivial automorphism group, found by random search:

Table 13.3. Highest minimal distance d in an $[[N, k, d]]$ additive binary quantum code. The symbols are explained in the text.

$N \setminus k$	0	1	2	3	4	5	6	7
3	2	1	1	1				
4	2	2	2	1	1			
5	3	$^h3$	2	1	1	1		
6	$^a4$	3^A	2	2	2	1	1	
7	3^B	$^s3$	2	2	2	1	1	1
8	$^b4$	$^s3$	$^s3$	$^f3$	2	2	2	1
9	4	$^s3$	$^s3$	$^s3$	2	2	2	1
10	$^b4$	4	4	$^s3$	$^s3$	2	2	2
11	5	5	4	$^s3$	$^s3$	$^s3$	2	2
12	$^e6$	5^A	4	4	$^i4$	$^s3$	$^s3$	2
13	5^B	5	4	4	4	$3-4$	$^s3$	$^s3$
14	$^b6$	5	$^g5$	$4-5$	4	4	$^i4$	$^s3$
15	$^c6$	5	5	5	$^g4^B$	4	4	$^s3^B$
16	$^b6$	6	6	5	$^k5$	$4-5$	4	$3-4$
17	7	7	6	$5-6$	5	$4-5$	$4-5$	4
18	$^b8$	7	6	$5-6$	$5-6$	5	$^g5$	4
19	7^B	7	6	$5-6$	$5-6$	$5-6$	5	$4-5$
20	$^b8$	7	$6-7$	$5-7$	$5-6$	$5-6$	$5-6$	$4-5$
21	$^c8$	7	$6-7$	$6-7$	$6-7$	$^c6$	$5-6$	$^c5-6^A$
22	$^b8$	$7-8$	$6-8$	$6-7$	$6-7$	$6-7$	$5-6$	$5-6$
23	$^c8-9$	$7-9$	$6-8$	$6-8$	$6-7$	$6-7$	$5-7$	$5-6$
24	$^b8-10$	$8-9^A$	$^y7-8$	$7-8$	$6-8$	$6-7$	$6-7$	$5-7$
25	$^c8-9^B$	$^d9$	$7-8$	$7-8$	$7-8$	$7-8$	$6-7$	$5-7$
26	$8-10$	9	$8-9$	$8-9$	8	$7-8$	$6-8$	$5-8$
27	$9-10$	9	9	9	$8-9$	$7-8$	$6-8$	$5-8$
28	10	10	10	9	$8-9$	$7-9$	$6-8$	$6-8$
29	11	11	10	$9-10$	$8-9$	$7-9$	$6-9$	$6-8$
30	$^b12$	11^A	10	$9-10$	$8-10$	$7-9$	$7-9$	$7-9$

Table 13.4. Highest minimal distance d in an $[[N, k, d]]$ additive binary quantum code, continued.

$N \backslash k$	8	9	10	11	12	13	14	15
8	1							
9	1	1						
10	2	1	1					
11	2	1	1	1				
12	2	2	2	1	1			
13	2	2	2	1	1	1		
14	$^s3$	2	2	2	2	1	1	
15	$^s3$	$^s3$	2	2	2	1	1	1
16	$^s3^B$	$^s3$	$^s3$	2	2	2	2	1
17	4	$^j4$	$^s3$	$^v3$	2	2	2	1
18	4	4	$^s3$	$^s3$	2^B	2	2	2
19	4^B	4	$3-4$	$^s3$	$^s3$	2^B	2	2
20	$4-5$	4	$^g4$	$3-4$	$^s3$	$^s3$	2	2
21	$4-5$	$4-5$	4	$^s4$	$3-4$	$^s3$	$^s3$	$^h3$
22	$4-6$	$4-5$	$4-5$	4	$^s4$	$3-4$	$^s3^B$	$^s3$
23	$4-6$	$4-6$	$4-5$	$4-5$	$^c4$	$^s4$	$3-4$	$^s3$
24	$^z5-6$	$4-6$	$4-6$	$4-5$	$4-5$	4	$^s4$	$3-4$
25	$5-7$	$4-6$	$4-6$	$4-6$	$4-5$	$4-5$	4	$^g4$
26	$5-7$	$5-7$	$5-6$	$5-6$	$^z5-6$	$4-5$	$4-5$	4
27	$5-8$	$5-7$	$5-7$	$5-6$	$5-6$	$4-5$	$4-5$	$4-5$
28	$^u6-8$	$6-8$	$^z6-7$	$5-7$	$5-6$	$5-6$	$^g5-6$	$4-5$
29	$6-8$	$6-8$	$6-7$	$5-7$	$5-6$	$5-6$	$5-6$	$4-5$
30	$^y7-8$	$6-8$	$6-8$	$^y6-7$	$^z6-7$	$5-6$	$5-6$	$5-6$

$N \backslash k$	16	17	18	19	20	21	22	23
16	1							
17	1	1						
18	2	1	1					
19	2	1	1	1				
20	2	2	2	1	1			
21	2	2	2	1	1	1		
22	2	2	2	2	2	1	1	
23	$^s3$	2	2	2	2	1	1	1
24	$^s3$	$^s3$	2	2	2	2	2	1
25	$3-4$	$^s3$	$^s3$	2	2	2	2	1
26	$^s4$	$3-4$	$^s3$	$^s3$	2	2	2	2
27	4	$^s4$	$3-4$	$^s3$	$^s3$	2	2	2
28	4	4	$^s4$	$3-4$	$^s3$	$^s3$	2	2
29	$4-5$	4	4	$3-4$	$3-4$	$^s3$	$^s3$	2
30	$^z5$	$4-5$	4	4	$^g4$	$3-4$	$^s3$	$^s3$

$$
\begin{bmatrix}
0\,0\,1\,0\,\omega\overline{\omega}\omega\overline{\omega}\,1\,1\,\omega\overline{\omega}\,0\,0\,1\,1\,\overline{\omega} \\
0\,0\,\omega\,1\,0\,\omega\,0\,\overline{\omega}\overline{\omega}\overline{\omega}\,1\,1\,\omega\overline{\omega}\overline{\omega}\,1\,1 \\
0\,1\,0\,0\,1\,\omega\,1\,\omega\overline{\omega}\overline{\omega}\overline{\omega}\,0\,\overline{\omega}\,1\,\omega\,0\,\overline{\omega} \\
0\,\omega\,0\,\omega\,\omega\,0\,\overline{\omega}\,1\,\overline{\omega}\,1\,\omega\overline{\omega}\omega\,1\,\omega\omega\,1 \\
1\,0\,0\,\omega\overline{\omega}\,0\,0\,1\,\omega\omega\overline{\omega}\,1\,\overline{\omega}\omega\,0\,\overline{\omega}\,1 \\
\omega\,0\,0\,1\,\overline{\omega}\overline{\omega}\overline{\omega}\,0\,\overline{\omega}\,0\,\overline{\omega}\,1\,0\,1\,1\,\omega\overline{\omega}
\end{bmatrix}.
$$

y. The following cyclic codes found by Grassl [198]: $[[24, 3, 7]]$, $[[30, 8, 7]]$, $[[30, 11, 6]]$.

z. The following quasi-cyclic codes found by Varbanov [198]: $[[24, 8, 5]]$, $[[26, 12, 5]]$, $[[30, 10, 6]]$, $[[30, 12, 6]]$, $[[30, 16, 5]]$.

Comparison of these tables of quantum codes with the existing tables of classical codes over $\mathbb{F}_4$ (e.g. Brouwer [84]) reveals a number of entries where it may be possible to improve the lower bounds. For example, classical linear $[30, 18, 8]_4$ codes certainly exist. If such a code can be found which contains its dual, we would obtain a $[[30, 6, 8]]$ quantum code.

References

The bibliography uses the following abbreviations:

arXiv = The arXiv eprint archive at http://arXiv.org/
DCC = *Designs, Codes and Cryptography*
DM = *Discrete Mathematics*
JCT = *Journal of Combinatorial Theory*
PGIT = *IEEE Transactions on Information Theory*

1. M. Aaltonen, Linear programming bounds for tree codes, *PGIT* **25** (1979), 85–90.
2. M. Aaltonen, A new upper bound on nonbinary block codes, *DM* **83** (1990), 139–160.
3. A. V. Alekseevskii, Finite commutative Jordan subgroups of complex simple Lie groups, *Functional Anal. Appl.* **8** (1974), 277–279.
4. O. Amrani and Y. Beéry, Reed-Muller codes: projections onto GF(4) and multilevel construction, *PGIT* **47** (2001), 2560–2565.
5. O. Amrani, Y. Beéry and A. Vardy, Bounded-distance decoding of the Leech lattice and the Golay code, in *Algebraic Coding (Paris, 1993)*, Lecture Notes Comput. Sci. **781** (1994), 236–248.
6. O. Amrani, Y. Beéry, A. Vardy, F.-W. Sun and H. C. A. van Tilborg, The Leech lattice and the Golay code: bounded-distance decoding and multilevel constructions, *PGIT* **40** (1994), 1030–1043.
7. J. B. Anderson, *Digital Transmission Engineering* , IEEE Press and Prentice-Hall, NY, 1998.
8. J. L. Anderson, On minimal decoding sets for the extended binary Golay code, *PGIT* **38** (1992), 1560–1561.
9. A. N. Andrianov, *Quadratic Forms and Hecke Operators*, Springer, 1987.
10. T. Aoki, P. Gaborit, M. Harada, M. Ozeki and P. Solé, On the covering radius of $\mathbb{Z}_4$-codes and their lattices, *PGIT* **45** (1999), 2162–2168.
11. K. T. Arasu and T. A. Gulliver, Self-dual codes over $\mathbb{F}_p$ and weighing matrices, *PGIT* **47** (2001) 2051–2055.

392   References

12. M. Araya and M. Harada, MDS codes over F_9 related to the ternary Golay code, *DM* **282** (2004) 233–237.
13. A. Ashikhmin and E. Knill, Nonbinary quantum stabilizer codes, *PGIT* **47** (2001) 3065–3072.
14. A. Ashikhmin and S. Litsyn, Upper Bounds on the size of quantum codes, *PGIT* **45** (1999), 1206–1216.
15. E. F. Assmus, Jr., H. F. Mattson, Jr. and R. J. Turyn, Research to develop the algebraic theory of codes, *Report AFCRL-67-0365*, Air Force Cambridge Res. Labs., Bedford, MA, June 1967.
16. E. F. Assmus, Jr. and V. S. Pless, On the covering radius of extremal self-dual codes, *PGIT* **29** (1983), 359–363.
17. A. O. L. Atkin and J. Lehner, Hecke operators on $\Gamma_0(m)$, *Math. Ann.* **185** (1970), 134–160.
18. A. Baartmans and V. Y. Yorgov, Some new extremal codes of lengths 76 and 78, *PGIT* **49** (2003), 1353–1354.
19. C. Bachoc, Applications of coding theory to the construction of modular lattices, *JCT* **A 78** (1997), 92–119.
20. C. Bachoc, On harmonic weight enumerators of binary codes, *DCC* **18** (1999), 11–28.
21. C. Bachoc, Harmonic weight enumerators of non-binary codes and MacWilliams identities, in *Codes and association schemes (Piscataway, NJ, 1999)*, DIMACS Ser. Discrete Math. Theoret. Comput. Sci. **56**, Amer. Math. Soc., Providence, RI, 2001; pp. 1–23.
22. C. Bachoc, Designs, groups and lattices, *J. Théorie Nombres Bordeaux*, to appear, 2005.
23. C. Bachoc and P. Gaborit, On extremal additive $\mathbb{F}_4$ codes of length 10 to 18, in *International Workshop on Coding and Cryptography (Paris, 2001), Electron. Notes Discrete Math.* **6** (2001), 10 pp.
24. C. Bachoc and P. Gaborit, Designs and self-dual codes with long shadows, JCT **A 105** (2004), 15–34.
25. B. Bajnok, Construction of spherical t-designs, *Geom. Dedicata*, **43** (1992), 167–179.
26. B. Bajnok, Construction of spherical 3-designs, *Graphs and Combinatorics*, **14** (1998), 97–107.
27. A. Bak, *K-Theory of Forms*, Princeton Univ. Press, 1981.
28. E. Bannai, S. T. Dougherty, M. Harada and M. Oura, Type II codes, even unimodular lattices and invariant rings, *PGIT* **45** (1999), 1194–1205.
29. E. S. Barnes and G. E. Wall, Some extreme forms defined in terms of Abelian groups, *J. Australian Math. Soc.* **1** (1959), 47–63.
30. H.-J. Bartels, Zur Galoiskohomologie definiter arithmetischer Gruppen, *J. Reine Angew. Math.* **298** (1978), 89–97.
31. G. F. M. Beenker, A note on extended quadratic residue codes over $GF(9)$ and their ternary images, *PGIT* **30** (1984), 403–405.
32. C. H. Bennett, D. DiVincenzo, J. A. Smolin and W. K. Wootters, Mixed state entanglement and quantum error correction, *Phys. Rev. A* **54** (1996), 3824–3851.
33. D. J. Benson, *Polynomial Invariants of Finite Groups*, Cambridge Univ. Press, 1993.
34. E. R. Berlekamp, F. J. MacWilliams and N. J. A. Sloane, Gleason's theorem on self-dual codes, *PGIT* **18**, (1972), 409–414.

35. K. Betsumiya, On the classification of Type II codes over $\mathbb{F}_{2^r}$ with binary length 32, Preprint, 2002.

36. K. Betsumiya and Y.-J. Choie, Jacobi forms over totally real fields and type II codes over Galois rings $GR(2^m, f)$, *European J. Combin.* **25** (2004), 475–486.

37. K. Betsumiya and Y.-J. Choie, Codes over $\mathbb{F}_4$, Jacobi forms and Hilbert-Siegel modular forms over $\mathbb{Q}(\sqrt{5})$, *European J. Combin.* **26** (2005), 629–650.

38. K. Betsumiya, S. Georgiou, T. A. Gulliver, M. Harada and C. Koukouvinos, On self-dual codes over some prime fields, *DM* **262** (2003), 37–58.

39. K. Betsumiya, T. A. Gulliver and M. Harada, Binary optimal linear rate 1/2 codes, in *Applied Algebra, Algebraic Algorithms and Error-Correcting Codes (Honolulu, HI, 1999)*, Lecture Notes Comput. Sci. **1719** (1999), pp. 462–471.

40. K. Betsumiya, T. A. Gulliver and M. Harada, On binary extremal formally self-dual even codes, *Kyushu J. Math.* **53** (1999), 421–430.

41. K. Betsumiya, T. A. Gulliver and M. Harada, Extremal self-dual codes over $\mathbb{F}_2 \times \mathbb{F}_2$, *DCC* **28** (2003), 171–186.

42. K. Betsumiya, T. A. Gulliver, M. Harada and A. Munemasa, On type II codes over $\mathbb{F}_4$, *PGIT* **47** (2001), 2242–2248.

43. K. Betsumiya and M. Harada, Binary optimal odd formally self-dual codes, *DCC* **23** (2001), 11–21.

44. K. Betsumiya and M. Harada, Classification of formally self-dual even codes of lengths up to 16, *DCC* **23** (2001), 325–332.

45. K. Betsumiya and M. Harada, Optimal self-dual codes over $\mathbb{F}_2 \times \mathbb{F}_2$ with respect to the Hamming weight, *PGIT* **50** (2004), 356–358.

46. K. Betsumiya, M. Harada and A. Munemasa, Type II codes over $\mathbb{F}_{2^r}$, in *Applied Algebra, Algebraic Algorithms and Error-Correcting Codes (Melbourne, 2001)*, Lecture Notes Comput. Sci. **2227** (2001), pp. 102-111.

47. K. Betsumiya, S. Ling and F. R. Nemenzo, Type II codes over $\mathbb{F}_{2^m} + u\mathbb{F}_{2^m}$, *DM* **275** (2004), 43–65.

48. V. K. Bhargava and C. Nguyen, Circulant codes based on the prime 29, *PGIT* **26** (1980), 363–364.

49. R. T. Bilous, Enumeration of binary self-dual codes of length 34, *J. Combin. Math. Combin. Computing*, to appear, 2005.

50. R. T. Bilous and G. H. J. van Rees, An enumeration of binary self-dual codes of length 32, *DCC* **26** (2002), 61–86.

51. I. F. Blake, Properties of generalized Pless codes, in *Proc. 12th Allerton Conf. Circuit and System Theory*, Univ. Ill., Urbana, 1974, pp. 787–789.

52. I. F. Blake, On a generalization of the Pless symmetry codes, *Inform Control* **27** (1975), 369–373.

53. M. Blaum and J. Bruck, Decoding the Golay code with Venn diagrams, *PGIT* **36** (1990), 906–910.

54. F. van der Blij, An invariant of quadratic forms mod 8, *Indag. Math.* **21** (1959), 291–293.

55. S. Böcherer, On the notion of extremal modular forms and analytically extremal lattices, Preprint, 2005.

56. S. Böcherer, Siegel modular forms and theta series, in *Theta functions (Bowdoin 1987)*, Proc. Sympos. Pure Math., **49**, Part 2, Amer. Math. Soc., Providence, RI, 1989, pp. 3–17.

57. B. Bolt, The Clifford collineation, transform and similarity groups III: generators and involutions, *J. Australian Math. Soc.* **2** (1961), 334–344.

58. B. Bolt, T. G. Room and G. E. Wall, On Clifford collineation, transform and similarity groups I, *J. Australian Math. Soc.* **2** (1961), 60–79.

59. B. Bolt, T. G. Room and G. E. Wall, On Clifford collineation, transform and similarity groups II, *J. Australian Math. Soc.* **2** (1961), 80–96.

60. A. Bonnecaze, A. R. Calderbank and P. Solé, Quaternary quadratic residue codes and unimodular lattices, *PGIT* **41** (1995), 366–377.

61. A. Bonnecaze, Y.-J. Choie, S. T. Dougherty and P. Solé, Splitting the shadow, *DM* **270** (2003), 43–60.

62. A. Bonnecaze, A. Desidiri Bracco, S. T. Dougherty, L. R. Nochefranca and P. Solé, Cubic self-dual binary codes, *PGIT* **49** (2003), 2253–2259.

63. A. Bonnecaze, P. Gaborit, M. Harada, M. Kitazume and P. Solé, Niemeier lattices and Type II codes over $\mathbb{Z}_4$, *DM* **205** (1999), 1–21.

64. A. Bonnecaze, B. Mourrain and P. Sol'e, Jacobi polynomials, type II codes, and designs, *DCC* **16** (1999), 215–234.

65. A. Bonnecaze, E. M. Rains and P. Solé, 3-colored 5-designs and Z_4-codes, *J. Statist. Plann. Inference* 86 (2000), 349–368.

66. A. Bonnecaze, P. Solé, C. Bachoc and B. Mourrain, Type II codes over $\mathbb{Z}_4$, *PGIT* **43** (1997), 969–976.

67. R. E. Borcherds, Automorphic forms with singularities on Grassmannians, *Invent. Math.* **132** (1998), 491–562 [arXiv: alg-geom/9609022].

68. W. Bosma, J. Cannon and C. Playoust, The Magma algebra system I: The user language, *J. Symb. Comp.*, **24** (1997), 235–265.

69. M. Bossert, On decoding binary quadratic residue codes, in *Applied algebra, algebraic algorithms and error-correcting codes (Menorca, 1987)*, Lecture Notes Comput. Sci. **356** (1989), 60–68.

70. I. Bouyukliev, S. Bouyuklieva, T. A. Gulliver and P. R. J. Östergård, Classification of optimal binary self-orthogonal codes, *J. Combin. Math. Combin. Computing*, to appear, 2005.

71. I. Bouyukliev and P. R. J. Östergård, Classification of self-orthogonal codes over $\mathbb{F}_3$ and $\mathbb{F}_4$, *SIAM J. Discrete Mathematics*, to appear, 2005.

72. S. Bouyuklieva, New extremal self-dual codes of lengths 42 and 44, *PGIT* **43** (1997), 1607–1612.

73. S. Bouyuklieva, On the binary self-dual codes with an automorphism of order 2, *DCC* **12** (1997), 39–48.

74. S. Bouyuklieva, On the automorphisms of order 2 with fixed points for the extremal self-dual codes of length $24m$, *DCC* **25** (2002), 5–13.

75. S. Bouyuklieva, On the automorphism group of a doubly-even (72, 36, 16) code, *PGIT* **50** (2004), 544–547.

76. S. Bouyuklieva, Some optimal self-orthogonal and self-dual codes, *DM* **287** (2004), 1–10.

77. S. Bouyuklieva and I. Bouyukliev, Extremal self-dual codes with an automorphism of order 2, *PGIT* **44** (1998), 323–328.

78. S. Bouyuklieva and M. Harada, On type IV self-dual codes over Z4, *DM* **247** (2002), 25–50.

79. S. Bouyuklieva and M. Harada, Extremal self-dual [50, 25, 10] codes with automorphisms of order 3 and quasi-symmetric 2-(49, 9, 6) designs, *DCC* **28** (2003), 163–169.

80. S. Bouyuklieva and V. Y. Yorgov, Singly-even self-dual codes of length 40, *DCC* **9** (1996), 131–141.

81. H. Braun, Geschlecter quadratischer Formen, *J. Reine Angew. Math.* **182** (1940), 32–49.

82. M. Broué and M. Enguehard, Polynômes des poids de certains codes et fonctions thêta de certains réseaux, *Ann. scient. Éc. Norm. Sup.* 4e série, **5** (1972), 157–181.

83. M. Broué and M. Enguehard, Une famille infinie de formes quadratiques entières; leurs groupes d'automorphismes, *Ann. scient. Éc. Norm. Sup.* 4e série, **6** (1973), 17–52. Summary in *C. R. Acad. Sc. Paris* **274** (1972), 19–22.

84. A. E. Brouwer, Bounds on the size of linear codes, Chapter 4 of [426], pp. 295–461.

85. A. E. Brouwer, A. M. Cohen and A. Neumaier, *Distance-Regular Graphs*, Springer, 1989.

86. R. A. Brualdi and V. S. Pless, Weight enumerators of self-dual codes, *PGIT* **37** (1991), 1222–1225.

87. J. H. Bruinier, Borcherds products on $O(2, l)$ and Chern classes of Heegner divisors, Lecture Notes Math. **1780**, 2002.

88. F. C. Bussemaker and V. D. Tonchev, New extremal doubly-even codes of length 56 derived from Hadamard matrices of order 28, *DM* **76** (1989), 45–49.

89. F. C. Bussemaker and V. D. Tonchev, Extremal doubly-even codes of length 40 derived from Hadamard matrices, *DM* **82** (1990), 317–321.

90. A. R. Calderbank, P. J. Cameron, W. M. Kantor and J. J. Seidel, $\mathbb{Z}_4$-Kerdock codes, orthogonal spreads and extremal Euclidean line-sets, *Proc. London Math. Soc.* **75** (1997), 436–480.

91. A. R. Calderbank, A. R. Hammons, Jr., P. V. Kumar, N. J. A. Sloane and P. Solé, A linear construction for certain Kerdock and Preparata codes, *Bull. Amer. Math. Soc.* **29** (1993), 218–222.

92. A. R. Calderbank, R. H. Hardin, E. M. Rains, P. W. Shor and N. J. A. Sloane, A group-theoretic framework for the construction of packings in Grassmannian spaces, *J. Algebraic Combin.* **9** (1999), 129–140 [arXiv: math.CO/0208002].

93. A. R. Calderbank and W. M. Kantor, The geometry of two-weight codes, *Bull. London Math. Soc.*, **118** (1986), 97–122.

94. A. R. Calderbank, W.-C. W. Li and B. Poonen, A 2-adic approach to the analysis of cyclic codes, *PGIT* **43** (1997), 977–986.

95. A. R. Calderbank, E. M. Rains, P. W. Shor and N. J. A. Sloane, Quantum error correction and orthogonal geometry, *Phys. Rev. Lett.* **78** (1997), 405–409 [arXiv: quant-ph/9605005].

96. A. R. Calderbank, E. M. Rains, P. W. Shor and N. J. A. Sloane, Quantum error correction via codes over GF(4), *PGIT* **44**, (1998), 1369–1387 [arXiv: quant-ph/9608006].

97. A. R. Calderbank and P. W. Shor, Good quantum error-correcting codes exist, *Phys. Rev. A*, **54**, pp. 1098–1105 (1996) [arXiv: quant-ph/9512032].

98. A. R. Calderbank and N. J. A. Sloane, Modular and p-adic cyclic codes, *DCC* **6** (1995), 21–35 [arXiv: math.CO/0311319].

99. A. R. Calderbank and N. J. A. Sloane, Double circulant codes over $\mathbb{Z}_4$ and even unimodular lattices, *J. Algebraic Combinatorics* **6** (1997), 119–131.

100. J. Cannon et al., *The Magma Computational Algebra System for Algebra, Number Theory and Geometry*, published electronically at http://magma.maths.usyd.edu.au/magma/.

101. J.-C. Carlach and A. Otmani, A systematic construction of self-dual codes, *PGIT* **49** (2003), 3005–3009.

102. C. Carlet, $\mathbb{Z}_{2^k}$-linear codes, *PGIT* **44** (1998), 1543–1547.

103. J. W. S. Cassels, *Rational Quadratic Forms*, Academic Press, NY, 1978.

104. R. Chapman, S. T. Dougherty, P. Gaborit and P. Solé, 2−modular lattices from ternary codes, *J. Théorie Nombres Bordeaux* **14** (2002), 1–13.

105. G. Chen and R. K. Brylinski, eds., *Mathematics of Quantum Computation*, Chapman and Hall, NY, 2002.

106. Y. Cheng and R. Scharlau, personal communication, Sept., 1987.

107. Y. Cheng and N. J. A. Sloane, The automorphism group of an [18,9,8] quaternary code, *DM* **83** (1990), 205–212.

108. C. Chevalley, Invariants of finite groups generated by reflections, *Amer. J. Math.* **67** (1955), 778–782.

109. Y.-J. Choie and S. T. Dougherty, Codes over Σ_{2m} and Jacobi forms over the quaternions, *Appl. Algebra Engrg. Comm. Comput.* **15** (2004), 129–147.

110. Y.-J. Choie and S. T. Dougherty, Codes over rings, complex lattices and Hermitian modular forms, *European J. Combin.* **26** (2005), 145–165.

111. Y.-J. Choie, S. T. Dougherty and H. Kim, Complete joint weight enumerators and self-dual codes, *PGIT* **49** (2003), 1275-1282.

112. Y.-J. Choie and E. Jeong, Jacobi forms over totally real fields and codes over $\mathbb{F}_p$, *Illinois J. Math.* **46** (2002), 627–643.

113. Y.-J. Choie and N. Kim, The complete weight enumerator of type II codes over $\mathbb{Z}_{2m}$ and Jacobi forms, *PGIT* **47** (2001), 396–399.

114. Y-J. Choie, B. Mourrain and P. Solé, Rankin-Cohen brackets and invariant theory, *J. Algebraic Combinatorics* **13** (2001), 5–13.

115. Y-J. Choie and P. Solé, A Gleason formula for Ozeki polynomials, *JCT* **A 98** (2002), 60–73.

116. Y-J. Choie and P. Solé, Self-dual codes over $\mathbb{Z}_4$ and half-integral weight modular forms, *Proc. Amer. Math. Soc.* **130** (2002), 3125–3131.

117. Y-J. Choie and P. Solé, Ternary codes and Jacobi forms, *DM* **282** (2004), 81–87.

118. R. Cleve, Quantum stabilizer codes and classical linear codes, *Phys. Rev. A* **55** (1997), 4054–4059.

119. H. Cohen, *A Course in Computational Algebraic Number Theory*, Springer, 1996.

120. J. H. Conway, R. H. Hardin and N. J. A. Sloane, Packing lines, planes, etc.: packings in Grassmannian space, *Experimental Math.* **5** (1996), 139–159 [arXiv: math.CO/0208004].

121. J. H. Conway and V. S. Pless, On the enumeration of self-dual codes, *JCT* **A 28** (1980), 26–53.

122. J. H. Conway and V. S. Pless, On primes dividing the group order of a doubly-even (72, 36, 16) code and the group of a quaternary (24, 12, 10) code, *DM* **38** (1982), 157–162.

123. J. H. Conway, V. S. Pless and N. J. A. Sloane, Self-dual codes over $GF(3)$ and $GF(4)$ of length not exceeding 16, *PGIT* **25** (1979), 312–322.

124. J. H. Conway, V. S. Pless and N. J. A. Sloane, The binary self-dual codes of length up to 32: A revised enumeration, *JCT* **A 60** (1992), 183–195.

125. J. H. Conway and N. J. A. Sloane, On the enumeration of lattices of determinant one, *J. Number Theory* **15** (1982), 83–94.

126. J. H. Conway and N. J. A. Sloane, Soft decoding techniques for codes and lattices, including the Golay code and the Leech lattice, *PGIT* **32** (1986), 41–50.

127. J. H. Conway and N. J. A. Sloane, Low-dimensional lattices I: Quadratic forms of small determinant, Proc. Royal Soc. **A 418** (1988), 17–41.

128. J. H. Conway and N. J. A. Sloane, Low-dimensional lattices II: Subgroups of $GL(n, \mathbb{Z})$, Proc. Royal Soc. **A 419** (1988), 29–68.

129. J. H. Conway and N. J. A. Sloane, A new upper bound for the minimum of an integral lattice of determinant one, *Bull. Amer. Math. Soc.* **23** (1990), 383–387; erratum: **24** (1991), 479.

130. J. H. Conway and N. J. A. Sloane, A new upper bound on the minimal distance of self-dual codes, *PGIT* **36** (1990), 1319–1333.

131. J. H. Conway and N. J. A. Sloane, Self-dual codes over the integers modulo 4, *JCT* **A 62** (1993), 30–45.

132. J. H. Conway and N. J. A. Sloane, On lattices equivalent to their duals, *J. Number Theory* **48** (1994), 373–382.

133. J. H. Conway and N. J. A. Sloane, *Sphere Packings, Lattices and Groups*, Springer, 1998, 3rd. ed., 1998.

134. D. Coppersmith and G. Seroussi, On the minimum distance of some quadratic residue codes, *PGIT* **30** (1984), 407–411.

135. H. S. M. Coxeter, *Regular Complex Polytopes* Cambridge Univ. Press, 2nd. ed., 1991.

136. D. B. Dalan, New extremal type I codes of lengths 40, 42 and 44, *DCC* **30** (2003), 151–157.

137. D. B. Dalan, New extremal binary [44, 22, 8] codes, *PGIT* **49** (2003), 747–748.

138. D. B. Dalan, Extremal binary self-dual codes of lengths 42 and 44 with new weight enumerators, *Kyushu J. Math.* **57** (2003), 333–345.

139. L. E. Danielsen and M. G. Parker, On the classification of all self-dual additive codes over GF(4) of length up to 12, Preprint, 2005.

140. F. N. David, M. G. Kendall and D. E. Barton, *Symmetric Function and Allied Tables*, Cambridge Univ. Press, 1966.

141. E. Dawson, Self-dual ternary codes and Hadamard matrices, *Ars Comb.* **19A** (1985), 303–308.

142. P. Delsarte, Bounds for unrestricted codes, by linear programming, *Philips Res. Reports* **27** (1972), 272–289.

143. P. Delsarte, J.-M. Goethals and J. J. Seidel, Spherical codes and designs, *Geom. Dedicata*, **6** (1977), 363–388.

144. H. Derksen and G. Kemper, *Computational Invariant Theory*, Springer, 2002.

145. S. M. Dodunekov, V. A. Zinoviev and J. E. M. Nilsson, On the algebraic decoding of some maximal quaternary codes and binary Golay codes (Russian), *Probl. Pered. Inform.* **35** (1999), no. 4, 59–67; *Problems Inform. Transmission* **35** (1999), no. 4, 338–345.

146. R. Dontcheva, New binary [70, 35, 12] self-dual and binary [72, 36, 12] self-dual doubly-even codes, *Serdica Math. J.* **27** (2001), 287–302.

147. R. Dontcheva and M. Harada, New extremal self-dual codes of length 62 and related extremal self-dual codes, *PGIT* **48** (2002), 2060–2064.

148. R. Dontcheva and M. Harada, Some extremal self-dual codes with an automorphism of order 7, *Appl. Algebra Engrg. Comm. Comput.* **14** (2003), 75–79.

149. R. Dontcheva and M. Harada, Extremal doubly-even [80,40,16] codes with an automorphism of order 19, *Finite Fields Appl.* **9** (2003), 157–167.

150. R. Dontcheva, A. van Zanten and S. Dodunekov, Binary self-dual codes with automorphisms of composite order, *PGIT* **50** (2004), 311–318.

151. S. T. Dougherty, Shadow codes and weight enumerators, *PGIT* **41** (1995), 762–768.

152. S. T. Dougherty, P. Gaborit, M. Harada, A. Munemasa and P. Solé, Type IV self-dual codes over rings, *PGIT* **45** (1999), 2345–2360.

153. S. T. Dougherty, T. A. Gulliver and M. Harada, Extremal binary self-dual codes, *PGIT* **43** (1997), 2036–2046.

154. S. T. Dougherty, T. A. Gulliver and M. Harada, Type II self-dual codes over finite rings and even unimodular lattices, *J. Alg. Combin.* **9** (1999), 233–250.

155. S. T. Dougherty and M. Harada, Shadow optimal self-dual codes, *Kyushu J. Math.* **53** (1999), 223–237.

156. S. T. Dougherty and M. Harada, New extremal self-dual codes of length 68, *PGIT* **45** (1999), 2133–2136.

157. S. T. Dougherty, M. Harada, P. Gaborit and P. Solé, Type II codes over $\mathbb{F}_2 + u\mathbb{F}_2$, *PGIT* **45** (1999), 32–45.

158. S. T. Dougherty, M. Harada and M. Oura, Note on the g-fold joint weight enumerators of self-dual codes over $\mathbb{Z}_k$, *Appl. Algebra Engrg. Comm. Comput.* **11** (2001), 437–445.

159. S. T. Dougherty, M. Harada and P. Solé, Shadow lattices and shadow codes, *DM* **219** (2000), 49–64.

160. S. T. Dougherty, M. Harada and P. Solé, Shadow codes over $\mathbb{Z}_4$, *Finite Fields Applic.* **7** (2001), 507–529.

161. W. Duke, On codes and Siegel modular forms, *Internat. Math. Res. Notices* **5** (1993), 125–136.

162. I. M. Duursma, M. Greferath and S. E. Schmidt, On the optimal $\mathbb{Z}_4$ codes of type II and length 16, *JCT* **A 92** (2000), 77–82.

163. H. Dym and H. P. McKean, *Fourier Series and Integrals*, Academic Press, NY, 1972.

164. W. Ebeling *Lattices and Codes*, Vieweg, Braunschweig/Wiesbaden, 2nd ed., 2002.

165. W. Eholzer and N. P. Skoruppa, Modular invariance and uniqueness of conformal characters, *Commun. Math. Phys.* **174** (1995) 117–136.

166. M. Eichler and D. Zagier, *The Theory of Jacobi Forms*, Birkhäuser, Boston, 1985.

167. N. D. Elkies, Lattices and codes with long shadows, *Math. Res. Lett.*, **2** (1995), 643–651.

168. N. D. Elkies, Lattices, linear codes, and invariants, *Notices Amer. Math. Soc.* **47** (2000), 1238–1245 and 1382–1391.

169. M. Esmaeili, T. A. Gulliver and A. K. Khandani, On the Pless-construction and ML decoding of the $(48, 24, 12)$ quadratic residue code, *PGIT* **49** (2003), 1527–1535.

170. C. Faith, *Algebra II: Ring Theory*, Springer, 1976.

171. F. Fekri, S. W. McLaughlin, R. M. Mersereau and R. W. Schafer, Double circulant self-dual codes using finite-field wavelet transforms, in *Applied Algebra, Algebraic Algorithms and Error-Correcting Codes (Honolulu, HI, 1999)*, Lecture Notes Comput. Sci. **1719** (1999), pp. 355–364.

172. X.-N. Feng and Z.-D. Dai, Notes on finite geometries and the construction of PBIB designs, V: Some "Anzahl" theorems in orthogonal geometry over finite fields of characteristic 2, *Sci. Sinica* **13** (1964), 2005-2008.

173. J. E. Fields, *Tables of Indecomposable Z/(4) Codes*, published electronically at http://www.math.uic.edu/~fields/z4/.

174. J. E. Fields, P. Gaborit, J. S. Leon and V. S. Pless,   All self-dual $\mathbb{Z}_4$ codes of length 15 or less are known, *PGIT* **44** (1998), 311–322.

175. G. D. Forney, Jr., N. J. A. Sloane and M. D. Trott, The Nordstrom-Robinson code is the binary image of the octacode, in *Coding and Quantization: DI-MACS/IEEE Workshop October 19–21, 1992*, ed. R. Calderbank, G. D. Forney, Jr. and and N. Moayeri, Amer. Math. Soc. (1993), pp. 19–26.

176. E. Freitag, *Siegelsche Modulfunktionen*, Springer, 1983.

177. A. Fröhlich and M. J. Taylor,   *Algebraic Number Theory*, Cambridge Univ. Press, 1991.

178. P. Gaborit, Mass formulas for self-dual codes over $\mathbb{Z}_4$ and $\mathbb{F}_q + u\mathbb{F}_q$ rings, *PGIT* **42** (1996), 1222–1228.

179. P. Gaborit, Quadratic double circulant codes over fields, *JCT* **A 97** (2002) 85–107.

180. P. Gaborit, *Tables of Self-Dual Codes*, published electronically at http://www.unilim.fr/pages_perso/philippe.gaborit/SD/, 2004.

181. P. Gaborit, A bound for certain s-extremal lattices and codes, Preprint, 2005.

182. P. Gaborit and M. Harada, Construction of extremal Type II codes over $\mathbb{Z}_4$, *DCC*, submitted.

183. P. Gaborit, W. C. Huffman, J.-L. Kim and V. S. Pless,   On additive GF(4) codes, in *Codes and Association Schemes (Piscataway, NJ, 1999)*, ed. A. Barg and S. Litsyn, Amer. Math. Soc., Providence, RI, 2001, pp. 135–149.

184. P. Gaborit, J.-L. Kim and V. S. Pless, Decoding binary $R(2,5)$ by hand, *DM* **264** (2003), 55–73.

185. P. Gaborit and A. Otmani, Experimental construction of self-dual codes, *Finite Fields Appl.* **9** (2003), 372–394.

186. P. Gaborit, V. S. Pless, P. Solé and A. O. L. Atkin,   Type II codes over $\mathbb{F}_4$, *Finite Fields Applic.* **8** (2002), 171–183.

187. M. Gaulter, Minima of odd unimodular lattices in dimension $24m$, *J. Number Theory* **91** (2001), 81–91.

188. S. Georgiou and C. Koukouvinos, Self-dual codes over $GF(7)$ and orthogonal designs, *Utilitas Math.* **60** (2001) 79–89.

189. S. Georgiou and C. Koukouvinos, MDS self-dual codes over large prime fields, *Finite Fields Appl.* **8** (2002), 455–470.

190. S. P. Glasby, On the faithful representations, of degree 2^n, of certain extensions of 2-groups by orthogonal and symplectic groups. *J. Australian Math. Soc. Ser. A* **58**, (1995), 232–247.

191. A. M. Gleason, Weight polynomials of self-dual codes and the MacWilliams identities, in *Actes, Congrés International de Mathématiques (Nice, 1970)*, Gauthiers-Villars, Paris, 1971, Vol. 3, pp. 211–215.

192. J.-M. Goethals and J. J. Seidel, Spherical designs, in *Relations Between Combinatorics and Other Parts of Mathematics*, ed. D. K. Ray-Chaudhuri, Proc. Symp. Pure Math. **34** (1979), 255–272.

193. J.-M. Goethals and J. J. Seidel, Cubature formulae, polytopes and spherical designs, in *The Geometric Vein: The Coxeter Festschrift*, ed. C. Davis et al., Springer, 1981, pp. 203–218.

194. J.-M. Goethals and J. J. Seidel, The football, *Nieuw Archief voor Wiskunde* **29** (1981), 50–58. Reprinted in *Geometry and Combinatorics: Selected Works of J. J. Seidel*, ed. D. G. Corneil and R. Mathon, Academic Press, NY, 1991, pp. 363–371.

195. I. J. Good, Generalizations to several variables of Lagrange's expansion, with applications to stochastic processes, *Proc. Camb. Phil. Soc.* **56** (1960), 367–380.

196. D. M. Gordon, Minimal permutation sets for decoding the binary Golay codes, *PGIT* **28** (1982), 541–543.

197. D. Gottesman, A class of quantum error-correcting codes saturating the quantum Hamming bound, *Phys. Rev. A* **54** (1996), 1862–1868 [arXiv: quant-ph/9604038].

198. M. Grassl, *Bounds on d_{min} for additive $[[n, k, d]]$ QECC*, published electronically at http://iaks-www.ira.uka.de/home/grassl/QECC/index.html.

199. M. Greferath and S. E. Schmidt, Linear codes and rings of matrices, in *Applied Algebra, Algebraic Algorithms and Error-Correcting Codes (Honolulu, HI, 1999)*, Lecture Notes Comput. Sci. **1719** (1999), pp. 160–169.

200. M. Greferath and S. E. Schmidt, Gray isometries for finite chain rings and a nonlinear ternary $(36, 3^{12}, 15)$ code, *PGIT* **45** (1999), 2522–2524.

201. M. Greferath and S. E. Schmidt, Finite-ring combinatorics and MacWilliams' equivalence theorem, *JCT* **A 92** (2000), 17–28.

202. M. Greferath and U. Vellbinger, Efficient decoding of $\mathbb{Z}_{p^k}$-linear codes, *PGIT* **44** (1998), 1288–1291.

203. M. Greferath and E. Viterbo, On $\mathbb{Z}_4$- and $\mathbb{Z}_9$-linear lifts of the Golay codes, *PGIT* **45** (1999), 2524–2527.

204. R. L. Griess, Automorphisms of extra special groups and nonvanishing degree 2 cohomology, *Pacific J. Math.* **48** (1973), 403–422.

205. P. Griffiths and J. Harris, *Principles of Algebric Geometry* Wiley, NY, 1978.

206. B. Gross and G. Nebe, Globally maximal arithmetic groups, *J. Algebra* **272** (2004), 625–642.

207. T. A. Gulliver, Optimal double circulant self-dual codes over $\mathbb{F}_4$, *PGIT* **46** (2000), 271–274.

208. T. A. Gulliver and V. K. Bhargava, Self-dual codes based on the twin prime product 35, *Appl. Math. Lett.* **5** (1992), 95–98.

209. T. A. Gulliver and M. Harada, Weight enumerators of extremal singly-even $[60, 30, 12]$ codes, *PGIT* **42** (1996), 658–659.

210. T. A. Gulliver and M. Harada, Classification of extremal double circulant formally self-dual even codes, *DCC* **11** (1997), 25–35.

211. T. A. Gulliver and M. Harada, Weight enumerators of double circulant codes and new extremal self-dual codes, *DCC* **11** (1997), 141–150.

212. T. A. Gulliver and M. Harada, Certain self-dual codes over $\mathbb{Z}_4$ and the odd Leech lattice, in *Proc. 12th Appl. Alg. Algorithms and Error-Correcting Codes*, Lect. Notes Comput. Sci. **1225** (1997), 130–137.

213. T. A. Gulliver and M. Harada, Classification of extremal double circulant self-dual codes of lengths 64 to 72, *DCC* **13** (1998), 257–269.

214. T. A. Gulliver and M. Harada, On the existence of a formally self-dual even $[70, 35, 14]$ code, *Appl. Math. Lett.* **11** (1998), 95–98.

215. T. A. Gulliver and M. Harada, Double circulant self-dual codes over $\mathbb{Z}_{2k}$, *PGIT* **44** (1998), 3105–3123.

216. T. A. Gulliver and M. Harada, New optimal self-dual codes over $GF(7)$, *Graphs Combin.* **15** (1999) 175–186.

217. T. A. Gulliver and M. Harada, Extremal double circulant Type II code over $\mathbb{Z}_4$ and construction of $5 - (24, 10, 36)$ designs, *DM* **194** (1999), 129–137.

218. T. A. Gulliver and M. Harada, Double circulant self-dual codes over $GF(5)$, *Ars Comb.*, **56** (2000), 3–13

219. T. A. Gulliver and M. Harada, On the minimum weight of codes over F_5 constructed from certain conference matrices, *DCC* **31** (2004) 139–145.

220. T. A. Gulliver and M. Harada, Classification of extremal double circulant self-dual codes of lengths 74 to 88, Preprint, 2005.

221. T. A. Gulliver, M. Harada and J.-L. Kim, Construction of new extremal self-dual codes *DM* **263** (2003), 81–91; erratum **289** (2004), 207.

222. T. A. Gulliver and J.-L. Kim, Circulant based extremal additive self-dual codes over $GF(4)$, *PGIT* **50** (2004), 359-366.

223. T. A. Gulliver, P. R. J. Östergård and N. I. Senkevitch, Optimal quaternary linear rate-1/2 codes of length ≤ 18, *PGIT* **49** (2003), 1540–1543.

224. R. C. Gunning, *Lectures on Modular Forms*, Princton Univ. Press, 1962.

225. R. M. Guralnick and P. H. Tiep, Decompositions of small tensor powers and Larsen's conjecture, *Represent. Theory* **9** (2005), 138–208.

226. A. J. Hahn and O. T. O'Meara, *The Classical Groups and K-Theory*, Springer, 1989.

227. A. R. Hammons, Jr., P. V. Kumar, A. R. Calderbank, N. J. A. Sloane and P. Solé, The $\mathbb{Z}_4$-linearity of Kerdock, Preparata, Goethals and related codes, *PGIT* **40** (1994), 301–319 [arXiv: math.CO/0207208].

228. M. Harada, Existence of new extremal doubly-even codes and extremal singly-even codes, *DCC* **8** (1996), 273–284.

229. M. Harada, The existence of a self-dual [70, 35, 12] code and formally self-dual codes, *Finite Fields Applic.* **3** (1997), 131–139.

230. M. Harada, Weighing matrices and self-dual codes, *Ars Comb.* **47** (1997), 65–73.

231. M. Harada, New extremal ternary self-dual codes, *Australas. J. Combin.*, **17** (1998), 133–145.

232. M. Harada, New extremal Type II codes over $\mathbb{Z}_4$, *DCC* **13** (1998), 271–284.

233. M. Harada, New 5-designs constructed from the lifted Golay code over $\mathbb{Z}_4$, *J. Combin. Designs* **6** (1998), 225–229.

234. M. Harada, Construction of an extremal self-dual code of length 62, *PGIT* **45** (1999), 1232–1233.

235. M. Harada, New extremal self-dual codes of lengths 36 and 38, *PGIT* **45** (1999), 2541–2543.

236. M. Harada, Self-orthogonal 3-(56, 12, 65) designs and extremal doubly-even self-dual codes of length 56, *DCC*, to appear, 2005.

237. M. Harada, On the self-dual F_5-codes constructed from Hadamard matrices of order 24, *J. Combin. Designs* **13** (2005), 152–156.

238. M. Harada, T. A. Gulliver and H. Kaneta, Classification of extremal double-circulant self-dual codes of length up to 62, *DM* **188** (1998), 127–136.

239. M. Harada and H. Kharaghani, Orthogonal designs, self-dual codes and the Leech lattice, *J. Combin. Designs* **13** (2005), 184–194.

240. M. Harada and H. Kharaghani, Orthogonal designs and MDS self-dual codes, *Australas. J. Combin.*, to appear, 2005.

241. M. Harada and H. Kimura, New extremal doubly-even [64, 33, 12] codes, *DCC* **6** (1995), 91–96.

242. M. Harada and H. Kimura, On extremal self-dual codes, *Math. J. Okayama Univ.* **37** (1995), 1–14.

243. M. Harada and M. Kitazume, $\mathbb{Z}_4$-code constructions for the Niemeier lattices and their embeddings in the Leech lattice, *European J. Combin.* **21** (2000), 473–485.

244. M. Harada and M. Kitazume, $\mathbb{Z}_6$-code constructions of the Leech lattice and the Niemeier lattices, *European J. Combin.* **23** (2002), 573–581.

245. M. Harada, M. Kitazume and A. Munemasa, On a 5-design related to an extremal doubly even self-dual code of length 72, *JCT* **A 107** (2004), 143–146.

246. M. Harada, M. Kitazume, A. Munemasa and B. B. Venkov, On some self-dual codes and unimodular lattices in dimension 48, *European J. Combin.* **26** (2005), 543–557.

247. M. Harada, M. Kitazume and M. Ozeki, Ternary code construction of unimodular lattices and self-dual codes over $\mathbb{Z}_6$, *J. Algebraic Combin.* **16** (2002), 209–223.

248. M. Harada and A. Munemasa, Classification of type IV self-dual Z_4-codes of length 16, *Finite Fields Appl.* **6** (2000), 244–254.

249. M. Harada and A. Munemasa, A quasi-symmetric 2-(49,9,6) design, *J. Combin. Des.* **10** (2002), 173–179.

250. M. Harada and A. Munemasa, Shadows, neighbors and covering radii of extremal self-dual codes, Preprint, 2005.

251. M. Harada, A. Munemasa and V. D. Tonchev, A characterization of designs related to an extremal doubly-even self-dual code of length 48, *Ann. Combin.*, to appear, 2005.

252. M. Harada and P. R. J. Östergård, Self-dual and maximal self-orthogonal codes over $\mathbb{F}_7$, *DM* **256** (2002), 471–477.

253. M. Harada and P. R. J. Östergård, On the classification of self-dual codes over F_5, *Graphs and Combinatorics* **19** (2003), 203–214.

254. M. Harada and M. Oura, On the Hamming weight enumerators of self-dual codes over $\mathbb{Z}_k$, *Finite Fields Appl.* **5** (1999), 26–34.

255. M. Harada and M. Ozeki, Extremal self-dual codes with the smallest covering radius, *DM* **215** (2000), 271–281.

256. M. Harada, M. Ozeki and K. Tanabe, On the covering radius of ternary extremal self-dual codes, *DCC* **33** (2004), 149-158.

257. M. Harada, P. Solé and P. Gaborit, Self-dual codes over $\mathbb{Z}_4$ and unimodular lattices: a survey, in *Algebras and combinatorics (Hong Kong, 1997)*, ed. K.-P. Shum, E. J. Taft and Z.-X. Wan, Springer, 1999, pp 255–275.

258. M. Harada and V. D. Tonchev, Singly-even self-dual codes and Hadamard matrices, in *Proc. Applied. Alg., Alg. Algorithms and Error-Correcting Codes*, ed. G. Cohen, M. Giusti and T. Mora, Lecture Notes Comput. Sci. **948** (1995), 279–284.

259. R. H. Hardin and N. J. A. Sloane, New spherical 4-designs, *DM* **106/107** (1992), 255–264. (Topics in Discrete Mathematics, vol. 7, "A Collection of Contributions in Honour of Jack Van Lint", ed. P. J. Cameron and H. C. A. van Tilborg, North-Holland, 1992.)

260. R. H. Hardin and N. J. A. Sloane, McLaren's improved snub cube and other new spherical designs in three dimensions, *Discrete Comput. Geometry* **15** (1996), 429–441.

261. S. Helgason, *Differential Geometry, Lie Groups and Symmetric Spaces*, Academic Press, NY, 1978.

262. T. Helleseth and P. V. Kumar, Sequences with low correlation, Chapter 21 of [426], pp. 1765–1854.

263. N. Herrmann, *Höhere Gewichtszähler von Codes und deren Beziehung zur Theorie der Siegelschen Modulformen*, Diplomarbeit, Bonn 1991

264. R. J. Higgs and J. F. Humphreys, Decoding the ternary Golay code, *PGIT* **39** (1993), 1043–1046.

265. F. Hirzebruch, The ring of Hilbert modular forms for real quadratic fields of small discriminant, in Lect. Notes Math., **627** (1977), 288–323; *Gesammelte Abhandlungen* **II** (1987), 501–536.

266. G. Höhn, Self-dual codes over the Kleinian four-group, *Math. Ann.* **327** (2003), 227–255 [arXiv: math.CO/0005266].

267. T. Honold and I. Landjev, Linear codes over finite chain rings, *Electronic J. Combin.* **7** (No. 1, 2000), #11.

268. H. Horimoto and K. Shiromoto, A Singleton bound for linear codes over finite quasi-Frobenius rings, in *Applied Algebra, Algebraic Algorithms and Error-Correcting Codes (Honolulu, HI, 1999)*, Lecture Notes Comput. Sci. **1719** (1999), pp. 51–52.

269. S. K. Houghten, C. W. H. Lam, L. H. Thiel and J. A. Parker, The extended quadratic residue code is the only (48, 24, 12) self-dual doubly-even code, *PGIT* **49** (2003), 53–59.

270. W. C. Huffman, The biweight enumerator of self-orthogonal codes, *DM* **26** (1978), 129–143.

271. W. C. Huffman, Automorphisms of codes with applications to extremal doubly even codes of length 48, *PGIT* **28** (1982), 511–521.

272. W. C. Huffman, Decomposing and shortening codes using automorphisms, *PGIT* **32** (1986), 833–836.

273. W. C. Huffman, On the [24,12,10] quaternary code and binary codes with an automorphism having two cycles, *PGIT* **34** (1988), 486–493.

274. W. C. Huffman, On the equivalence of codes and codes with an automorphism having two cycles, *DM* **83** (1990), 265–283.

275. W. C. Huffman, On extremal self-dual quaternary codes of lengths 18 to 28, I, *PGIT* **36** (1990), 651–660.

276. W. C. Huffman, On 3-elements in monomial automorphism groups of quaternary codes, *PGIT* **36** (1990), 660–664.

277. W. C. Huffman, On extremal self-dual quaternary codes of lengths 18 to 28, II, *PGIT* **37** (1991), 1206–1216.

278. W. C. Huffman, On extremal self-dual ternary codes of lengths 28 to 40, *PGIT* **38** (1992), 1395–1400.

279. W. C. Huffman, On the classification of self-dual codes, *Proc. 34th Allerton Conf. Commun. Control and Computing*, Univ. Ill., Urbana, 1996, pp. 302–311.

280. W. C. Huffman, Characterization of quaternary extremal codes of lengths 18 and 20, *PGIT* **43** (1997), 1613–1616.

281. W. C. Huffman, Decompositions and extremal type II codes over $\mathbb{Z}_4$, *PGIT* **44** (1998), 800–809.

282. W. C. Huffman, On the classification and enumeration of self-dual codes, *Finite Fields Applic.* **11** (2005), 451–490.

283. W. C. Huffman and N. J. A. Sloane, Most primitive groups have messy invariants, *Advances in Math.* **32** (1979), 118–127.

284. W. C. Huffman and V. D. Tonchev, The existence of extremal [50, 25, 10] codes and quasi-symmetric 2-(49,9,6) designs, *DCC* **6** (1995), 97–106.

285. W. C. Huffman and V. D. Tonchev, The [52, 26, 10] binary self-dual codes with an automorphism of order 7, *Finite Fields Appl.* **7** (2001), 341–349.

286. W. C. Huffman and V. Y. Yorgov, A [72, 36, 16] doubly even code does not have an automorphism of order 11, *PGIT* **33** (1987), 749–752.

287. J. E. Humphreys, *Reflection Groups and Coxeter Groups*, Cambridge Univ. Press, 1990.

288. B. Huppert, *Endliche Gruppen I*, Springer, 1967.

289. J.-I. Igusa, On Siegel modular forms of genus II, *Amer. J. Math.* **84** (1962), 175–200.

290. N. Ito, Symmetry codes over $GF(3)$, *JCT* **A 29** (1980), 251–253.

291. N. Jacobson, *Basic Algebra II*, Freeman, New York, 1980.

292. D. B. Jaffe, *Binary Linear Codes: New Results on Nonexistence*, published electronically at http://www.math.unl.edu/~djaffe/codes/code.ps.gz.

293. D. B. Jaffe, Optimal binary linear codes of length ≤ 30, *DM* **223** (2000), 135–155.

294. G. A. Kabatiansky and V. I. Levenshtein, Bounds for packings on a sphere and in space, *Probl. Pered. Inform.* **14** (1978), no. 1, 3–25; *Problems Inform. Transmission* **14** (1978), no. 1, 1–17.

295. S. N. Kapralov and V. D. Tonchev, Extremal doubly-even codes of length 64 derived from symmetric designs, *DM* **83** (1990), 285–289.

296. M. Karlin, New binary coding results by circulants, *PGIT* **15** (1969), 81–92.

297. F. Kasch, *Moduln und Ringe*, Teubner, Stuttgart, 1977.

298. L. S. Kazarin, On certain groups defined by Sidelnikov (in Russian), *Mat. Sb.* **189** (No. 7, 1998), 131–144; English translation in *Sb. Math.* **189** (1998), 1087–1100.

299. G. T. Kennedy, Weight distributions of linear codes and the Gleason-Pierce theorem, *JCT* **A 67** (1994), 72–88.

300. I. L. Kheifets, Extension theorem for linear codes over finite quasi-Frobenius modules (in Russian), *Fundam. Prikl. Mat.* **7** (No. 4, 2001), 1227–1236.

301. D. K. Kim, H. K. Kim and J.-L. Kim, Type I codes over GF(4), Preprint, 2004.

302. J.-L. Kim, New extremal self-dual codes of lengths 36, 38, and 58, *PGIT* **47** (2001), 386–393.

303. J.-L. Kim, New self-dual codes over $GF(4)$ with the highest known minimum weights, *PGIT* **47** (2001), 1575–1580.

304. J.-L. Kim and Y. Lee, Euclidean and Hermitian self-dual MDS codes over large finite fields, *JCT* **A 105** (2004), 79–95.

305. J.-L. Kim, K. E. Mellinger and V. S. Pless, Projections of binary linear codes onto larger fields, *SIAM J. Discrete Math.* **16** (2003), 591–603.

306. J.-L. Kim and V. S. Pless, Decoding some doubly-even self-dual [32, 16, 8] codes by hand, in *Codes and Designs (Ohio State University, May 2000, the Ray-Chaudhuri Festschrift)*, ed. K. T. Arasu and Á. Seress, de Gruyter, Berlin, 2002, pp. 165–178.

307. J.-L. Kim and V. S. Pless, Designs in additive codes over $GF(4)$, *DCC* **30** (2003), 187–199.

308. H. Kimura, Extremal doubly even $(56, 28, 12)$ codes and Hadamard matrices of order 28, *Australas. J. Combin.* **10** (1994), 171–180.

309. O. D. King, The mass of extremal doubly-even self-dual codes of length 40, *PGIT* **47** (2001), 2558–2560.

310. A. Y. Kitaev, Quantum computations: algorithms and error correction (in Russian), *Uspekhi Mat. Nauk.* **52** (No. 6, 1997), 53–112; English translation in *Russian Math. Surveys* **52** (1997), 1191–1249.

311. A. Y. Kitaev, A. H. Shen and M. N. Vyalyi, *Classical and Quantum Computation*, Amer. Math. Soc., Providence, RI, 2002.

312. Y. Kitaoka, *Arithmetic of Quadratic Forms*, Cambridge Univ. Press, 1993.

313. M. Kitazume, T. Kondo and I. Miyamoto, Even lattices and doubly even codes, *J. Math. Soc. Japan* **43** (1991), 67–87.

314. P. B. Kleidman and M. W. Liebeck, *The Subgroup Structure of the Finite Classical Groups*, Cambridge Univ. Press, 1988.

315. M. Klemm, Selbstduale Codes über dem Ring der ganzen Zahlen modulo 4, *Arch. Math. (Basel)* **53** (1989), 201–207.

316. M. Klemm, Eine Invarianzgruppe für die vollständige Gewichtsfunktion selbstdualer Codes, *Arch. Math. (Basel)* **53** (1989), 332–336.

317. H. Klingen, *Introductory Lectures on Siegel Modular Forms* , Cambridge Univ. Press, 1990.

318. M. Kneser, *Quadratische Formen*, Springer, 2002.

319. E. Knill, Non-binary unitary error bases and quantum codes, *Technical Report LAUR-96-2717*, Los Alamos National Laboratory, 1996 [arXiv: quant-ph/9608048].

320. E. Knill, Group representations, error bases and quantum codes, *Technical Report LAUR-96-2807*, Los Alamos National Laboratory, 1996 [arXiv: quant-ph/9608049].

321. E. Knill and R. Laflamme, A theory of quantum error correcting codes, *Phys. Rev.* **A 55** (1997), 900–911 [arXiv: quant-ph/9604034].

322. E. Knill, R. Laflamme and W. Zurek, Resilient quantum computation: Error models and thresholds, *Proc. R. Soc. Lond.* **A 454** (1998), 365–384 [arXiv: quant-ph/9702058].

323. M.-A. Knus, *Quadratic and Hermitian Forms over Rings*, Springer, 1991.

324. M.-A. Knus, A. Merkurjev, M. Rost and J.-P. Tignol, *The Book of Involutions*, Amer. Math. Soc., Providence, RI, 1998.

325. H. Koch, Unimodular lattices and self-dual codes, in *Proc. Intern. Congress Math. Berkeley 1986*, Amer. Math. Soc., Providence, RI, Vol. **1**, 1987, 457–465.

326. H. Koch, On self-dual, doubly even codes of length 32, *JCT* **A 51** (1989), 63–76.

327. H. Koch, On self-dual doubly-even extremal codes, *DM* **83** (1990), 291–300.

328. H. Koch, The 48-dimensional analogues of the Leech lattice, *Trudy Mat. Inst. Steklov.* **208** (1995), 193–201.

329. H. Koch and G. Nebe, Extremal even unimodular lattices of rank 32 and related codes, *Math. Nachr.* **161** (1993), 309–319.

330. H. Koch and B. B. Venkov, Über ganzzahlige unimodulare euklidische Gitter, *J. Reine Angew. Math.* **398** (1989), 144–168.

331. H. Koch and B. B. Venkov, Über gerade unimodulare Gitter der Dimension 32, III, *Math. Nachr.* **152** (1991), 191–213.

332. W. Kohnen and R. Salvati Manni, Linear relations between theta series, *Osaka J. Math.* **41** (2004), 353–356.

333. T. Kogiso and K. Tsushima, On an algebra of Siegel modular forms associated with the theta group $\Gamma_2(1,2)$, *Tsukuba J. Math.* **22** (1998), 645–656.

334. A. I. Kostrikin and P. H. Tiep, *Orthogonal Decompositions and Integral Lattices*, de Gruyter, Berlin, 1994.

335. I. Krasikov and S. Litsyn, Linear programming bounds for doubly-even self-dual codes, *PGIT* **43** (1997), 1238–1244.

336. I. Krasikov and S. Litsyn, An improved upper bound on the minimum distance of doubly-even self-dual codes, *PGIT* **46** (2000), 274–278.

406   References

337. F. R. Kschischang and S. Pasupathy, Some ternary and quaternary codes and associated sphere packings, *PGIT* **38** (1992), 227–246.

338. C. W. H. Lam, The search for a finite projective plane of order 10, *Amer. Math. Monthly* **98** (1991) 305–318.

339. C. W. H. Lam and V. S. Pless, There is no (24,12,10) self-dual quaternary code, *PGIT* **36** (1990), 1153–1156.

340. C. W. H. Lam, L. Thiel and S. Swiercz, The non-existence of finite projective planes of order 10, *Canad. J. Math.* **41** (1989), 1117–1123.

341. T. Y. Lam, *Lectures on Modules and Rings*, Springer, 1999.

342. T. Y. Lam, *A First Course in Noncommutative Rings*, Springer, 2nd. ed., 2001.

343. S. Lang, *Algebra*, 3rd. ed., Addison-Wesley, Reading, MA, 1993.

344. J. Leech and N. J. A. Sloane, Sphere packing and error-correcting codes, *Canadian J. Math.*, **23** (1971), 718–745. (See also Chapter 5 in [133].)

345. J. S. Leon, J. M. Masley and V. S. Pless, Duadic codes, *PGIT* **30** (1984), 709–714.

346. J. S. Leon, V. S. Pless and N. J. A. Sloane, On ternary self-dual codes of length 24, *PGIT* **27** (1981), 176–180.

347. J. S. Leon, V. S. Pless and N. J. A. Sloane, Self-dual codes over GF(5), *JCT* **A 32** (1982), 178–194.

348. V. I. Levenshtein, Universal bounds for codes and designs, Chapter 6 of [426], pp. 499–648.

349. S. Ling and P. Solé, Type II Codes over $\mathbb{F}_4 + u\mathbb{F}_4$, *European J. Combinatorics* **22** (2001), 983–997.

350. J. H. van Lint, *Introduction to Coding Theory*, Springer, 1982.

351. J. H. van Lint and F. J. MacWilliams, Generalized quadratic residue codes, *PGIT* **24** (1978), 730–737.

352. S. Litsyn, An updated table of the best binary codes known, Chapter 5 of [426], pp. 463–498.

353. O. Loos, Bimodule-valued Hermitian and quadratic forms, *Arch. Math. (Basel)* **62** (1994), 134–142.

354. X. Ma and L. Zhu, Nonexistence of extremal doubly even self-dual codes, unpublished manuscript, 1997.

355. S. MacLane, *Categories for the Working Mathematician*, Springer, 2nd ed., 1998.

356. F. J. MacWilliams, A theorem on the distributionm of weights in a systematic code, *Bell Syst. Tech. J.* **42** (1963), 79–94.

357. F. J. MacWilliams, Orthogonal matrices over finite fields, *Amer. Math. Monthly* **76** (1969), 152–164.

358. F. J. MacWilliams, Orthogonal circulant matrices over finite fields, and how to find them, *JCT* **10** (1971), 1–17.

359. F. J. MacWilliams, C. L. Mallows and N. J. A. Sloane, Generalizations of Gleason's theorem on weight enumerators of self-dual codes, *PGIT* **18** (1972), 794–805.

360. F. J. MacWilliams, A. M. Odlyzko, N. J. A. Sloane and H. N. Ward, Self-dual codes over GF(4), *JCT* **A 25** (1978), 288–318

361. F. J. MacWilliams and N. J. A. Sloane, *The Theory of Error-Correcting Codes*, North-Holland, Amsterdam, 1977; 11th impression 2003.

362. F. J. MacWilliams, N. J. A. Sloane and J. G. Thompson, Good self-dual codes exist, *DM* **3** (1972), 153–162.

363. C. L. Mallows, A. M. Odlyzko and N. J. A. Sloane, Upper bounds for modular forms, lattices and codes, *J. Algebra* **36** (1975), 68–76.

364. C. L. Mallows, V. S. Pless and N. J. A. Sloane, Self-dual codes over $GF(3)$, *SIAM J. Appl. Math.* **31** (1976), 649–666.

365. C. L. Mallows and N. J. A. Sloane, An upper bound for self-dual codes, *Information and Control* **22** (1973), 188–200.

366. C. L. Mallows and N. J. A. Sloane, Weight enumerators of self-orthogonal codes, *DM* **9** (1974), 391–400.

367. C. L. Mallows and N. J. A. Sloane, Weight enumerators of self-orthogonal codes over $GF(3)$, *SIAM J. Algebraic and Discrete Methods* **2** (1981), 425–460.

368. J. Martinet, *Perfect Lattices in Euclidean Spaces*, Springer, 2003.

369. B. R. McDonald, *Finite Rings with Identity*, Dekker, NY, 1974.

370. R. J. McEliece, E. R. Rodemich, Jr., H. Rumsey and L. R. Welch, New upper bounds on the rate of a code via the Delsarte-MacWilliams inequalities, *PGIT* **23** (1977), 157–166.

371. J. Milnor and D. Husemoller, *Symmetric Bilinear Forms*, Springer, 1973.

372. T. Miyake, *Modular Forms*, Springer, 1989.

373. T. Molien, Ueber die invarianten der linear Substitutionsgruppe, *Sitzungsber König. Akad. Wiss.*, (1897), 1152–1156.

374. E. H. Moore, *Double Circulant Codes and Related Algebraic Structures*, Ph.D. Dissertation, Dartmouth College, July 1976.

375. D. Mumford, *Tata Lectures on Theta I*, Birkhäuser, Boston, 1983.

376. D. Mumford, *Tata Lectures on Theta III*, Birkhäuser, Boston, 1983.

377. A. Munemasa, A mass formula for Type II codes over finite fields of characteristic two, in *Codes and Designs (Ohio State University, May 2000, the Ray-Chaudhuri Festschrift)*, ed. K. T. Arasu and Á. Seress, de Gruyter, Berlin, 2002, pp. 207-214.

378. T. Nakayama, On Frobeniusean algebras I, *Annals Math.* **40** (1939), 611–633.

379. G. Nebe, The normaliser action and strongly modular lattices, *L'Enseign. Math.* **43** (1997), 67–76.

380. G. Nebe, Some cyclo-quaternionic lattices, *J. Alg.* **199** (1998), 472–498.

381. G. Nebe, An analogue of Hecke operators in coding theory, Preprint, 2005 [arXiv: math.NT/0509474].

382. G. Nebe, H.-G. Quebbemann, E. M. Rains and N. J. A. Sloane, Complete weight enumerators of generalized doubly-even self-dual codes, *Finite Fields Applic.*, **10** 2004, 540–550 [arXiv: math.NT/0311289].

383. G. Nebe, E. M. Rains and N. J. A. Sloane, The invariants of the Clifford groups, *DCC* **24** (2001), 99–121 [arXiv: math.CO/0001038].

384. G. Nebe, E. M. Rains and N. J. A. Sloane, A simple construction for the Barnes-Wall lattices, in *Codes, Graphs and Systems: A Celebration of the Life and Career of G. David Forney, Jr. on the Occasion of his Sixtieth Birthday*, ed. R. E. Blahut and R. Koetter, Kluwer, Boston, 2002, pp. 333–342. [arXiv: math.CO/0207186].

385. G. Nebe, E. M. Rains and N. J. A. Sloane, Codes and invariant theory, *Math. Nachr.* **274–275** (2004), 104–116. [math.NT/0311046]

386. G. Nebe and B. B. Venkov, Nonexistence of extremal lattices in certain genera of modular lattices, *J. Number Theory* **60** (2) (1996) 310–317.

387. W. K. Nicholson and M. F. Yousif, *Quasi-Frobenius Rings*, Cambridge Univ. Press, 2003.

408   References

388. M. A. Nielsen and I. L. Chuang, *Quantum Computation and Quantum Information*, Cambridge Univ. Press, 2000.
389. H.-V. Niemeier, Definite quadratische Formen der Dimension 24 und Diskriminante 1, *J. Number Theory* **5** (1973), 142–178.
390. T. Nishimura, A new extremal self-dual code of length 64, *PGIT* **50** (2004), 2173–2174.
391. A. Nobs, Die irreduziblen Darstellungen der Gruppen $SL_2(\mathbb{Z}_p)$, insbesondere $SL_2(\mathbb{Z}_2)$. I. Teil. *Comm. Math. Helvetii* **51** (1976), 456–489.
392. P. R. J. Östergård, There exists no Hermitian self-dual quaternary $[26, 13, 10]_4$ code, *PGIT* **50** (2004), 3316–3317.
393. M. Oura, The dimension formula for the ring of code polynomials in genus 4, *Osaka J. Math.* (1997), **34**, pp. 53–72.
394. M. Ozeki, On the basis problem for Siegel modular forms of degree 2, *Acta Arith.* **31** (1976), 17–30.
395. M. Ozeki, Hadamard matrices and doubly even self-dual error-correcting codes, *JCT* **A 44** (1987), 274–287.
396. M. Ozeki, On a class of self-dual ternary codes, *Science Reports Hirosaki Univ.* **36** (1989), 184–191.
397. M. Ozeki, Quinary code construction of the Leech lattice, *Nihonkai Math. J.* **2** (1991), 155–167.
398. M. Ozeki, On the notion of Jacobi polynomials for codes, *Math. Proc. Camb. Phil. Soc.* **121** (1997), 15–30.
399. M. Ozeki, On covering radius and coset weight distributions of extremal binary self-dual codes of length 40, *Theoret. Comp. Sci.* **235** (2000), 283-308.
400. M. Ozeki, On covering radii and coset weight distributions of extremal binary self-dual codes of length 56, *PGIT* **46** (2000), 2359-2372.
401. M. Ozeki, Jacobi polynomials for singly even self-dual codes and the covering radius problems, *PGIT* **48** (2002), 547–557.
402. M. Ozeki, Notes on the shadow process in self-dual codes, *DM* **264** (2003), 187–200.
403. B. Pareigis, Non-additive ring and module theory I: General theory of monoids, *Publ. Math. Debrecen* **24** (1977), 189–204.
404. G. Pasquier, Binary self-dual codes construction from self-dual codes over a Galois field $\mathbb{F}_{2^m}$, in *Combinatorial Mathematics (Luminy, 1981)*, ed. C. Berge et al., Annals Discrete Math. **17** (1983), 519–526.
405. N. J. Patterson, personal communication, 1980.
406. A. Peres, *Quantum Theory: Concepts and Methods*, Kluwer, Dordrecht, 1995.
407. L. Ping and K. L. Yeung, Symbol-by-symbol APP decoding of the Golay code and iterative decoding of concatenated Golay codes, *PGIT* **45** (1999), 2558–2562.
408. P. M. Piret, Algebraic construction of cyclic codes over $\mathbb{Z}_8$ with a good Euclidean minimum distance, *PGIT* **41** (1995), 815–817.
409. A. O. Pittenger, *An Introduction to Quantum Computing Algorithms*, Birkhäuser, Boston, 2000.
410. W. Plesken, Lattices of covariant quadratic forms, *L'Enseign. Math.* **47** (2001), 21–56.
411. V. S. Pless, The number of isotropic subspaces in a finite geometry, *Rend. Cl. Scienze fisiche, matematiche e naturali, Acc. Naz. Lincei* **39** (1965), 418–421.
412. V. S. Pless, On the uniqueness of the Golay codes, *JCT* **5** (1968), 215–228.

413. V. S. Pless, On a new family of symmetry codes and related new five-designs, *Bull. Amer. Math. Soc.* **75** (1969), 1339–1342.

414. V. S. Pless, A classification of self-orthogonal codes over $GF(2)$, *DM* **3** (1972), 209–246.

415. V. S. Pless, Symmetry codes over $GF(3)$ and new five-designs, *JCT* **A 12** (1972), 119–142.

416. V. S. Pless, The children of the (32,16) doubly even codes, *PGIT* **24** (1978), 738–746.

417. V. S. Pless, 23 does not divide the order of the group of a (72,36,16) doubly even code, *PGIT* **28** (1982), 113–117.

418. V. S. Pless, A decoding scheme for the ternary Golay code, in *Proc. 20th Allerton Conf. Comm. Control*, Univ. Ill., Urbana, 1982, pp. 682–687.

419. V. S. Pless, On the existence of some extremal self-dual codes, in *Enumeration and Design*, ed. D. M. Jackson and S. A. Vanstone, Academic Press, NY, 1984, pp. 245–250.

420. V. S. Pless, Q-codes, *JCT* **A 43** (1986), 258–276.

421. V. S. Pless, Decoding the Golay codes, *PGIT* **32** (1986), 561–567.

422. V. S. Pless, Extremal codes are homogeneous, *PGIT* **35** (1989), 1329–1330.

423. V. S. Pless, More on the uniqueness of the Golay code, *DM* **106** (1992), 391–398.

424. V. S. Pless, Parents, children, neighbors and the shadow, *Contemporary Math.* **168** (1994), 279–290.

425. V. S. Pless, Coding constructions, Chapter 2 of [426], pp. 141–176.

426. V. S. Pless and W. C. Huffman, eds., *Handbook of Coding Theory*, Elsevier, Amsterdam, 2 vols., 1998.

427. V. S. Pless, W. C. Huffman and R. A. Brualdi, An introduction to algebraic codes, Chapter 1 of [426], pp. 3–139.

428. V. S. Pless, J. S. Leon and J. E. Fields, All $\mathbb{Z}_4$ codes of Type II and length 16 are known, *JCT* **A 78** (1997), 32–50.

429. V. S. Pless and J. N. Pierce, Self-dual codes over $GF(q)$ satisfy a modified Varshamov-Gilbert bound, *Information and Control* **23** (1973), 35–40.

430. V. S. Pless and Z. Qian, Cyclic codes and quadratic residue codes over $\mathbb{Z}_4$, *PGIT* **42** (1996), 1594–1600.

431. V. S. Pless and N. J. A. Sloane, On the classification and enumeration of self-dual codes, *JCT* **A 18** (1975), 313–335.

432. V. S. Pless, N. J. A. Sloane and H. N. Ward, Ternary codes of minimum weight 6 and the classification of self-dual codes of length 20, *PGIT* **26** (1980), 305–316.

433. V. S. Pless, P. Solé and Z. Qian, Cyclic self-dual $\mathbb{Z}_4$-codes, *Finite Fields Appl.* **3** (1997), 48–69.

434. V. S. Pless and J. G. Thompson, 17 does not divide the order of a group of a (72,36,16) code, *PGIT* **28** (1982), 537–541.

435. V. S. Pless and V. D. Tonchev, Self-dual codes over $GF(7)$, *PGIT* **33** (1987), 723–727.

436. V. S. Pless, V. D. Tonchev and J. S. Leon, On the existence of a certain $(64, 32, 12)$ extremal code, *PGIT* **39** (1993), 214–215.

437. A. Poli and C. Rigoni, Enumeration of self-dual $2k$ circulant codes, in *Applied Algebra, Algorithmics and Error-Correcting Codes (Toulouse, 1984)*, Lect. Notes. Comput. Sci. **228** (1986), 61–70.

438. H.-G. Quebbemann, On even codes, *DM* **98** (1991), 29–34.
439. H.-G. Quebbemann, Modular lattices in euclidean spaces, *J. Number Th.* **54** (1995), 190–202.
440. H.-G. Quebbemann, Atkin-Lehner eigenforms and strongly modular lattices. *L'Ens. Math.* **43** (1997), 55–65.
441. H.-G. Quebbemann and E. M. Rains, On the involutions fixing the class of a lattice, *J. Number Theory*, **101** (2003), 185–194.
442. E. M. Rains, Shadow bounds for self-dual codes, *PGIT* **44** (1998), 134–139.
443. E. M. Rains, Quantum weight enumerators, *PGIT* **44** (1998), 1388–1394 [arXiv: quant-ph/9612015].
444. E. M. Rains, Quantum codes of minimum distance two, *PGIT* **45** (1999), 266–271 [arXiv: quant-ph/9704043].
445. E. M. Rains, Nonbinary quantum codes, *PGIT* **45** (1999), 1827–1832 [arXiv: quant-ph/9703048].
446. E. M. Rains, Optimal self-dual codes over Z_4, *DM* **203** (1999), 215–228.
447. E. M. Rains, Quantum shadow enumerators, *PGIT* **45** (1999), 2361–2366 [arXiv: quant-ph/9611001].
448. E. M. Rains, Monotonicity of the quantum linear programming bound, *PGIT* **45** (1999), 2489–2492 [arXiv: quant-ph/9802070].
449. E. M. Rains, Polynomial invariants of quantum codes, *PGIT* **46** (2000), 54–59 [arXiv: quant-ph/9704042].
450. E. M. Rains, Bounds for self-dual codes over Z_4, *Finite Fields Appl.* **6** (2000), 146–163.
451. E. M. Rains, New asymptotic bounds for self-dual codes and lattices, *PGIT* **49** (2003), 1261–1274 [arXiv: math.CO/0104145].
452. E. M. Rains, R. H. Hardin, P. W. Shor and N. J. A. Sloane, A nonadditive quantum code, *Phys. Rev. Lett.* **79** (1997), 953–954 [arXiv: quant-ph/9802061].
453. E. M. Rains and N. J. A. Sloane, The shadow theory of modular and unimodular lattices, *J. Number Theory*, **73** (1998), 359–389 [arXiv: math.CO/0207294].
454. E. M. Rains and N. J. A. Sloane, Self-dual codes, Chapter 1 of [426], pp. 177-294 [arXiv: math.CO/0208001].
455. D. K. Ray-Chaudhuri, Some results on quadrics in finite projective geometry based on Galois fields, *Canad. J. Math.* **14** (1962), 129–138.
456. I. S. Reed, X. Yin and T. K. Truong, Algebraic decoding of the $(32, 16, 8)$ quadratic residue code, *PGIT* **36** (1990), 876–880.
457. I. Reiner, *Maximal Orders*, Academic Press, NY, 1975.
458. B. Reznick, Some constructions of spherical 5-designs, *Linear Algebra and Its Applications*, **226/228** (1995), 163–196.
459. J. Rifà, A new algebraic algorithm to decode the ternary Golay code, *Inform. Process. Lett.* **68** (1998), no. 6, 271–274.
460. L. H. Rowen, *Ring Theory*, Academic Press, San Diego, Student ed., 1991.
461. B. Runge, On Siegel modular forms I, *J. Reine Angew. Math.* **436** (1993), 57–85.
462. B. Runge, On Siegel modular forms II, *Nagoya Math. J.* **138** (1995), 179–197.
463. B. Runge, The Schottky ideal, in *Abelian Varieties (Egloffstein, 1993)*, de Gruyter, Berlin, 1995, pp. 251–272.
464. B. Runge, Codes and Siegel modular forms, *DM* **148** (1996), 175–204.
465. R. P. Ruseva, Uniqueness of the [36, 18, 8] double circulant code, in *Proc. Internat. Workshop on Optimal Codes and Related Topics*, May 26–June 1, 1995, Sozopol, Bulgaria, 126–129.

466. R. P. Ruseva, New extremal self-dual codes of length 36, in *Proc. of the* 25th *Spring Conf. of the Union of Bulgarian Mathematicians*, Bulgarian Academy Sci., 1996, 150–153.

467. R. P. Ruseva, Self-dual [24, 12, 8] codes with a non-trivial automorphism of order 3, *Finite Fields Appl.* **8** (2002), 34–51.

468. R. A. Sack, Interpretation of Lagrange's expansion and its generalization to several variables as integration formulas, *J. SIAM* **13** (1965), 47–59.

469. R. A. Sack, Generalization of Lagrange's expansion for functions of several implicitly defined variables, *J. SIAM* **13** (1965), 913–926.

470. R. A. Sack, Factorization of Lagrange's expansion by means of exponential generating functions, *J. SIAM* **14** (1966), 1–15.

471. C. H. Sah, Cohomology of split group extensions: II, *J. Algebra* **45** (1977), 17–68.

472. A. Samorodnitsky, On linear programming bounds for spherical codes and designs, *Discrete Comput. Geom.*, **31** (2004), 385–394.

473. R. Scharlau and R. Schulze-Pillot, Extremal lattices, in *Algorithmic Algebra and Number Theory*, ed. B. H. Matzat, G. M. Greuel and G. Hiss, Springer, 1999, pp. 139–170.

474. W. Scharlau and D. Schomaker, personal communication, April 1991.

475. W. Scharlau, *Quadratic and Hermitian Forms*, Springer, 1985.

476. P. Schmid, On the automorphism group of extraspecial 2-groups, *J. Algebra* **234** (2000), 492–506.

477. B. Schoeneberg, *Elliptic Modular Functions*, Springer, 1974.

478. B. Segre, Le geometrie di Galois, *Ann. Mat. Pura Appl.* **(4) 48** (1959), 1–96.

479. J.-P. Serre, *Linear Representations of Finite Groups*, Springer, 1977.

480. J.-P. Serre, *Cours d'arithmétique*, Presses Universitaires de France, 3rd. ed., Paris, 1988. English translation of 1st edition published by Springer, 1977.

481. P. D. Seymour and T. Zaslavsky, Averaging sets: a generalization of mean values and spherical designs, *Advances in Math.*, **52** (1984), 213–240.

482. G. C. Shephard and J. A. Todd, Finite unitary reflection groups, *Canad. J. Math.* **6** (1954), 274–304.

483. G. Shimura, On modular forms of half integral weight, *Annals Math.* **97** (1973), 440–481.

484. D.-J. Shin, P. V. Kumar and T. Helleseth, 5-designs from the lifted Golay code over Z_4 via an Assmus-Mattson type approach, *DM* **241** (2001), 479–487.

485. K. Shiromoto and L. Storme, A Griesmer bound for linear codes over finite quasi-Frobenius rings, *Discrete Applied Math.* **128** (2003), 263–274 [cage.rug.ac.be/~ls/artgriesmerwcc2001-35final.pdf].

486. P. W. Shor, "Fault-tolerant quantum computation," *Proc. 37th Sympos. Foundations of Computer Science*, IEEE Computer Society Press, 1996, pp. 56–65 [arXiv: quant-ph/9605011].

487. P. W. Shor and R. Laflamme, Quantum analog of the MacWilliams identities for classical coding theory, *Phys. Rev. Lett.* **78** (1997), 1600–1602 [arXiv: quant-ph/9610040].

488. P. W. Shor and N. J. A. Sloane, A family of optimal packings in Grassmannian manifolds, *J. Algebraic Combin.* **7** (1998), 157–163 [arXiv: math.CO/0208003].

489. I. Siap, Linear codes over $\mathbb{F}_2 + u\mathbb{F}_2$ and their complete weight enumerators, in *Codes and Designs (Ohio State University, May 2000, the Ray-Chaudhuri Festschrift)*, ed. K. T. Arasu and Á. Seress, de Gruyter, Berlin, 2002, pp. 259–271.

490. V. M. Sidelnikov, On a finite group of matrices and codes on the Euclidean sphere (in Russian), *Probl. Pered. Inform.* **33** (1997), 35–54 (1997); English translation in *Problems Inform. Transmission* **33** (1997), 29–44 .

491. V. M. Sidelnikov, On a finite group of matrices generating orbit codes on the Euclidean sphere, in *Proceedings IEEE Internat. Sympos. Inform. Theory, Ulm, 1997*, IEEE Press, 1997, p. 436.

492. V. M. Sidelnikov, Spherical 7-designs in 2^n-dimensional Euclidean space, *J. Algebraic Combin.* **10** (1999), 279–288.

493. V. M. Sidelnikov, Orbital spherical 11-designs in which the initial point is a root of an invariant polynomial (in Russian), *Algebra i Analiz* **11** (No. 4, 1999), 183–203.

494. C. L. Siegel, Einfürung in die Theorie der Modulfunktionen n-ten Grades, *Math. Ann.* **116** (1939), 617–657; *Gesammelte Abhandlungen* **II** (1966), pp. 97–137.

495. C. L. Siegel, Berechnung von Zetafunktionen an ganzzahligen Stellen, *Nachr. Akad. Wiss. Göttingen* **10** (1969), 87–102.

496. J. Simonis, The [18, 9, 6] code is unique, *DM* **106** (1992), 439–448.

497. N. P. Skoruppa, *MODI: A modular forms dimension calculator*, published electronically at wotan.algebra.math.uni-siegen.de/~modi/, 2005.

498. N. J. A. Sloane, Is there a (72, 36) $d = 16$ self-dual code?, *PGIT* **19** (1973), 251.

499. N. J. A. Sloane, Weight enumerators of codes, in *Combinatorics*, ed. M. Hall Jr. and J. H. van Lint, Mathematical Centre, Amsterdam and Reidel Publishing Co., Dordrecht, Holland, 1975, pp. 115–142.

500. N. J. A. Sloane, Error-correcting codes and invariant theory: New applications of a nineteenth-century technique, *Amer. Math. Monthly* **84** (1977), 82–107.

501. N. J. A. Sloane, Binary codes, lattices and sphere-packings, in *Combinatorial Surveys: Proceedings of the Sixth British Combinatorial Conference*, ed. P. J. Cameron, Academic Press, NY, 1977, pp. 117–164.

502. N. J. A. Sloane, Codes over $GF(4)$ and complex lattices, *J. Algebra* **52** (1978), 168–181.

503. N. J. A. Sloane, Self-dual codes and lattices, in *Relations Between Combinatorics and Other Parts of Mathematics*, Proc. Symp. Pure Math., Vol 34, Amer. Math. Soc., Providence, RI, 1979, pp. 273–308.

504. N. J. A. Sloane, *The On-Line Encyclopedia of Integer Sequences*, published electronically at www.research.att.com/~njas/sequences/, 2005.

505. N. J. A. Sloane, R. H. Hardin and P. Cara, Spherical Designs in Four Dimensions (Extended Abstract) in *Proceedings Information Theory Workshop (Paris, April 2003)*, IEEE Press, 2003, pp. 253–257.

506. L. Smith, *Polynomial Invariants of Finite Groups*, Peters, Wellesley, MA, 1995.

507. S. L. Snover, *The Uniqueness of the Nordstrom-Robinson and the Golay Binary Codes*, Ph.D. Dissertation, Department of Mathematics, Michigan State Univ., 1973.

508. S. L. Sobolev, Cubature formulae on the sphere invariant under finite groups of rotations (Russian), *Doklady Akademii Nauk SSR*, **146** (No. 2, 1962), 310–313; translation in *Soviet Mathematics Doklady*, **3** (1962), 1307–1310.

509. G. Solomon, Golay encoding/decoding via BCH-Hamming, *Comput. Math. Appl.*, **39** (2000), 103–108.

510. E. Spence and V. D. Tonchev, Extremal self-dual codes from symmetric designs, *DM* **110** (1992), 265–268.

511. R. P. Stanley, Relative invariants of finite groups generated by pseudoreflections, *J. Algebra* **49** (1977), 134–148.

512. R. P. Stanley, Invariants of finite groups and their application to combinatorics, *Bull. Amer. Math. Soc.* **1** (1979), 475–511.

513. A. M. Steane, Multiple particle interference and quantum error correction, *Proc. Roy. Soc. London A*, **452** (1996), 2551–2577 [arXiv: quant-ph/9601029].

514. A. M. Steane, Simple quantum error correcting codes, *Phys. Rev. Lett.* **77** (1996), 793–797 [arXiv: quant-ph/9605021].

515. J. Stolze and D. Suter, *Quantum Computing: A Short Course from Theory to Experiment*, Wiley-VCH, Weinheim, Germany, 2004.

516. B. Sturmfels, *Algorithms in Invariant Theory*, Springer, 1993.

517. K. Tanabe, An Assmus-Mattson theorem for Z_4-codes, *PGIT* **46** (2000), 48–53.

518. K. Tanabe, A new proof of the Assmus-Mattson theorem for non-binary codes, *DCC* **22** (2001), 149–155.

519. H. Tapia-Recillas and G. Vega, On $\mathbb{Z}_{2^k}$-linear and quaternary codes, *SIAM J. Discrete Math.* **17** (2003), 103–113.

520. A. Terras, *Fourier Analysis on Finite Groups and Applications*, Cambridge Univ. Press, 1999.

521. J. G. Thompson, Weighted averages associated to some codes, *Scripta Math.* **29** (1973), 449–452.

522. V. D. Tonchev, Self-orthogonal designs and extremal doubly-even codes, *JCT A* **52** (1989), 197–205.

523. V. D. Tonchev, Self-dual codes and Hadamard matrices, *Discr. Appl. Math.* **33** (1991), 235–240.

524. V. D. Tonchev, Codes and designs, Chapter 15 of [426], pp. 1229–1268.

525. V. D. Tonchev and R. V. Raev, Cyclic 2-(17, 8, 7) designs and related doubly even codes, *Comp. Rend. Acad. Bulg. Sci.* **35** (1982), 1367–1370.

526. V. D. Tonchev and V. Y. Yorgov, The existence of certain extremal [54, 27, 10] self-dual codes, *PGIT* **42** (1996), 1628–1631.

527. H.-P. Tsai, Existence of some extremal self-dual codes, *PGIT* **38** (1992), 1829–1833.

528. H.-P. Tsai, The covering radius of extremal self-dual code D11 and its application, *PGIT* **43** (1997), 316–319.

529. H.-P. Tsai, Extremal self-dual codes of lengths 66 and 68, *PGIT* **45** (1999), 2129–2133.

530. H.-P. Tsai and Y.-J. Jiang, Some new extremal self-dual [58,29,10] codes, *PGIT* **44** (1998), 813–814.

531. J. V. Uspensky, *Theory of Equations*, McGraw-Hill, NY, 1948.

532. A. Vardy, The Nordstrom-Robinson code: representation over GF(4) and efficient decoding, *PGIT* **40** (1994), 1686–1693.

533. B. B. Venkov, The classification of integral even unimodular 24-dimensional quadratic forms, *Trudy Matemat. Inst. Steklova* **148**, 65–76; *Proc. Steklov Inst. Math.* (1980), 63–74; [133, Chap. 18].

534. B. B. Venkov, Réseaux et designs sphériques, in *Réseaux euclidiens, designs sphériques et formes modulaires*, ed. J. Martinet, Monogr. Enseign. Math., 37, Geneva, 2001, pp. 10–86.

535. M. Ventou and C. Rigoni, Self-dual doubly circulant codes, *DM* **56** (1985), 291–298.

536. G. E. Wall, On Clifford collineation, transform and similarity groups IV: an application to quadratic forms, *Nagoya Math. J.* **21** (1962), 199–222.

414   References

537. Z.-X. Wan, Studies in finite geometries and the construction of incomplete block designs, I: Some "Anzahl" theorems in symplectic geometry over finite fields (Chinese), *Acta Math. Sinica* **15** 354-361; *Chinese Math.–Acta* **7** (1965) 55-62.

538. Z.-X. Wan, *Geometry of Classical Groups Over Finite Fields*, Studentlitteratur, Lund; Chartwell-Bratt Ltd., Bromley, 1993.

539. Z.-X. Wan, *Quaternary Codes*, World Scientific, Singapore, 1997.

540. H. N. Ward, A restriction on the weight enumerator of self-dual codes, *JCT* **21** (1976), 253–255.

541. H. N. Ward, Divisible codes, *Arch. Math. (Basel)* **36** (1981), 485–494.

542. H. N. Ward, personal communication.

543. H. N. Ward, A bound for divisible codes, *PGIT* **38** (1992), 191–194.

544. H. N. Ward, Quadratic residue codes and divisibilty, Chapter 9 of [426], pp. 827–870.

545. H. N. Ward and J. A. Wood, Characters and the equivalence of codes, *JCT* **A 73** (1996), 348–352.

546. A. Weil, Sur certaines groupes d'opérateurs unitaires, *Acta Math.* **111** (1964), 143–211; *Oeuvres Scientifiques* III, Springer, 1979, pp. 1–69.

547. E. T. Whittaker and G. N. Watson, *A Course of Modern Analysis*, Cambridge Univ. Press, 4th ed., 1963.

548. D. L. Winter, The automorphism group of an extraspecial p-group, *Rocky Mtn. J. Math.* **2** (1972), 159–168.

549. J. Wolfmann, New decoding methods [for] the Golay code $(24, 12, 8)$, in *Combinatorial mathematics (Marseille-Luminy, 1981)*, North-Holland, Amsterdam, 1983, pp. 651–656.

550. J. Wolfmann, Nouvelles méthodes de décodage du code de Golay $(24, 12, 8)$, *Rev. CETHEDEC* no. 2, (1982), 79-88.

551. J. Wolfmann, A class of doubly even self-dual binary codes, *DM* **56** (1985), 299–303.

552. J. A. Wood, Extension theorems for linear codes over finite rings, in *Applied Algebra, Algebraic Algorithms and Error-Correcting Codes, Proc. 12th Internat. Symp., AAECC-12, Toulouse, June, 1997*, ed. T. Mora and H. Mattson, Lect. Notes Comput. Sci. **1255** (1997), pp. 329–340.

553. J. A. Wood, Weight functions and the extension theorem for linear codes over finite rings, in *Finite Fields: Theory, Applications and Algorithms, Proc. Fourth Internat. Conf. Finite Fields, Waterloo, August 1997*, ed. R. C. Mullin and G. L. Mullen, Contemp. Math. **225**, Amer. Math. Soc., Providence, RI, 1999, pp. 231–243.

554. J. A. Wood, Duality for modules over finite rings and applications to coding theory, *Amer. J. Math.* **121** (1999), 555–575.

555. V. Y. Yorgov, Binary self-dual codes with automorphisms of odd order, *Probl. Pered. Inform.* **19** (1983); English translation in *Prob. Inform. Trans.* **19** (1983), 11–24.

556. V. Y. Yorgov, A method for constructing inequivalent self-dual codes with applications to length 56, *PGIT* **33** (1987), 77–82.

557. V. Y. Yorgov, Doubly-even codes of length 64, *Probl. Pered. Inform.* **22** (1986), 35–42; English translation in *Prob. Inform. Trans.* **22** (1986), 277–284.

558. V. Y. Yorgov, The extremal codes of length 42 with an automorphism of order 7, *DM* **190** (1998), 210–213.

559. V. Y. Yorgov, On the minimal weight of some singly-even codes, *PGIT* **45** (1999), 2539–2541.
560. V. Y. Yorgov, New self-dual codes of length 106, *Congr. Numer.* **162** (2003), 111-117.
561. V. Y. Yorgov and R. P. Ruseva, Two extremal codes of length 42 and 44, *Probl. Pered. Inform.* **29** (1993), 99–103; English translation in *Prob. Inform. Trans.* **29** (1994), 385–388.
562. V. Y. Yorgov and N. Yankov, On the extremal binary codes of lengths 36 and 38 with an automorphism of order 5, *Proc. of the* 5th *International Workshop on Algebraic and Combinatorial Coding Theory*, June 1–7, 1996, Sozopol, Bulgaria, 307–312.
563. V. Y. Yorgov and N. P. Ziapkov, Doubly-even self-dual [40, 20, 8] codes with an automorphism of odd order, *Probl. Pered. Inform.* **32** (1996), 41–46; English translation in *Prob. Inform. Trans.* **32** (1996), 253–257.
564. J. Yuan and C.-M. Leung, Two-level decoding of the (32, 16, 8) quadratic residue code, *Southeast Asian Bull. Math.* **18** (1994), 173–182.
565. H. Zassenhaus, Über eine Verallgemeinerung des Henselschen Lemmas, *Arch. Math. (Basel)* **V** (1954), 317–325.
566. S. Zhang, On the nonexistence of extremal self-dual codes, *Discrete Appl. Math.* **91** (1999), 277–286.
567. S. Zhang and S. Li, Some new extremal self-dual codes with lengths 42, 44, 52, and 58, *DM* **228** (2001), 147–150.

Index